12 CHEMISTRY

PNG UPPER SECONDARY

Reddy Kuama
Suzanne Boniface
Terry Bunn
Alex Eames
Douglas Barns-Graham
Jan Giffney
Mark Sayes

OXFORD

OXFORD
UNIVERSITY PRESS

Oxford University Press is a department of the University of Oxford. It furthers the University's objective of excellence in research, scholarship, and education by publishing worldwide. Oxford is a registered trademark of Oxford University Press in the UK and in certain other countries.

Published in Australia by
Oxford University Press
Level 8, 737 Bourke Street, Docklands, Victoria 3008, Australia

First published 2013
Reprinted 2014, 2015, 2016, 2017, 2018, 2019, 2021, 2022, 2024

This book was originally published by ESA Publications, Auckland, New Zealand. This edition, specially adapted for the Grade 12 syllabus in Papua New Guinea, is published by arrangement with ESA Publications. Authors of the original work were Suzanne Boniface, Terry Bunn, Alex Eames, Douglas Barns-Graham, Jan Giffney and Mark Sayes; and this edition has been has been produced by Reddy Kuama.

ISBN 978 0 19 5578805

Edited by Catherine Greenwood
Illustrated by Guy Holt and Ian Laver
Typeset by diacriTech, India
Printed in China by Golden Cup Printing Co. Ltd

Oxford University Press Australia & New Zealand is committed to sourcing paper responsibly.

Contents

Preliminary Unit

Unit 12.1 Masses, Moles and Concentrations

Unit 12.2 Acids, Bases and Salts

Unit 12.3 Electrochemistry

Unit 12.4 Carbon Compounds

Unit 12.5 Natural Resources and Chemical Industries in Papua New Guinea

Introduction

This book has been published to provide the information required by students in order to successfully complete the Upper Secondary course in Chemistry in Grade 12 in Papua New Guinea.

The book is written in a manner that best develops an overall understanding of the Chemistry concepts and processes set out in the PNG Grade 12 Syllabus. The order of units in this book follows the order of units presented in the Chemistry syllabus for Grade 12:

Unit 12.1 Masses, Moles and Concentrations

Unit 12.2 Acids, Bases and Salts

Unit 12.3 Electrochemistry

Unit 12.4 Carbon Compounds

Unit 12.5 Natural Resources and Chemical Industries in Papua New Guinea

Within each unit there are a number of topics that follow the main subheadings and bullet points set out within each unit in the syllabus. The aim is to provide structure and content in a concise and compact format for students to use as an effective resource to support the classroom experience. It is acknowledged that Chemistry is more interesting and meaningful when students are exposed to a variety of resources and materials and we encourage students and teachers not to rely on this book as a sole source of information.

If you have any suggestions about how this book might be improved in future editions, please make contact with Oxford University Press:

Fax: 00 61 3 9934 9100

Customer Service Email: exportsales.au@oup.com

We wish you every success in your studies.

The authors

Acknowledgements

There are many people to be thanked for their help and assistance in enabling the publication of this series to happen. First, we acknowledge the cooperation and generosity of Mark Sayes at ESA Publications in New Zealand, who responded with interest and support when the proposal to adapt his Study Guide series was put to him.

We would also like to acknowledge many other individuals who have been happy to advise and assist in different ways: Joy Sahumlal, Greg Kapanombo, Anne Sangi, Safak Deliismail, Frank Griffin, Peter Petsul and Justin Narimbi; also Catherine Greenwood, who edited the new material.

Authorship

Authors of the original editions were Suzanne Boniface, Terry Bunn, Alex Eames, Douglas Barns-Graham, Jan Giffney and Mark Sayes.

This edition has been produced with the assistance of Reddy Kuama, lecturer in Chemistry at the University of Papua New Guinea. An administrator and teacher for over 20 years for public and private institutions at secondary and tertiary levels, Mr Kuama has a passion for education and teaching young people. He has published widely, served as chief examiner and marker of the PNGHSC Year 12 Chemistry Examinations since 1998 and contributed to various school and government boards. In addition he is one of the co-founding members and senior fellow of the Institute of Chemists PNG (ICPNG). Mr Kuama holds a bachelor's degree (BSc) and master's degree (MSc) in Chemistry from the University of Newcastle, Australia (1994, 1996), along with qualifications in Administration (DipAdmin) and Education (BEd, DipEd) from Pacific Adventist University (1989, 1987) and Fulton College, Fiji (1981). He resides with his wife and four children in Port Moresby.

The aim of this book is to provide Grade 12 Chemistry students in PNG with a book that they can use as a compact summary of content and skills related to the PNG Grade 12 syllabus.

Preliminary Unit
Practical investigations in chemistry

This chapter provides information for carrying out an extended practical investigation involving quantitative analysis by:

- Developing and carrying out a plan for the investigation.
- Collecting and processing quantitative experimental data.
- Interpreting the results to reach a conclusion.
- Presenting a report on the investigation.

Planning an investigation

The practical chemistry investigation in Grade 12 involves individually investigating variations in the amount of a chemical substance in some context within which there may be a possible trend in the amount present. The investigation will involve:

- deciding a purpose for the investigation.
- identifying and/or developing an appropriate analytical procedure.
- background research about the substance being investigated and/or the method used.
- preparing solutions, and standardising these where necessary.
- collecting quantitative experimental data.
- processing that data.
- interpreting results to reach a conclusion based on the processed data and relevant to the purpose.
- presenting a report.

Throughout the investigation, you will record in a logbook:

- all details relating to the methods used.
- how the procedure developed during the investigation.
- the raw data obtained.
- examples of the calculations using this data.
- any problems arising.

The logbook will be submitted with the final report.

You must allow enough time for the investigation for satisfactory results to be obtained.

Choosing the method of analysis

The analytical technique used may be:

- **Volumetric analysis**, ie a **titration** (**acid–base**, **oxidation–reduction**, formation of a complex, or precipitation).
- **Gravimetric analysis** (determination of a mass of product).
- Colorimetry (determination of depth of colour in reactions involving either a coloured reactant or product).

All of these techniques involve **quantitative analysis**. At the end of this chapter are some sample contexts for the above analytical methods.

Dependent and independent variables

The investigation is intended to measure how amounts or concentrations *vary* (rather than, for example, to determine the concentration of a single product). This can be done as a function of time, place, source or method of treatment (eg exposure to heat or light).

Example A

An investigation may compare the effects of different methods of treatment (the independent variable) on the amount of vitamin C in orange juice (the dependent variable).

The analytical technique selected to measure the dependent variable should be manageable and feasible. Investigations should include as wide a range of the independent variable as possible – this will assist in drawing a valid conclusion.

The scope of the investigation and the analytical technique selected may be limited by:

- availability of equipment.
- access to the chemicals needed to complete the analysis.
- the time needed to complete the investigation.

In designing the investigation, take care to ensure that all significant variables are controlled.

Example B

Significant variables could be the pH of the solution (eg in an EDTA titration for Mg^{2+} or Ca^{2+}) or the amount of time the solutions are left before taking a measurement (eg in the colorimetric determination of Fe^{3+} using the formation of the $[FeSCN]^{2+}$complex ion).

It is not always possible to control environmental conditions but these must be noted and/or considered where relevant.

Example C

When collecting samples of seawater in an estuary, is it high or low tide?

Trialling the method

The method you select will need to be trialled, and adjusted where necessary, to ensure that:

- The method is workable.

Example D

The presence of coloured species in the raw materials containing the substance being investigated can partially obscure the endpoint of a titration. These coloured species may be removed by filtration through activated charcoal.

Example E

Fruit juices being analysed for vitamin C commonly contain other natural products that are also oxidised by the iodine added in the analytical method. This will result in an inflated value for the calculated amount of vitamin C present.

- An appropriate range of data values can be obtained.

Example F

Initial trials may give **titres** that are either too small (eg less than a few mL) to be valid in terms of accuracy, ie the percentage uncertainty is too high, or too large to be practical (eg more than 50 mL). In both cases, the method needs to be modified.

Trials can be done on a small scale, and may involve only approximate volumes, and/or concentrations. This may mean using a measuring cylinder rather than a pipette or burette and solutions that have not been standardised. Having determined the appropriate concentration of solutions needed for the investigation, you will prepare and standardise these solutions where necessary.

Checking reliabilty and validity of data

If the investigation involves a titration, it is important to repeat the titration on each sample until you obtain sufficient concordant data. The level of concordancy that can be expected may vary depending on the levels of accuracy of the analytical method.

In addition, you must repeat the whole procedure for sample preparation to check on the reliablity of the method. When obtaining samples for analysis, the reliablity of the method, and hence validity of the results, is also checked by doing a full analysis of a second sample taken from the same site or equivalent source (such as a second bottle of the same brand).

Example G

In the investigation of the variation in chloride content of water in an estuary, more than one sample should be collected for analysis from any given site. For each sample, analysis should involve repeating the titration until **concordant** titres are obtained.

Analysis and reporting of results

When analysing the results, the amount or **concentration** may be expressed in **moles** or grams, moles per litre or grams per litre, or as a percentage. The results will need to be clearly presented in tables and/or graphs so that a valid conclusion can be drawn.

Calculated values should be given with no more significant figures than the least precise value.

The report should include:

- a statement of the purpose of the investigation.
- relevant background information.
- a description of the procedure used, including details of collection and/or preparation of samples and solutions used, control of significant variables, and the analytical technique used. Reasons for any modifications to the procedure should be given.
- a summary of processed data using graphs or tables where appropriate. A sample calculation using data collected should be included.
- a conclusion that links the experimental results to the purpose of the investigation.
- a discussion of the relationship between the practical outcomes of the investigation and the background research.
- an evaluation of the investigation, which considers the significance of the results in relation to the background information, the validity of the conclusion, the reliability of the data and sources of error.

Analytical methods

Volumetric analysis

Volumetric procedures (including acid–base, redox, compleximetric and precipitation titrations) can be used for determination of species such as vitamin C, ethanol, **hypochlorite** ion (ClO^-), dissolved oxygen, chloride ion, iodide ion, copper and the metal ions Ca^{2+}, Mg^{2+} and Al^{3+}.

In some cases, the species whose concentration is being investigated is present in the sample to be titrated.

Example H

Determination of chloride ion

The concentration of chloride ions in aqueous solution can be determined by reacting a sample of the solution with aqueous silver nitrate ($AgNO_3$) to form a white precipitate of AgCl.

$Ag^+(aq) + Cl^-(aq) \rightarrow AgCl(s)$

The indicator in this titration is potassium chromate (K_2CrO_4) solution, which reacts with a slight excess of silver ions to form brown, insoluble silver chromate (Ag_2CrO_4). Since Ag_2CrO_4 is more soluble than AgCl, the brown coloration appears only after most of the chloride ions have reacted with the silver ions.

In some volumetric procedures, the species in the sample must be reacted to produce another species, whose concentration is determined by titration.

Example I

Determination of hypochlorite ion

The concentration of hypochlorite in bleaching agents can be determined by adding acid and excess iodide:

$ClO^- + 2I^- + 2H^+ \rightarrow Cl^- + I_2 + H_2O$

The liberated iodine is then determined by titration with a **standard solution** of sodium thiosulfate:

$I_2 + 2S_2O_3^{2-} \rightarrow 2I^- + S_4O_6^{2-}$

Example J

Determination of dissolved oxygen in water

The method (called the Winkler method) for determining dissolved oxygen in water depends on the oxidation of manganese(II) ions, Mn^{2+}, by the dissolved oxygen to produce manganese(III):

$O_2 + 4Mn^{2+} + 8OH^- \rightarrow 2Mn_2O_3 + 4H_2O$

The Mn_2O_3 is dissolved by the addition of acid, and the Mn^{3+} ions react with added iodide to produce iodine:

$Mn_2O_3 + 6H^+ \rightarrow 2Mn^{3+} + 3H_2O$

$2Mn^{3+} + 2I^- \rightarrow 2Mn^{2+} + I_2$

The amount of iodine produced is determined by titration with sodium thiosulfate, using starch as the indicator:

$I_2 + 2S_2O_3^{2-} \rightarrow 2I^- + S_4O_6^{2-}$

The ratio of the number of moles of dissolved oxygen originally present in the water sample to the number of moles of thiosulfate used in the titration is:

$n(O_2) : n(S_2O_3^{2-}) = 1 : 4$

Back titrations

To be suitable for use in titrations, a reaction must:

- be fast.
- be free from competing reactions.
- have a suitable colour change – either inherently or resulting from the addition of a suitable indicator.

Example K

These criteria can *not* be met in the direct titration of ethanol as there is no fast-acting oxidant available with a sharp colour change occurring at the endpoint. The reaction of the alcohol with potassium dichromate needs heating and takes time to ensure the alcohol is fully oxidised through to the carboxylic acid, rather than being converted to a mixture of the aldehyde and carboxylic acid.

Where reactions are not suitable for direct titrations, a 'back-titration' method may be used:

- To the sample, a measured amount of a **reagent** (known to be more than is needed to completely react with the sample) is added all at once.
- When the reaction is complete, the amount of excess reagent remaining is determined by titration with an appropriate standard solution.
- From the difference between the amount of reagent added and that remaining, the amount of the unknown substance in the sample can be calculated using the stoichiometric relationship given by the balanced equation.
- If a back-titration method is being used, a 'blank' titration should be carried out to determine the titre volume in the absence of the species whose concentration is being determined. The difference between the titre volume for the blank titration and the titre volume when the sample for analysis is added should be at least 5–10 mL. If this is not the case, then either the method of sample preparation or the concentration of solutions used in the analysis will need to be changed.

Example L

Analysis of ethanol in wine

The alcohol content of wine can be determined by back titration. The ethanol is oxidised to ethanoic acid by heating with an excess of acidified potassium dichromate solution:

$$3C_2H_5OH + 2Cr_2O_7^{2-} + 16H^+ \rightarrow 3CH_3COOH + 4Cr^{3+} + 11H_2O$$

Excess iodide ions, in the form of solid potassium iodide, are added and react with any remaining $Cr_2O_7^{2-}$ to produce iodine:

$$Cr_2O_7^{2-} + 14H^+ + 6I^- \rightarrow 2Cr^{3+} + 7H_2O + 3I_2$$

The iodine produced is titrated with standard sodium thiosulfate solution, using starch as the indicator:

$$I_2 + 2S_2O_3^{2-} \rightarrow 2I^- + S_4O_6^{2-}$$

Example M

Analysis of vitamin C

Vitamin C (ascorbic acid) is a reducing agent and reacts with iodine, I_2, according to the following equation:

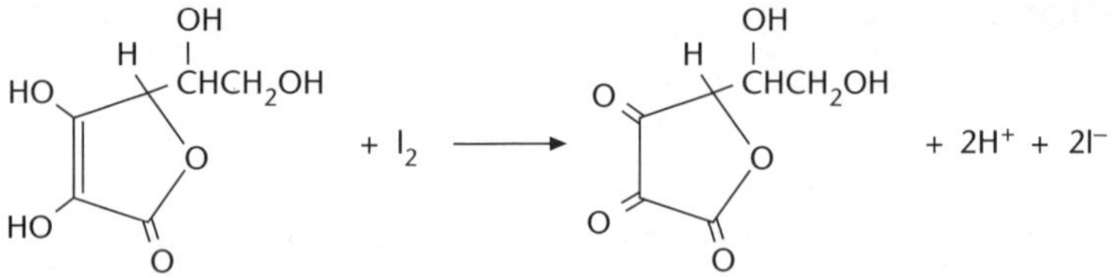

$$\text{(ascorbic acid)} + I_2 \rightarrow \text{(dehydroascorbic acid)} + 2H^+ + 2I^-$$

The structures in the above equation are line formulae – each corner where lines meet represents a carbon atom.

An iodine solution prepared from solid iodine can be 'messy' to handle and its concentration is uncertain. A preferred technique is to react a known amount of iodate with excess iodide to produce a solution containing a known amount of iodine:

$IO_3^- + 5I^- + 6H^+ \rightarrow 3I_2 + 3H_2O$

When a known amount of a solution containing vitamin C is added to the solution containing iodine, the vitamin C reduces the equivalent amount of I_2 (they react in a 1 : 1 ratio).

The excess iodine remaining in the solution is determined by back titration with sodium thiosulfate, using starch as the indicator:

$2S_2O_3^{2-} + I_2 \rightarrow 2I^- + S_4O_6^{2-}$

Example N

Analysis of aspirin

The amount of aspirin in a tablet can be found by hydrolysing the aspirin by **boiling** gently with excess aqueous sodium hydroxide solution.

$$\text{(aspirin)} + 2NaOH \rightarrow \text{(sodium salicylate)} + CH_3COONa + H_2O$$

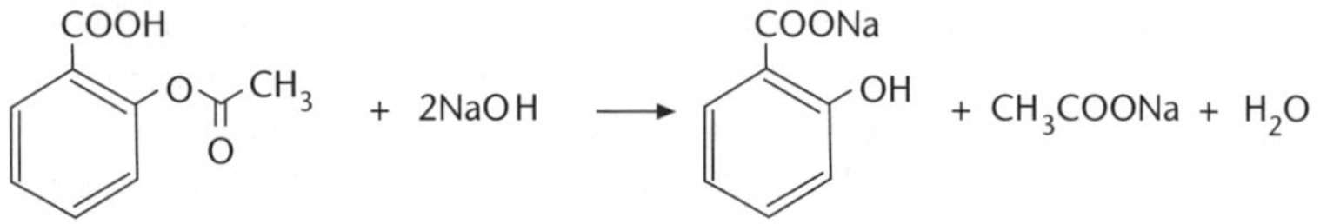

The amount of *unreacted* sodium hydroxide is determined by titrating the sodium hydroxide solution containing the aspirin with a standard solution of dilute hydrochloric acid, using phenolphthalein as the indicator.

Gravimetric methods

Gravimetric methods are suitable for the determination of species such as phosphate and sulfate.

Example O

Determination of amount of sulfate

To determine the amount of sulfur present as sulfate in a fertiliser, the sulfate is precipitated as barium sulfate from a solution containing a known mass of fertiliser. This precipitation is done by adding excess barium chloride solution to a solution containing the dissolved fertiliser:

$Ba^{2+}(aq) + SO_4^{2-}(aq) \rightarrow BaSO_4(s)$

The mass of sulfate present in the fertiliser is calculated from the mass of $BaSO_4$ precipitated.

Colorimetric methods

If a **colorimeter** or **spectrophotometer** is available, investigations may be carried out involving species that are coloured or that form coloured complexes, eg iron, lead, fluoride and copper. Colorimetric techniques use a prepared series of standard solutions of known concentrations to determine the concentration of the unknown solutions.

Unit 12.1 Masses, Moles and Concentrations

Topic 1: Qualitative analysis

Topic 1 covers reagents used in qualititive analysis and procedures used to determine what ions are present in solution. Qualitative analysis is important knowledge that you should understand berfore proceeding with your Year 12 studies.

Qualititative chemical analysis involves the identification of chemicals present in a sample. It is the study of what reacts or is produced in a reaction.

Reagents used in qualitative analysis

A **reagent** is a chemical compound or solution used to bring about a reaction. Reagents commonly used in qualitative analysis are solutions of sodium hydroxide, ammonia and **dilute** acid. When a reagent is being added to a solution, it should be added dropwise, with mixing, and the process should be carefully observed.

Sodium hydroxide solution

Precipitates containing hydroxide ions are formed by adding dilute sodium hydroxide solution to a solution containing a suitable cation. The colour of the precipitate is often quite distinct. The sodium ions are **spectator ions** and are not shown in the **ionic equation**.

Example A

Sodium hydroxide and iron(III) ions

When sodium hydroxide solution is added to a solution containing iron(III) ions, an insoluble rust-brown precipitate of iron(III) hydroxide forms.

$$Fe^{3+}(aq) + 3OH^-(aq) \rightarrow Fe(OH)_3(s)$$

orange colourless brown

Some hydroxides are **amphoteric** (they react with both acids and bases). In this case, the initial hydroxide precipitate dissolves if further sodium hydroxide solution is added, due to the formation of a complex ion.

Example B

Aluminium hydroxide

If sodium hydroxide is added to a solution containing Al^{3+} ions, a white precipitate of $Al(OH)_3$ initially forms:

$$Al^{3+}(aq) + 3OH^-(aq) \rightarrow Al(OH)_3(s)$$

With the addition of further hydroxide ions, aluminium hydroxide forms **soluble** aluminate ions, $[Al(OH)_4]^-(aq)$, so the precipitate dissolves:

$$Al(OH)_3(s) + OH^-(aq) \rightarrow [Al(OH)_4]^-(aq)$$

white precipitate — colourless solution of aluminate ions

Ammonia solution

Solutions of ammonia contain ammonia molecules, $NH_3(aq)$, a large amount of water, a few hydroxide ions, $OH^-(aq)$, and ammonium ions, $NH_4^+(aq)$. This is because ammonia is a **weak base**:

$$NH_3(aq) + H_2O(\ell) \rightleftharpoons NH_4^+(aq) + OH^-(aq)$$

The ⇌ arrows indicate that the solution is at **equilibrium**, so both reactants (NH_3(aq) and H_2O(ℓ)) and products (NH_4^+(aq) and OH^-(aq)) are present in the solution.

Small volumes of ammonia solution behave in the same way as solutions containing hydroxide ions and can precipitate suitable cations as insoluble hydroxides.

Example C

Small volumes of ammonia solution

When a small volume of dilute ammonia solution is added dropwise to copper sulfate solution, a light blue precipitate of copper hydroxide forms:

$$Cu^{2+}(aq) + 2OH^-(aq) + NH_4^+(aq) \rightarrow \underset{\text{light blue precipitate}}{Cu(OH)_2(s)} + NH_4^+(aq)$$

When large volumes of ammonia solution or concentrated NH_3(aq) are used, the large amount of ammonia molecules sometimes results in the formation of complex ions.

Example D

Excess ammonia solution

When an excess of concentrated ammonia solution is added to a solution containing copper ions, a deep blue-coloured solution of tetraammine copper(II) ions, $[Cu(NH_3)_4]^{2+}$(aq), forms:

$$Cu^{2+}(aq) + 4NH_3(aq) \rightarrow \underset{\text{deep blue tetraammine copper(II) ion}}{[Cu(NH_3)_4]^{2+}(aq)}$$

or $$Cu(OH)_2 + 4NH_3(aq) \rightarrow [Cu(NH_3)_4]^{2+}(aq) + 2OH^-(aq)$$

The two different reactions between copper ions and ammonia solution can be combined:

Copper ions in solution: Cu^{2+}(aq)	Add a few drops of dilute ammonia →	Pale blue precipitate of copper hydroxide forms: $Cu(OH)_2$(s)	Add excess ammonia →	Precipitate dissolves to give a deep blue soluble complex: $[Cu(NH_3)_4]^{2+}$(aq)

Similarly, Zn^{2+} ions first precipitate as white zinc hydroxide, $Zn(OH)_2$, but excess ammonia gives the soluble tetraammine zinc ions:

$$Zn^{2+}(aq) + 2OH^-(aq) \rightarrow Zn(OH)_2(s)$$

$$Zn(OH)_2(s) + 4NH_3(aq) \rightarrow \underset{\text{tetraammine zinc(II) ion}}{[Zn(NH_3)_4]^{2+}(aq)} + 2OH^-(aq)$$

Ag^+ ions first precipitate as brown silver oxide, Ag_2O, but ammonia gives the soluble diammine silver(I) ion $[Ag(NH_3)_2]^+$(aq):

$$2Ag^+(aq) + 2OH^-(aq) \rightarrow Ag_2O(s) + H_2O(\ell)$$

$$H_2O(\ell) + 4NH_3(aq) + \underset{\text{brown precipitate}}{Ag_2O(s)} \rightarrow \underset{\text{colourless soluble diammine silver(I) complex ion}}{2[Ag(NH_3)_2]^+(aq)} + 2OH^-(aq)$$

Similarly:

$$AgCl(s) + 2NH_3(aq) \rightarrow [Ag(NH_3)_2]^+(aq) + Cl^-(aq)$$

Dilute acid solutions

Dilute acid solutions will react with precipitates containing carbonate ions or hydroxide ions. These precipitates are basic (or amphoteric) and will dissolve in dilute acid solutions.

Dilute acid solutions and carbonate ions

Carbonates dissolve in dilute acid solutions to form carbon dioxide gas and water.

Example E

Dilute acid and carbonates

When dilute nitric acid is added to a (white) precipitate of calcium carbonate, the precipitate dissolves, forming carbon dioxide gas and water:

$$CaCO_3(s) + 2H^+(aq) \rightarrow Ca^{2+}(aq) + CO_2(aq) + H_2O(\ell)$$

Nitrate ions (from the nitric acid) are spectator ions.

Similarly, solutions containing **hydrogen carbonate** ions will react with acid to release bubbles of CO_2 gas.

Dilute sulfuric acid and Ba^{2+} or Pb^{2+} ions

When dilute sulfuric acid (or any solution containing sulfate ions) is added to a solution containing Ba^{2+} or Pb^{2+} ions, a white precipitate forms:

$$Ba^{2+}(aq) + SO_4^{2-}(aq) \rightarrow \underset{\text{white}}{BaSO_4(s)}$$

$$Pb^{2+}(aq) + SO_4^{2-}(aq) \rightarrow \underset{\text{white}}{PbSO_4(s)}$$

All other sulfate compounds except $CaSO_4$ are soluble.

Dilute acid solutions and hydroxide ions

Hydroxides dissolve in dilute acid solutions to form water.

Example F

Dilute acid and hydroxides

When dilute hydrochloric acid is added to a (blue) precipitate of copper(II) hydroxide, the precipitate dissolves, forming copper ions and water:

$$Cu(OH)_2(s) + 2H^+(aq) \rightarrow Cu^{2+}(aq) + 2H_2O(\ell)$$

The chloride ions are spectator ions.

Silver nitrate solution

Precipitates form when a solution containing Ag^+ ions is added to a solution containing chloride, Cl^-, bromide, Br^-, or iodide, I^-, ions.

Example G

Silver nitrate solution

When silver nitrate solution is added to a sodium chloride solution, a white precipitate of silver chloride is formed. This precipitate will darken on standing.

$$Ag^+(aq) + Cl^-(aq) \rightarrow AgCl(s)$$

Identifying ions in solution

Grade 12 Chemistry requires identification of the following cations and anions.

Cations to be identified		
Ag^+	Ba^{2+}	Al^{3+}
*Na^+	Cu^{2+}	Fe^{3+}
NH_4^+	Fe^{2+}	
	Mg^{2+}	
	Pb^{2+}	
	Zn^{2+}	

Anions to be identified	
Cl^-	CO_3^{2-}
I^-	SO_4^{2-}
OH^-	
*NO_3^-	

*Identified by a process of elimination

A knowledge of suitable reagents makes identification of all required ions possible.

Equations to explain the formation of precipitates

$Ag^+(aq) + Cl^-(aq) \rightarrow AgCl(s)$
$Ag^+(aq) + I^-(aq) \rightarrow AgI(s)$
$2Ag^+(aq) + 2OH^-(aq) \rightarrow Ag_2O(s) + H_2O(\ell)$
$2Ag^+(aq) + CO_3^{2-}(aq) \rightarrow Ag_2CO_3(s)$

$Ba^{2+}(aq) + 2OH^-(aq) \rightarrow Ba(OH)_2(s)$
$Ba^{2+}(aq) + CO_3^{2-}(aq) \rightarrow BaCO_3(s)$
$Ba^{2+}(aq) + SO_4^{2-}(aq) \rightarrow BaSO_4(s)$

$Cu^{2+}(aq) + 2OH^-(aq) \rightarrow Cu(OH)_2(s)$
$Cu^{2+}(aq) + CO_3^{2-}(aq) \rightarrow CuCO_3(s)$

$Al^{3+}(aq) + 3OH^-(aq) \rightarrow Al(OH)_3(s)$

$Mg^{2+}(aq) + 2OH^-(aq) \rightarrow Mg(OH)_2(s)$
$Mg^{2+}(aq) + CO_3^{2-}(aq) \rightarrow MgCO_3(s)$

$Zn^{2+}(aq) + 2OH^-(aq) \rightarrow Zn(OH)_2(s)$
$Zn^{2+}(aq) + CO_3^{2-}(aq) \rightarrow ZnCO_3(s)$

$Pb^{2+}(aq) + 2OH^-(aq) \rightarrow Pb(OH)_2(s)$
$Pb^{2+}(aq) + CO_3^{2-}(aq) \rightarrow PbCO_3(s)$
$Pb^{2+}(aq) + SO_4^{2-}(aq) \rightarrow PbSO_4(s)$
$Pb^{2+}(aq) + 2I^-(aq) \rightarrow PbI_2(s)$

$Fe^{2+}(aq) + 2OH^-(aq) \rightarrow Fe(OH)_2(s)$
$Fe^{2+}(aq) + CO_3^{2-}(aq) \rightarrow FeCO_3(s)$
$Fe^{3+}(aq) + 3OH^-(aq) \rightarrow Fe(OH)_3(s)$

Equations to explain the formation of complex ions

$Ag_2O(s) + 4NH_3(aq) + H_2O(\ell) \rightarrow 2[Ag(NH_3)_2]^+(aq) + 2OH^-(aq)$
$AgCl(s) + 2NH_3(aq) \rightarrow [Ag(NH_3)_2]^+(aq) + Cl^-(aq)$

$Zn(OH)_2(s) + 4NH_3(aq) \rightarrow [Zn(NH_3)_4]^{2+}(aq) + 2OH^-(aq)$
$Zn(OH)_2(s) + 2OH^-(aq) \rightarrow [Zn(OH)_4]^{2-}(aq)$

$Cu(OH)_2(s) + 4NH_3(aq) \rightarrow [Cu(NH_3)_4]^{2+}(aq) + 2OH^-(aq)$

$Al(OH)_3(s) + OH^-(aq) \rightarrow [Al(OH)_4]^-(aq)$

Identifying cations

The presence of Ag^+, Na^+, NH_4^+, Ba^{2+}, Cu^{2+}, Fe^{2+}, Mg^{2+}, Pb^{2+}, Zn^{2+}, Al^{3+} or Fe^{3+} ions can be detected using the following procedures:

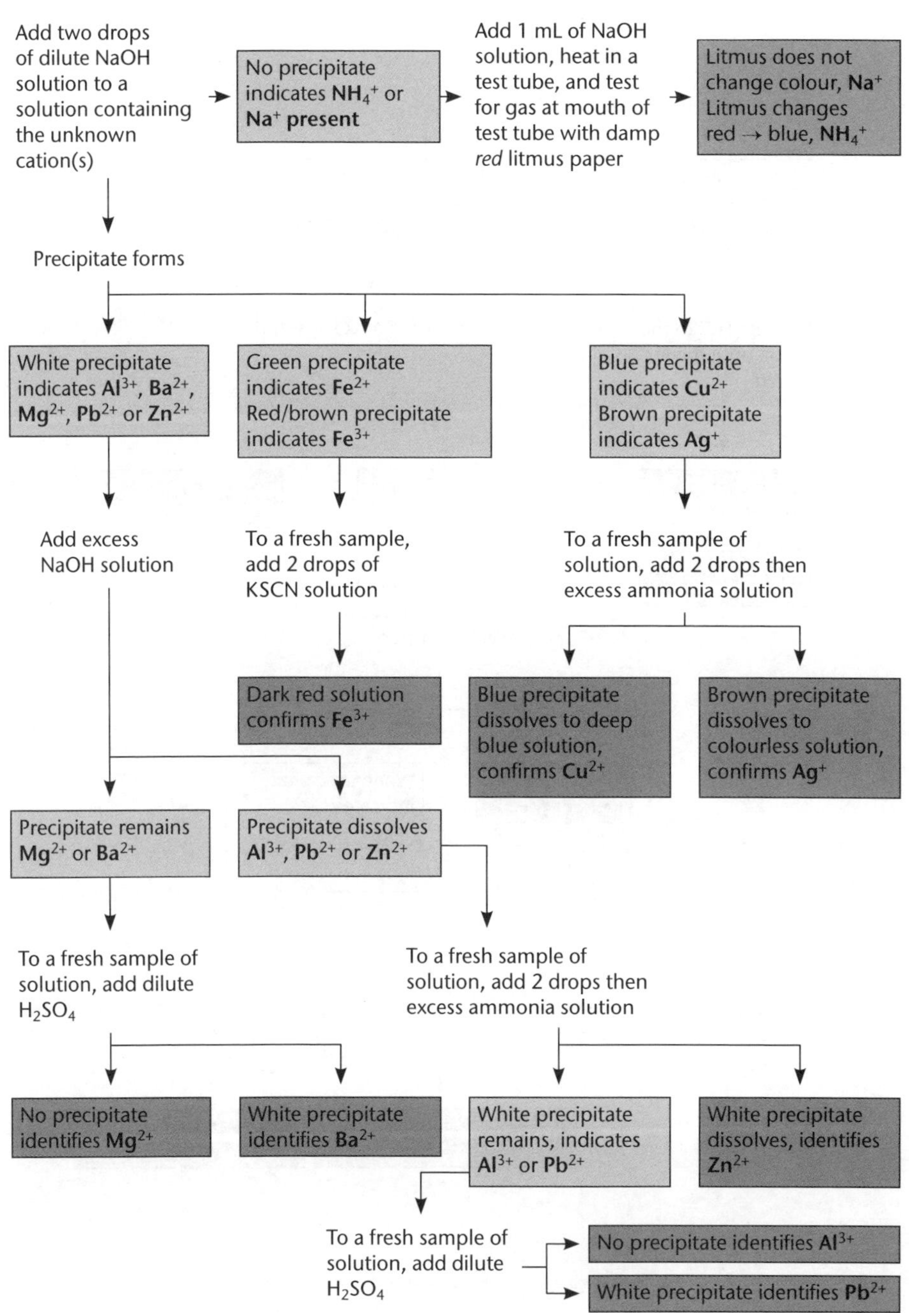

Identifying anions

The presence of OH^-, CO_3^{2-}, SO_4^{2-}, Cl^-, NO_3^-, and I^- can be detected using the following procedures:

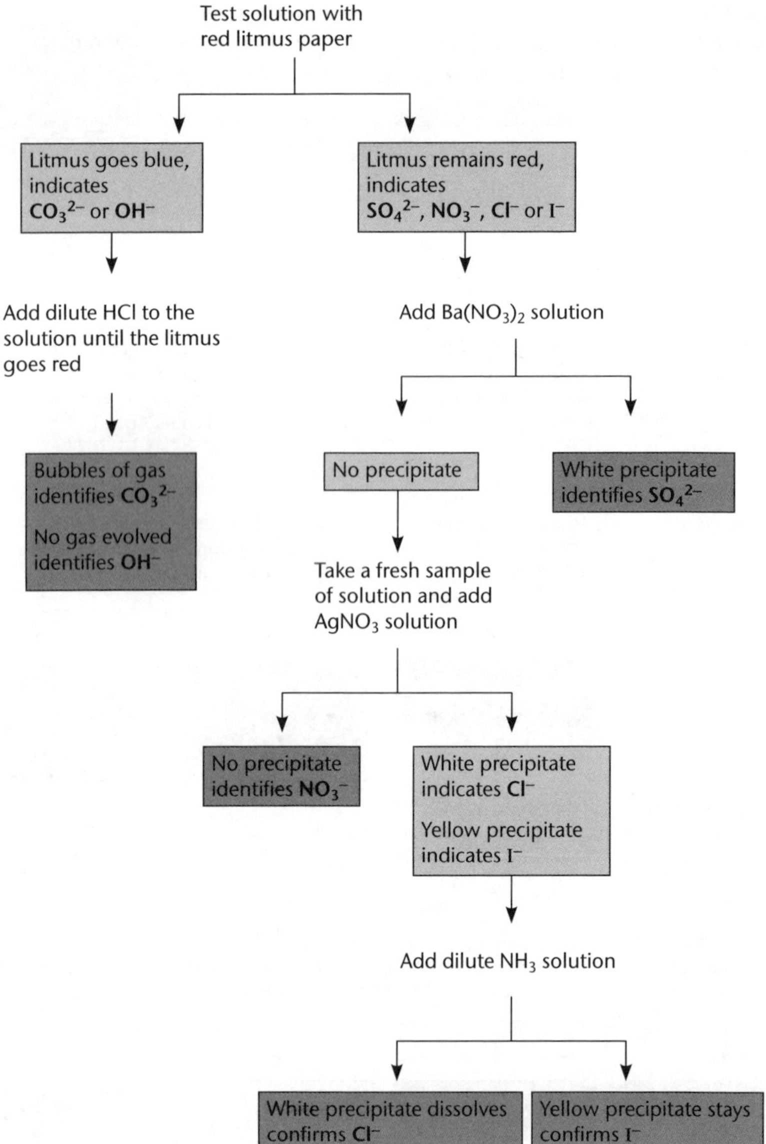

Unit 12.1 Activity 1A: Testing for cations and anions

1. When red litmus is added to a solution, it turns blue. On the addition of dilute hydrochloric acid, bubbles of gas are observed. Name the anion in the original solution.
2. A colourless solution formed a white precipitate when barium chloride solution was added. The solution was filtered, and when dilute hydrochloric acid was added to the residue, no reaction took place. Identify the anion present in the colourless solution. Write an ionic equation for the formation of the precipitate.
3. An unknown solution X is pale blue in colour. When excess ammonia is added to some of solution X, a light-blue precipitate forms, which dissolves to give a deep-blue solution. When silver nitrate solution is added to some of solution X, nothing happens. When barium chloride solution is added to some of solution X, nothing happens. Identify the ions present in the solution. Write balanced ionic equations for the reactions that occur.
4. Some silver nitrate is added to a solution containing chloride ions. Later, dilute ammonia is added to the reaction mixture.
   - **a.** What observations would be made?
   - **b.** If, instead of chloride ions, iodide ions were used, what observations would be made? Write balanced ionic equations for any reactions that occur.
5. Four common test solutions are:

   Test W Barium nitrate.

   Test X Silver nitrate followed by aqueous ammonia.

   Test Y Sodium hydroxide (a small quantity and then excess reagent).

   Test Z Aqueous ammonia (a small quantity and then excess reagent).

   Choose the test (W, X, Y or Z) that distinguishes between the ions in each of the following pairs. Give the observation that would be made as each ion pair is tested, with balanced equations for expected reactions.

   Pairs of ions:
   - **a.** Cl^- (chloride ions) and I^- (iodide ions).
   - **b.** SO_4^{2-} (sulfate ions) and NO_3^- (nitrate ions).
   - **c.** Al^{3+} (aluminium ions) and Zn^{2+} (zinc ions).
   - **d.** Cu^{2+} (copper(II) ions) and Fe^{2+} (iron(II) ions).
6. When sodium hydroxide is added dropwise to a solution of zinc nitrate, a white precipitate forms that dissolves in excess reagent. A similar reaction occurs when aqueous ammonia is added to aqueous zinc nitrate, with the white precipitate that forms dissolving in excess aqueous ammonia.

   Write equations for each reaction described.
7. Sodium hydroxide is added to a colourless solution containing only one type of metal ion. A white precipitate forms.
   - **a.** Give the formulae for five ions that could be present in the solution, and write equations for the precipitation reactions.
   - **b.** Outline further reactions that could be carried out to distinguish between these five ions. Write equations for any reactions that would occur.

8. A white powder was dissolved in water and analysed in the following way:

Step 1: Aqueous silver nitrate was added and a white precipitate formed.

Step 2: Excess aqueous ammonia was added to the white precipitate and it disappeared.

Step 3: Aqueous sodium hydroxide was added to a fresh sample of the dissolved powder and no precipitate formed. When the resulting solution was heated, a gas was evolved, which turned moist red litmus blue.

a. Identify the ions in the white powder.

b. Write equations for the reactions that occurred.

9. A blue powder was dissolved in water and analysed in the following way:

Step 1: Aqueous barium nitrate was added and a precipitate formed.

Step 2: Aqueous ammonia was added to a fresh sample and a blue precipitate formed.

Step 3: Excess aqueous ammonia was added to the blue precipitate, which dissolved, and a deep blue solution was observed.

a. Identify the ions in the blue powder.

b. Write equations for the reactions that occur.

10. Four colourless solutions are unlabelled. They are known to be Na_2CO_3, Na_2SO_4, $MgSO_4$, and NaOH.

Use the reaction schemes for identifying cations and anions on pages 11 and 12 to devise a method that could be used to identify each solid.

Describe any observations you would expect to make and write balanced equations for any reactions you would expect to occur.

11. A colourless solution is analysed to determine the cation and anion present. To separate samples of this solution, various tests were carried out and the following observations were made.

Identify the cation and anion present. Write balanced equations for the precipitation reactions that occurred and for the formation of any complex ions.

Observations: No reaction to litmus. On addition of barium nitrate, no reaction occurred. On addition of silver nitrate, no reaction occurred. A white precipitate formed when a small volume of aqueous sodium hydroxide was added, and this precipitate disappeared when excess sodium hydroxide was added. No precipitate formed with dilute sulfuric acid. A white precipitate formed when a small volume of aqueous ammonia was added, and this remained in the presence of excess ammonia.

12. An orange-brown coloured solution is analysed to determine the cation and anion present. To separate samples of this solution, various tests were carried out and the observations that follow were made.

Identify the cation and anion present. Write balanced equations for the precipitation reactions that occurred and for the formation of any complex ions.

Observations: No reaction to litmus. On addition of barium nitrate, a white precipitate occurs. A red-brown precipitate is formed when a small volume of aqueous sodium hydroxide is added. When potassium cyanate is added to a sample of the solution, a blood-red solution is formed.

13. A colourless solution is analysed to determine the cation and anion present. To separate samples of this solution, various tests were carried out and the observations that follow were made.

Identify the cation and anion present. Write balanced equations for the precipitation reactions that occurred and for the formation of any complex ions.

Observations: No reaction to litmus. On addition of barium nitrate, no reaction occurred. On addition of silver nitrate, a yellow precipitate formed that remained when excess ammonia solution was added. A white precipitate formed when a small volume of aqueous sodium hydroxide was added and this disappeared when excess sodium hydroxide was added. A white precipitate formed when a small volume of aqueous ammonia was added, and this disappeared when excess ammonia was added.

Unit 12.1 Activity 1B: Qualitative analysis – multiple choice

1. When adding a reagent to a solution, it should be:

A. added dropwise with mixing.

B. poured in quickly.

C. added slowly and the mixture allowed to settle.

D. none of the above.

2. A chemical that is amphoteric reacts with:

A. acids only.

B. bases only.

C. both acids and bases.

D. metals.

3. In solution, ammonia, $NH_3(aq)$, acts as a:

A. strong base.

B. weak acid.

C. weak base.

D. strong acid.

4. When sodium hydroxide is added to a solution of aluminium chloride, a white precipitate forms. The precipitate has the formula:

A. $[Al(OH)_4]^-$

B. HCl

C. $NaCl$

D. $Al(OH)_3$

Unit 12.1 Masses, Moles and Concentrations

Topic 2: Quantitative chemistry

Topic 2 covers quantitative chemical problems and how to solve them. This topic will introduce you to:

- Determining molar masses of molecules from given molar masses of atoms and formulae.
- Determing mole ratios from given formulae and equations.
- Using the formulae $n = \frac{m}{M}$ and $c = \frac{n}{V}$.
- Calculating percentage composition of a compound.
- Determining empirical and molecular formulae.
- Solving stoichiometric problems involving two-step and three-step calculations.
- Determining the number of waters of cyrstallisation in the formula of a hydrated salt.

Introduction

Quantitative chemistry, or **quantitative analysis**, is the study of *how much* reacts or is produced.

As we saw in Topic 1, qualitative chemistry is the study of *what* reacts or is produced.

(The difference between the two types is easy to remember – *quantit*ative chemistry deals with *quantit*ies.)

Even small amounts of substances contain *very* large numbers of **particles**.

Example A

Large numbers of particles

One teaspoon of copper sulfate contains about 2×10^{22} copper ions and 2×10^{22} sulfate ions.
A cup of water contains about 8.3×10^{24} water molecules.
An 'empty' 500 mL flask contains about 2.7×10^{21} oxygen molecules and 1.1×10^{22} nitrogen molecules.

The numbers in Example A are written in **scientific notation**, which is a convenient way of writing very large or very small numbers.

- 3×10^{22} means 30 000 000 000 000 000 000 000 or 3 followed by 22 zeros.
- The decimal number 0.000 000 000 02 is written 2×10^{-11}.

When large quantities are involved, the number of particles is never counted. Weighing is one method commonly used to specify a large number of items.

Example B

'Counting' large quantities

In a bank, large quantities of one-kina coins are weighed instead of being counted.
A farmer needing staples to wire up a fence will order a 25 kg box of staples rather than specifying a particular number of staples.

Although the number of particles handled in chemistry is very large, the **masses** of each of these particles is usually very small. The mass of a chlorine molecule is about 1.18×10^{-22} g. It would take about 10^{22} chlorine molecules to equal a mass of just 1 g.

All the measurements made in science use the same set of units for convenience, called the **Système International d'Unités**, abbreviated to **SI**.

- The SI unit for mass is the **kilogram** (symbol kg).
- The unit **gram** (symbol g), which is 0.001 or $\frac{1}{1000}$ kg, is used for small quantities.
- In industry, where chemicals are used in large quantities, the unit **tonne** (1000 kg) may be used.

The mole

When particularly large numbers of objects are needed, quantities are defined to suit these large numbers.

Example C

Large quantities

A baker refers to a *dozen* doughnuts rather than 12 doughnuts or to a *gross* of cream buns rather than 144 buns.

A physicist refers to 0.5 *coulomb* of electrons rather than 3.12×10^{18} electrons.

In the school office, sheets of paper for the photocopier are ordered by the *ream*, which is about 500 sheets.

In chemistry, the quantity chosen to describe the amount of substance is the **mole**, abbreviated to **mol** when used as a unit. The amount of substance (in mol) is represented by the symbol n.

$n(H_2O)$ means 'the amount of water in moles'.

The mole is defined as the amount of substance that contains the same number of particles (atoms, ions or molecules) as there are atoms in exactly 12 g of carbon-12 (^{12}C or C-12).

Example D

The mole

12 g of carbon is 1 mol of carbon.
The number of carbon atoms in 12 g of carbon is: 602 000 000 000 000 000 000 000, or 6.02×10^{23} atoms.

58.5 g of sodium chloride is 1 mol of sodium chloride, NaCl.
The number of sodium chloride ion pairs in 58.5 g is 6.02×10^{23}.

Avogadro's number

A mole always contains 6.02×10^{23} particles. This is called **Avogadro's number** (N_A).

$$\text{Avogadro's number} = 6.02 \times 10^{23} \text{ particles per mole}$$

$$N_A = 6.02 \times 10^{23} \text{ mol}^{-1}$$

Avogadro's number is sometimes approximated to 6×10^{23} mol^{-1}.

It is important to be able to change quantities between amount (in moles) and number of particles.

Example E

Moles and particles

Approximating Avogadro's number to 6×10^{23} means:

- 1 mol of carbon contains 6×10^{23} carbon atoms.
- 0.5 mol of carbon dioxide contains $0.5 \times 6 \times 10^{23} = 3 \times 10^{23}$ carbon dioxide molecules.
- 18×10^{24} sodium ions is $\frac{18 \times 10^{24}}{6 \times 10^{23}}$ = 30 mol of sodium ions.

The mole, a suitable quantity for chemists, may be quite unsuitable for other people. 0.5 mol of 10 cm nails is 3×10^{23} nails; laid end-to-end this is enough to stretch 750 000 000 000 000 times around the Earth. Hardly a suitable quantity for a builder!

In this book, the *amount* of substance will be measured in *moles*.

The definition of the mole is useful when dealing with both molecular substances and **ionic compounds**.

Example F

Moles and particles

Since one molecule of methane, CH_4, contains one carbon atom and four hydrogen atoms, 1 mol of methane contains 1 mol of carbon atoms and 4 mol of hydrogen atoms.

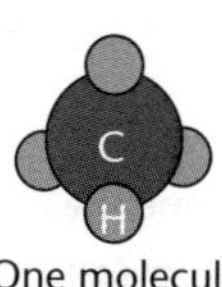

One molecule of CH_4

Many molecules of CH_4; 6.02×10^{23} molecules would make up one mole of CH_4

In 1 mol of calcium chloride, $CaCl_2$, there is 1 mol of calcium ions, Ca^{2+}, and 2 mol of chloride ions, Cl^-.

Molar mass

The mass of one mole of a substance (element or compound) is known as its **molar mass** (*M*). Molar mass has the unit 'grams per mole' or 'g mol^{-1}'.

Molar masses are defined relative to the molar mass of carbon-12 (the isotope of carbon with 6 protons and 6 neutrons), which is given a molar mass of 12 g mol^{-1} (ie 12 g of C-12 contains 1 mole, 6×10^{23}, C-12 atoms). By comparing the masses of all elements and compounds with carbon-12, the masses of all elements and compounds can be compared with each other (eg atomic hydrogen, H-1, has a molar mass of 1 g mol^{-1}, so hydrogen atoms are 12 times lighter than C-12 atoms).

The molar mass for each element appears on the periodic table on page 341.

Molar mass values are rarely whole numbers, because molar masses take into account the relative abundances of all the isotopes of an element.

Example G

Molar mass of chlorine

Chlorine has two naturally occurring forms, or isotopes:

- Chlorine-35, which makes up about 75% of all chlorine atoms.
- Chlorine-37, which makes up about 25% of all chlorine atoms.

Compared to carbon-12, a chlorine atom on average has a molar mass of:

$$\frac{75}{100} \times 35 + \frac{25}{100} \times 37 = 35.5$$

Thus, atomic chlorine has a molar mass of 35.5 g mol^{-1}, written $M(Cl) = 35.5$ g mol^{-1}. On average, a chlorine atom is $\frac{35.5}{12.0} \approx 3.0$ times heavier than an atom of carbon-12.

The molar mass of a compound can be calculated by adding together the molar mass of each atom in the compound.

Example H

Molar mass of ammonia

The molar mass, M, of ammonia, NH_3, is:

$$\begin{aligned} M(NH_3) &= M(N) + 3 \times M(H) \\ &= 14.0 + 3 \times 1.0 \\ &= 17.0 \\ \therefore \quad M(NH_3) &= 17.0 \text{ g mol}^{-1} \end{aligned}$$

$$\begin{aligned} N &= 14.0 \\ 3 \times H &= \underline{\ \ 3.0} \\ \therefore \quad M(NH_3) &= 17.0 \end{aligned}$$

(An ammonia molecule is thus $\frac{17}{12}$ times heavier than an atom of carbon-12.)

The loss or gain of electrons to form ions makes very little difference to the mass of an atom, so the mass of an ion is taken to be the mass of the atom it is derived from:

$$M(\text{ion}) = M(\text{atom(s) derived from})$$

Example I

Molar mass of sodium chloride

To find the molar mass of sodium chloride, NaCl, assume NaCl to be made from a sodium atom and a chlorine atom.

$$\begin{aligned} M(NaCl) &= M(Na) + M(Cl) \\ &= 23.0 + 35.5 \\ &= 58.5 \text{ g mol}^{-1} \end{aligned}$$

$$\begin{aligned} Na &= 23.0 \\ Cl &= 35.5 \\ \therefore \quad M(NaCl) &= 58.5 \text{ g mol}^{-1} \end{aligned}$$

Significant figures

Significant figures indicate the level of accuracy of the data and/or equipment used. When recording measurements, the correct number of significant figures for each piece of equipment used must be quoted.

Example J

25.2 g implies a one-decimal point balance has been used for weighing; 25.20 g implies a two-decimal point balance had been used for weighing.

When carrying out calculations, the final answer should be quoted to the number of significant figures of the least accurate measurement.

Most of the measurements used in calculations in this book are quoted to three significant figures.

Unit 12.1 Activity 2A: Molar mass

Determine the molar mass of each of the following. Use molar mass values from the periodic table on p. 341. Give answers to three significant figures where appropriate.

1. HCl	**2.** O_2	**3.** NO_2	**4.** HNO_3
5. H_2SO_4	**6.** $C_{12}H_{22}O_{11}$	**7.** C_3H_8	**8.** CH_3OH
9. MgO	**10.** $CaCl_2$	**11.** $CaSO_4$	**12.** $CuSO_4.5H_2O$
13. $Na_2CO_3.10H_2O$	**14.** CH_3COO^-	**15.** PO_4^{3-}	

Mole calculations

Chemical calculations often require that the mass of a substance be converted into an amount (moles). Conversions are done using the following relationship:

$$\text{amount of substance} = \frac{\text{mass of substance}}{\text{molar mass of substance}}$$

$$n = \frac{m}{M}\left(\frac{\text{g}}{\text{g mol}^{-1}}\right)$$

m is measured in grams; molar mass, M, is measured in grams per mole (g mol^{-1})

If the three variables are displayed in a triangle, it is easy to see how the formula can be rearranged:

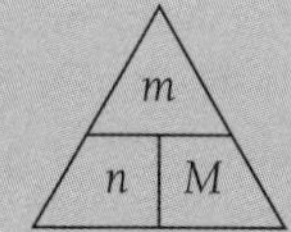

$$m = nM; \quad n = \frac{m}{M}; \quad M = \frac{m}{n}$$

Example K

Mole calculations

The mass of 3 mol of neon atoms is:

$m = nM$

$= 3 \text{ mol} \times 20.2 \text{ g mol}^{-1}$ $\quad$ [M(Ne) = 20.2 g mol^{-1}]

$= 60.6 \text{ g}$

The amount of copper atoms in 5 kg of copper is:

$n = \frac{m}{M}$

$= \frac{5 \times 1000 \text{ g}}{63.6 \text{ g mol}^{-1}}$ $\quad$ [M(Cu) = 63.6 g mol^{-1}]

$= 78.6 \text{ mol}$

The mass of 2.8 mol of ammonia gas is:

$m = nM$

$= 2.8 \text{ mol} \times 17.0 \text{ g mol}^{-1}$ $[M(NH_3) = 17.0 \text{ g mol}^{-1}]$

$= 47.6 \text{ g}$

The amount of sodium chloride in 409.5 g of sodium chlorie is:

$n = \frac{m}{M}$

$= \frac{409.5 \text{ g}}{58.5 \text{ g mol}^{-1}}$ $[M(NaCl) = 58.5 \text{ g mol}^{-1}]$

$= 7.00 \text{ mol}$

Unit 12.1 Activity 2B: Mole calculations

1. For the compound methane, CH_4, find the:
   a. molar mass.
   b. mass of two moles.
   c. amount of compound in 64.0 g.
   d. mass of 3.00 mol.
2. For the ionic compound calcium carbonate, $CaCO_3$, find the:
   a. molar mass.
   b. mass of 0.0500 mol.
   c. amount of calcium carbonate in 10.0 g of calcium carbonate.
   d. number of calcium ions in 50.0 g of calcium carbonate.
3. Calculate the molar mass of carbon dioxide.
4. Sulfur dioxide has the molecular formula SO_2 and a molar mass of 64.1 g mol^{-1}. Calculate how much SO_2 (in moles) is present in 6.4 kg of SO_2.
5. The molar mass of sulfur trioxide, SO_3, is 80.1 g mol^{-1}. Calculate the number of moles of SO_3 contained in 8.0 kilograms of sulfur trioxide.
6. Calculate the amount of carbon in 88.0 g of propane, C_3H_8.
7. Calculate the mass of the following substances:
   a. 5.50 mol of NO_2 — $M(NO_2) = 46.0 \text{ g mol}^{-1}$
   b. 0.150 mol of $NaHCO_3$ — $M(NaHCO_3) = 84.0 \text{ g mol}^{-1}$
   c. 10.7 mol of C_4H_{10} — $M(C_4H_{10}) = 58.0 \text{ g mol}^{-1}$
   d. 1.25 mol of AgCl — $M(AgCl) = 143.5 \text{ g mol}^{-1}$
   e. 0.600 mol of $C_6H_{12}O_6$ — $M(C_6H_{12}O_6) = 180 \text{ g mol}^{-1}$
   f. 25.0 mol of I_2 — $M(I_2) = 254 \text{ g mol}^{-1}$
   g. 0.50 mol of CCl_4
   h. 0.250 mol of CuO
8. A sample of aluminium oxide, Al_2O_3, has a mass of 1.02 g. Find the:
   a. molar mass of the compound.
   b. amount of aluminium oxide in the sample.
   c. amount of aluminium ions in the sample.
   d. amount of oxide ions in the sample.

9. For a sample of ammonium sulfate, $(NH_4)_2SO_4$, with a mass of 1320 g, calculate the:

a. molar mass of the ammonium sulfate.
b. amount of ammonium sulfate.
c. amount of ammonium ions.
d. amount of sulfate ions.
e. amount of sulfur atoms.
f. amount of nitrogen atoms.

10. Calculate the number of moles in each of the following substances:

a. NaCl in 15.2 g — M(NaCl) = 58.5 g mol^{-1}
b. CF_2Cl_2 in 10.9 g — $M(CF_2Cl_2)$ = 121 g mol^{-1}
c. $Zn(OH)_2$ in 25.0 g — $M(Zn(OH)_2)$ = 99.4 g mol^{-1}
d. Na_2CO_3 in 21.5 g — $M(Na_2CO_3)$ = 106 g mol^{-1}
e. NH_3 in 1.95 g — $M(NH_3)$ = 17.0 g mol^{-1}
f. $CuSO_4$ in 10.0 g — $M(CuSO_4)$ = 159.7 g mol^{-1}
g. $ZnCO_3$ in 282 g
h. $(NH_4)_2CO_3$ in 34 g

Chemical equations

A **chemical equation** is a brief, accurate way of summarising what happens in a chemical reaction. Chemical equations are written in the form:

reactants → products

Reactants are the substances that *react* and are written before the arrow (ie on the *left* side of the equation).

Products are *produced* in the reaction and are written after the arrow (ie on the *right* side of the equation).

A chemical equation has these features:

- The formulae of reactants and products are shown.
- The equation must be **balanced** (ie there must be the same number of each kind of atom on each side of the equation, and the total charge must be the same on each side of the equation).

Sometimes the **state** (or **phase**) of the reactants and products is shown, using the symbols:

(g) gas, (ℓ) liquid, (s) solid, (aq) dissolved in water

Writing balanced chemical equations

The following steps are used to write balanced chemical equations.

Step 1
Identify the reactant(s).

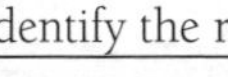

Step 2
Identify the product(s).

Example L

A piece of magnesium burns in air, forming a white powder.

Magnesium and oxygen

The white powder is magnesium oxide.

Step 3 Write the word equation.	Magnesium + oxygen → magnesium oxide

Step 4 Write the equation using correct symbols, formulae and states of both reactants and products.	$Mg(s) + O_2(g) \rightarrow MgO(s)$ (unbalanced because there are 2 oxygen atoms on the left and 1 oxygen atom on the right)

Step 5 Balance atoms and charge if necessary, starting with atoms that are not H or O. This is done by 'trial and error'; put whole numbers in front of reactants and products until the equation balances.	Put 2 in front of MgO to balance O: $Mg(s) + O_2(g) \rightarrow 2MgO(s)$ Then put 2 in front of Mg to balance Mg: $2Mg(s) + O_2(g) \rightarrow 2MgO(s)$

When ions are involved, follow the method illustrated in Example L.

Example M

When zinc metal reacts with dilute hydrochloric acid, the zinc metal dissolves and hydrogen gas is produced.

The reactants are zinc metal, Zn, and hydrogen ions, H^+ (from the dilute hydrochloric acid). (Hydrochloric acid, HCl, exists as hydrogen ions, H^+, and chloride ions, Cl^-.) The products are hydrogen gas, H_2, and zinc ions, Zn^{2+}.

The word equation is:

zinc metal + hydrogen ions → zinc ions + hydrogen gas

The unbalanced equation is:

$$Zn(s) + H^+(aq) \rightarrow Zn^{2+}(aq) + H_2(g)$$

The balanced equation is:

$$Zn(s) + 2H^+(aq) \rightarrow Zn^{2+}(aq) + H_2(g)$$

The last equation is balanced because:

- There are two Hs and one Zn on each side.
- The charge on the left is +2 (due to the two H^+) and the charge on the right is +2 (due to the one Zn^{2+}).

The chloride ions, Cl^-, in the hydrochloric acid are not involved in the reaction. They are called spectator ions. It is not necessary to show spectator ions, since they are neither reactants nor products.

Unit 12.1 Activity 2C: Balancing equations

Balance the following equations:

1. $H_2(g) + Cl_2(g) \rightarrow HCl(g)$
2. $H_2(g) + O_2(g) \rightarrow H_2O(\ell)$
3. $Al(s) + O_2(g) \rightarrow Al_2O_3(s)$
4. $Fe_2O_3(s) + C(s) \rightarrow Fe(s) + CO(g)$
5. $Na(s) + O_2(g) \rightarrow Na_2O(s)$
6. $N_2(g) + H_2(g) \rightarrow NH_3(g)$
7. $SO_2(g) + O_2(g) \rightarrow SO_3(g)$

8. $CH_4(g) + O_2(g) \rightarrow CO_2(g) + H_2O(g)$
9. $C(s) + O_2(g) \rightarrow CO(g)$
10. $C_3H_8(g) + O_2(g) \rightarrow CO_2(g) + H_2O(g)$
11. $C_3H_8(g) + O_2(g) \rightarrow C(s) + H_2O(g)$
12. $Fe(s) + Cl_2(g) \rightarrow FeCl_3(s)$
13. $PbO(s) + C(s) \rightarrow Pb(s) + CO_2(g)$

The mole and chemical equations

A balanced chemical equation is used to look at the relative amounts of reactants and products involved in a reaction.

Example N

Information from balanced equations

	$2\,Mg(s)$	$+$	$O_2(g) \rightarrow 2\,MgO(s)$		
	2 mol Mg	reacts with	1 mol O_2	to produce	2 mol MgO
or	20 mol Mg	reacts with	10 mol O_2	to produce	20 mol MgO
or	48.6 g Mg	reacts with	32.0 g O_2	to produce	80.6 g MgO
or	12.15 g Mg	reacts with	8.0 g O_2	to produce	20.15 g MgO

[M(Mg) = 24.3 g mol^{-1}; $M(O_2)$ = 32.0 g mol^{-1}]

Simple proportion allows information to be manipulated as required.

Chemists and chemical engineers are often involved in calculating amounts of product and reactant for chemical reactions.

Example O

Steel production

At the Steel Industries plant at Port Moresby, iron can be produced from iron oxide, Fe_2O_3, using carbon monoxide, CO.

*The actual ore used is Fe_3O_4 – a mixture of Fe_2O_3 and FeO.

An engineer may be interested in finding out the amount of iron that could be produced from 16 tonnes of iron oxide.

The balanced equation for the reaction is:

$Fe_2O_3(s) + 3CO(g) \rightarrow 2Fe(s) + 3CO_2(g)$

This reaction shows that:

	produces	
1 mol Fe_2O_3	produces	2 mol Fe
$\therefore$ 160.0 g Fe_2O_3	produces	111.8 g Fe
$\therefore$ 16 tonnes Fe_2O_3	produces	11.2 tonnes Fe

$2 \times Fe = 111.8$
$3 \times O = 48.0$
$\therefore M(Fe_2O_3) = 159.8$ g mol^{-1}

Mole ratios

For the general equation:

$$aA + bB \rightarrow cC + dD$$

a is the amount (in moles) of reactant A; b is the amount (in moles) of reactant B; c is the amount (in moles) of product C; d is the amount (in moles) of product D.

If the amount (moles) of any species is known (ie a of A is known, *or* b of B is known, *or* c of C is known, *or* d of D is known), the amount of any other reactant or product is known.

The ratio of amounts is:

$$\frac{n(\mathrm{A})}{a} = \frac{n(\mathrm{B})}{b} = \frac{n(\mathrm{C})}{c} = \frac{n(\mathrm{D})}{d}$$

For the reaction in Example N:

$$Fe_2O_3(s) + 3CO(g) \rightarrow 2Fe(s) + 3CO_2(s)$$

the mole ratio is:

$$\frac{n(Fe_2O_3)}{1} = \frac{n(CO)}{3} = \frac{n(Fe)}{2} = \frac{n(CO_2)}{3}$$

Using equations to calculate masses of products or reactants

The following steps are used to calculate masses of products or reactants.

Step 1
Write a balanced equation for the reaction.

Step 2
Convert the mass of the known reactant or product to moles.

Step 3
From the equation, deduce the relationship between the moles of known and unknown.

Step 4
Convert the moles of unknown to mass.

Example P

1.5 g of magnesium ribbon is burnt in air to produce magnesium oxide, MgO.
What mass of MgO wll be produced?

$$2Mg + O_2 \rightarrow 2MgO$$

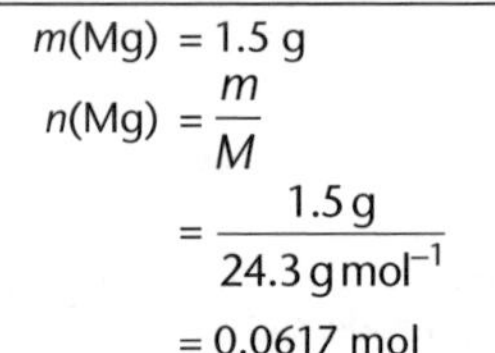

$$\begin{aligned} m(\mathrm{Mg}) &= 1.5\ \mathrm{g} \\ n(\mathrm{Mg}) &= \frac{m}{M} \\ &= \frac{1.5\,\mathrm{g}}{24.3\,\mathrm{g\,mol^{-1}}} \\ &= 0.0617\ \mathrm{mol} \end{aligned}$$

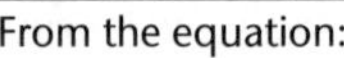

From the equation:

$$\begin{aligned} \frac{n(\mathrm{MgO})}{2} &= \frac{n(\mathrm{Mg})}{2} \\ n(\mathrm{MgO}) &= n(\mathrm{Mg}) \\ &= 0.0617\ \mathrm{mol} \end{aligned}$$

$$\begin{aligned} n(\mathrm{MgO}) &= 0.0617\ \mathrm{mol} \\ m(\mathrm{MgO}) &= nM \\ &= 0.0617\ \mathrm{mol} \times 40.3\ \mathrm{g\ mol^{-1}} \\ &= 2.5\ \mathrm{g} \end{aligned}$$

Chemical equations do *not* give information about:

- how *fast* a reaction proceeds.
- how *far* the reaction proceeds, ie how much of the reactants is used up.
- the way the reaction occurs.

Occasionally, fractional numbers are used in balanced equations.

Example Q

Using fractions in balanced equations

$2Mg(s) + O_2(g) \rightarrow 2MgO(s)$ can be written:

$Mg(s) + \frac{1}{2}O_2(g) \rightarrow MgO(s)$ and means that:

1 mol Mg reacts with $\frac{1}{2}$mol O_2 to make 1 mol MgO.

Note: $\frac{1}{2}O_2$ does *not* refer to half a molecule of O_2; it refers to half a mole of O_2.

Unit 12.1 Activity 2D: The mole and chemical equations

1. Zinc and iodine react to form zinc iodide, according to the equation:

$$Zn(s) + I_2(s) \rightarrow ZnI_2(s)$$

$M(Zn) = 65.4$ g mol^{-1}; $M(ZnI_2) = 319.4$ g mol L^{-1}

A student weighs out exactly 1.65 g of zinc and allows it to react with excess iodine. Calculate the amount (mol) of zinc used. What mass of zinc iodide would be formed?

2. The reaction between magnesium carbonate and dilute hydrochloric acid is represented by the equation:

$$MgCO_3(s) + 2HCl(aq) \rightarrow MgCl_2(aq) + H_2O(\ell) + CO_2(g)$$

$M(MgCO_3) = 84.3$ g mol^{-1}; $M(MgO) = 40.3$ g mol^{-1}

a. Calculate the molar mass of carbon dioxide.

b. Calculate the mass of magnesium carbonate that will be needed to produce 8.80 g of CO_2.

3. A student weighed out 2.40 g of magnesium and burnt it in air. Magnesium burns in air to form magnesium oxide. The equation for the reaction is:

$$2Mg(s) + O_2(g) \rightarrow 2MgO(s)$$

Calculate the mass of magnesium oxide produced in the reaction.

4. When copper carbonate is heated, it decomposes. The reaction occurring is:

$$CuCO_3(s) \rightarrow CuO(s) + CO_2(g)$$

a. Calculate the mass of copper carbonate that must be decomposed to provide 11.0 g of carbon dioxide.

b. Calculate the mass of copper oxide that forms when 247 g of copper carbonate is completely decomposed.

c. If 318 g of copper oxide is produced, what mass of carbon dioxide is produced?

5. The reaction for the decomposition of calcium carbonate is:

$$CaCO_3 \rightarrow CaO + CO_2$$

If 100 kg of calcium carbonate is heated, what mass of calcium oxide will form?

6. Calcium burns in air according to the equation:

$$2Ca + O_2 \rightarrow 2CaO$$

How much calcium has reacted when 8.00 kg of oxygen is used up?

7. When sulfur trioxide dissolves in water, the reaction occurring is:

$$SO_3 + H_2O \rightarrow H_2SO_4$$

Find the mass of H_2SO_4 formed when 8.0 tonnes of sulfur trioxide dissolves in water.

8. Iron oxide is converted into iron by carbon monoxide according to the equation:

$$Fe_2O_3 + 3CO \rightarrow 2Fe + 3CO_2$$

Calculate the mass of iron that could be obtained from 1.60 tonnes of iron oxide.

9. $CH_4(g) + 2O_2(g) \rightarrow 2H_2O(g) + CO_2(g)$

How many moles of water vapour form when 32 g of methane burns?

10. Calculate the mass of water that will react completely with 4.00 g of pure calcium metal according to the following equation:

$$Ca(s) + 2H_2O(\ell) \rightarrow Ca(OH)_2(s) + H_2(g)$$

11. A substance X reacts with oxygen according to the equation:

$$4X + O_2 \rightarrow 2X_2O$$

a. How many moles of oxygen molecules react with one mole of X?

b. 4.6 g of X burns completely to produce 6.2 g of X oxide (X_2O). How much oxygen (in moles) has reacted in this experiment?

12. Calculate the mass of ammonia that is required to produce 182 kg of urea, $CO(NH_2)_2$, according to the following equation:

$$CO_2(g) + 2NH_3(g) \rightarrow CO(NH_2)_2(s) + H_2O(\ell)$$

13. Calculate the mass of HF that can be prepared from 15.6 g of CaF_2 when treated with excess H_2SO_4, according to the following equation:

$$CaF_2(s) + H_2SO_4(aq) \rightarrow 2HF(g) + CaSO_4(s)$$

14. Calculate the mass of CO_2 produced in the complete combustion of 21.2 g of butene, C_4H_8, according to the following equation:

$$C_4H_8 + 6O_2 \rightarrow 4CO_2 + 4H_2O$$

15. Calculate the mass of H_2O produced in the complete combustion of 9.90 g of cyclohexene, C_6H_{10}, according to the following equation:

$$2C_6H_{10} + 17O_2 \rightarrow 12CO_2 + 10H_2O$$

16. Calculate the mass of sulfur trioxide, SO_3, produced from 100 kg of sulfur dioxide, SO_2, according to the following equation:

$$2SO_2 + O_2 \rightarrow 2SO_3$$

Percentage composition

Percentage composition is a measure of the mass of the different elements in a compound expressed as a percentage of the total mass. Compounds can be identified through their percentage composition.

Calculating percentage composition

The following steps are used in calculating percentage compositions.

Example R

Percentage composition of iron oxide

Find the percentage composition of iron oxide, Fe_2O_3.

Step 1
From the chemical formula, find the amount of each element present (in moles).

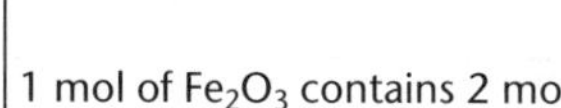

1 mol of Fe_2O_3 contains 2 mol of Fe and 3 mol of O.

Step 2
Use the molar masses of each element to find the mass of each element present.

1 mol of Fe_2O_3 has a mass of 159.8 g and contains:
$2 \times 55.9 = 111.8$ g of Fe and
$3 \times 16.0 = 48.0$ g of O.

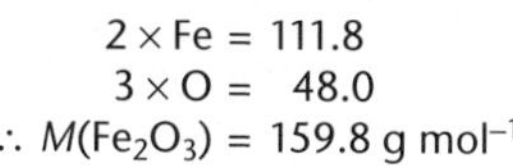

$$2 \times \text{Fe} = 111.8$$
$$3 \times \text{O} = 48.0$$
$$\therefore M(Fe_2O_3) = 159.8 \text{ g mol}^{-1}$$

Step 3
Express the amount of each element as a percentage.

There is 111.8 g of Fe in 159.8 g of Fe_2O_3, so the percentage composition is:

$$\frac{111.8}{159.8} \times \frac{100}{1}\% = 70\% \text{ Fe}$$

and similarly there is 48.0 g of O in 159.6 g of Fe_2O_3, so there is:

$$\frac{48.0}{159.8} \times \frac{100}{1}\% = 30\% \text{ O}$$

Step 4
Check that adding the percentages gives 100%.

70% Fe + 30% O = 100% (checked)

Empirical formulae

When **analysis** is carried out to find what and/or how much of a particular element is present in a compound, the result is often expressed as a percentage composition. From the percentage composition, an **empirical formula** can be calculated. The empirical formula is the *simplest whole number ratio* of the atoms or ions in a compound.

Finding an empirical formula

The following steps are used in calculating percentage compositions.

Step 1
Find the mass of each element present in 100 g of the compound.

Step 2
Change these masses to amounts, using $n = \frac{m}{M}$.

Step 3
Divide each result by the smallest number of moles.

Step 4
Make small approximations or multiply all results by a constant to get a simple whole number ratio.

Step 5
This is the empirical formula.

Example S

Analysis of a clear liquid with bleaching properties showed that it contained 5.9% hydrogen and 94.1% oxygen.

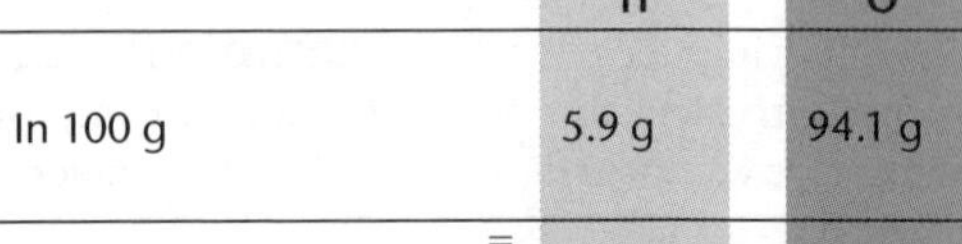

	H	O
In 100 g	5.9 g	94.1 g
Number of moles	$= \frac{5.9}{1.0}$ $= 5.9$	$= \frac{94.1}{16.0}$ $= 5.88$
Mole ratio	$= \frac{5.9}{5.88}$ $= 1.003$	$= \frac{5.88}{5.88}$ $= 1.0$
Whole number mole ratio	≈ 1.0	1.0
Empirical formula	HO	

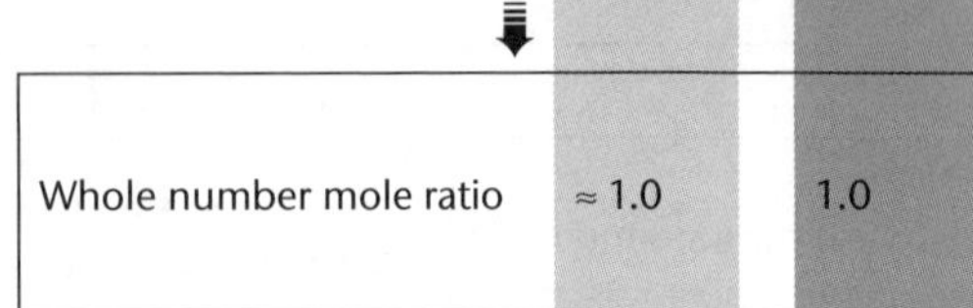

Example T

Finding an empirical formula from masses

In an experiment, a sample of an unknown organic compound was found to contain 0.18 g of carbon and 0.03 g of hydrogen.

	C	H
Masses	0.18 g	0.03 g
Number of moles	$\frac{0.18}{12.0}$ $= 0.015$	$\frac{0.03}{1.0}$ $= 0.03$
Ratio of moles	$\frac{0.015}{0.015}$ $= 1$	$\frac{0.030}{0.015}$ $= 2$

Empirical formula is CH_2

Example U

Determing the empirical formula of an oxide of magnesium by experiment

Step 1
Weigh accurately and record the mass of a clean, dry crucible and lid (eg 29.10 g).

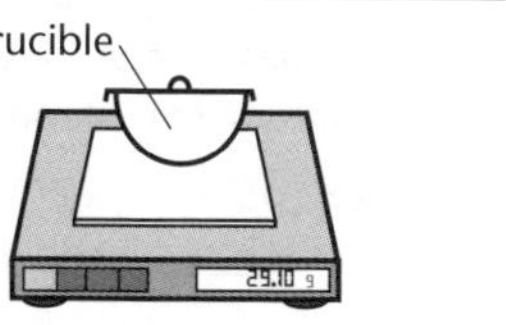

Step 2
Add approximately 2.5 g of magnesium to the crucible and record the mass of the crucible and lid plus magnesium.
(The mass of magnesium needs to be known accurately, but the amount used is not critical. For the purpose of this experiment, the amount should be between 2.4 g and 2.6 g.)

Add solid to crucible, then return to the balance

Step 3
Heat the magnesium in the crucible with the lid on. (The lid may be lifted in the early stages of heating to allow oxygen to react with the magnesium, but the lid should be replaced once the reaction begins.) Heat strongly for 15 minutes.
(If the lid is left off once the reaction has commenced, some of the product may be lost.)

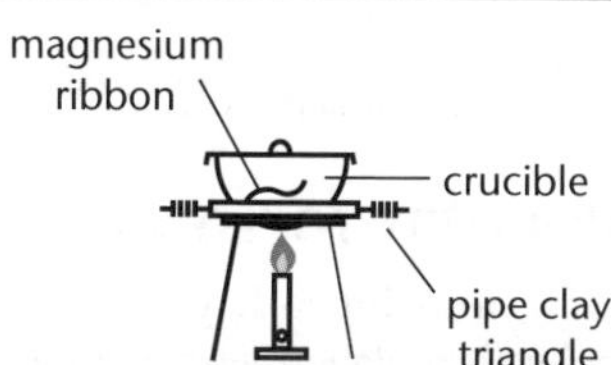

Crucible in pipe clay triangle over Bunsen burner

Step 4
Remove the crucible from the heat and allow it to cool. Weigh the crucible and sample.

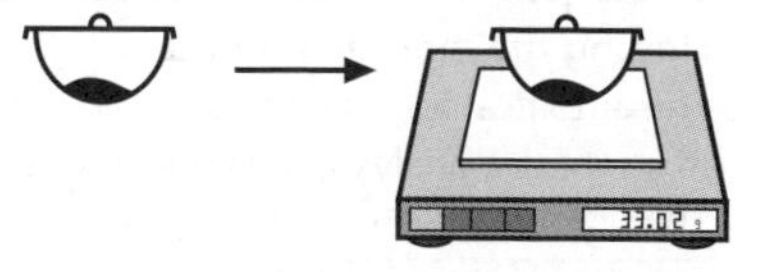

Step 5
Heat the sample for a further 10 minutes, remove the crucible from the heat, and cool and weigh it.

Step 6
Repeat Step 5 until there is no change in the mass. (If there is no change in the mass after further heating and weighing, then it can be assumed that all the magnesium has reacted.)

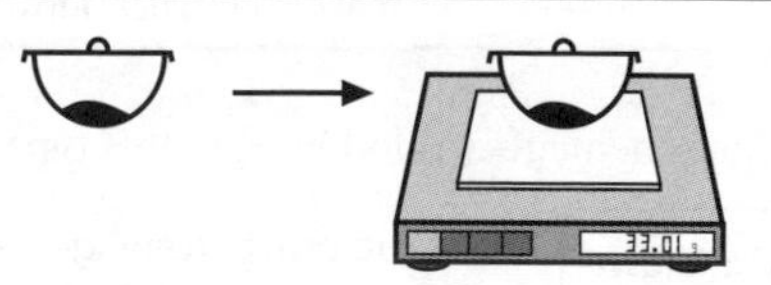

Results

Mass of a clean, dry crucible = 29.10 g [Step 1]
Mass of crucible + magnesium = 31.45 g [Step 2]
Mass of crucible and sample after heating = 33.02 g [Step 4]
Constant mass of crucible and sample after re-heating = 33.01 g [Step 6]

Calculations

Mass of Mg = 31.45 – 29.10 g = 2.35 g [Mass of the crucible with magnesium – mass of the empty crucible.]

$$n(\text{Mg}) = \frac{m}{M} = \frac{2.35 \text{ g}}{24.3 \text{ g mol}^{-1}} = 0.0967 \text{ mol}$$

Mass of final product = 33.01 – 29.10 g = 3.91 g [Mass of crucible and sample after heating – mass of the empty crucible.]

Mass of oxygen in compound = 3.91 – 2.35 = 1.56 g [Mass of the product – mass of magnesium used.]

$$n(\text{O}) = \frac{1.56}{16.0} = 0.0975 \text{ mol}$$

$n(\text{Mg}) : n(\text{O}) = 0.0967 : 0.0975 = 1 : 1$ [Divide through by 0.0967 (ie the smaller number of moles) and then make approximations to get a whole number ratio.]

Empirical formula is MgO.

Molecular formulae

The **molecular formula** gives the *actual* numbers of each atom in a molecule. The molecular formula is a whole number multiple of the empirical formula.

The molecular formula of a compound can be calculated by using both the empirical formula and the molar mass, *M*.

Example V

Finding the molecular formula

The empirical formula of the liquid in Example S was found to be HO.
In a separate analysis, the liquid was shown to have a molar mass of 34 g mol^{-1}.
The possible molecular formulae are shown with their molar masses:

Possible molecular formulae	HO	H_2O_2	H_3O_3	H_4O_4
Molar mass (g mol^{-1})	17	34	51	68

The molecular formula is thus H_2O_2.
Note: H_2O_2 is called hydrogen peroxide.
Generally: Molecular formula = x × empirical formula,

$$\text{where } x = \frac{\text{molar mass}}{\text{mass of empirical formula}}$$

The scheme that is followed in this type of analysis is:

Unknown formula — Analysis to give percentage composition
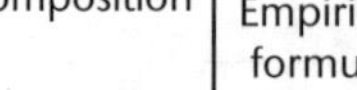
Empirical formula — Molecular mass ➡ Molecular formula

Unit 12.1 Activity 2E: Percentage composition, empirical and molecular formulae

1. What is the percentage composition of each element in potassium nitrate, KNO_3?

2. Calculate the percentage of magnesium in magnesium oxide.

3. Find the percentage of carbon in acetylene, C_2H_2.

4. Find the empirical formulae of the substances with the following percentage compositions:

a. 80% copper, 20% oxygen.

b. 53% aluminium, 47% oxygen.

c. 1.6% hydrogen, 22.2% nitrogen, 76.2% oxygen.

5. The empirical formula for magnesium chloride is $MgCl_2$.

a. What is the ratio of ions in this formula?

b. Calculate the percentage, by weight, of magnesium in magnesium chloride.

6. Calculate the empirical formula of the oxide of sulfur that is 40% sulfur by weight.

7. A hydrocarbon contains 90% carbon. Calculate the empirical formula of the hydrocarbon.

8. An oxide of silicon was produced from 0.28 g of silicon. The mass of the oxide was 0.60 g. Calculate the empirical formula of the oxide.

9. 10.2 g of vanadium is combined with 21.3 g of chlorine to make vanadium chloride. Calculate the empirical formula of the vanadium chloride.

10. The empirical formula of a substance is CH_2. Its molar mass is 84 g mol^{-1}. Find the molecular formula of the substance.

11. Analysis gave the following data for three unknown substances:

a. 85.7% C and 14.3% H, M = 28 g mol^{-1}

b. 30.4% N and 69.6% O, M = 92 g mol^{-1}

c. 2% H, 33% S and 65% O, M = 98 g mol^{-1}

Find the molecular formula of each.

12. A hydrocarbon contains 82.7% carbon and 17.3% hydrogen by weight.

a. Work out the empirical formula.

b. The molar mass of the compound is 58 g mol^{-1}. What is its molecular formula?

13. A gaseous hydrocarbon was found to contain 80% carbon and 20% hydrogen by mass.

a. Calculate the empirical formula for the hydrocarbon.

b. The hydrocarbon was found to have a molar mass of 30 g mol^{-1}. Use this value to work out the molecular formula of the hydrocarbon.

14. a. An oxide of nitrogen contains 30.3% nitrogen. Calculate the empirical formula of the oxide.

b. If the molar mass of the oxide is 92 g mol^{-1}, use your answer in **a.** to determine the molecular formula of the oxide.

15. a. Vitamin C has a mass composition of 40.92% carbon, 4.58% hydrogen and 54.50% oxygen. Calculate the empirical formula of vitamin C.

b. If the molar mass of vitamin C is 176 g mol^{-1}, use your answer in **a.** to determine the molecular formula of vitamin C.

16. a. An organic compound has a mass composition of 26.67% carbon, 2.22% hydrogen and 71.11% oxygen. Calculate the empirical formula of the compound.

b. If the molar mass of the compound is 90 g mol^{-1}, use your answer in **a.** to determine its molecular formula.

17. A lead sulfide compound was made by heating 2.95 g of lead with excess sulfur. Once all the lead had reacted, the excess sulfur was burnt off. The mass of the final product was 3.42 g.
M(Pb) = 207 g mol^{-1}, M(S) = 32.1 g mol^{-1}

a. Calculate the moles of lead used.

b. Calculate the mass and hence the moles of sulfur that reacted.

c. Calculate the empirical formula of the compound.

Water of crystallisation

The crystals of some salts contain molecules of water. They are known as **hydrates**. These molecules fit into the crystal **lattice**. The ratio of water to salt is always a whole number. The water present in **hydrated** salts is known as **water of crystallisation**.

When a hydrated salt is heated, the water is driven off and the salt is said to become **anhydrous**.

Example W

Determining the formula of a sample of hydrated copper sulfate, $CuSO_4.xH_2O$

Step 1: Weigh a clean, dry crucible accurately and record its mass.

Step 2: Add approximately 2.0 g of the hydrated salt to the crucible and record the mass. (The mass of salt taken needs to be known exactly, but the actual amount taken is not critical. For the purpose of this experiment, the amount should be between 1.9 g and 2.1 g.)

Step 3: Heat the sample in the crucible with the lid on for 10 minutes.

Step 4: Remove the crucible from the heat, allow it to cool, and weigh the sample and crucible.

Step 5: Heat the crucible strongly for another 5 minutes, and cool and reweigh it.

Step 6: Repeat Step 5 until there is no change in the mass of the sample.

Results

Mass of a clean, dry crucible = 25.20 g [Step 1]
Mass of crucible + hydrated salt = 27.28 g [Step 2]
Constant mass of crucible and sample after re-heating = 26.54 g [Step 6]

Calculations

Mass of hydrated salt ($CuSO_4.xH_2O$) = 27.28 – 25.20 g = 2.08 g

Mass of anhydrous salt ($CuSO_4$) = 26.54 – 25.20 g = 1.34 g

[$M(CuSO_4)$ = 159.6 g mol^{-1}]

$$\text{Moles of anhydrous salt} = \frac{m}{M} = \frac{1.34\text{ g}}{159.6\text{ g mol}^{-1}} = 0.008\,40\text{ mol}$$

Mass of water removed = 2.08 g – 1.34 g = 0.74 g

[$M(H_2O)$ = 18.0 g mol^{-1}]

$$\text{Moles of water removed} = \frac{0.74\text{ g}}{18.0\text{ g mol}^{-1}} = 0.0411\text{ mol}$$

$CuSO_4 : H_2O$ = 0.008 40 : 0.0411 = 1 : 4.89 ≈ 1 : 5 [Divide through by 0.008 40 (ie the smaller number of moles) and then make approximations to get a whole number ratio.]

Formula is $CuSO_4.5H_2O$.

Unit 12.1 Activity 2F: Water of crystallisation

1. Nickel chloride is a bright green crystalline solid when it contains water of crystallisation. On heating, the water is driven off and a yellow powder remains.

The following results were obtained when a sample of nickel chloride was heated in order to determine the formula of the hydrated salt $NiCl_2.xH_2O$.

Mass of crucible and lid	21.53 g
Mass of crucible, lid and hydrated nickel chloride	24.09 g
Mass of crucible, lid and nickel chloride after heating and cooling twice	22.84 g

a. Calculate the mass of anhydrous nickel chloride.

b. Calculate the moles of anhydrous nickel chloride. $M(NiCl_2) = 129.7\ g\ mol^{-1}$

c. Calculate the mass and hence the moles of water lost. $M(H_2O) = 18.0\ g\ mol^{-1}$

d. Calculate the formula of the hydrated salt.

2. A 2.07 g sample of hydrated sodium carbonate was heated strongly to determine the water of crystallisation. The final mass of the anhydrous salt was 0.77 g.

a. Calculate the moles of sodium carbonate left after the water was driven off. $M(Na_2CO_3) = 106.0\ g\ mol^{-1}$

b. Calculate the moles of water lost and hence the formula of the hydrated salt.

3. Calcium chloride is often used as a drying agent, since it can absorb water from the atmosphere to become hydrated (ie forms $CaCl_2.xH_2O$).

5.00 g of anhydrous calcium chloride was used as a drying agent until it could absorb no more water. The hydrated crystals had a mass of 9.86 g. Calculate the formula of the hydrated salt. $M(CaCl_2) = 111.1\ g\ mol^{-1}$, $M(H_2O) = 18.0\ g\ mol^{-1}$

4. To find the formula of the hydrated salt $Na_2S_2O_3.xH_2O$, the following data were collected:

- Mass of crucible and lid 20.26 g
- Mass of crucible, lid and hydrated salt 25.58 g
- Mass after heating to constant mass 23.66 g

Calculate the formula of the hydrated salt.

$M(Na_2S_2O_3) = 158\ g\ mol^{-1}$, $M(H_2O) = 18.0\ g\ mol^{-1}$

5. A 5.00 g sample of hydrated barium chloride, $BaCl_2.xH_2O$, was heated to drive off the water. After heating, 4.26 g of anhydrous barium chloride remains. What is the value of x in the formula of $BaCl_2.xH_2O$?

$M(BaCl_2) = 208\ g\ mol^{-1}$, $M(H_2O) = 18.0\ g\ mol^{-1}$

Unit 12.1 Activity 2G: Quantitative analysis – multiple choice

1. 1 mol of ethane, C_2H_6, contains:

A. 1 mol of carbon atoms and 3 mol of hydrogen atoms.

B. 1 mol of carbon atoms and 1 mol of hydrogen atoms.

C. 2 mol of carbon atoms and 6 mol of hydrogen atoms.

D. 2 mol of carbon atoms and 3 mol of hydrogen atoms.

2. What is the mass of 1 mole of C_4H_{10}?

A. 48 g
B. 58 g
C. 14 g
D. 36 g

3. What is the mass of 1 mole of AgCl?

A. 143.5 g
B. 136.5 g
C. 97 g
D. 43 g

4. 118 g of NaCl contains how many moles?

A. 1
B. 2
C. 2.5
D. 4

5. 4 moles of H_2O have what mass?

A. 4 g
B. 40 g
C. 72 g
D. 36 g

6. 0.075 mol of H_2O contains how many molecules?

A. 1.2×10^{-25}
B. 75
C. 6.02×10^{23}
D. 4.5×10^{22}

7. Which of the following equations is balanced correctly?

A. $H_2(g) + Cl_2(g) \rightarrow HCl(g)$
B. $H_2(g) + Cl_2(g) \rightarrow 2HCl(g)$
C. $H_2(g) + \frac{1}{2}Cl_2(g) \rightarrow HCl(g)$
D. $\frac{1}{2}H_2(g) + Cl_2(g) \rightarrow HCl(g)$

8. A compound contains 6.0 g of carbon and 1.0 g of hydrogen. The percentage composition of the compound is:

A. 14% hydrogen and 86% carbon.
B. 86% hydrogen and 14% carbon.
C. 17% hydrogen and 83% carbon.
D. 83% hydrogen and 17% carbon.

9. Which of the following could be an empirical formula?

A. N_2O_4
B. N_2H_6
C. N_2O_5
D. None of the above.

Unit 12.1 Masses, Moles and Concentrations

Topic 3: Concentration and standard solutions

Topic 3 covers solution chemistry:

- Using $c = \frac{n}{V}$.
- Calculating the mass of solid needed to prepare a given volume of standard solution.
- Changing concentration by dilution.

Concentration of solutions

A **solution** is formed when a **solute** is dissolved in a **solvent**. When the solvent is water, the solution is described as an **aqueous solution**. The solute is usually a solid.

The **concentration** of a solution is a measure of the *quantity* of dissolved solute in a given *volume* of solution. The concentration of a chemical solution is often measured in moles per litre, calculated using the following equation:

$c = \frac{n}{V}$ where c = concentration (mol L^{-1})
n = number of moles of dissolved solute (mol)
V = volume of solution (L)

If the three variables are displayed in a triangle, it is easy to see how the formula can be rearranged:

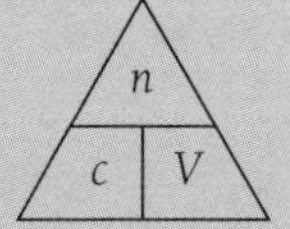

$$n = cV; \quad V = \frac{n}{c}; \quad c = \frac{n}{V}$$

Note: Volume, *V*, must be in litres; convert volume from mL to L by dividing by 1000.

Example A

Concentration of a salt solution

Find the concentration of a salt solution made up in the following way:

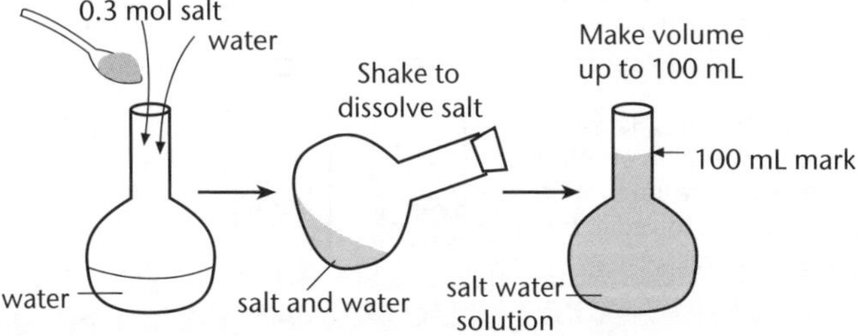

The concentration is $c = \frac{n}{V}$

$= \frac{0.3\ \text{mol}}{0.1\ \text{L}}$ [substituting n = 0.3 mol, V = 0.1 L]

$= 3\ \text{mol L}^{-1}$

The following example shows the relationship between the relative amounts of solute, solvent and a concentration.

Example B

Volumes and concentrations

The flasks below show three solutions of glucose, $C_6H_{12}O_6$, being prepared. The concentration of each solution is calculated.

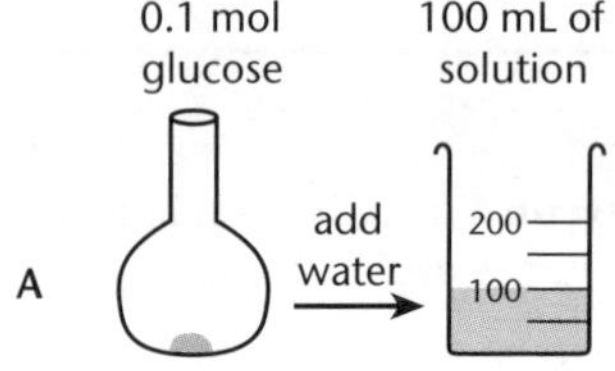

$n = 0.1$ mol
$V = 100$ mL $= 0.1$ L
$c = \frac{n}{V} = \frac{0.1\,\text{mol}}{0.1\,\text{L}} = 1$ mol L^{-1}

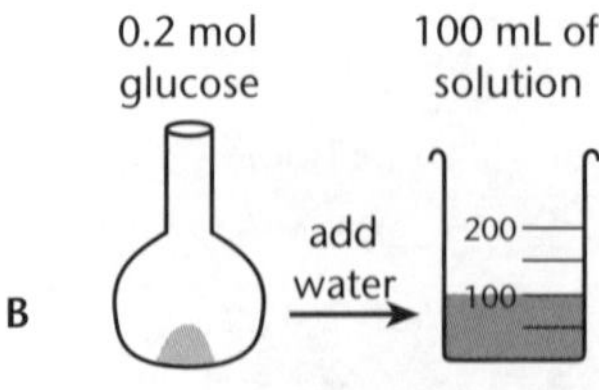

$n = 0.2$ mol
$V = 100$ mL $= 0.1$ L
$c = \frac{n}{V} = \frac{0.2\,\text{mol}}{0.1\,\text{L}} = 2$ mol L^{-1}

$n = 0.2$ mol
$V = 200$ mL $= 0.2$ L
$c = \frac{n}{V} = \frac{0.2\,\text{mol}}{0.2\,\text{L}} = 1$ mol L^{-1}

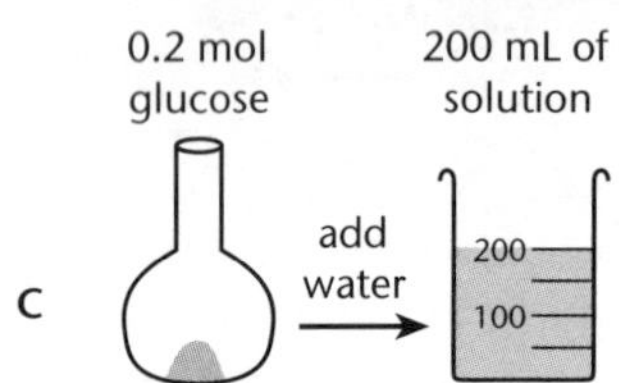

The calculations show:

- The concentration of solution B is *twice* the concentration of solution A because *twice* the amount of solute has been dissolved to give the *same* volume of solution.

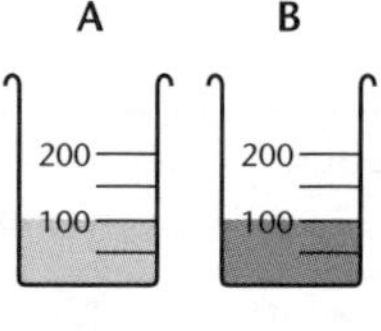

- The concentration of solution B is *twice* the concentration of solution C, because the *same* amount of solute has been dissolved in *half* the volume of water.

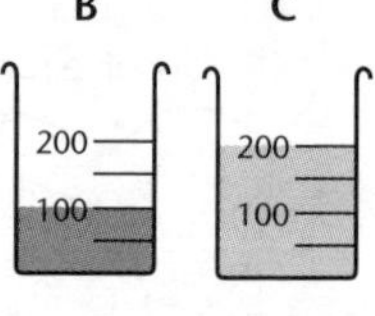

- Solution A and solution C have the *same* concentration although they have been made using *different* amounts of solute and solvent.

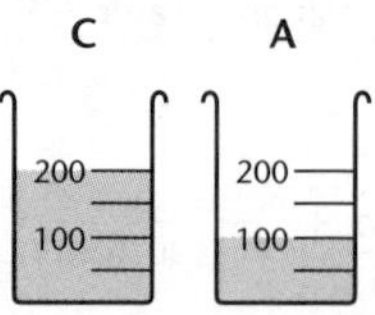

Diluting solutions

An aqueous solution is diluted by adding more water to it.

Example C

Diluting solutions

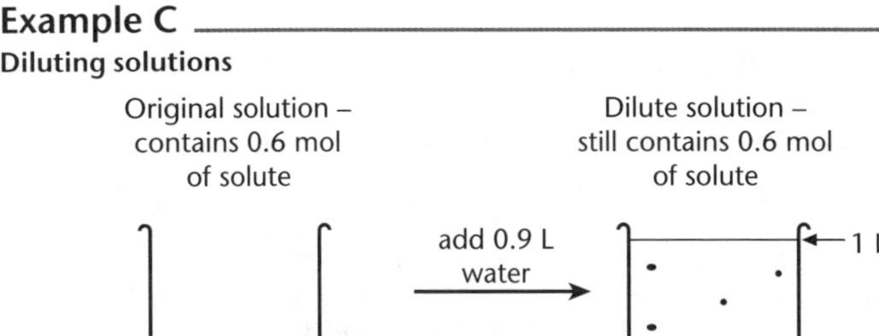

The solution contains 0.6 mol of solute in 0.1 L of solution.

The solution is diluted by adding 900 mL of water. The amount of solute stays the same, but the volume becomes 1 L.

$$c = \frac{n}{V} = \frac{0.6\text{ mol}}{0.1\text{ L}} = 6\text{ mol L}^{-1}$$

$$c = \frac{n}{V} = \frac{0.6\text{ mol}}{1.0\text{ L}} = 0.6\text{ mol L}^{-1}$$

By *increasing* the volume of water × 10, the concentration has *decreased* × 10.

If additional solute is added or some water is removed (by evaporation), a solution becomes more concentrated.

Mass concentration

The concentration of a solution can be measured by using the *mass* of dissolved solute rather than the *amount* of solute. The concentration of the solution, sometimes known as the **mass concentration**, is then calculated using the relationship:

$c = \frac{m}{V}$ where c = concentration (g L^{-1})
m = mass of dissolved solute (g)
V = volume of solution (L)

Example D

Mass concentration

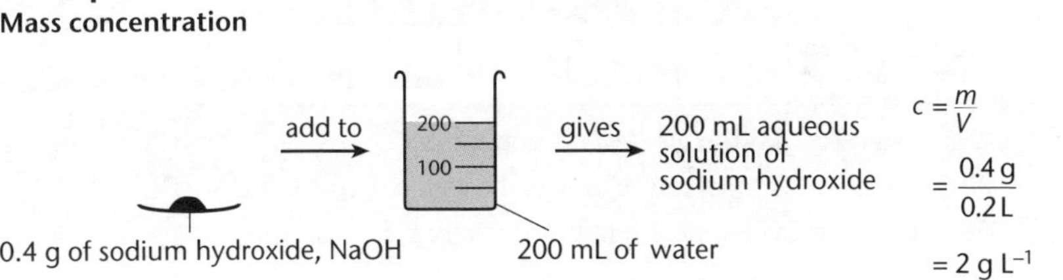

$$c = \frac{m}{V} = \frac{0.4\text{ g}}{0.2\text{ L}} = 2\text{ g L}^{-1}$$

Concentrations in g L^{-1} are easily converted to mol L^{-1} using $n = \frac{m}{M}$.

Example E

Converting concentrations

For the solution in Example D, 0.4 g of sodium hydroxide was dissolved to give 200 mL of aqueous solution.

- Change mass of solute to number of moles of solute:

  mass = 0.4 g

$$n = \frac{m}{M} = \frac{0.4\,\text{g}}{40.0\,\text{g mol}^{-1}} = 0.01\ \text{mol}$$

Na = 23.0
O = 16.0
H = 1.0
∴ M(NaOH) = 40.0

- Convert concentration to moles per litre:

$$c = \frac{n}{V} = \frac{0.01\ \text{mol}}{0.2\ \text{L}} = 0.05\ \text{mol L}^{-1}$$

So, the concentration of sodium hydroxide is 0.05 mol L^{-1}.

Unit 12.1 Activity 3A: Concentrations

1. When cobalt chloride is dissolved in water, it produces a deep-pink-coloured solution. The following solutions of cobalt chloride were prepared:

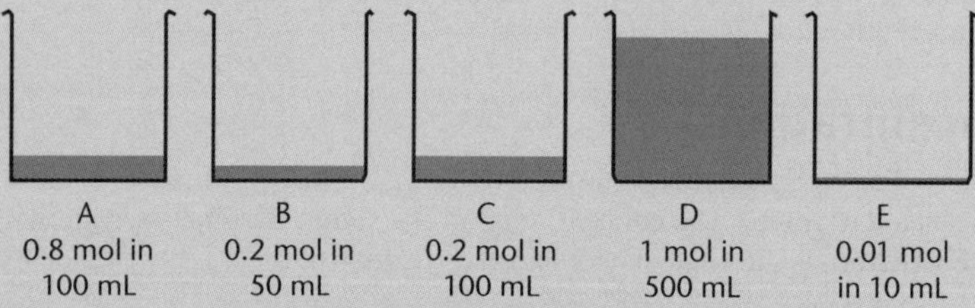

A	B	C	D	E
0.8 mol in 100 mL	0.2 mol in 50 mL	0.2 mol in 100 mL	1 mol in 500 mL	0.01 mol in 10 mL

a. Which solution contains the most dissolved solute?

b. Which solution has the greatest volume?

c. Copy and complete the following table:

	Solutions				
	A	B	C	D	E
Concentration (mol L^{-1})					

d. Which solution has the greatest concentration?

e. Which solution is most dilute (ie has the smallest concentration)?

f. Which solution will appear lightest in colour?

2. Calculate the concentrations of the following solutions in mol L^{-1}.

a. 0.004 mol sodium chloride in 20 mL of solution.

b. 0.050 mol potassium iodide in 40 mL of solution.

c. 0.15 mol of sodium carbonate in 250 mL of solution.

3. Calculate the concentrations of the following solutions in g L^{-1}.

a. **i.** 18 g of glucose, $C_6H_{12}O_6$, in 200 mL of solution.

ii. 180 g of glucose in 400 mL of solution.

b. 23.0 g of ethanol, C_2H_5OH, in 200 mL of solution.

c. 39.9 g of copper sulfate, $CuSO_4$, in 0.500 L of solution.

4. Find the concentration of the following solutions in both g L^{-1} and mol L^{-1}.

a. 58.5 g of sodium chloride is dissolved to make 2.00 L of solution.

b. 10.6 g of sodium carbonate, Na_2CO_3, is dissolved to give a 100.0 mL aqueous solution.

c. 0.2 g sodium hydroxide, NaOH, is dissolved in water to make a 10 mL solution.

d. 1.7 g of silver nitrate, $AgNO_3$, is dissolved to produce a 500 mL solution.

e. 5.3 g of ammonium chloride, NH_4Cl, is dissolved to produce a 1.00 L solution.

5. Find the concentrations of the following solutions, before and after dilution.

a. 0.5 mol NaCl in 500 mL solution is diluted by adding 500 mL of water.

b. 0.1 mol sodium hydroxide in 250 mL is diluted to 1000 mL.

c. 0.02 mol sodium chloride in 5 mL of solution is diluted to 500 mL by adding 495 mL of water.

d. 10 mL of a solution of hydrochloric acid containing 0.04 mol of solute is diluted to 100 mL by adding 90 mL of water.

6. Calculate the mass of solid needed to prepare the following solutions.

a. 250 mL of 0.0100 mol L^{-1} $AgNO_3$. $M(AgNO_3)$ = 170 g mol^{-1}.

b. 500 mL of 0.500 mol L^{-1} $K_2Cr_2O_7$. $M(K_2Cr_2O_7)$ = 294.2 g mol^{-1}.

c. 100 mL of 0.100 mol L^{-1} $BaCl_2$. $M(BaCl_2)$ = 208 g mol^{-1}.

d. 200 mL of 1.00 mol L^{-1} $FeSO_4.7H_2O$. $M(FeSO_4.7H_2O)$ = 277.9 g mol^{-1}.

7. A colorimetric analysis of a solution containing Co^{2+}(aq) is to be carried out.

Calculate the mass of $CoCl_2.6H_2O$ that would be needed to prepare 250 mL of a stock solution of concentration 0.00100 mol L^{-1}.

$M(CoCl_2.6H_2O)$ = 237.8 g mol^{-1}

8. A colorimetric analysis of a solution containing Fe^{3+}(aq) is to be carried out.

Calculate the mass of $Fe(NO_3)_3.9H_2O$ that would be needed to prepare 500 mL of a stock solution of concentration 0.0125 mol L^{-1}.

$M(Fe(NO_3)_3.9H_2O)$ = 403.9 g mol^{-1}

Standard solutions

A **standard solution** is a solution for which the concentration is accurately known. Standard solutions are used in quantitative work for finding the concentration of other solutions.

A distinction is made between **primary standard** and **secondary standard** solutions:

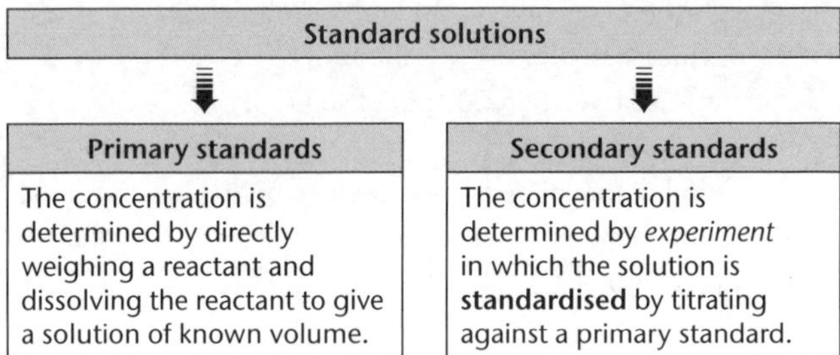

For a substance to be suitable as a primary standard, it must:

- be readily available in a pure form.
- resist absorbing water.
- be **stable** (unreactive) when stored.
- have a high molar mass, so weighing errors are less significant.

Example F

Sodium carbonate

Anhydrous sodium carbonate, Na_2CO_3, is a common acid–base primary standard, since it is pure, readily available, *not* **hygroscopic** (ie not water absorbing), forms stable solutions and has a relatively high molar mass (106 g mol^{-1}).

Neither sodium hydroxide, NaOH, nor hydrochloric acid, HCl, make suitable primary standards, because:

- NaOH rapidly absorbs both moisture and carbon dioxide from the air, thus the mass changes during and after weighing.
- HCl is not sufficiently pure – variation occurs in the product when it is commercially prepared.

Solutions of sodium hydroxide and hydrochloric acid can be *standardised* against a primary standard and then used as secondary standards.

For both primary and secondary standard solutions, the concentration is found using the equation:

$$c = \frac{n}{V}$$

where: n is the amount of dissolved solute (moles)
c is the concentration of the standard solution (in mol L^{-1})
V is the volume of standard solution (in L)

Where possible:

- All volumetric analysis measurements should be to two decimal places.
- All concentrations should be quoted to three significant figures.

Example G

Standard sodium carbonate solution

A standard solution of sodium carbonate, Na_2CO_3, is prepared by dissolving 5.30 g of Na_2CO_3 in water to give a volume of 250.0 mL.

$M(Na_2CO_3) = 106.0 \text{ g mol}^{-1}$

$$n = \frac{\text{mass } Na_2CO_3 \text{ used}}{\text{molar mass } Na_2CO_3}$$

$$= \frac{5.30 \text{ g}}{106.0 \text{ g mol}^{-1}}$$

$$= 0.0500 \text{ mol}$$

The concentration of the sodium carbonate solution is given by:

$$c = \frac{n}{V}$$

$$= \frac{0.0500 \text{ mol}}{0.2500 \text{ L}} \quad \text{[substituting } V = 250.0 \text{ mL} = 0.2500 \text{ L]}$$

$$= 0.200 \text{ mol L}^{-1}$$

Preparing a primary standard soution

Step 1

Accurately weigh the required solute (ie primary standard reagent) into a clean, dry, glass container (eg a small beaker). Record the mass of solute used.

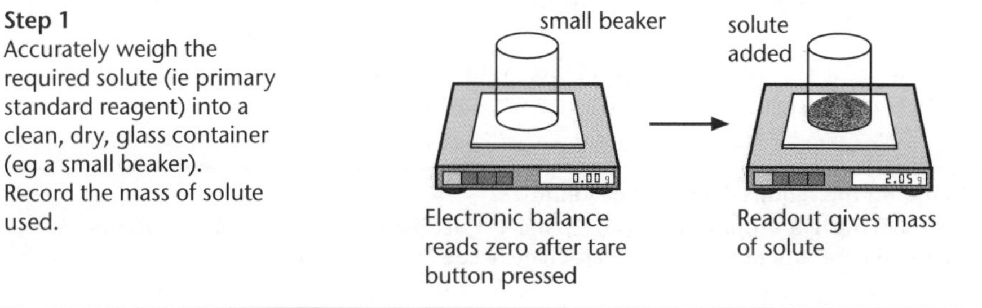

Step 2

Transfer the solute to a volumetric flask using a funnel and wash-bottle. Take care to transfer all the solute to the **volumetric flask**. Use deionised/ distilled water if available.

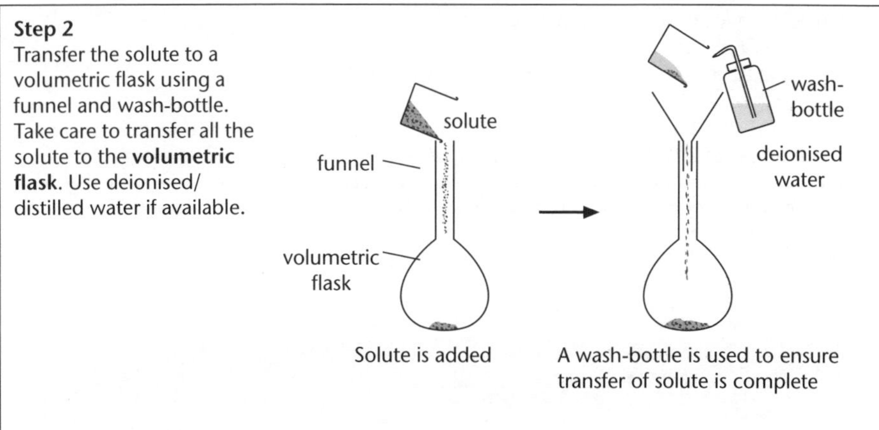

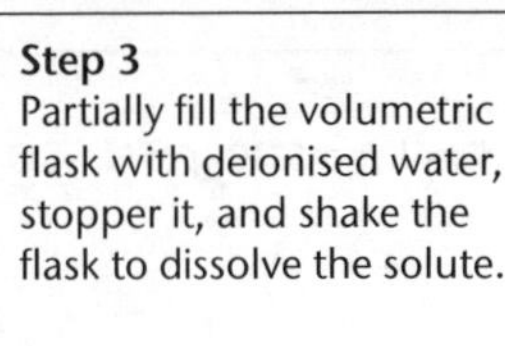

Step 3
Partially fill the volumetric flask with deionised water, stopper it, and shake the flask to dissolve the solute.

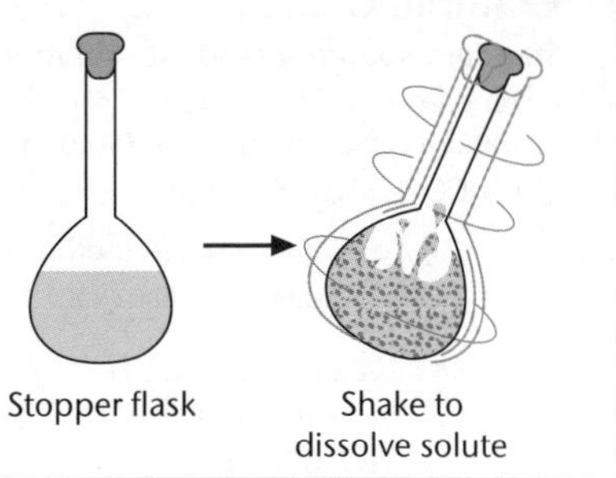

Step 4
Remove the stopper and let any solution run down from the top of the neck of the volumetric flask. Add deionised water to fill the volumetric flask up to the volume mark. Ensure the *bottom* of the meniscus sits on the volume mark. Stopper the flask, and invert it a number of times to mix well. Label the flask, stating the name and concentration of the solution.

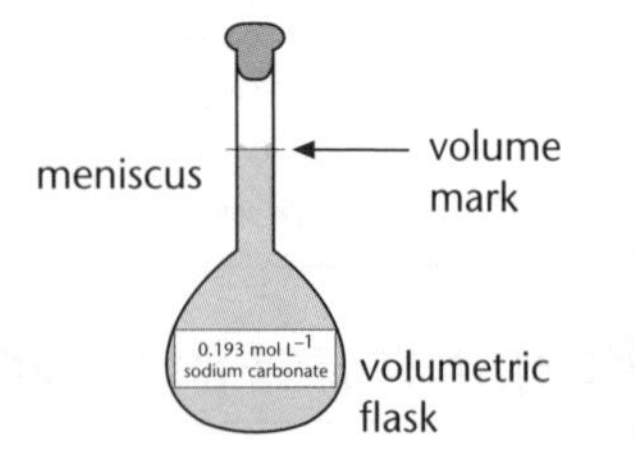

In practice, it is hard to weigh exactly the required amount, so an amount close to the required amount, but *accurately known*, is weighed out.

Example H

Finding the required amount of solute

If 100.0 mL of an approximately 0.200 mol L^{-1} standard solution of sodium carbonate is required, the amount of sodium carbonate needed is:

$n = cV$
$= 0.200 \times 0.100$ [substituting and changing mL to L]
$= 0.0200$ mol

Hence, the mass of sodium carbonate needed is:

$m = nM$
$= 0.0200 \times 106.0$ [substituting $M(Na_2CO_3) = 106.0$ g mol^{-1}]
$= 2.12$ g

Usually, the solution needs to be close to, but not exactly, the required concentration. An amount close to 2.12 g will be weighed out, ie between 2.02 g and 2.22 g. The sodium carbonate is then transferred to a 100 mL volumetric flask and dissolved in deionised water and the volume of the solution is brought up to 100 mL and the exact concentration of the solution is calculated.

Example I

Calculating the concentration of the primary standard

If 2.05 g of sodium carbonate was weighed out, rather than the 2.12 g calculated in Example H, the concentration of the standard solution would be 'close enough'.

Amount of sodium carbonate:

$$n = \frac{m}{M}$$

$$= \frac{2.05\ \text{g}}{106.0\ \text{g mol}^{-1}} \quad \text{[substituting } M(Na_2CO_3) = 106.0\ \text{g mol}^{-1}\text{]}$$

$$= 0.0193\ \text{mol}$$

The concentration of the sodium carbonate standard solution is therefore actually:

$$c = \frac{n}{V}$$

$$= \frac{0.0193\ \text{mol}}{0.100\ \text{L}} \quad \text{[substituting 100 mL = 0.100 L]}$$

$$= 0.193\ \text{mol L}^{-1}$$

Note that the concentration is *known* to be 0.193 mol L^{-1} – this concentration is 'close enough' to 0.200 mol L^{-1}.

Preparing secondary standard solutions

Secondary standards are standard solutions prepared by making up a solution to the approximate concentration required, then finding the concentration *exactly* by **titrating** against a primary standard solution. Finding the concentration is called **standardising** the solution.

Example J

Standard sodium hydroxide

Sodium hydroxide cannot be used to make a primary standard solution because it is **deliquescent** (solid NaOH dissolves in water it has absorbed from the air), thus NaOH does not have a constant composition and its mass can change. Standard sodium hydroxide solution must therefore be prepared as a secondary standard.

Most acid solutions, such as HCl(aq), must be prepared as secondary standards.

Unit 12.1 Activity 3B: Standard solutions

1. Name the particles present and calculate the amount (in mol) of each in the following aqueous solutions. (**Note:** All solutions contain water molecules. Do not include these in the answers.)

a. 5 mL of 0.1 mol L^{-1} sodium chloride solution.

b. 25 mL of 0.5 mol L^{-1} hydrochloric acid.

c. 10 mL of 0.2 mol L^{-1} sodium carbonate solution.

d. 100 mL of 1 mol L^{-1} sodium hydroxide solution.

e. 50 mL of 2 mol L^{-1} nitric acid.

f. 35 mL of 0.25 mol L^{-1} iron(III) chloride solution.

g. 0.5 mL of 0.30 mol L^{-1} sodium sulfate solution.

h. 30 mL of 0.01 mol L^{-1} calcium hydroxide solution.

i. 20 mL of 0.25 mol L^{-1} sulfuric acid.

2. You are provided with a 14.32 g sample of hydrated sodium carbonate, $Na_2CO_3.10H_2O$, in a beaker, a 250 mL volumetric flask, a funnel, a spatula, a pipette, a wash-bottle and a supply of distilled water.
   a. Describe the steps that you would take to prepare a standard solution of sodium carbonate. Include the equipment that you would use in each step.
   b. Calculate the concentration of this standard solution.
   $M(Na_2CO_3.10H_2O) = 286.1 \text{ g mol}^{-1}$
   c. Is this a primary or a secondary standard solution?
3. A student was asked to prepare 250 mL of standard 0.200 mol L^{-1} sodium carbonate solution.
   An empty beaker was weighed (116.48 g) and *anhydrous* sodium carbonate, Na_2CO_3, was added until the combined mass of beaker and sodium carbonate was 122.05 g. The solid was transferred to a 250 mL volumetric flask. The sodium carbonate was dissolved in deionsed water and the volume was made up to the mark.
   $M(Na_2CO_3) = 106.0 \text{ g mol}^{-1}$
   a. **i.** What mass of sodium carbonate was weighed out?
   ii. What amount of sodium carbonate is this?
   b. What is the concentration of the standard solution in mol L^{-1}?
   c. If a 20.00 mL sample of this solution was taken, what amount (mol) of carbonate ions would be present in the sample?
4. 5.30 g of sodium carbonate, Na_2CO_3, is dissolved in water to make 250.0 mL of solution.
   $M(Na_2CO_3) = 106.0 \text{ g mol}^{-1}$
   a. Find the amount of Na_2CO_3 in 5.30 g of the solid.
   b. Find the concentration in mol L^{-1} of the sodium carbonate solution.
   c. What amount of CO_3^{2-} is present in 20.0 mL of this solution?
5. 10.00 g of anhydrous sodium hydroxide is dissolved in water to make 500.0 mL of solution.
   a. Determine the molar mass of NaOH.
   b. Calculate the amount of NaOH in 10.00 g.
   c. Calculate the concentration of the NaOH(aq) in mol L^{-1}.
   d. What amount of OH^-(aq) is present in 75.00 mL of this solution?
6. **a.** 0.16 g of anhydrous sodium hydroxide is dissolved in enough distilled water to form 50.0 mL of solution. Find the concentration of this solution.
   $M(NaOH) = 40.0 \text{ g mol}^{-1}$
   b. Explain why this solution would not make a good primary standard.

Unit 12.1 Masses, Moles and Concentrations

Topic 4: Atomic structure

Topic 4 covers atomic structure:

- The structure of atoms and ions.
- Mass number and atomic number.
- Isotopes.

Introduction

Atoms are very small particles made up of three main **subatomic particles – protons**, **neutrons** and **electrons**. Protons and neutrons are found in the **nucleus**, a very small central region.

The nucleus is about $\frac{1}{10\,000}$ the size of an atom.

Protons and neutrons have approximately equal masses (measured in atomic mass units, amu), and are much heavier than electrons, so the nucleus contains over 99% of the mass of the atom. Electrons move rapidly in a wide region around the nucleus.

Subatomic particle	Symbol	Mass compared to a proton	Charge	Location
Proton	p	1	+1	In the nucleus
Neutron	n	1	0	In the nucleus
Electron	e	$\frac{1}{1840}$	–1	Moving outside the nucleus

The nucleus is positively charged because it contains positively charged protons and neutral neutrons.

All atoms are *neutral* because they contain equal numbers of protons and negatively charged electrons.

Example A

Atomic structure

A sodium atom has 11 protons, 12 neutrons and 11 electrons.

Structure of the sodium atom

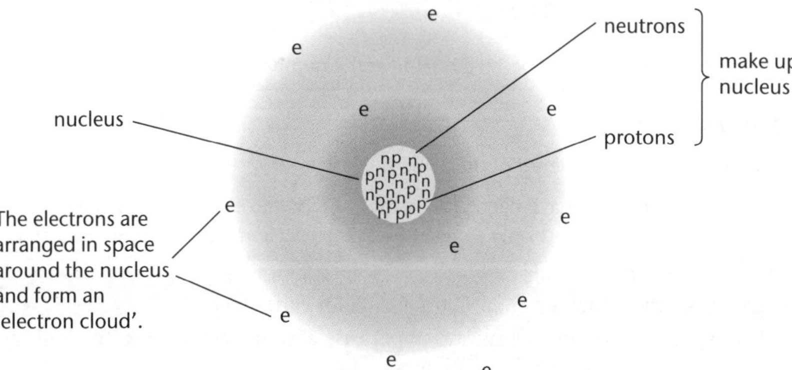

The nucleus is drawn much larger than scale.

The New Zealand scientist Ernest Rutherford first suggested the structure of the atom (shown in Example A) following a series of experiments involving **alpha** (α) **particles** (the nuclei of helium atoms) fired at gold foil a few hundred atoms thick:

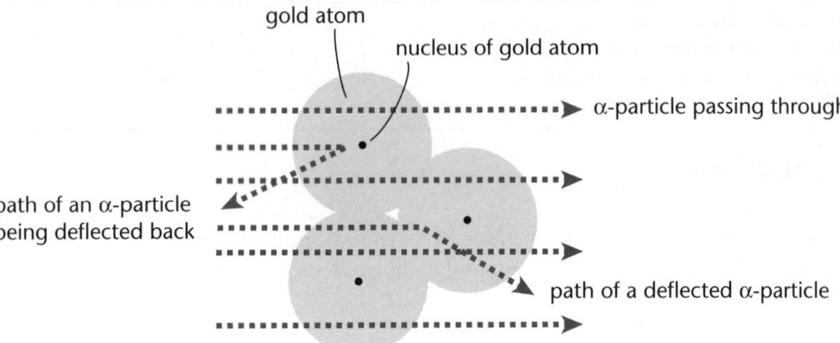

Nearly all the α-particles passed right through the foil, which suggested to Rutherford that the atom was mainly empty space. However, a few (about 1 in 20 000) were deflected. Rutherford concluded that these deflections were due to α-particles 'hitting' the nuclei of gold atoms.

Mass number and atomic number

The **atomic number** (symbol **Z**) of an atom is the number of protons in the nucleus. Because atoms are neutral, the number of electrons in an atom is equal to the atomic number.

The **mass number** (symbol **A**) of an atom is the number of protons *plus* the number of neutrons in the nucleus.

> For any atom, the number of neutrons is $A - Z$.

Example B

Mass number and atomic number

A chlorine atom has 17 protons and 18 neutrons, so:

- Atomic number $Z = 17$.
- Mass number A = number of protons + number of neutrons
  $= 17 + 18$
  $= 35$
- Number of *electrons* must be 17, the same as the number of protons.

Elements

Elements are substances made up of only one type of atom.

Over 118 different elements have been 'discovered'. A few have yet to be studied completely. Some elements exist only transiently before decaying into a different element. The symbol of an element is *either*:

- a single upper case letter, eg H for hydrogen, C for carbon; *or*
- two letters, the first of which is always upper case and the second of which is always lower case, eg He for helium, Cu for copper.

The symbol for an element, X, can be written together with the mass number, A, and atomic number, Z, as ${}^{A}_{Z}\mathrm{X}$.

Example C

Using *A* and *Z*

$^{23}_{11}Na$ is the symbol for a sodium atom.

The 11 shows there are 11 protons (and 11 electrons).

The number of neutrons is: $A - Z = 23 - 11$
$= 12$

Isotopes

Isotopes are atoms with the *same* number of protons (ie same atomic number) but *different* numbers of neutrons (ie different mass numbers). Isotopes of an element have different masses due to the different numbers of neutrons in their nuclei. Isotopes of an element behave the same way in chemical reactions since the number of *electrons* determines the way an atom reacts. An element can be made up of a mixture of isotopes.

Example D

Isotopes of oxygen

For the element oxygen there are three isotopes:

Isotope	$^{16}_{8}O$	$^{17}_{8}O$	$^{18}_{8}O$
Number of electrons	8	8	8
Number of protons	8	8	8
Number of neutrons	8	9	10

Since all atoms of the same element differ only in their number of neutrons, a particular isotope of an element can be described by writing the symbol and mass number without the atomic number.

Example E

Describing isotopes

The isotope $^{16}_{8}O$ can be written as ^{16}O or O-16 and is called 'oxygen-16'.

$^{18}_{8}O$ can be written as ^{18}O or O-18 and is called 'oxygen-18'.

Some isotopes have special names.

Example F

Naming isotopes

$^{2}_{1}H$ or hydrogen-2 is called **deuterium.**

$^{3}_{1}H$ or hydrogen-3 is called **tritium.**

Unit 12.1 Activity 4A: Atoms, elements and isotopes

1. For a carbon atom with six protons and eight neutrons:
   - **a.** Find the number of electrons.
   - **b.** Write down the atomic number.
   - **c.** Write down the mass number.
2. What is the mass number of an atom containing 16 protons, 18 neutrons and 16 electrons?
3. Consider atom Y with the symbol $^{39}_{19}Y$.
   - **a.** What is the mass number of the atom?
   - **b.** How many neutrons does this atom have?
4. Copy the following table and identify *a–r*.

Symbol	Mass number	Atomic number	Number of neutrons	Number of protons
$^{3}_{1}H$	3	1	*a*	*b*
$^{c}_{d}C$	13	*e*	*f*	6
$^{g}_{h}Fe$	*i*	26	30	*j*
$^{k}_{17}Cl$	*l*	*m*	20	*n*
$^{16}_{o}O$	*p*	*q*	*r*	8

5. Consider the atom with the symbol $^{42}_{20}X$.
   - **a.** What is the mass number of the atom?
   - **b.** How many neutrons does the atom have?
6. What is the difference, in terms of subatomic particles, between $^{1}_{1}H$ and $^{3}_{1}H$?
7. Name the following elements:
   - **a.** Be **b.** Na
   - **c.** Pb **d.** Fe
   - **e.** Cu **f.** Li
   - **g.** Ag **h.** Au
   - **i.** K **j.** Si
8. Atom A has 10 protons, 10 neutrons and 10 electrons.
   Atom B has 10 protons, 12 neutrons and 10 electrons.
   What is the relationship between atoms A and B?
9. **a.** Define the term 'isotope'. Using one element, give two examples.
   b. Explain why isotopes of the same element show the same chemical properties.

Electron arrangement

The energy of electrons is **quantised**, ie electrons can only have certain energies with no energy values in between. Electrons occupy certain **energy levels** around the nucleus.

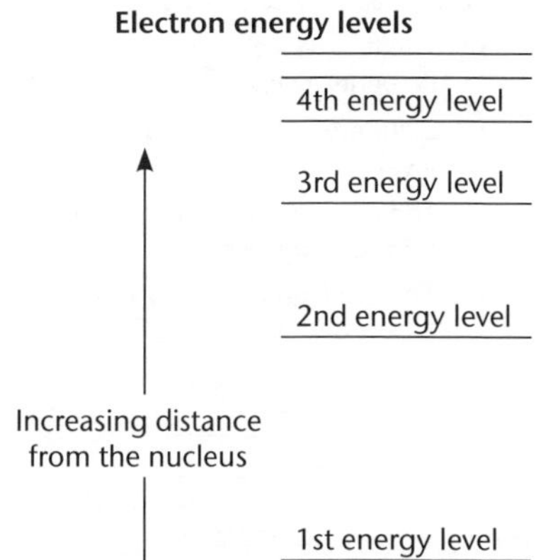

Electrons in atoms occupy lower energy levels *before* they occupy higher energy levels. There is a limit to the number of electrons that can be present in any one energy level, and the following rules hold for the first 20 elements:

- The first energy level holds a maximum of *two* electrons.
- The second level holds a maximum of *eight* electrons.
- The third level holds a maximum of *eight* electrons.
- The fourth level holds the remaining electrons.

For elements with more than 20 protons, the third level holds a maximum of 18 electrons.

The way electrons are arranged is called the **electron configuration** (or electron arrangement) of the atom.

Electron configuration can be written in brackets after the element symbol with the number of electrons in each level separated by commas.

Example G

Electron configurations

Carbon

The atomic number of carbon is 6.
Carbon has six electrons and six protons:

- The first two electrons are in the first energy level.
- The remaining four are in the second energy level.
- The electron configuration for carbon is written C (2, 4).

Argon

The atomic number of argon is 18.
Argon has 18 electrons:

- The first two electrons are in the first energy level.
- Eight are in the second energy level.
- The remaining eight are in the third energy level.
- The electron configuration for argon is written Ar (2, 8, 8).

C (2, 4)		Ar (2, 8, 8)
	3rd energy level	••••••••
••••	2nd energy level	••••••••
••	1st energy level	••
carbon atom		argon atom

Key
• electron

Valence electrons

Electrons in the highest occupied energy level are called **valence electrons** or outer-shell electrons, and the energy level they are found in is the **valence level** or *outer level*.

Example H

Valence electrons

For a carbon atom (see Example G), the *second* energy level is the valence level and carbon has four valence electrons.

For an argon atom (see Example G), the *third* energy level is the valence level and argon has eight valence electrons.

Valence electrons for a particular atom can be represented using **electron dot diagrams** (also called **Lewis diagrams** or **Lewis structures**). These show the symbol for the atom together with dots (•) for each valence electron.

Electron dot diagrams are usually drawn according to the following conventions:

- Imagine four possible sites for the valence electrons around the symbol for the atom:

symbol for the atom

one of four possible valence electron sites

- Valence electrons are first placed *unpaired* around the symbol, until all sites are occupied, eg:

Na Mg Al Si

- Additional valence electrons then pair up with the single electrons, eg:

P S Cl Ar

It is *only* valence electrons that are involved in chemical reactions.

Elements whose atoms have the same number of electrons in their outer shell, ie atoms with the same number of valence electrons, tend to have similar chemical properties.

Example I

Valence electrons

Li (2, 1), Na (2, 8, 1) and K (2, 8, 8, 1) all have one valence electron. These elements are all metals and all have similar chemical properties, eg they all react violently with water to produce hydrogen gas.

Unit 12.1 Activity 4B: Electron arrangements

1. Write the electron arrangements for:

a. Lithium. **b.** Nitrogen. **c.** Sodium-23. **d.** Argon.

e. Potassium. **f.** $^{16}_{8}O$ **g.** $^{17}_{8}O$ **h.** $^{40}_{20}Ca$

2. Name the element that has atoms with the electron configuration of 2, 8, 7.

3. Draw electron dot diagrams for the following atoms:

a. Hydrogen. **b.** Nitrogen. **c.** Fluorine. **d.** Calcium.

e. Sulfur. **f.** Potassium. **g.** Neon.

4. Answer questions **a.** to **e.** by choosing from the key list A–E:

A. $^{1}_{1}H$ **B.** $^{4}_{2}He$ **C.** $^{12}_{6}C$ **D.** $^{23}_{11}Na$ **E.** $^{37}_{17}Cl$

a. Which element has four valence electrons?

b. Which two elements have one valence electron?

c. Which atom has its valence electrons in the second energy level?

d. Give an example of an element with seven valence electrons.

e. Which element has a completely filled valence electron level?

Valence electrons and the periodic table

When elements are arranged in order of the increasing number of protons in the nucleus of their atoms, *and* with atoms with the same number of valence electrons in the same column, they form the **periodic table**. Elements in the same row will have the valence electrons in their atoms in the same energy level.

- Elements in the same row *across* the periodic table are said to be in the same **period**.
- Elements within the same column *down* the periodic table are said to be in the same **group**.

The number of valence electrons in the atoms of the first 20 elements of the periodic table is determined as follows:

- For groups 1 and 2:
  Number of valence electrons = group number
  (eg group 2 elements all have 2 valence electrons).
- For groups 13–18:
  Number of valence electrons = group number – 10
  (eg group 17 elements have 17 – 10 = 7 valence electrons).

The number of valence electrons *increases* across a period.

Example J

Valence electrons across a period

The number of valence electrons increases across the second period.

H							He	Period 1
Li	Be	B	C	N	O	F	Ne	Period 2
Na	Mg	Al	Si	P	S	Cl	Ar	Period 3

Element	Li	Be	B	C	N	O	F	Ne
Number of valence electrons	1	2	3	4	5	6	7	8

Elements in the same group have the same number of valence electrons, but their valence electrons exist in different valence levels.

The number of valence electrons remains the *same* down a group.

Example K

Valence electrons down a group

Group 1 elements each have one valence electron.

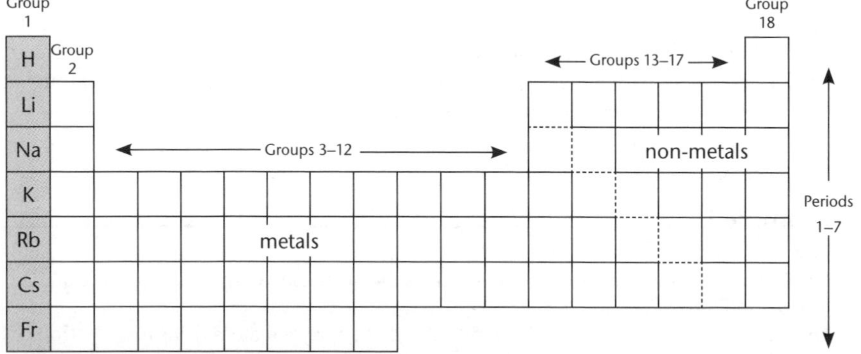

Element	Group	Period	Number of valence electrons
H	1	1	1
Li	1	2	1
Na	1	3	1
K	1	4	1
Rb	1	5	1
Cs	1	6	1
Fr	1	7	1

Metals are found on the left-hand side of the periodic table – all of groups 1–12 – and some are found in groups 13–16.

Non-metals are found on the right-hand side of the periodic table – mainly in groups 13–17 and all of group 18.

Unit 12.1 Activity 4C: The periodic table

1. Describe the relationship of the number of valence electrons in the atoms of the elements in the periodic table to the *groups* and *periods* in the table.
2. Compare the number of valence electrons of atoms of metal elements to atoms of non-metal elements for the first 20 elements of the periodic table.

Ions

The gain or loss of electrons by atoms produces **ions**. When atoms gain or lose electrons they produce i*ons*.

Ions are either cations or anions:

- **Cations**,or *positively* charged ions, form when atoms or groups of (*covalently bonded*) atoms lose electrons. Cations have *fewer* electrons than protons, eg, magnesium ions,

Mg^{2+}, have 10 electrons and 12 protons whereas the atoms of the element, Mg, have 12 electrons and 12 protons; aluminium ions, Al^{3+}, have 10 electrons and 13 protons, whereas atoms of the element, Al, have 13 electrons and 13 protons.

- **Anions**, or *negatively* charged ions, form when atoms or groups of (*covalently bonded*) atoms gain electrons. Anions have *more* electrons than protons, eg, chloride ions, Cl^-, have 18 electrons and 17 protons, whereas atoms of the element, Cl, have 17 electrons and 17 protons; oxide ions, O^{2-}, have 10 electrons and 8 protons, whereas atoms of the element, O, have 8 electrons and 8 protons.

Ions are written with the charge shown on the *right* and *above* the symbol for the atom or group of atoms.

Example L

Ions

The sodium ion is written Na^+, which indicates that a sodium atom loses one electron when it forms a sodium *cation.*

The oxygen ion is written O^{2-}, which indicates that an oxygen atom gains two electrons when it forms an oxide *anion.*

Atoms gain or lose electrons to produce stable (full) valence shells.

Cations

Metals are in groups 1–12 and 13–16, and tend to *lose* electrons to form **monatomic** ('one atom') *cations.*

Example M

Sodium atoms and sodium ions

Sodium is element 11 and has 11 protons and 11 electrons. Sodium ions form when sodium atoms lose an electron from the third energy level.

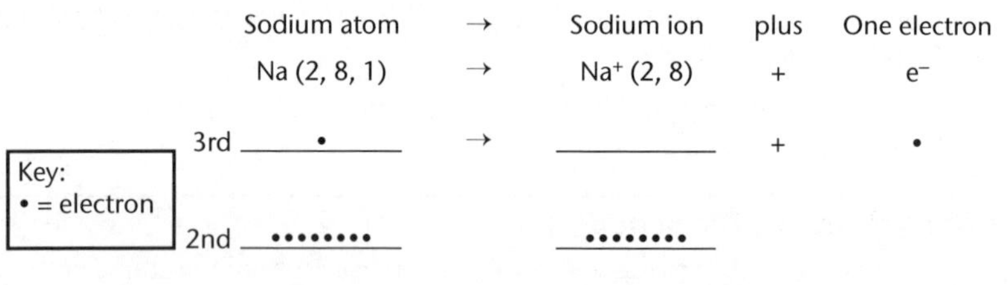

Sodium atom

- Valence level is *third* energy level and has *one* electron.
- Has no charge, because has 11 protons and 11 electrons.

Sodium ion

- Valence level is *second* energy level and has *eight* electrons.
- Has a charge of +1, because has 11 protons but only 10 electrons.

Anions

Non-metals tend to *gain* electrons to form monatomic *anions*. These ions are named using the suffix *-ide*, eg chlor*ide*.

Example N

Chlorine atoms and chloride ions

Chlorine atoms have 17 protons and 17 electrons. A chlorine atom can fill the third valence level by gaining one electron to form a chloride ion, Cl^-.

	Chlorine atom	plus	One electron	→	Chloride ion
	Cl (2, 8, 7)	+	e^-	→	Cl^- (2, 8, 8)
3rd	•••••••	+	•	→	••••••••
2nd	••••••••				••••••••
1st	••				••
	chlorine atom				**chloride ion**

Chlorine atom

- Valence level is *third* energy level and has *seven* electrons.
- Has no charge, because it has 17 protons and 17 electrons.

Chloride ion

- Valence level is *third* energy level and has *eight* electrons.
- Has a charge of –1, because it has 17 protons but 18 electrons.

The group 18 elements – such as neon and argon – are atoms with stable valence levels so they are unlikely to form ions.

Polyatomic ions

Ions that consist of a stable *group* of atoms with an overall charge are called **polyatomic ions**.

A polyatomic ion with two atoms is called a **diatomic ion**.

Example O

Polyatomic ions

The ammonium ion, NH_4^+, is polyatomic. It consists of one atom of nitrogen and four hydrogen atoms, bonded together, and overall has one *less* electron than it has protons, so it has a charge of +1.

The hydroxide ion, OH^-, is diatomic. It consists of one oxygen atom and one hydrogen atom, and overall has one *more* electron than it has protons, so it has a charge of –1.

Valency of ions

The **valency** of an ion refers to the magnitude or size of the charge of the ion, not whether it is positive or negative.

Example P

Valency

The sodium ion, Na^+, chloride ion, Cl^-, and ammonium ion, NH_4^+, all have a valency of 1. The aluminium ion, Al^{3+}, has a valency of 3 and the oxide ion, O^{2-}, has a valency of 2.

The **transition metals**, such as copper, Cu, manganese, Mn, and iron, Fe, form cations that can have more than one valency. A roman numeral is used to indicate the valency of a transition metal ion.

Example Q

Transition metal ions

Copper can form two copper ions, Cu^+ and Cu^{2+}; and iron, Fe, can form two ions, Fe^{2+} and Fe^{3+}:

- Cu^+ is called copper(I) ion; Cu^{2+} is called copper(II) ion.
- Fe^{2+} is called iron(II) ion; Fe^{3+} is called iron(III) ion.

Unit 12.1 Activity 4D: Ions

1. Write the symbols for each of the following ions:

a. Lithium ion. **b.** Bromide ion. **c.** Barium ion.
d. Sulfide ion. **e.** Silver ion. **f.** Fluoride ion.
g. Iron(II) ion. **h.** Copper(I) ion. **i.** Nitrate ion.
j. Hydroxide ion. **k.** Hydrogen carbonate ion. **l.** Sulfate ion.

2. Write the names and give the valency for each of the following ions:

a. Mg^{2+} **b.** Zn^{2+} **c.** Pb^{2+} **d.** O^{2-} **e.** Cl^-
f. Fe^{3+} **g.** Al^{3+} **h.** I^- **i.** Cu^{2+} **j.** F^-

3. For each of the ions:

a. K^+ **b.** O^{2-} **c.** F^- **d.** Al^{3+}

i. Name the ion and write down its electron arrangement.
ii. State whether the ion formed from the corresponding atom by gain or loss of electrons.

4. Write the electron arrangement for the:

a. Calcium atom.
b. Calcium ion.
c. Sulfide ion.

Unit 12.1 Masses, Moles and Concentrations

Topic 5: Molecules

Topic 5 covers molecules:

- Describing bonding using Lewis structures.
- Describing atoms or molecules and the attractive forces between them (weak intermolecular forces).
- Predicting polarity of molecules from bond polarity and shape.
- Understanding polar solvents – water and ethanol; non-polar hydrocarbon solvents (eg hexane, cyclohexane, mineral turpentine).

Introduction

Most natural substances do not consist of isolated atoms; they consist of elements or compounds, in which the atoms are bound strongly to other atoms by **chemical bonds**. A chemical bond is an electrostatic force of attraction between positively and negatively charged species. There are two general ways that atoms combine to form compounds – either electrons are shared between atoms, or they are transferred between atoms.

- When electrons are shared, the force of attraction between the atoms is called a **covalent bond** and electrically neutral **molecules** are formed.
- When electrons are transferred between atoms, electrically charged particles called ions are formed and the force of attraction between the ions is known as an **ionic bond**.

Example A

Chemical bonds

Water, H_2O, consists of an oxygen atom linked by chemical bonds to two hydrogen atoms.

Sodium chloride, NaCl, consists of sodium ions, Na^+, attracted strongly to chloride ions, Cl^-. These forces of strong **electrostatic attraction** are a chemical (*ionic*) bond.

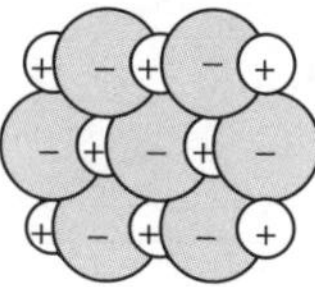

When chemical bonding occurs, there is a tendency for most atoms to acquire the electron arrangement of the closest **inert gas** (ie an outer shell with eight electrons). This is sometimes known as the **octet rule**.

Covalent bonding

Covalent bonds form when electrons are *shared*. The shared electrons, called **bonding electrons**, 'belong' to both atoms *simultaneously*. Shared electrons come from the valence shells of the atoms sharing electrons and are found between the nuclei:

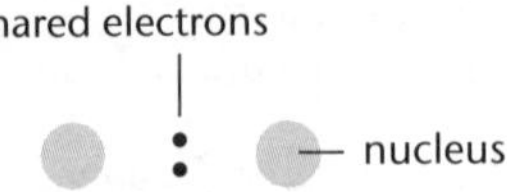

Covalent bonds usually form between atoms of the non-metals, such as those in groups 14–16.
When groups of atoms bond covalently, molecules are formed, such as hydrogen gas, H_2, methane, CH_4, and carbon dioxide, CO_2.

Example B

Chlorine, Cl_2

Chlorine is element 17. Its atom has the electron configuration Cl (2, 8, 7).

Two atoms of chlorine

valence electrons

Cl

Cl

nucleus (inner-shell electrons *not* shown)

chlorine atoms move together

Two atoms of chlorine covalently bonded together

bonding electrons

Cl

Cl

Outer shells overlap and a pair of electrons are shared.

Each atom now has a full valence shell of eight electrons.

In a covalent bond, the negatively charged bonding electrons are attracted to the positively charged nuclei of *both* bonding atoms – this attraction holds the atoms together strongly. This force of attraction between two atoms as a result of sharing a pair of electrons is a covalent bond.

Lewis structures (electron dot diagrams)

Covalent bonds and the molecules they exist in can be shown using electron dot diagrams, known as **Lewis structures**. The covalent bonds are shown as : or —.

The number of valence electrons of an element can be determined from its group in the periodic table.

Example C

Lewis structure for chlorine

Chlorine is in group 17, so has 17 – 10 = 7 valence electrons.

Seven valence electrons is '1 short' of an inert gas electron arrangement – each chlorine atom must share one electron.

:C̈l• + •C̈l: —covalent bonding→ :C̈l:C̈l: (2 bonding electrons)

Chlorine, Cl_2, is thus represented by the Lewis structures:

:C̈l:C̈l: or :C̈l—C̈l:

In Example C, the two shared electrons (one pair) are most likely to be found between the two chlorine nuclei.

For a hydrogen atom, the inert gas electron arrangement is reached when it contains two electrons.

Example D

Hydrogen and hydrogen chloride

For both hydrogen, H_2, and hydrogen chloride, HCl, molecules, hydrogen achieves an inert gas arrangement of two electrons.

H:H or H—H

Each hydrogen atom contributes one valence electron to form the bonding pair.

:C̈l:H or H—C̈l:

The hydrogen and chlorine atoms contribute one valence electron each.

Multiple covalent bonds

Molecules with **double bonds** have four electrons (two pairs) shared by two atoms simultaneously.

In a **triple bond**, six electrons (three pairs) are shared.

Example E

Single, double and triple bonds

In Cl_2, O_2 and N_2, the atoms in each molecule achieve an inert gas arrangement by sharing two, four and six electrons, respectively:

Chlorine, Cl_2 2 electrons shared	**Oxygen, O_2** 4 electrons shared	**Nitrogen, N_2** 6 electrons shared
:C̈l:C̈l:	:Ö::Ö:	:N:::N:
or	or	or
:C̈l—C̈l:	:Ö=Ö:	:N≡N:

Rules for drawing Lewis structures

The following steps show how to draw Lewis structures.

Example F

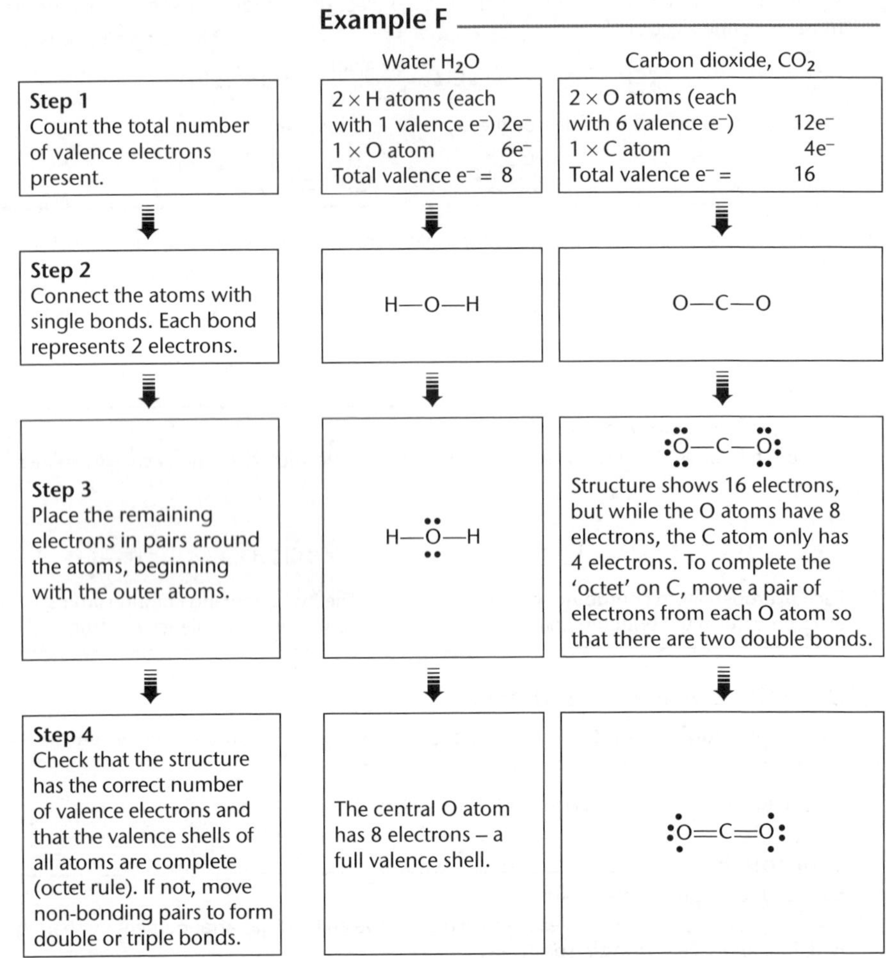

	Water H_2O	Carbon dioxide, CO_2
Step 1 Count the total number of valence electrons present.	2 × H atoms (each with 1 valence e^-) $2e^-$ 1 × O atom $6e^-$ Total valence e^- = 8	2 × O atoms (each with 6 valence e^-) $12e^-$ 1 × C atom $4e^-$ Total valence e^- = 16
Step 2 Connect the atoms with single bonds. Each bond represents 2 electrons.	H—O—H	O—C—O
Step 3 Place the remaining electrons in pairs around the atoms, beginning with the outer atoms.	H—Ö—H	:Ö—C—Ö: Structure shows 16 electrons, but while the O atoms have 8 electrons, the C atom only has 4 electrons. To complete the 'octet' on C, move a pair of electrons from each O atom so that there are two double bonds.
Step 4 Check that the structure has the correct number of valence electrons and that the valence shells of all atoms are complete (octet rule). If not, move non-bonding pairs to form double or triple bonds.	The central O atom has 8 electrons – a full valence shell.	:O=C=O:

There are some molecules that have fewer than eight electrons around the central atom. Common examples are molecules containing boron and beryllium.

Example G

BF_3 only has 6 electrons around the central B atom, and $BeCl_2$ only has 4 electrons around Be:

:F̈—B—F̈: (with :F: below B) and :C̈l— Be— C̈l:

Unit 12.1 Activity 5A: Lewis structures

1. Draw Lewis structures for each of the following molecules.

a.	**i.** H_2	**ii.** N_2	**iii.** HOCl
b.	**i.** HCl	**ii.** CO_2	**iii.** C_2H_6
c.	**i.** CH_4	**ii.** O_2	**iii.** H_2O_2
d.	**i.** CCl_4	**ii.** H_2S	**iii.** CH_3Cl

2. Draw Lewis structures for:

a. PH_3 **b.** CH_2Cl_2 **c.** H_2CO

d. F_2O **e.** $COCl_2$ (C is the central atom).

Polar and non-polar bonds

Covalent bonds can be either *polar* or *non-polar*.

If the atoms involved in a bond are different, they will have different attractions for the bonded pair of electrons. This causes an unequal sharing of electrons and results in a **polar bond**.

Example H

The non-polar bond of oxygen

Oxygen, O_2, contains **non-polar covalent bonds** since the atoms involved in the bond are identical.

O=O

Identical atoms bonding

The polar bond of hydrogen fluoride

Hydrogen fluoride, HF, contains a **polar covalent bond**, since the atoms involved in the bond are not identical. The electrons are closer to the fluorine atom.

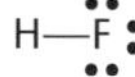

Non-identical atoms bonding

In a polar covalent bond, one atom is more **electronegative** than the other.

Electronegativity is a measure of how strongly bonding electrons are attracted to the nucleus of an atom in a bond. The Pauling electronegativity scale ranges from about 0.7 to 4.0.

Electronegativities increase across a period.

Electronegativities decrease down a group.

2.1 H						
1.0 Li	1.5 Be	2.0 B	2.5 C	3.0 N	3.5 O	4.0 F
0.9 Na	1.2 Mg	1.5 Al	1.8 Si	2.1 P	2.5 S	3.0 Cl
0.8 K	1.0 Ca					

Inert gases have zero electronegativity. For all other elements, the trend of increasing electronegativity is in the direction of the arrows

Polar bonds occur because the most electronegative atom of a bond has the greater pull on the shared electrons, which will result in a slight negative charge (shown by $\delta-$ and called 'delta negative'), because the shared electrons will be held closer to it. The less electronegative atom has a slight positive charge, written $\delta+$ (called 'delta positive'). This means that a charge separation occurs across the bond, resulting in a **dipole**.

Electronegativities can be used to predict which atom in a polar covalent bond will carry the slight negative charge and which will carry the slight positive charge.

Example I

Bonding electrons in a polar bond

Hydrogen chloride, HCl, contains a **polar covalent bond** because electrons are more strongly attracted to the chlorine atom.

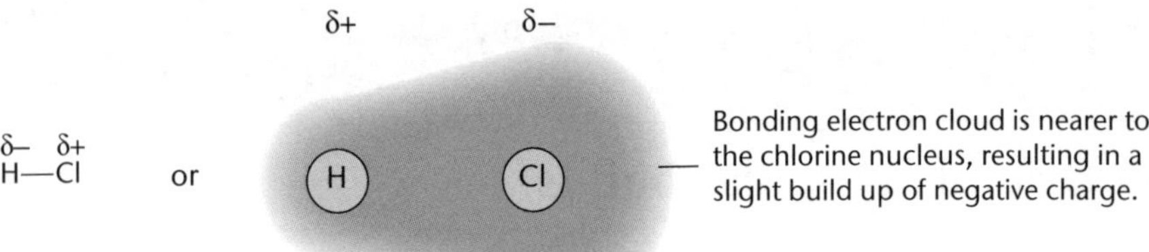

Bonding electrons are pulled closer to the chlorine nucleus because a chlorine atom is more electronegative than a hydrogen atom. The chlorine atom gains a small negative charge, so takes the sign δ–. Since electron density is removed from hydrogen, it acquires a slight positive charge, so takes the sign δ+.

In a polar covalent bond between nitrogen and hydrogen, the bonding electrons will be closer to the nitrogen, because nitrogen is more electronegative than hydrogen. The nitrogen atom will carry a slight negative charge: δ– δ+ N—H

Bonding as a continuum

Ionic bonds form when the electrons involved in the bond are transferred *entirely* to another (different) atom involved in the bond. In non-polar covalent bonds, bonding electrons are not transferred between atoms but are evenly *shared* between the (identical) atoms.

It is useful to consider non-polar covalent bonding (also called pure covalent bonding) and ionic bonding as two extremes. In between are polar covalent bonds, where the electrons are shared but the atoms in the bond are different. The difference in electronegativity between atoms indicates where, in such a continuum, the bonding type lies.

Differences in electronegativity between the atoms in a bond can be used to predict the type of bond that forms:

- If the two atoms have similar electronegativities, then the bond is non-polar.
- If the electronegativity difference is between 0.4 and 1.6, then the bond will be polar covalent.
- If the electronegativity difference is greater than 1.6, then the covalent transfer of electrons is complete and the bonding is ionic.

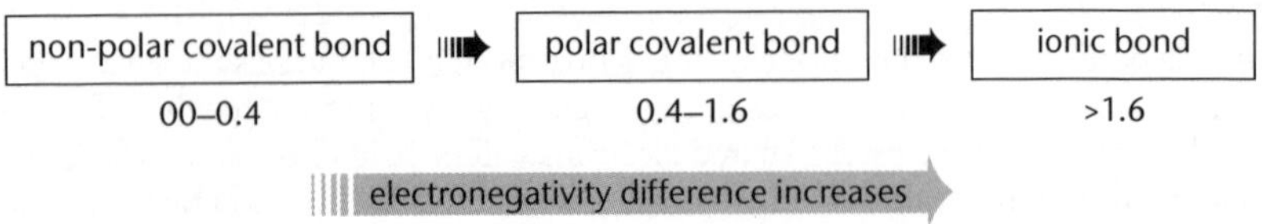

The bonding continuum

Non-polar covalent bonding	Polar covalent bonding	Ionic bonding
Atoms have similar tendency to attract bonding electrons (similar electronegativity values). Electrons are shared equally.	Atoms have small differences in their tendency to attract bonding electrons (small electronegativity differences). Electrons shared unequally between atoms.	Atoms have large differences in their tendency to attract bonding electrons. One atom donates its valence electron(s) to the other, ions are formed, resulting in ionic bonding.
Present when identical non-metal atoms bond, eg oxygen, O_2; hydrogen, H_2.	Commonly between two different non-metal atoms, eg oxygen–hydrogen bonds, carbon–chlorine bonds.	Common in combinations of bonding between the atoms of metals and non-metals, eg sodium and chlorine, magnesium and fluorine.

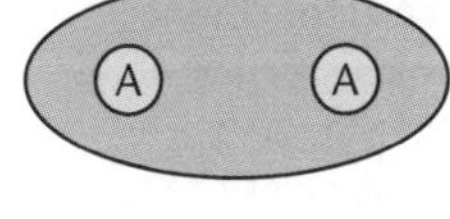

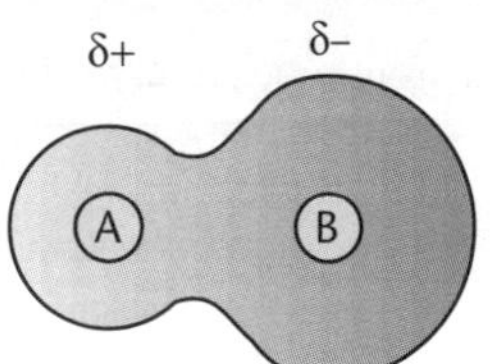

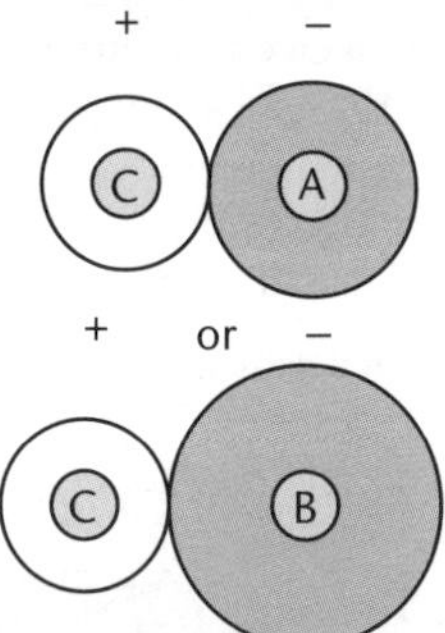

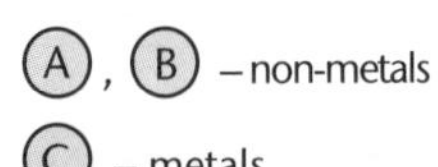

Unit 12.1 Activity 5B: Polar bonds and bonding as a continuum

1. If a polar covalent bond formed between each atom in the pairs of atoms below, which atom would have the slightly negative charge?

a. N, H **b.** H, O **c.** C, H **d.** C, O **e.** H, F

2. a. Draw a diagram of the bonding continuum.

b. Place the following at an appropriate place on the continuum:

i. A bond between P and Cl. **ii.** A bond between H and S.

iii. A bond between N atoms. **iv.** A bond between K and Cl atoms.

3. For each of the following pairs of elements, predict the type of bonding – ionic, polar covalent or covalent – that could occur between atoms of the elements.

Note: This can be completed using 'trends' in electronegativity – it is *not* necessary to learn the eletronegativity values.

a. Magnesium and chlorine. **b.** Sulfur and chlorine.

c. Phosphorus and oxygen. **d.** Aluminium and oxygen.

e. Calcium and oxygen. **f.** Oxygen and carbon.

g. Sulfur and sulfur.

h. The element whose atoms have an electron configuration (2, 8, 1) and the element whose atoms have an electron configuration (2, 6).

i. The element whose atoms have an electron configuration (2, 8, 6) and the element whose atoms have an electron configuration (2, 6).

j. The element whose atoms have an electron configuration (2, 5) and the element whose atoms have an electron configuration (2, 5).

Shapes of molecules

The position of each atom in a molecule gives the molecule its particular shape. The shape can be predicted from the Lewis structures by counting the number of *sets* of electrons around the central atom.

A *set* of electrons is considered a region of electron density, sometimes known as an electron cloud. A set may be a **bonding pair** (ie single bond), two bonding pairs (ie double bond), three bonding pairs (ie triple bond) or an unbonded (ie **lone**) **pair of electrons**.

Each set of electrons around the central atom repels every other set of electrons, since the electrons are all negatively charged and 'like charges repel'. Molecules are most stable when the repulsive forces are at a *minimum* – this occurs when the electron sets are as far apart as possible around an atom. The repulsion between electron sets gives rise to three-dimensional molecular shapes.

For 3D structures, the following conventions are used:

- X —— Y Both atoms are in the plane of the page.
- X ◄ Y X is in the plane of the page, Y is pointing 'out'.
- X ⁞⁞⁞ Y X is in the plane of the page, Y is pointing 'into' the page.

The shape of the molecule depends on the number of:

- regions of electron density or sets of electrons (bonded and non-bonded) around the central atom.
- atoms bonded to the central atom.

The three basic shapes for molecules with up to 8 electrons on the central atom are:

- Tetrahedral – 4 sets of electrons arranged as far apart as possible take up a tetrahedral arrangement. The **bond angle** between the electron clouds is 109°.

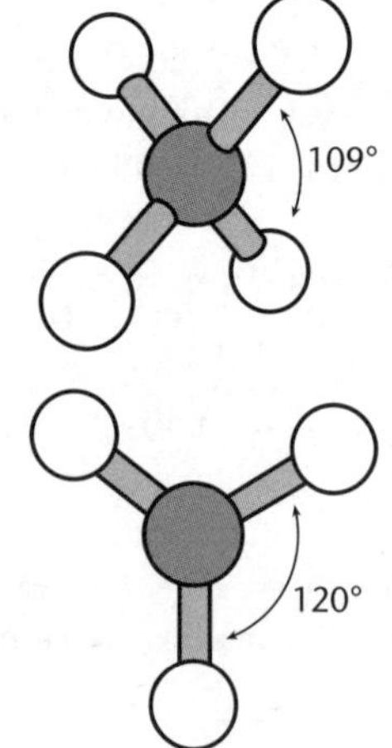

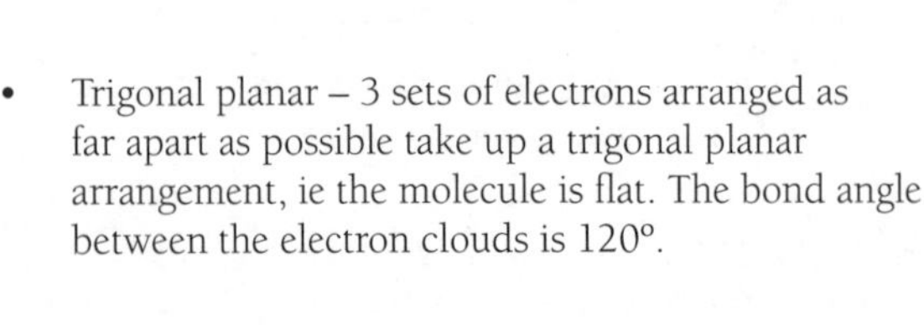

- Trigonal planar – 3 sets of electrons arranged as far apart as possible take up a trigonal planar arrangement, ie the molecule is flat. The bond angle between the electron clouds is 120°.

- Linear – 2 sets of electrons arranged as far apart as possible take up a linear arrangement. The bond angle between the electron clouds is 180°.

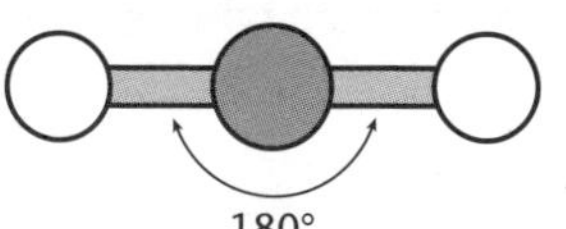

Trigonal pyramids and V-shaped (or bent) molecules result when there are non-bonded pairs of electrons.

To determine the shape of a molecule (and the resulting angles about the central atom):

- Draw the Lewis structure for the molecule.
- Count the number of regions of electron density or sets of electrons on the central atom. This will determine the arrangement of electrons about the central atom.
- Count the number of bonded atoms (bonding sets of electrons).
  This will determine the shape of the molecule. Non-bonding electron pairs are not drawn in the final shape diagram.

Number of sets of electrons on central atom	Shape for sets of electrons	Number of bonded atoms (bonding sets of electrons)	Shape of molecule	Examples	
4	Tetrahedral Bond angle 109°	4	Tetrahedral Bond angle 109°	CF_4	
				4 sets of e⁻ around the C atom. All are bonded.	Shape of the molecule is tetrahedral.
		3	Trigonal pyramid Bond angle aproximately 109°	NH_3	
				4 sets of e⁻, 3 are bonded, 1 is non-bonded.	Shape of the molecule is a trigonal pyramid.
		2	Bent or V-shaped Bond angle approximately 109°	H_2O	
				4 sets of e⁻, 2 are bonded, 2 are non-bonded.	Molecule is V-shaped or bent, with an angle of approximately 109°.

<table>
<tr><th>Number of sets of electrons on central atom</th><th>Shape for sets of electrons</th><th>Number of bonded atoms (bonding sets of electrons)</th><th>Shape of molecule</th><th colspan="2">Examples</th></tr>
<tr><td rowspan="4">3</td><td rowspan="4">Trigonal planar
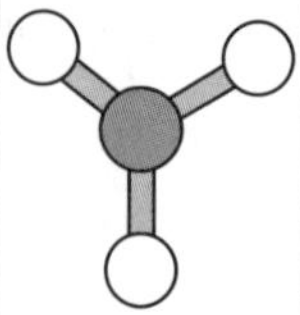
Bond angle 120°</td><td rowspan="2">3</td><td rowspan="2">Trigonal planar</td><td colspan="2">BF_3</td></tr>
<tr><td>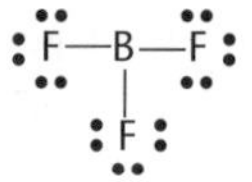
3 sets of e⁻, all are bonded.</td><td>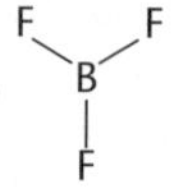
Shape of the molecule is trigonal planar.</td></tr>
<tr><td rowspan="2">2</td><td rowspan="2">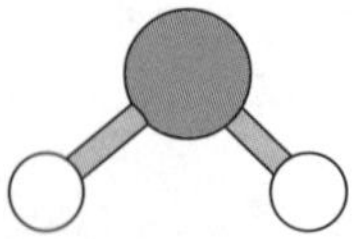
Bent or V-shaped
Bond angle approximately 120°</td><td colspan="2">SO_2</td></tr>
<tr><td>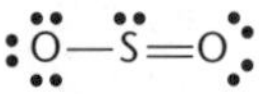
3 sets of e⁻, 2 are bonded, 1 is non-bonded.</td><td>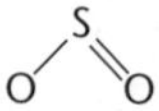
Molecule is V-shaped or bent with an angle of 120°.</td></tr>
<tr><td rowspan="2">2</td><td rowspan="2">Linear
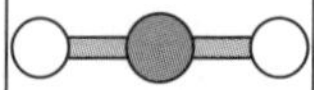
Bond angle 180°</td><td rowspan="2">2</td><td rowspan="2">Linear</td><td colspan="2">CO_2</td></tr>
<tr><td>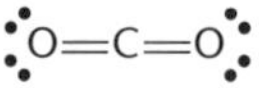</td><td>O=C=O</td></tr>
</table>

Example J

Shape of the OF_2 molecule

The OF_2 molecule has four regions of negative charge. Two are formed by bonding pairs of electrons; two are formed by the two **non-bonding pairs** from oxygen. The molecule is V-shaped, with a bond angle of approximately 109°.

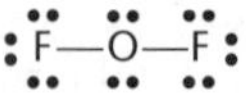

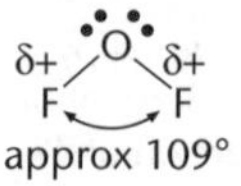

Shape of the methanal molecule, CH_2O

Methanal has three regions of negative charge around the central C atom. All are bonding. The shape of the molecule is trigonal planar.

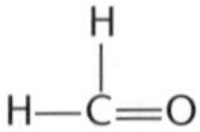

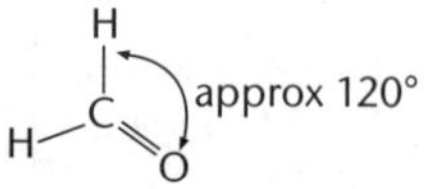

Non-bonding pairs of valence electrons are found *closer* to the central atom than bonding pairs of electrons, and thus exert a slightly *larger* repulsive force than do bonding pairs of electrons. This slightly affects the angles between atoms in trigonal pyramidal and V-shaped molecules. This effect can be seen in the ammonia molecule, where the N–H bond angles are reduced to 107° and in the water molecule, where the bond angle is reduced to 105°.

Unit 12.1 Activity 5C: Shapes of molecules

1. Copy and complete the following table.

	Molecule		
	Methane	**Ammonia**	**Water**
Molecular formula	**a.**	**b.**	**c.**
Number of pairs of electrons around central atom	**d.**	**e.**	**f.**
Number of pairs of non-bonding electrons around central atom	**g.**	**h.**	**i.**
Shape of molecule	**j.**	**k.**	**l.**

2. Describe the shape of each of the following molecules, giving reasons for your answers.

a. H_2S **b.** CF_4

c. NF_3 **d.** CO_2

3. Copy and complete the following table.

Compound	Formula	Lewis structure	Shape	Reason for shape
Tetrachloromethane	CCl_4	**a.**	**b.**	**c.**
Nitrogen trichloride	NCl_3	**d.**	**e.**	Three bonding pairs and one lone pair of electrons around the central atom.
Chlorine fluoride	ClF	**f.**	**g.**	**h.**

4. One of the following molecules is linear:

C_3H_8, CH_4, NH_3, CO_2.

Which molecule is linear? Justify your choice.

5. Draw Lewis structures for the following molecules and use these to predict the shape of each molecule.

a. HCl **b.** PCl_3 **c.** CCl_4

d. OCl_2 **e.** HOCl

6. Draw the Lewis structure for phosgene, CCl_2O, and predict the shape of the phosgene molecule.

7. The shapes of SO_2 and H_2O are both described as V-shaped. The bond angle in SO_2 is 120°, and in H_2O it is 105°. Explain why these bond angles are different.

The polarity of molecules

In diatomic molecules, if the bond joining the two atoms is polar, the molecule as a whole will be polar.

Polar molecule

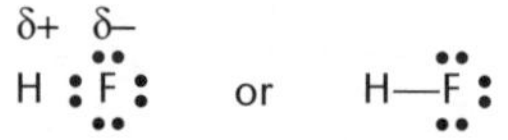

δ+ δ–
H : F : or H—F :

HF, contains a polar covalent bond.
Hydrogen fluoride is a **polar** molecule.

Non-polar molecule

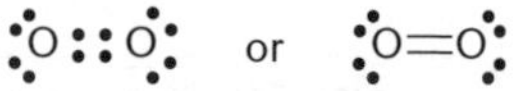

:O::O: or :O=O:

Oxygen, O_2, contains only **non-polar covalent bonds,** so oxygen is a non-polar molecule.

For molecules containing more than two atoms and polar bonds, the molecule may be polar. When the polar bonds are arranged symmetrically around the central atom, the **polarity** of the bonds cancel each other out, and the molecule overall is non-polar.

Example K

Polar and non-polar molecules

The water molecule contains polar bonds. It is V-shaped. The spread of charge is uneven due to the polar bonds being asymmetrically arranged around the O atom. *Water is a polar molecule.*

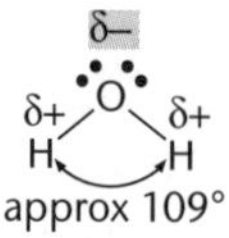

δ–
δ+ O δ+
H H
approx 109°

The carbon dioxide molecule contains polar bonds. It is a linear shape. The spread of negative charge is even around the central carbon atom, because the bonds are arranged symmetrically around the carbon atom. *Carbon dioxide is a non-polar molecule.*

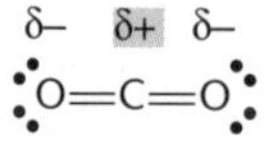

δ– δ+ δ–
O=C=O

The spread of charge in a molecule is uneven when the central atom is bonded to more than one type of atom.

Example L

Tetrachloromethane and trichloromethane

Tetrachloromethane, CCl_4, is a non-polar molecule.
The four polar C–Cl bonds are all equal and are spread symmetrically around the carbon. The even spread of charge makes the molecule non-polar.

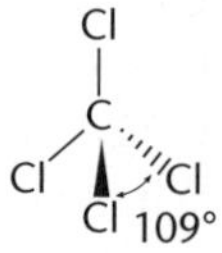

Cl
C
Cl Cl
Cl 109°

Trichloromethane, $CHCl_3$, is a polar molecule.
The polarity of the C–H bond differs from the polarity of the C–Cl bonds. This destroys the symmetrical spread of charge and makes the molecule polar.

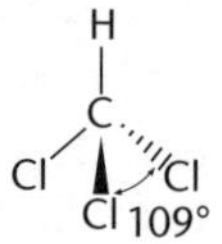

H
C
Cl Cl
Cl 109°

Polar molecules can thus be identified by:

- polar bonds in the molecule arranged asymmetrically around the central atom (eg water, H_2O, and ammonia, NH_3).
- more than one type of atom bonded to the central atom (eg trichloromethane, $CHCl_3$).

Substances that contain polar molecules		Substances that contain non-polar molecules	
Hydrogen chloride	Ammonia	Carbon dioxide	Oxygen
Phosphorus trichloride	Water	Hydrogen	Methane
Dichloromethane		Nitrogen	

Testing for polarity in liquids

Molecules in liquids have mobility. If placed near an electrically charged object, polar molecules *align* and are attracted to the charged object ('unlike charges attract'). This causes a stream of a liquid polar substance (eg water) to be deflected *towards* a charged object. This provides a test for polarity. Non-polar liquids flow without deflection past charged objects.

A polar liquid deflects towards a charged object.

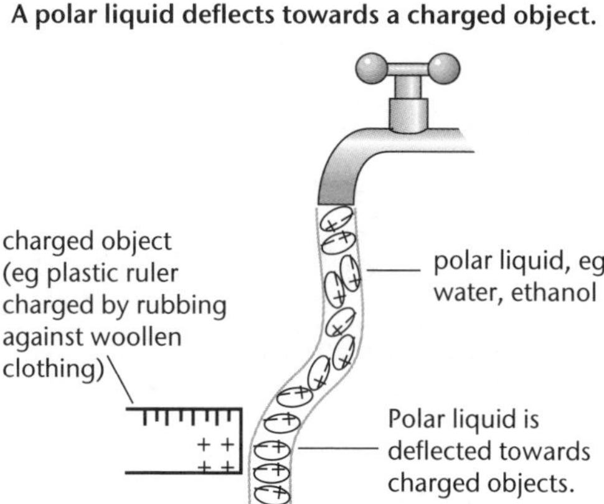

Polarity of water molecules

Water molecules are polar, since they have polar (O–H) bonds and a bent (non-symmetrical) shape (due to the two **lone pairs of electrons** on the central oxygen atom).

Water has high **melting** and **boiling points** for a molecule with such low mass.

The relatively large electronegativity difference between hydrogen and oxygen, and the small size of the molecules, produce strong attractive forces between the molecules. Considerable energy is required to overcome these forces and separate the molecules, explaining the high melting and boiling points.

Strong intermolecular bonding in water

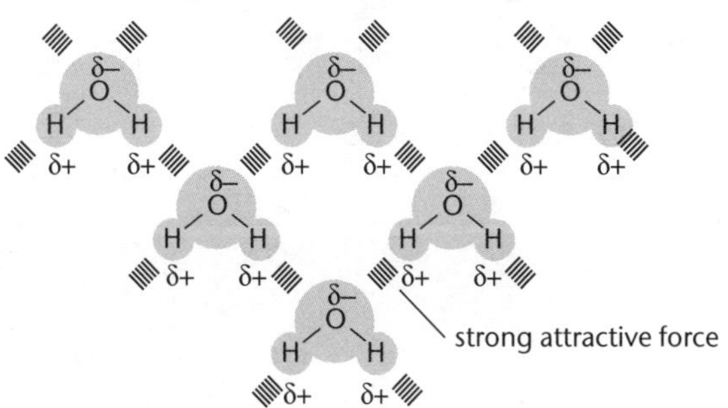

The polarity of water molecules causes some ionic compounds to dissolve.

Dissolving is a two-step process.

The strong ionic bonds holding ions together are overcome by attraction between the ions and the polar water molecules. The ions separate.

The separated ions are surrounded by water molecules to form **hydrated** ions. These hydrated ions are written with (aq) after their symbol.

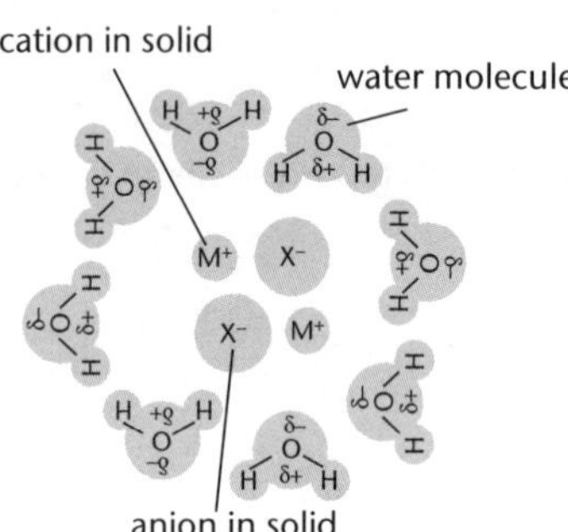

Polar water molecules are attracted to the ions; the ions separate.

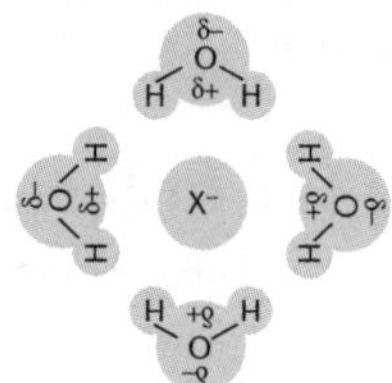

hydrated anion, $X^-(aq)$

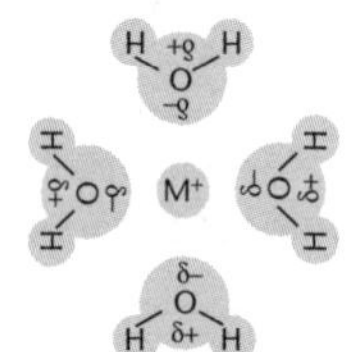

hydrated cation, $M^+(aq)$

Example M

Sodium chloride dissolving in water

$$NaCl(s) \xrightarrow{H_2O} Na^+(aq) + Cl^-(aq)$$

Solid sodium chloride **Solution of sodium choride**

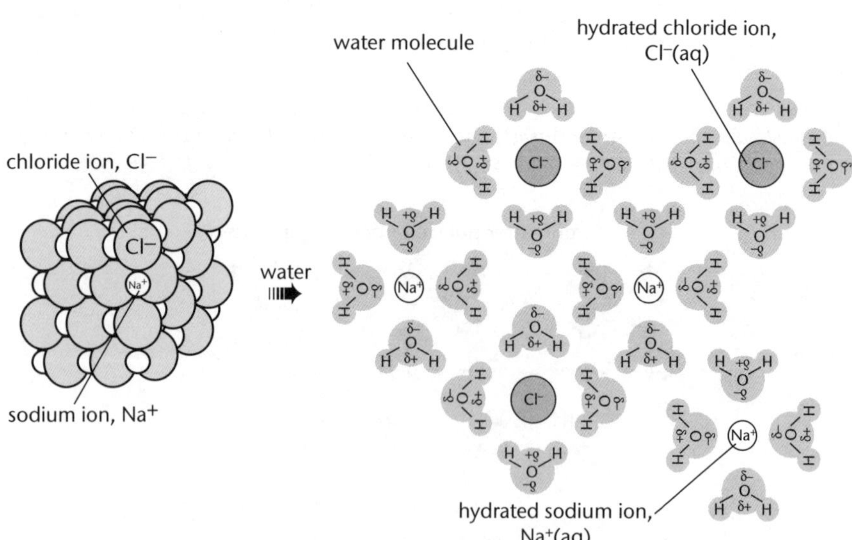

The amount of energy (called the **lattice energy**) required to break up an ionic lattice so that hydrated ions can be produced varies for different ionic compounds. Ionic compounds are **insoluble** if *more* energy is required to separate the ions than can be gained by forming hydrated ions.

Unit 12.1 Activity 5D: Polarity of molecules

1. Which one of the following molecules is non-polar yet contains polar bonds? Explain your answer.

A. H_2O

B. NH_3

C. CH_2Cl_2

D. CCl_4

2. Label the molecules below as polar or non-polar, giving reasons for your answers.

a. F_2 **b.** CH_3Cl **c.** $CHCl_3$ **d.** CF_4

e. HCl **f.** PCl_3 **g.** H_2S

3. Answer questions **a.** and **b.** by choosing from the key list A–D:

A. $CHCl_3$ **B.** NO **C.** NH_3 **D.** O_2

a. Select from the list those molecules that are linear.

b. For each linear molecule, state and explain its polarity or non-polarity.

4. Determine the polarity of NCl_3 and CCl_4. Explain your answers.

5. a. Draw a diagram to show how potassium iodide dissolves in water.

b. Give symbols for a hydrated:

i. Potassium ion.

ii. Iodide ion.

6. The table below shows the Lewis structures and shapes of two molecules.

Molecule	Lewis structure	Diagram to show shape
CO_2	$\ddot{\underset{..}{O}}=C=\ddot{\underset{..}{O}}$	O—C—O
SO_2	$\ddot{\underset{..}{O}}=\ddot{S}—\ddot{\underset{..}{O}}:$	O S O (bent)

Using the information in this table, *describe* CO_2 and SO_2 molecules as either *polar* or *non-polar*, and *discuss* the reasons for your choice.

7. When volcanoes erupt, a number of gases are released, including sulfur dioxide, SO_2. After the eruption, lakes and streams become more acidic, due to SO_2 dissolving in H_2O. Use the structure and bonding in H_2O and SO_2 to explain why SO_2 is soluble in H_2O.

8. Two substances that damage Earth's ozone layer are CCl_4 and CH_2Cl_2.

State whether the molecules are *polar* or *non-polar*, and discuss the reasons for your choice. Include a Lewis structure of the molecules with your answer.

Unit 12.1 Activity 5E: Molecules – multiple choice

1. Which of the following identifies the bonding in sodium chloride?
   A. Covalent.
   B. Hydrogen.
   C. Ionic.
   D. None of the above.

2. Which of the following identifies the bonding in chlorine molecules?
   A. Covalent.
   B. Hydrogen.
   C. Ionic.
   D. None of the above.

3. How many electrons are shared in a single bond?
   A. 1
   B. 2
   C. 6
   D. 4

4. How many electrons are shared in a double bond?
   A. 1
   B. 2
   C. 6
   D. 4

5. How many electrons are shared in a triple bond?
   A. 1
   B. 2
   C. 6
   D. 4

6. In the periodic table:
   A. electronegativity increases across a period and down a group.
   B. electronegativity decreases across a period and down a group.
   C. electronegativity increases across a period and decreases down a group.
   D. electronegativity decreases across a period and increases down a group.

7. The shape of the carbon tetrachloride molecule is:
   A. trigonal pyramid.
   B. tetrahedral.
   C. bent or V-shaped
   D. linear.

8. The shape of the carbon dioxide molecule is:
   A. trigonal pyramid.
   B. tetrahedral.
   C. bent or V-shaped.
   D. linear.

Unit 12.2 Acids, Bases and Salts

Topic 1: Acids and bases

Topic 1 covers acids and bases:

- Interpreting and explaining the nature of acid–base solutions.
- Writing equations for proton transfer reactions between acids and bases.
- Explaining the behaviour of strong and weak acids and bases in terms of the concentration of H_3O^+ or OH^- ions.
- pH.
- Identifing conjugate acid–base pairs.
- Accounting for the acidic or basic properties of some salts.

Introduction

The reactions of acids and bases in aqueous solutions involve the transfer of **hydrogen ions**, H^+. Hydrogen ions are often referred to as protons, and are formed when a hydrogen atom loses an electron:

$$H \rightarrow H^+ + e^-$$

The scientists Brønsted and Lowry defined acids and bases in terms of **proton transfer**.

- A Brønsted **acid** is a substance whose molecules or ions can *donate protons*. Acid particles contain a hydrogen atom bonded polar covalently to a non-metal atom.

Structures of some acids

HCl	H_2O	CH_3COOH	NH_4^+
$\overset{\delta+}{H}-\overset{\delta-}{Cl}$	$\overset{\delta-}{O}$ bonded to two $\overset{\delta+}{H}$	$CH_3C(=O)-\overset{\delta-}{O}-\overset{\delta+}{H}$	$[\overset{\delta-}{N}$ bonded to four $\overset{\delta+}{H}]^+$

- A Brønsted **base** is a substance whose molecules or ions can *accept protons*. Bases are species with particles containing a non-metal atom with a lone pair of electrons.

Structures of some bases

OH^-	H_2O	CH_3COO^-	NH_3	CO_3^{2-}
$[:\ddot{\underset{..}{O}}-H]^-$	$\ddot{\underset{..}{O}}$ bonded to H and H	$[CH_3C(=O)-:\ddot{\underset{..}{O}}:]^-$	$\ddot{N}$ bonded to H, H, H	$[C$ bonded to $:\ddot{O}$ (double), $:\ddot{\underset{..}{O}}:$ and $:\ddot{\underset{..}{O}}:]^{2-}$

- An acid–base reaction is one in which a proton is *transferred* from an acid to a base.

Hydrogen ions do not exist on their own in water but are always attached to water molecules and exist as **hydronium ions**, $\mathbf{H_3O^+(aq)}$. Hydronium ions are sometimes written as aqueous hydrogen ions, $H^+(aq)$.

Acids

Molecules such as hydrochloric acid, HCl, sulfuric acid, H_2SO_4, nitric acid, HNO_3, and ethanoic acid, CH_3COOH, act as acids by donating protons. The protons are transferred to water molecules to form hydronium ions, $H_3O^+(aq)$. The resulting aqueous solutions are **acidic** solutions because of the $H_3O^+(aq)$ ions that have been added.

Acidic substances and their reactions with water

Acidic substance	Reaction with water (in each case water is acting as a base)	Products
Ethanoic acid, CH_3COOH	$CH_3COOH(\ell) + H_2O(\ell) \rightleftharpoons H_3O^+(aq) + CH_3COO^-(aq)$ proton transfer	Hydronium ion, $H_3O^+(aq)$; ethanoate ion, $CH_3COO^-(aq)$
Ammonium ion, NH_4^+	$NH_4^+(g) + H_2O(\ell) \rightleftharpoons H_3O^+(aq) + NH_3(aq)$ proton transfer	Hydronium ion, $H_3O^+(aq)$; ammonia molecule, $NH_3(aq)$

Example A

Hydrochloric acid

When hydrogen chloride gas, HCl(g), is dissolved in water, the reaction occurring is:

acid base

$$HCl(g) + H_2O(\ell) \rightarrow Cl^-(aq) + H_3O^+(aq)$$

proton transfer

HCl acts as an acid because it *donates* a proton. Water acts as a base because it *accepts* a proton (from the HCl molecule). The proton has been *transferred* from the HCl to the H_2O. The resultant solution, called hydrochloric acid, is acidic because of the $H_3O^+(aq)$ ions present. The solution also contains chloride ions, $Cl^-(aq)$, and water molecules which did not accept protons.

The transfer of hydrogen ions can also be described as ionisation (since ions have formed) or **dissociation** (since the HCl molecule has, in effect, dissociated or been 'split up').

An ionisation or dissociation process (like the reaction in Example A), is sometimes written as:

$$HCl(g) \xrightarrow{H_2O} H^+(aq) + Cl^-(aq)$$

Acids are described as monoprotic if one proton can be transferred or diprotic if two protons can be transferred. HCl is a monoprotic acid because each HCl molecule donates only one proton. Sulfuric acid is a diprotic acid because each H_2SO_4 molecule has 2 hydrogen ions that can be donated, one at a time:

1st transfer acid base

$$H_2SO_4(\ell) + H_2O(\ell) \rightarrow HSO_4^-(aq) + H_3O^+(aq)$$

proton transfer

2nd transfer acid base

$HSO_4^-(aq) + H_2O(\ell) \rightarrow SO_4^{2-}(aq) + H_3O^+(aq)$

proton transfer

Overall $H_2SO_4(\ell) + 2H_2O(\ell) \rightarrow 2H_3O^+(aq) + SO_4^{2-}(aq)$

Strong and weak acids

Acids that completely dissociate in water are known as **strong acids**. Common strong acids are HCl, HNO_3 and H_2SO_4.

Example B

Hydrochloric acid is a strong acid because it completely dissociates in water:

$HCl(aq) + H_2O(\ell) \rightarrow H_3O^+(aq) + Cl^-(aq)$

Organic acids such as ethanoic acid, CH_3COOH, act as acids because the hydrogen of the –COOH group can be transferred to a base. Not all of the organic acid molecules donate their protons, so these acids are said to be only *partially* ionised or dissociated. The reaction of an organic acid with water is written as an equilibrium reaction using arrows $\rightleftharpoons$, showing that not all the acid molecules exist in a dissociated form in the solution.

Acids that are partially ionised or dissociated are known as **weak acids**.

Example C

Ethanoic acid

When ethanoic acid reacts with water, the following equilibrium situation occurs:

acid base

$CH_3COOH(aq) + H_2O(\ell) \rightleftharpoons CH_3COO^-(aq) + H_3O^+(aq)$

proton transfer

This can also be written as:

$CH_3COOH(aq) \overset{H_2O}{\rightleftharpoons} CH_3COO^-(aq) + H^+(aq)$

Note: The reaction is in equilibrium. Only a few CH_3COOH molecules react, so the resulting solution contains mostly $CH_3COOH(aq)$ and H_2O molecules and a few $CH_3COO^-(aq)$ and $H_3O^+(aq)$ ions.

Some *ions* are acidic because they donate protons.

Example D

The ammonium ion

The ammonium ion, NH_4^+, is an acid because, in water, *some* of the ammonium ions present transfer their protons to water molecules:

$NH_4^+(aq) + H_2O(\ell) \rightleftharpoons NH_3(aq) + H_3O^+(aq)$

proton transfer

The resulting solution is acidic because of the hydronium ions produced.

Note: The reaction is an equilibrium with only a few NH_4^+ ions reacting, so the resulting solution contains ammonium ions, $NH_4^+(aq)$; ammonia molecules, $NH_3(aq)$; hydronium ions, and water molecules.

Acid salts

Some salts produce an acidic solution when they dissolve in water.

Example E

Dissolving ammonium chloride

When ammonium chloride, NH_4Cl, dissolves in water, it dissociates completely to form ammonium ions and chloride ions:

$$NH_4Cl(aq) \xrightarrow{H_2O} NH_4^+(aq) + Cl^-(aq)$$

Some of the ammonium ions behave as acids and donate protons to water molecules to form hydronium ions (see Example D):

$$NH_4^+(aq) + H_2O(\ell) \rightarrow NH_3(aq) + H_3O^+(aq)$$

Note: Ammonium chloride produces an acidic solution. The chloride ions are *spectator ions* and have no effect on the acidity of the solution.

Bases

Some molecules can act as bases by accepting protons. In aqueous solutions, protons are donated by water molecules and hydroxide ions, $OH^-(aq)$, form. Aqueous solutions formed from these molecules are called **basic** or **alkaline** solutions. An **alkali** is a soluble base. Common alkalis are sodium hydroxide solution and ammonia solution.

Basic substances and their reactions with water

Basic substance	Reaction with water (in each case water is acting as an acid)	Products
Ethanoate ion, CH_3COO^-	$CH_3COO^-(aq) + H_2O(\ell) \rightleftharpoons OH^-(aq) + CH_3COOH(aq)$ proton transfer	Hydroxide ions, $OH^-(aq)$; ethanoic acid, $CH_3COOH(aq)$
Ammonia molecule, NH_3	$NH_3(g) + H_2O(\ell) \rightleftharpoons OH^-(aq) + NH_4^+(aq)$ proton transfer	Hydroxide ions, $OH^-(aq)$; ammonium ions, $NH_4^+(aq)$

Strong and weak bases

Strong bases completely dissociate in water.

Example F

NaOH is a strong base because it completely dissociates in water:

$$NaOH(s) \xrightarrow{H_2O(\ell)} Na^+(aq) + OH^-(aq)$$

Weak bases only react to a small extent with water.

Example G

Ammonia solution

When ammonia, NH_3, dissolves in water, a proton transfer reaction occurs to form ammonium ions, $NH_4^+(aq)$, and hydroxide ions, $OH^-(aq)$. Only a few of the ammonia molecules accept protons from the water molecules. The reaction is written as an equilibrium because the reaction of those few ammonia molecules that accept protons from water is a **reversible reaction**:

acid base

$$H_2O(\ell) + NH_3(g) \rightleftharpoons NH_4^+(aq) + OH^-(aq)$$

proton transfer

Water acts as an acid by donating protons to the ammonia molecules. Protons are transferred *from* water molecules *to* ammonia molecules.

Because ammonia is only partially ionised and only a few ammonium ions and hydroxide ions form, the resulting solution contains *mainly* water and unreacted ammonia molecules, together with small numbers of ammonium ions, $NH_4^+(aq)$, and hydroxide ions, $OH^-(aq)$; hence ammonia is an example of a *weak base*.

Some ions are bases because they accept protons, donated by water molecules. The water molecule forms a hydroxide ion. So the water molecule donates a proton and acts as an acid.

Example H

Solutions containing the carbonate ion

The carbonate ion, CO_3^{2-}, is a base because carbonate ions can accept protons from water molecules to form hydrogen carbonate ions, $HCO_3^-(aq)$, and hydroxide ions, $OH^-(aq)$:

acid base

$$H_2O(\ell) + CO_3^{2-}(aq) \rightleftharpoons HCO_3^-(aq) + OH^-(aq)$$

proton transfer

This is an equilibrium reaction, and the resulting solution will contain unreacted carbonate ions, $CO_3^{2-}(aq)$, water molecules, hydrogen carbonate ions, $HCO_3^-(aq)$, and hydroxide ions, $OH^-(aq)$.

Basic salts

Some salts produce basic solutions when they dissolve in water.

Example I

Dissolving sodium carbonate

When sodium carbonate dissolves in water, it dissociates completely to form sodium and carbonate ions:

$$Na_2CO_3(s) \xrightarrow{H_2O} 2Na^+(aq) + CO_3^{2-}(aq)$$

Some of the carbonate ions behave as bases and accept protons from water molecules to form OH^- ions (see Example H).

$$CO_3^{2-}(aq) + H_2O(\ell) \rightleftharpoons HCO_3^-(aq) + OH^-(aq)$$

A basic solution is produced, with the Na^+ ions acting as spectator ions.

Other **basic salts** include CH_3COONa and $NaHCO_3$.

pH of solutions

Acidity can be measured by the **pH** of the aqueous solution. pH is a number and has no units. The pH values of different solutions form a **pH scale** that ranges between 0 and 14.

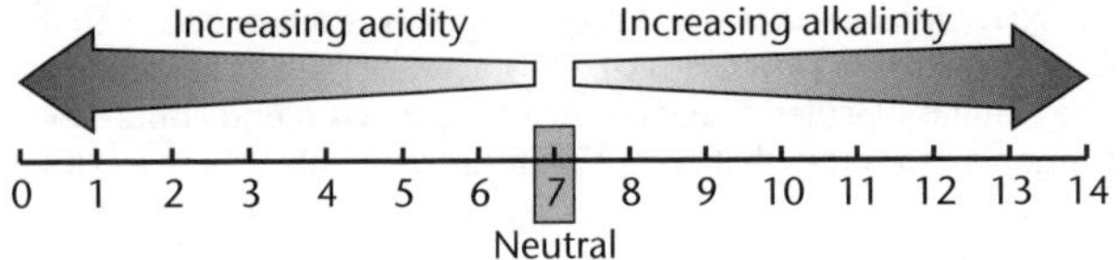

The pH scale

pH is linked to the concentration of hydronium ions in the solution. A *high* concentration of hydronium ions is registered as a *low* pH, which indicates high acidity of the solution. Alkalis have *low* concentrations of hydronium ions, and consequently have *high* pH values.

Testing for acidity and alkalinity

Acid solutions will:

- turn blue **litmus** paper red.

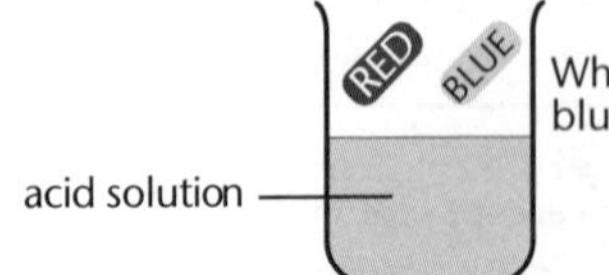

- indicate a pH of less than 7 with **universal indicator**, which produces a series of colour changes across the whole pH range.

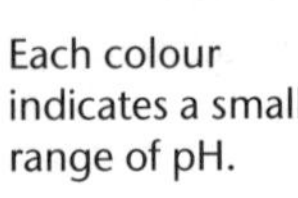

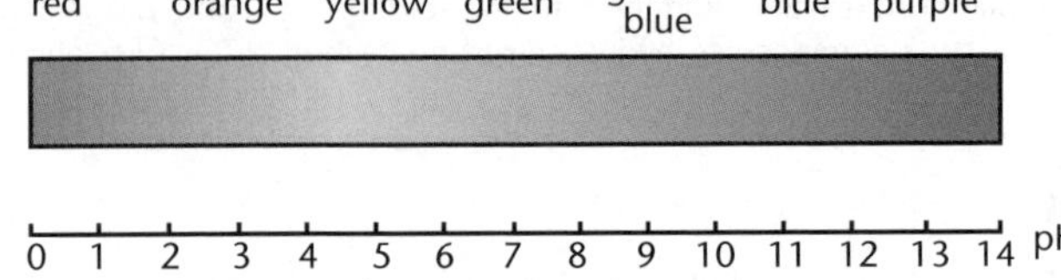

Effect of pH on universal indicator

Alkali solutions will:

- turn red litmus paper blue.
- indicate a pH of more than 7 with universal indicator.

Amphiprotic and neutral substances

Earlier examples show that water molecules can act as either:

- an acid by donating protons – see Examples G and H, where *bases* react with water.
- a base by accepting protons – see Examples C, D and E, where *acids* react with water.

Substances such as water, which can either accept or donate protons, are said to be **amphiprotic**. Other substances that exhibit this behaviour are the hydrogen carbonate ion, $HCO_3^-(aq)$, and the hydrogen sulfate ion, $HSO_4^-(aq)$.

Amphiprotic substances contain a:

- hydrogen atom with a polar bond to a non-metal atom, so they can act as proton donors or acids.
- non-metal atom with a lone pair of electrons, so they can act as proton acceptors or bases.

Substances that neither accept nor donate protons are said to be **neutral**. In the reaction between NH_4Cl and water in Example E, the spectator chloride ions, Cl^-, are neutral.

Example J

The amphiprotic nature of water

Water is an amphiprotic molecule. Water contains hydrogen atoms linked by polar bonds to oxygen and lone pairs of electrons, so a water molecule can act as both a proton donor and a proton acceptor.

- Acting as an acid, water is a proton donor:

$$H_2O \rightarrow H^+ + OH^-$$

- Acting as a base, water is a proton acceptor:

$$H_2O + H^+ \rightarrow H_3O^+$$

Example K

Amphiprotic ions

Hydrogen carbonate and hydrogen sulfate are amphiprotic ions.

Amphiprotic ion	Hydrogen carbonate, HCO_3^-	Hydrogen sulfate, HSO_4^-
Acting as an acid	$HCO_3^- + H_2O \rightarrow H_3O^+ + CO_3^{2-}$	$HSO_4^- + H_2O \rightarrow H_3O^+ + SO_4^{2-}$
Acting as a base	$HCO_3^- + H_2O \rightarrow OH^- + H_2CO_3$	$HSO_4^- + H_2O \rightarrow OH^- + H_2SO_4$
Dominant reaction in solution	$HCO_3^- + H_2O \rightarrow OH^- + H_2CO_3$	$HSO_4^- + H_2O \rightarrow H_3O^+ + SO_4^{2-}$
Nature of solution	Basic	Acidic

Amphoteric substances are solids that show:

- basic properties by reacting with acids.
- acidic properties by reacting with bases.

Example L

Aluminium hydroxide and zinc oxide are amphoteric substances.

Solution	Aluminium hydroxide, $Al(OH)_3$	Zinc oxide, ZnO
Acidic	$Al(OH)_3 + 3H^+ \rightarrow Al^{3+} + 3H_2O$	$ZnO + 2H^+ \rightarrow Zn^{2+} + H_2O$
Basic	$Al(OH)_3 + OH^- \rightarrow [Al(OH)_4]^-$ aluminate ion	$ZnO + 2OH^- + H_2O \rightarrow [Zn(OH)_4]^{2-}$ zincate ion

Conjugate acid–base pairs

When an acid loses a proton, it forms its **conjugate base**.

When a base gains a proton, it forms its **conjugate acid**.

Conjugate acid–base pairs are therefore molecules and ions that differ in their structure by a single proton, H^+:

acid $\rightleftharpoons$ conjugate base + proton

Example M

Conjugate acid–base pairs

HCl/Cl^- acid–conjugate base pair	CO_3^{2-}/HCO_3^- base–conjugate acid pair
The hydrogen chloride molecule, HCl, differs from the chloride ion, Cl^-, by one proton. The hydrogen chloride molecule and the chloride ion are thus considered to be an acid–conjugate base pair: $HCl \rightleftharpoons H^+ + Cl^-$ acid — conjugate base	The carbonate ion, CO_3^{2-}, differs from the hydrogen carbonate ion, HCO_3^-, by one proton. The carbonate ion and the hydrogen carbonate ion are thus considered to be a base–conjugate acid pair: $CO_3^{2-} + H^+ \rightleftharpoons HCO_3^-$ base — conjugate acid

Other examples of conjugate acid–base pairs are:

- CH_3COOH/CH_3COO^-
- NH_4^+/NH_3
- H_2O/OH^-
- H_3O^+/H_2O

Unit 12.2 Activity 1A: Acids and bases

1. Define the following:

a. Acid.

b. Base.

c. Amphiprotic.

2. Group the following into conjugate acid–base pairs, listing the acid first:

a. HCOOH **b.** SO_4^{2-}

c. OH^- **d.** HCl

e. NH_3 **f.** H_2O

g. Cl^- **h.** NH_2^-

i. $HCOO^-$ **j.** HSO_4^-

3. Give the formulae for the conjugate bases of:

a. H_2SO_4

b. $CH_3NH_3^+$

4. Give the formulae for the conjugate acids of:

a. $C_2H_5NH_2$

b. CH_3COO^-

c. NO_3^-

5. a. Write an equation for the reaction that occurs when the molecule HNO_3 is dissolved in water.

b. Name the ions that form.

c. What is the resultant solution called?

6. When sulfuric acid dissolves in water, the following reaction occurs:

$$H_2SO_4(\ell) + 2H_2O(\ell) \rightarrow 2H_3O^+(aq) + SO_4^{2-}(aq)$$

a. How does the reaction indicate that sulfuric acid acts as an acid?

b. If 0.2 mol H_2SO_4 dissolved, what amount of H_3O^+ would be present in the solution?

7. a. Write an equation showing potassium hydroxide dissolving in water.

b. Use the equation in **a**. to explain why potassium hydroxide is a *strong base*.

8. Write equations for the reactions occurring when:

a. The strong acid HNO_3 dissolves in water.

b. The weak acid CH_3CH_2COOH dissolves in water.

c. Ammonium chloride dissolves in water.

d. The strong base KOH dissolves in water.

e. The weak base CH_3NH_2 dissolves in water.

9. Give the formula for the conjugate acids of:

a. Cl^- **b.** NH_3

c. H_2O **d.** HCO_3^-

e. CO_3^{2-} **f.** OH^-

10. Give the formula for the conjugate bases of:

a. H_2CO_3 **b.** HCl

c. NH_4^+ **d.** H_2O

e. HNO_3 **f.** HSO_4^-

g. H_3O^+

11. Classify the following substances as basic, amphoteric or amphiprotic. Give reasons for your answers, including equations.

a. $Zn(OH)_2$

b. NaOH

c. $NaHCO_3$

12. Give two examples of water acting as:

a. a base.

b. an acid.

For each example, state whether the resulting solution is acidic, basic or neutral.

13. Identify the acid–base reactions from the list below. Give reasons for your answers in terms of proton transfers.

a. $CuO(s) + 2NH_4^+(aq) \rightarrow Cu^{2+}(aq) + 2NH_3 + H_2O$

b. $ZnCl_2 + 2NaOH \rightarrow Zn(OH)_2 + 2NaCl(aq)$

c. $Na_2CO_3(aq) + 2HCl(aq) \rightarrow 2NaCl(aq) + H_2O(\ell) + CO_2(g)$

14. a. Identify the acid in the following reactions. Give a reason for each answer.

i. $HCOOH + H_2O \rightleftharpoons HCOO^- + H_3O^+$

ii. $CuO + 2NH_4^+ \rightarrow Cu^{2+} + 2NH_3 + H_2O$

iii. $HCO_3^- + H_2O \rightarrow H_2CO_3 + OH^-$

b. Identify two conjugate acid–base pairs in **a. i.** and **a. iii.**

15. a. A sample of solid ammonium chloride, NH_4Cl, is dissolved in water. The solution formed is found to be acidic. Explain why the solution is acidic. Include appropriate equation(s) in your answer.

b. Explain why a solution of sodium carbonate is basic.

16. $HCO_3^-(aq)$ is a species that may act as an acid or a base. Consider the following equilibrium system:

$$HCO_3^-(aq) + H_2O(\ell) \rightleftharpoons H_2CO_3(aq) + OH^-(aq)$$

Is $HCO_3^-(aq)$ acting as an acid or a base? Justify your answer.

17. Account for the fact that solutions of $NaHCO_3$ are basic, while solutions of NaCl are neutral.

18. Phosphoric acid, H_3PO_4, is a weak acid that reacts with water according to the following equations:

Equation 1: $H_3PO_4(aq) + H_2O(\ell) \rightleftharpoons H_2PO_4^-(aq) + H_3O^+(aq)$

Equation 2: $H_2PO_4^-(aq) + H_2O(\ell) \rightleftharpoons HPO_4^{2-}(aq) + H_3O^+(aq)$

Equation 3: $HPO_4^{2-}(aq) + H_2O(\ell) \rightleftharpoons PO_4^{3-}(aq) + H_3O^+(aq)$

a. Identify four conjugate acid–base pairs in the equations above.

b. HPO_4^{2-} can act as both an acid and a base. Specify which equation above shows HPO_4^{2-} acting as an *acid*. Give a reason for your answer.

Unit 12.2 Acids, Bases and Salts
Topic 2: Acid–base titrations

Topic 2 covers acid–base titrations:
- Using the correct procedure to collect accurate titration data.
- Using titration data to calculate the concentration of a solution.

Introduction

An **acid–base titration** is an experimental procedure in which a reaction between an acid and a base is used to determine the concentration of the acid or the base. The titration depends on accurately measuring the volumes of reactants used – hence the procedure is also known as a **volumetric analysis**.

Acid–base titrations involve:

- A knowledge of the reaction occurring so that mole ratios can be determined.
- The preparation of standard solutions.
- The use of volumetric apparatus, eg pipette and burette.

Acid–base titrations

An acid–base titration uses a reaction between an acidic solution and basic solution to determine the concentration of one of the solutions where the concentration of the other is known.

Essential items used in an acid–base titration are a burette, pipette, conical flask, wash bottle and acid–base indicator.

- A **burette** is a long vertical tube with a tap at the bottom. It has graduations on the side so the volume of solution in the burette can be read. The burette usually holds the standard solution and is held in a clamp during use.

> The volume in the burette should be recorded to two decimal places, eg 10.00 mL, 21.96 mL.

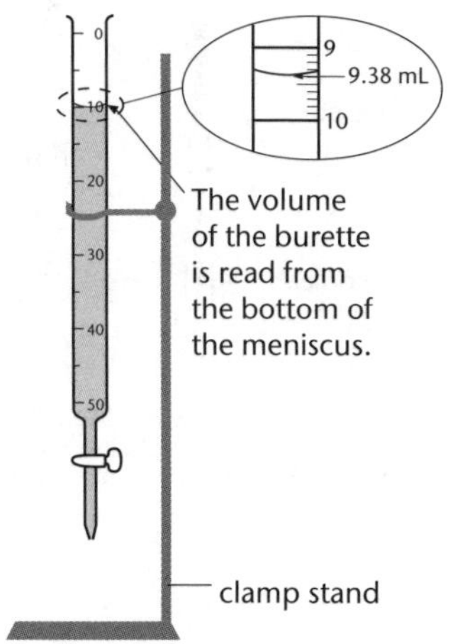

- A **pipette** is a glass tube **calibrated** to dispense an exact volume of solution, typically 10.00 mL, 20.00 mL or 25.00 mL. A pipette filler is used to fill and dispense liquid from the pipette. The pipette usually dispenses the solution with the *unknown* concentration.

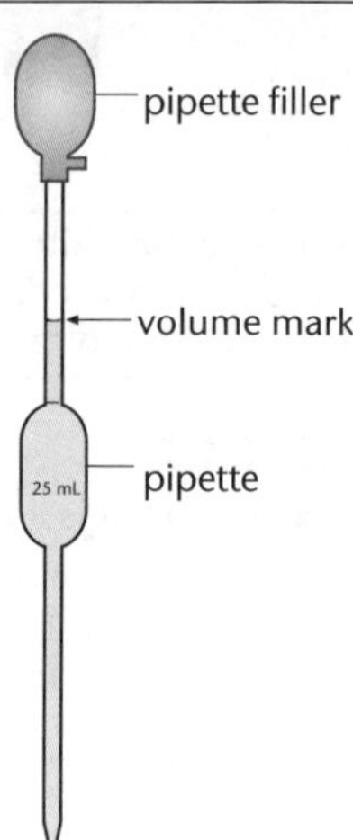

The reaction is carried out in a **conical flask**. One of the solutions is pipetted into the conical flask, a few drops of acid–base **indicator** are added, and the conical flask is placed under the burette. The other solution is slowly added from the burette, while the flask is constantly swirled to ensure the two solutions fully mix. A **wash-bottle** is also used to wash *splashes* from the sides of the flask back into the solution.

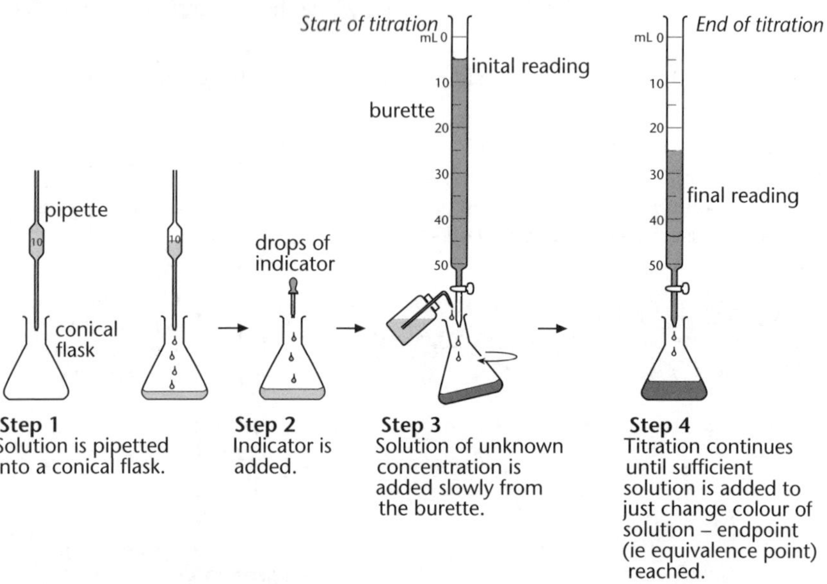

The point at which *all* the acid or base in the flask has reacted with the added base or acid is called the **equivalence point**.

*The equivalence point is not necessarily at pH = 7. (If the reaction produces an **acidic salt**, such as ammonium chloride, NH_4Cl, the equivalence point will be at pH < 7; if the reaction produces a basic salt, such as sodium ethanoate, CH_3COONa, the equivalence point will be at pH > 7.)

An acid–base indicator is used to indicate when the equivalence point has been reached. The pH at which an indicator changes colour is called the **endpoint**.

*An indicator should be chosen so that the equivalence point of the titration reaction lies close to, or within, the endpoint range of the indicator.

Indicators and endpoint ranges

Name	Colour change	pH range for endpoint
Methyl orange	Red to yellow	3.0 to 4.5
Bromothymol blue	Yellow to blue	6.0 to 7.6
Litmus solution	Red to blue	Below 7 to above 7
Phenolphthalein	Colourless to pink	8.3 to 10.0

Example A

Standardising hydrochloric acid

10.00 mL of a standard 0.200 mol L^{-1} sodium hydroxide solution is placed in a conical flask using a pipette. Three drops of phenolphthalein indicator are added. Hydrochloric acid is then added dropwise from a burette. The volume of hydrochloric acid required to make the phenolphthalein change from pink to colourless is 20.00 mL (30.00 mL – 10.00 mL).

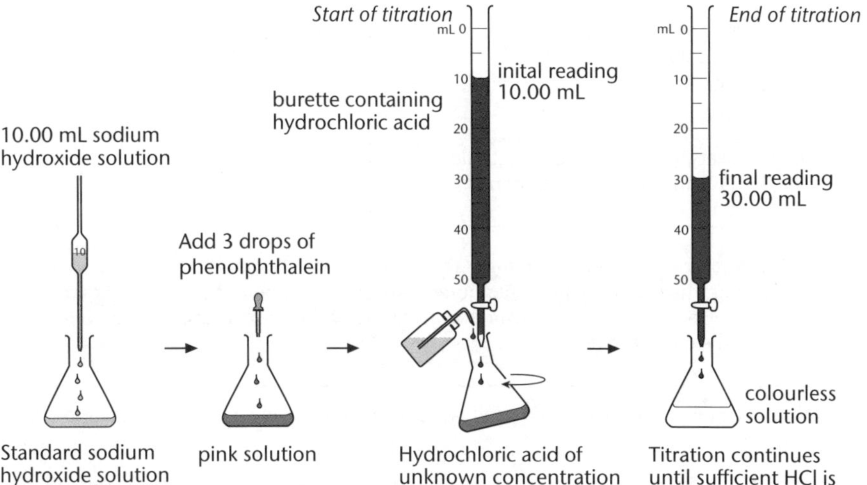

Reaction

The reaction occurring in the conical flask, as the contents of the burette are added, is:

$$HCl(aq) + NaOH(aq) \rightarrow H_2O(\ell) + Na^+(aq) + Cl^-(aq)$$

or $$H^+(aq) + OH^-(aq) \rightarrow H_2O(\ell)$$

Amount of known substance

For the standard sodium hydroxide, NaOH:

$c = 0.200$ mol L^{-1} and $V = 10.00$ mL $= 0.01000$ L

Amount of hydroxide ions, $OH^-(aq)$, originally in the conical flask is:

$n = cV$

$= 0.200 \times 0.01000$ [substituting and changing 10 mL to 0.0100 L]

$= 0.00200$ mol

Equivalence point

At the moment that all the solution in the flask turns colourless, it will be **neutral**. This is the equivalence point, where the amount of $H^+(aq)$ (in moles) added from the burette is exactly the same as the amount of $OH^-(aq)$ (in moles) initially in the flask.

From the balanced equation for the reaction:

1 mol H^+ (HCl)	reacts with	1 mol OH^- (NaOH)
$\therefore$ 0.00200 mol H^+	reacts with	0.00200 mol OH^-

Calculation of the unknown concentration

The volume of HCl used in the titration is given by:

volume used = final reading – initial reading on burette

= 30.00 mL – 10.00 mL

= 20.00 mL (0.02000 L)

Since the amount of HCl used is 0.00200 mol, the concentration of HCl can be calculated:

$$c = \frac{n}{V}$$

$$[H^+] = \frac{0.00200}{0.02000} \quad \text{[substituting } n = 0.00200 \text{ mol, } V = 20.00 \text{ mL} = 0.02000 \text{ L]}$$

$= 0.100$ mol L^{-1}

ie [HCl] $= 0.100$ mol L^{-1}

Square brackets, [ ], indicate 'concentration'. Hence, $[H^+]$ means 'the concentration of hydrogen ions'.

Note: The final answer is quoted to three significant figures. The measurements used in the calculations have three significant figures (concentration) or four significant figures (volumes). The accuracy of the answer is determined by the least accurate measurement, in this case, the concentration.

Titration procedures accompanied by *good practice* techniques ensure that accurate results are achieved.

- The pipette is rinsed with a small amount of the solution to be used in the pipette. This prevents dilution when the pipette is filled. The fixed volumes of liquid delivered by a pipette are called **aliquots**.
- The burette, like the pipette, is rinsed with a small volume of the solution it will contain.
- If a funnel is used for the burette, the funnel is removed from the burette so that the funnel does not drip, which would lead to inaccuracies.
- Before each titration, the conical flask is rinsed with deionised water. The conical flask does not need to be dried because any water left in it does not change the *amount* of dissolved solute added to the flask from the pipette.
- The burette should be read to two decimal places with an expected error of ±0.02 mL.

- Only a *few drops* (2 or 3) of indicator are added to the conical flask.
- A **rough titration** is done first. This is a titration done rapidly to determine the *approximate* volume of solution needed from the burette to reach endpoint. This volume is used to guide later titrations.
- During the titration, the flask is *swirled* to mix the solutions from the burette and pipette. Near the endpoint, deionised water from a wash-bottle is used to rinse any solution on the side of the flask into the acid–base mixture.
- A white background under the flask and behind the burette make for ease of reading results.
- At least two, but often three, titrations are needed to achieve **titres** (titration volumes that lie within a predetermined range, eg within 0.1 mL of each other) that are **concordant**. The results of concordant titres are *averaged* to provide the data for calculations.

Titres are the individual volumes of liquid delivered by a burette.

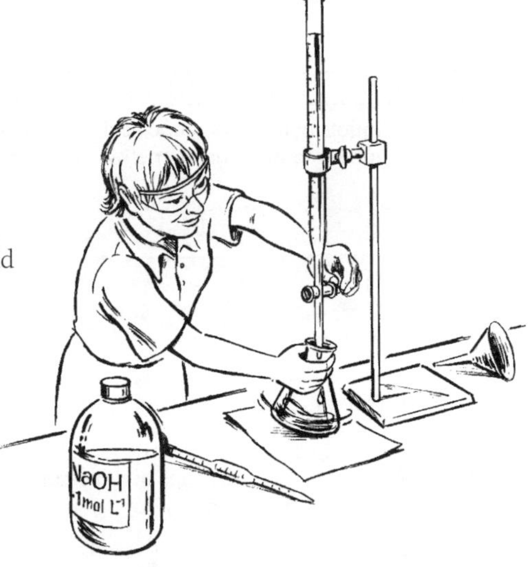

Some important titrations

Using a primary standard solution of sodium carbonate to standardise hydrochloric acid

Methyl orange is usually the indicator chosen to indicate the endpoint in the titration of sodium carbonate with hydrochloric acid. The hydrochloric acid solution is placed in the conical flask. The methyl orange will initially be *red* (in the acid) and turn *yellow* at the endpoint.

The reaction between the carbonate ion and acid is:

$$Na_2CO_3\,(aq) + 2HCl(aq) \rightarrow H_2O(\ell) + CO_2(g) + 2NaCl(aq)$$

or

$$CO_3^{2-}(aq) + 2H^+(aq) \rightarrow H_2O(\ell) + CO_2(g)$$

The balanced equation at the equivalence point shows that 2 mol $H^+(aq)$ will have reacted with 1 mol CO_3^{2-}:

ie $\frac{n(H^+)}{2} = \frac{n(CO_3^{2-})}{1}$; $n(H^+) = 2n(CO_3^{2-})$.

> *n* stands for 'number of moles'.
> $n(H^+)$ means 'number of moles of hydrogen ions'; $n(CO_3^{2-})$ means 'number of moles of carbonate ions'.
>
> $n(H^+) = 2n(CO_3^{2-})$ means 'the number of moles of hydrogen ions is twice the number of moles of carbonate ions'.

Example B

Standardising hydrochloric acid with sodium carbonate

20.00 mL aliquots of a hydrochloric acid solution were titrated with a 0.160 mol L^{-1} standard sodium carbonate solution using methyl orange indicator.

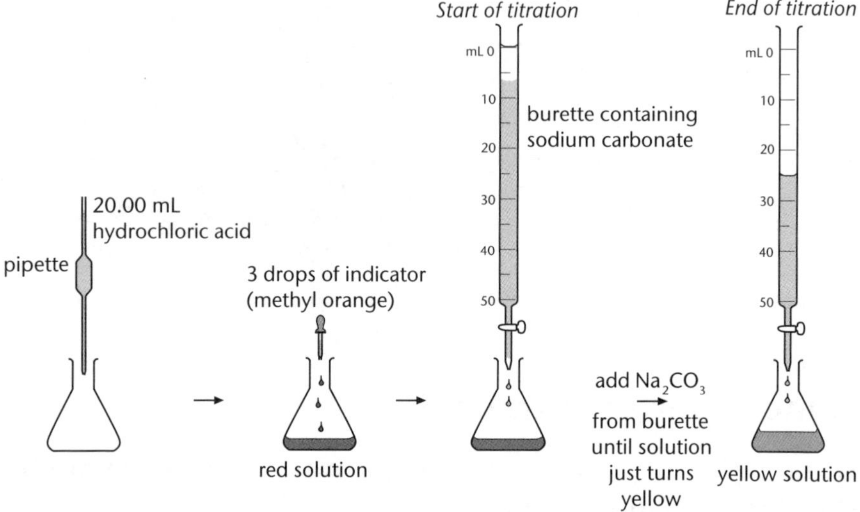

The titration results were:

Titration	Burette reading (mL)		Titre (mL)
	Initial	Final	
1 (rough)	0.90	26.32	25.42
2	26.32	49.58	23.26
3	1.16	24.48	23.32
4	24.48	47.68	23.20

Three concordant titres (titrations 2–4)

The average titre (volume Na_2CO_3) is $\frac{23.26 + 23.32 + 23.20}{3} = 23.26$ mL.

The amount of carbonate ion is:

$n(Na_2CO_3) = cV$

$= 0.160 \times 0.02326$ [substituting and changing 23.26 mL to 0.023 26 L]

$= 0.00372$ mol

$n(H^+) = 2n(CO_3^{2-}) = 2 \times 0.00372$

$= 0.00744$ mol

Hence, 0.007 44 mol H^+ is present in 20.00 mL of acid solution.

$c = \frac{n}{V}$

$[HCl] = \frac{0.00744}{0.0200}$ [substituting and changing mL to L]

$= 0.372 \text{ mol L}^{-1}$

Note: The rough titration acts only as a guide and is *not* used in the calculations.

Using standard hydrochloric acid to standardise calcium hydroxide solution

The calcium hydroxide solution of unknown concentration is pipetted into a conical flask. Phenolphthalein indicator is used, giving a colour change from pink (basic) to colourless (acidic).

The balanced equation for the reaction occurring is:

$$2HCl(aq) + Ca(OH)_2(aq) \rightarrow CaCl_2(aq) + 2H_2O(\ell)$$

Example C

Calcium hydroxide solution

50.00 mL of calcium hydroxide solution is added to a rinsed conical flask and titrated with 0.0500 mol L^{-1} HCl. The endpoint is detected by phenolphthalein at an average volume of 20.00 mL HCl (obtained from three concordant titres).

The amount of HCl used in the titration is:

$n = cV$

$= 0.0500 \times 0.0200$ mol [substituting and changing mL to L]

$= 0.00100$ mol

From the equation for the reation:

$$\frac{n(HCl)}{2} = \frac{n(Ca(OH)_2)}{1}$$

$$nCa(OH)_2) = \frac{1}{2}n(HCl)$$

$$= \frac{1}{2} \times 0.00100$$

$$= 0.000500 \text{ mol} = 5.00 \times 10^{-4} \text{ mol}$$

Concentration of $Ca(OH)_2 = \frac{n}{V}$

$= \frac{0.000500}{0.05000}$ [5.00×10^{-4} mol present in 50.00 mL]

$= 0.0100 \text{ mol L}^{-1}$

Using standard sodium hydroxide to standardise ethanoic acid (acetic acid) solution

The indicator chosen is usually phenolphthalein. The reaction occurring is:

$$CH_3COOH(aq) + OH^-(aq) \rightarrow CH_3COO^-(aq) + H_2O(\ell)$$

This reaction can be used to analyse the ethanoic acid content of vinegar.

Example D

Vinegar concentration

10.00 mL of vinegar is diluted to 100.00 mL. 10.00 mL aliquots of this diluted vinegar are titrated, using phenolphthalein indicator, with a standard solution of sodium hydroxide of concentration 0.0500 mol L^{-1}. The results are:

Titration	Burette reading (mL)		Titre (mL)
	Initial	Final	
1 (rough)	0.80	18.92	18.12
2	18.92	33.96	15.04
3	2.25	17.35	15.10
4	17.62	32.66	15.04

The average titre of NaOH is 15.06 mL.

The amount of OH^- used in the titration is:

$$\begin{aligned} n(OH^-) &= cV \\ &= 0.0500 \times 0.01506 \quad \text{[substituting and changing mL to L]} \\ &= 0.000\,753 \text{ mol} \end{aligned}$$

From the balanced equation, 1 mol of ethanoic acid will react with 1 mol of hydroxide, ie $n(CH_3COOH) = 0.000\,753$ mol. The concentration of ethanoic acid in the *diluted* ethanoic acid is:

$$\begin{aligned} c(CH_3COOH) &= \frac{n}{V} \\ &= \frac{0.000\,753}{0.0100} \quad \text{[substituting and changing mL to L]} \\ &= 0.0753 \text{ mol L}^{-1} \end{aligned}$$

The original vinegar has been diluted by a factor of 10, hence the original vinegar solution is ten times *more* concentrated:

$$\begin{aligned} \text{original concentration} &= 0.0753 \times 10 \\ &= 0.753 \text{ mol L}^{-1} \end{aligned}$$

In commercial preparations of chemicals, it is common to express the concentration as 'weight per volume', ie as the *mass* of dissolved solute in *100 mL* of solution. This is often given as a percentage.

Example E

Percentage weight/volume

The % weight/volume for the vinegar in Example D is given by the mass of ethanoic acid in 100 mL:

amount of ethanoic acid (in mol) in 100 mL = cV
= 0.753×0.100 [substituting; 100 mL = 0.1 L]
= 0.0753 mol

mass of ethanoic acid in 100 mL solution = nM
= 0.0753×60.0 [M(ethanoic acid) = 60.0 g mol^{-1}]
= 4.52 g

So, in 100 mL of solution there is 4.52 g of ethanoic acid, hence the vinegar is:

$\frac{4.52}{100} \times 100 = 4.52\%$ ethanoic acid.

Using standard hydrochloric acid to standardise sodium hydroxide solution

This is a similar process to that of standardising calcium hydroxide solution (Example C). The main difference is that for every 1 mol of sodium hydroxide dissolved, only *one* mole of OH^- ions are produced. Phenolphthalein is the indicator chosen. The reaction is:

$$H^+(aq) + OH^-(aq) \rightarrow H_2O(\ell)$$

or $$HCl(aq) + NaOH(aq) \rightarrow H_2O(\ell) + NaCl(aq)$$

The standardised sodium hydroxide solution is a secondary standard.

Using standard hydrochloric acid to standardise ammonia solution

This is another example of the preparation of a secondary standard solution. The indicator chosen is usually methyl orange. The reaction occurring is:

$$NH_3(aq) + H^+(aq) \rightarrow NH_4^+(aq)$$

or $$NH_3(aq) + HCl(aq) \rightarrow NH_4Cl(aq)$$

Unit 12.2 Activity 2A: Acid–base titrations

1. 25.00 mL of 0.400 mol L^{-1} sodium hydroxide solution is placed in a freshly rinsed conical flask. Some phenolphthalein is added. A rough titration with HCl showed that *about* 17 mL of HCl was needed for neutralisation. Three accurate titrations gave the volume readings: 14.96 mL, 15.02 mL and 15.00 mL.
   - **a.** Write a balanced ionic equation for the reaction occurring during the titration.
   - **b.** Calculate the average titre.
   - **c.** Calculate the amount (in mol) of hydroxide ions used in the titration.
   - **d.** From the balanced equation, what amount of $H^+(aq)$ has reacted in the titration?
   - **e.** Calculate the concentration of the hydrochloric acid.
2. A 10.00 mL aliquot of a standard potassium carbonate solution, concentration 0.0500 mol L^{-1}, is pipetted into a flask. This sample is titrated with hydrochloric acid of unknown concentration. The indicator (methyl orange) showed that the endpoint had been reached after an average titre of 25.00 mL of hydrochloric acid was added.
   - **a.** What is the function of the indicator?
   - **b.** How many moles of potassium carbonate are there in the 10.00 mL sample?
   - **c.** Write a balanced equation for the reaction between hydrochloric acid and potassium carbonate.
   - **d.** How many moles of hydrochloric acid are there in the 25.00 mL titre?
   - **e.** What is the concentration of the hydrochloric acid?
3. Calcium hydroxide dissolves in water to form a saturated solution (limewater). To determine the concentration of limewater in mol L^{-1}, four separate 10.00 mL aliquots of the solution were titrated with 0.125 mol L^{-1} standardised hydrochloric acid using bromophenol blue as the indicator. The four titre values were 28.00 mL, 23.92 mL, 24.10 mL and 24.06 mL.
   - **a.** What piece of equipment would be used to measure out the aliquots of limewater?
   - **b.** Why is the value for the first titration higher than the other three?
   - **c.** Use the average of the three concordant titre values to calculate the amount of hydrochloric acid that reacted.
   - **d.** Write a balanced equation for the reaction.
   - **e.** Calculate the amount of $Ca(OH)_2$ in the 10.00 mL aliquots.
   - **f.** Calculate the concentration of the limewater solution.
4. 20.00 mL of a 0.100 mol L^{-1} solution of ethanoic acid was neutralised by 16.00 mL of sodium hydroxide. The ionic equation for this reaction is:

   $$CH_3COOH + NaOH \rightarrow CH_3COONa + H_2O$$

   - **a.** Calculate the number of moles of ethanoic acid in the 20.00 mL sample.
   - **b.** How many moles of sodium hydroxide are required to neutralise the acid?
   - **c.** Calculate the concentration of the sodium hydroxide solution.

5. A solution of NaOH was titrated with 15.00 mL aliquots of 0.252 mol L^{-1} ethanoic acid. An average titre of 24.96 mL of NaOH was obtained. Find the concentration of the NaOH.

6. When 25.00 mL of sulfuric acid, H_2SO_4, was titrated with 0.0820 mol L^{-1} sodium hydroxide solution, the endpoint was detected (with phenolphthalein) at 22.5 mL. Calculate the concentration of the sulfuric acid (in mol L^{-1}).

7. During the preparation of a standard solution of anhydrous sodium carbonate, Na_2CO_3, a student obtained the following results:

- Mass of beaker and anhydrous sodium carbonate = 131.10 g.
- Mass of empty beaker = 128.45 g.

The student then dissolved the sodium carbonate in enough water to form 100.00 mL of solution.

a. Calculate the mass of sodium carbonate that was weighed out.

b. Calculate the concentration of the solution prepared in:

i. g L^{-1}

ii. mol L^{-1}

$M(Na_2CO_3) = 106.0$ g mol^{-1}

The standard solution was then titrated against a hydrochloric acid solution of unknown concentration, using methyl orange indicator. It was found that 20.00 mL of the sodium carbonate solution was neutralised by 5.16 mL of the acid.

c. What piece of apparatus would be used to measure the 20.00 mL of standard solution into a conical flask for the titration?

d. What piece of apparatus would be used to measure the volume(s) of acid necessary to neutralise the standard solution?

e. Briefly describe how it was known when the two solutions were neutralised.

f. Write an equation for the reaction which occurred between the hydrochloric acid and sodium carbonate.

g. Calculate the concentration of the hydrochloric acid solution in mol L^{-1}.

8. 20.00 mL of 0.208 mol L^{-1} sodium carbonate was titrated with hydrochloric acid. When the endpoint was reached, an average titre of 31.22 mL HCl had been added. What is the concentration of HCl?

9. A student wished to find the concentration of a saturated solution of calcium hydroxide (limewater) by titrating it with standard hydrochloric acid. The hydrochloric acid was standardised against a standard sodium carbonate, Na_2CO_3, solution.

$M(Na_2CO_3) = 106.0 \text{ g mol}^{-1}$

a. Calculate the amount of sodium carbonate used, given that 3.02 g was dissolved to produce a 500.0 mL solution.

b. Calculate the concentration of the standard sodium carbonate solution.

c. Give an ionic equation for the reaction between sodium carbonate and hydrochloric acid.

d. 10.00 mL aliquots of the standard Na_2CO_3 were titrated with HCl, using methyl orange indicator. The following results were obtained:

Titration	Burette reading (mL)		Titre (mL)
	Initial	Final	
1 (rough)	0.80	18.90	18.10
2	18.90	33.92	15.02
3	2.54	17.66	15.12
4	17.66	32.66	15.00

Calculate the concentration of the HCl.

e. Give an equation for the reaction between limewater and hydrochloric acid.

f. Calculate the concentration of the limewater from the following data:

- 10.00 mL aliquots of limewater were titrated with the standard hydrochloric acid.
- Three titrations gave titres of 13.72 mL, 13.82 mL and 13.80 mL.

Unit 12.2 Acids, Bases and Salts
Topic 3: More about acids and bases

Topic 3 covers:

- Equilibrium constant expressions, K_a and K_w.
- Correlation between acid or base strength, K_a and pH.
- Calculations involving K_a, K_w and pH for solutions of bases and monoprotic acids.

Acidity and basicity

Definitions of acidity and basicity are shown below.

Definitions of acidity	Definitions of basicity
The acidity of an acid refers to the relative number of protons it donates per mole of acid (in aqueous solution).	Basicity refers to the relative number of protons accepted per mole of base (in aqueous solution).
Monoprotic acids release one mole of hydrogen ions per mole of acid. Hydrochloric acid, HCl, is a monoprotic acid: $HCl \rightarrow H^+ + Cl^-$	Monobasic substances, such as potassium hydroxide, KOH, accept one mole of protons per mole of substance: $KOH + H^+ \rightarrow K^+ + H_2O$
Diprotic acids release two moles of hydrogen ions per mole of acid. Sulfuric acid, H_2SO_4, is a diprotic acid: $H_2SO_4 \rightarrow 2H^+ + SO_4^{2-}$	Dibasic substances accept two moles of protons per mole of substance. Calcium hydroxide, $Ca(OH)_2$, is dibasic: $Ca(OH)_2 + 2H^+ \rightarrow Ca^{2+} + 2H_2O$

Acid–base reactions

The reaction between aqueous acids and aqueous bases involves proton transfer. The hydronium ions released by the acid donate protons to the hydroxide ions released by the base:

$H_3O^+(aq) + OH^-(aq) \rightarrow 2H_2O(\ell)$

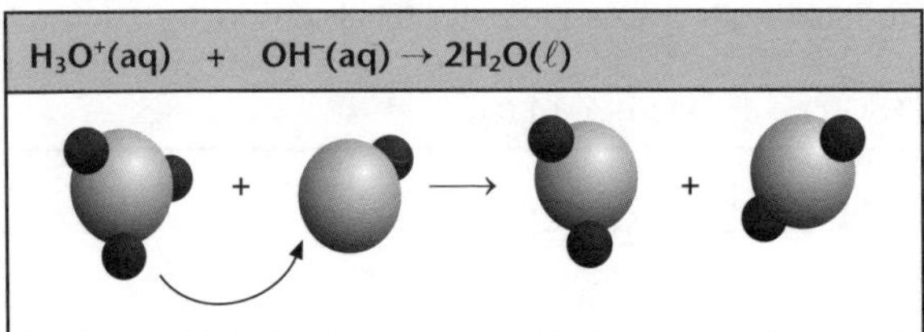

Acid–base reaction

Example A

Reaction between HCl and NaOH

In the reaction between hydrochloric acid and sodium hydroxide:

$$HCl(aq) + NaOH(aq) \rightarrow NaCl(aq) + H_2O(\ell)$$

- The hydronium ions from the HCl react with the hydroxide ions from the NaOH:

$$H_3O^+(aq) + OH^-(aq) \rightarrow 2H_2O(\ell)$$

- The chloride ions Cl^- from the HCl(aq) and the sodium ions Na^+ from the NaOH(aq) are spectator ions.

Example B

Acid–base reaction between hydrogen chloride and ammonia

In the reaction between hydrogen chloride and ammonia:

$$HCl(g) + NH_3(g) \rightarrow NH_4Cl(s)$$

the hydrogen chloride donates protons to the ammonia.

$$HCl(g) + NH_3(g) \rightarrow NH_4^+(s) + Cl^-(s)$$

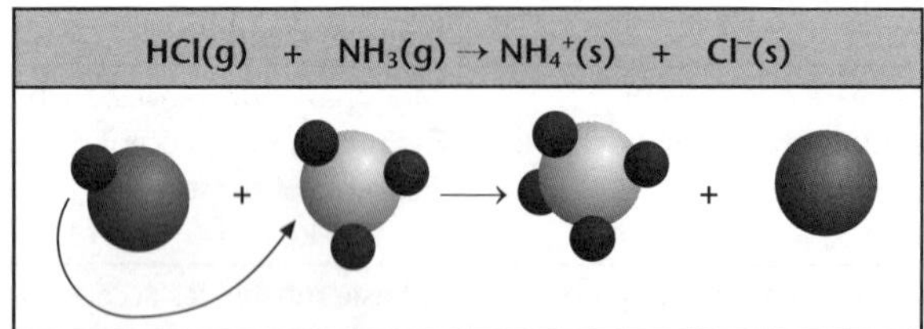

Strong acids

A **strong acid** is virtually 100% dissociated in water to form hydronium ions and the conjugate base.

Nitric acid, $HNO_3(aq)$, and hydrochloric acid, HCl(aq), are strong acids.

Example C

When hydrogen chloride gas dissolves in water, it fully dissociates to form $H_3O^+(aq)$ and $Cl^-(aq)$:

$$HCl(g) + H_2O(\ell) \rightarrow H_3O^+(aq) + Cl^-(aq)$$

Calculating pH of strong acids

$$pH = -\log[H_3O^+]$$

The pH of a strong monoprotic acid of concentration c(HX) is calculated thus:

$$pH = -\log c(HX)$$

Example D

The pH of 0.006 00 mol L^{-1} HCl is:

$$pH = -\log 0.006\,00 = -(-2.22) = 2.22$$

Weak acids

A **weak acid** only partially dissociates in water – often less than 5% of the acid molecules present dissociate. This means that most of the acid molecules do not release their protons.

The equilibrium position for the acid dissolving in water lies towards the left-hand side of the equilibrium equation.

For the acid HA:

$$HA(aq) + H_2O(\ell) \rightleftharpoons H_3O+(aq) + A^-(aq)$$

Example E

Ethanoic acid

Ethanoic acid, $CH_3COOH(aq)$, is a weak acid:

$CH_3COOH(aq)$ + $H_2O(\ell)$ ⇌	$H_3O^+(aq)$ +	$CH_3COO^-(aq)$
0.1 mol L^{-1} only a small percentage dissociates	≈ 0.001 mol L^{-1}	≈ 0.001 mol L^{-1}

In this equilibrium the:

- concentration of ethanoic acid remains high.
- concentrations of both the hydronium ions and the ethanoate ions (the conjugate base of the ethanoic acid) are identical and small.

Example F

Ammonium ion

The ammonium ion, $NH_4^+(aq)$, is an even weaker acid than ethanoic acid:

$NH_4^+(aq)$ + $H_2O(\ell)$ ⇌	$H_3O^+(aq)$ +	$NH_3(aq)$
0.1 mol L^{-1} only a very small percentage dissociates	≈ 0.00001 mol L^{-1}	≈ 0.00001 mol L^{-1}

Example G

Water

Water is an even weaker acid than the ammonium ion:

$H_2O(\ell)$ +	$H_2O(\ell)$ ⇌	$H_3O^+(aq)$ +	$OH^-(aq)$
55.6 mol L^{-1}	55.6 mol L^{-1}	1×10^{-7} mol L^{-1}	1×10^{-7} mol L^{-1}

Although the four species HCl, CH_3COOH, NH_4^+ and H_2O can all react as acids (proton donors), they react differently with water molecules because:

- water is amphiprotic.
- HCl, CH_3COOH, NH_4^+ and H_2O have different acid strengths.

The acidity constant, K_a

The percentage dissociation of a weak acid depends not only upon the chemical make-up of the acidic substance, but also upon:

- concentration.
- temperature.

The **equilibrium equation** for a weak acid HA with water is:

$$HA(aq) + H_2O(\ell) \rightleftharpoons H_3O^+(aq) + A^-(aq)$$

The **equilibrium constant** expression, $\mathbf{K_c}$, is:

$$K_c = \frac{[H_3O^+][A^-]}{[HA][H_2O]}$$

Rearranging the equilibrium constant formula gives:

$$K_c \times [H_2O] = \frac{[H_3O^+][A^-]}{[HA]}$$

The concentration of water is so large (55.6 mol L^{-1}) that it does not effectively change upon addition of HA molecules. The water concentration can therefore be regarded as constant. Since K_c is constant and $[H_2O]$ is constant, $K_c \times [H_2O]$ will also be constant.

For an acid, $K_c \times [H_2O]$ is given the name the **acid dissociation constant** or the **acidity constant** and the symbol $\mathbf{K_a}$, where:

$$K_a = \frac{[H_3O^+][A^-]}{[HA]}$$

pH calculations for weak acids

For the reaction of a weak acid with water,

$$HA(aq) + H_2O(\ell) \rightleftharpoons A^-(aq) + H_3O^+(aq)$$

the K_a expression is:

$$K_a = \frac{[H_3O^+][A^-]}{[HA]}$$

From the reaction equation:

$$[H_3O^+] = [A^-]$$

Substituting into the K_a expression:

$$K_a = \frac{[H_3O^+]^2}{[HA]}$$

If K_a is small, then $[H_3O^+]$ will be much smaller than [HA], ie $[H_3O^+] \lll [HA]$.

This means the concentration of the acid at equilibrium can be assumed to be the same as the initial concentration of the acid, $c(HA)$, ie:

$$K_a = \frac{[H_3O^+]^2}{c(HA)}$$

Rearranging: $[H_3O^+]^2 = K_a \times c(HA)$

$$[H_3O^+] = \sqrt{K_a \times c(HA)}$$

and $pH = -\log [H_3O^+]$

Example H

Calculating the pH of a weak acid solution

Calculate the pH of a 0.200 mol L^{-1} solution of methanoic acid, HCOOH.

$K_a(HCOOH) = 1.82 \times 10^{-4}$.

$$HCOOH(aq) + H_2O(\ell) \rightleftharpoons HCOO^-(aq) + H_3O^+(aq)$$

$$K_a = \frac{[H_3O^+][HCOO^-]}{[HCOOH]}$$

From the reaction equation: $[H_3O^+] = [HCOO^-]$

Since K_a is small: $[HCOOH] = 0.200$ mol L^{-1}

and so $K_a = 1.82 \times 10^{-4} = \frac{[H_3O^+]^2}{0.200}$

Rearranging: $[H_3O^+]^2 = 1.82 \times 10^{-4} \times 0.200$

$[H_3O^+] = \sqrt{1.82 \times 10^{-4} \times 0.200}$

$= 6.03 \times 10^{-3}$

and pH $= -\log(6.03 \times 10^{-3})$

$= 2.22$ (3 significant figures)

Unit 12.2 Activity 3A: Strong and weak acids

1. Calculate the pH of:

- **a.** 0.015 mol L^{-1} HCl.
- **b.** 0.500 mol L^{-1} HCl.
- **c.** 0.00213 mol L^{-1} HCl.

2. Explain the differences between the terms:

- **a.** Strong and weak acids.
- **b.** Concentrated and dilute acids.

3. Give the K_a expressions for the following dissociations of weak acids:

- **a.** $CH_3COOH(aq) + H_2O(\ell) \rightleftharpoons CH_3COO^-(aq) + H_3O^+(aq)$
- **b.** $NH_4^+(aq) + H_2O(\ell) \rightleftharpoons NH_3(aq) + H_3O^+(aq)$
- **c.** $HF(aq) + H_2O(\ell) \rightleftharpoons F^-(aq) + H_3O^+(aq)$

4. Explain the assumption made when determining the pH of a weak acid.

5. Calculate the pH of the following solutions. The K_a values are given below.

- **a.** 0.125 mol L^{-1} ethanoic acid (CH_3COOH).
- **b.** 0.500 mol L^{-1} ethanoic acid (CH_3COOH).
- **c.** 0.00250 mol L^{-1} propanoic acid (C_2H_5COOH).
- **d.** 0.0100 mol L^{-1} ammonium chloride (NH_4Cl).

$K_a(CH_3COOH) = 1.74 \times 10^{-5}$, $K_a(C_2H_5COOH) = 1.35 \times 10^{-5}$, $K_a(NH_4^+) = 5.75 \times 10^{-10}$

6. Vitamin C is an organic acid. The pH of an aqueous solution of vitamin C is 3.20. Calculate the concentration of vitamin C in the solution.

$K_a(\text{vitamin C}) = 7.94 \times 10^{-5}$

K_a and pK_a

As pH is defined as –log [H_3O^+], **pK_a** is similarly defined as:

$$pK_a = -\log K_a$$

Just as [H_3O^+] = 10^{-pH}, K_a can similarly be determined from:

$$K_a = 10^{-pK_a}$$

The relationship between acid strength, K_a and pK_a, is shown in the following table.

Acid strength, K_a and pK_a

Acid	Formula	K_a	pK_a
Strong acids	**HX**	**Very large**	**Negative**
Hydrochloric acid	HCl	1×10^7	–7
Weak acids	**HA**	**10^{-3} to 10^{-13}**	**3 to 13**
Hydrofluoric acid	HF	6.76×10^{-4}	3.17
Methanoic acid	HCOOH	1.82×10^{-4}	3.74
Hypochlorous acid	HClO	2.95×10^{-8}	7.53
Ammonium ions	NH_4^+	5.75×10^{-10}	9.24

Conjugate base strengths and K_a values

As the strength (ie the degree of dissociation) of acids decreases, then:

- K_a values become smaller.
- pK_a values become larger.
- The equilibrium concentration of both hydronium ions, [H_3O^+], and the conjugate base of the acid, [A^-], decreases.
- The equilibrium concentration of the undissociated acid molecules, [HA], increases.
- The strengths (ie the ability to accept protons) of the conjugate bases of the acids increase.

If the acid of a conjugate pair is weak, the conjugate base will be more likely to accept protons. The *weaker* the acid (smaller K_a and larger pK_a), the *stronger* the conjugate base.

Example I

For $NH_4^+(aq) + H_2O(\ell) \rightleftharpoons NH_3(aq) + H_3O^+(aq)$

K_a is calculated by:

$$K_a = \frac{[NH_3]\,[H_3O^+]}{[NH_4^+]}$$

$$= 5.75 \times 10^{-10}$$

The very small size of the K_a value indicates that the:

- ammonium ion NH_4^+ is a very weak acid.
- conjugate base of NH_4^+, which is ammonia NH_3, will be a less weak base.

Unit 12.2 Activity 3B: K_a and pK_a

1. Place the following substances in order of decreasing acid strength:
   a. HCOOH, K_a 1.78×10^{-4}
   b. H_2O_2, K_a 5.01×10^{-12}
   c. HF, K_a 6.76×10^{-4}
   d. $[Cu(H_2O)_4]^{2+}(aq)$, K_a 4.50×10^{-8}
   e. C_3H_7COOH, K_a 1.35×10^{-5}
   f. CH_3COOH, K_a 1.74×10^{-5}
2. For the table that follows:
   a. Calculate the missing pK_a or K_a value.
   b. List the species in order of decreasing acid strength.

Substance	K_a	pK_a	Substance	K_a	pK_a
HCOOH	1.82×10^{-4}	i.	NH_4^+	vi.	9.24
HF	6.76×10^{-4}	ii.	HNO_3	vii.	−1.30
$[Al(H_2O)_6]^{3+}$	1.10×10^{-5}	iii.	HNO_2	viii.	3.15
$[Cu(H_2O)_4]^{2+}$	4.57×10^{-8}	iv.	$[Fe(H_2O)_6]^{3+}$	ix.	2.17
H_2O_2	2.24×10^{-12}	v.	HCN	x.	9.21

3. Calculate the pH of the following:
   a. 0.100 mol L^{-1} lactic acid solution (HLac). $pK_a(HLac) = 3.86$.
   b. 0.0200 mol L^{-1} hydrocyanic acid. $pK_a(HCN) = 9.21$.

The ionic product, K_w, for water

The equilibrium reaction for water dissociating is:

$$2H_2O(\ell) \rightleftharpoons H_3O^+(aq) + OH^-(aq)$$

Numerically, the equilibrium constant, K_c, equals:

$$K_c = \frac{[H_3O^+][OH^-]}{[H_2O]^2}$$

Rearranging this formula gives:

$$K_c \times [H_2O]^2 = [H_3O^+]\,[OH^-]$$

The concentration of water is so large, 55.6 mol L^{-1}, that it does not effectively change as such a small amount dissociates. The water concentration can therefore be regarded as constant. Since $\mathbf{K_c}$ is constant and $[H_2O]$ is constant, $K_c \times [H_2O]^2$ will also be constant.

For water, $K_c \times [H_2O]^2$ is given the name the **ionic product for water** and the symbol $\mathbf{K_w}$ where:

$$K_w = K_c \times [H_2O]^2 = [H_3O^+][OH^-]$$

Pure water is neutral. In any neutral aqueous solution, the hydronium ion concentration equals the hydroxide ion concentration.

At 25°C, $[H_3O^+] = 1 \times 10^{-7}$ mol L^{-1}

and $[OH^-] = 1 \times 10^{-7}$ mol L^{-1}

Therefore, $K_w = [H_3O^+][OH^-] = (1 \times 10^{-7})^2 = 1 \times 10^{-14}$ at 25°C.

The effect of temperature on K_w

The dissociation of water is:

$$2H_2O(\ell) \rightleftharpoons H_3O^+(aq) + OH^-(aq) \quad \Delta H = +ve$$

The dissociation of water is an endothermic process. Increasing the temperature favours the **forward reaction**, which results in:

- more dissociation.
- more hydrogen ions and hydroxide ions.
- an increase in K_w.

Therefore, values for K_w alter with temperature.

Example J

At 25°C, $K_w = 1 \times 10^{-14}$. At 100°C, $K_w = 5.12 \times 10^{-13}$.

The increase in K_w does not mean water is more acidic at higher temperatures – the concentrations of hydrogen and hydroxide ions still remain identical.

So, since $[H_3O^+]$ still equals $[OH^-]$, and since $[H_3O^+][OH^-] = K_w$, it follows that:

$$[H_3O^+] = \sqrt{K_w}$$
$$= \sqrt{5.12 \times 10^{-13}}$$
$$= 7.16 \times 10^{-7}$$

$$pH = -\log [H_3O^+]$$
$$= -\log (7.16 \times 10^{-7})$$
$$= 6.1$$

As the water temperature has risen, the pH scale has changed with the central, neutral point now at pH 6.1.

Strong bases

A **strong base** completely dissociates in water.

Example K

NaOH is a strong base:

$$NaOH(s) \xrightarrow{H_2O(\ell)} Na^+(aq) + OH^-(aq)$$

The pH of a strong base is calculated using:

$K_w = [H_3O^+][OH^-]$

$[H_3O^+] = \frac{K_w}{[OH^-]}$

$pH = -\log [H_3O^+]$

Example L

To calculate the pH of 0.0200 mol L^{-1} NaOH:

$$[H_3O^+] = \frac{K_w}{[OH^-]}$$

$$= \frac{1 \times 10^{-14}}{0.0200}$$

$$= 5.00 \times 10^{-13}$$

$$pH = -\log (5.00 \times 10^{-13}) = 12.3$$

Weak bases

A **weak base** reacts only to a small extent with water.

Example M

Ammonia is a weak base, ie only a very small percentage reacts:

$$NH_3(aq) + H_2O(\ell) \rightleftharpoons NH_4^+(aq) + OH^-(aq)$$

0.1 mol L^{-1} 0.001 mol L^{-1} 0.001 mol L^{-1}

In aqueous ammonia, there will be a mixture of $NH_3(aq)$, $OH^-(aq)$, $NH_4^+(aq)$ and $H_2O(\ell)$.

Example N

The ethanoate ion, CH_3COO^-, is an even weaker base than ammonia:

$$CH_3COO^-(aq) + H_2O(\ell) \rightleftharpoons CH_3COOH(aq) + OH^-(aq)$$

0.1 mol L^{-1} 0.00001 mol L^{-1} 0.00001 mol L^{-1}

In aqueous ethanoic acid, there will be a mixture of $CH_3COOH(aq)$, $H^+(aq)$, $CH_3COO^-(aq)$ and $H_2O(\ell)$.

Calculating the pH of a weak base solution

Calculations that involve determining the hydronium ion concentration and/or the pH of a solution of a weak base use the K_a value for the conjugate acid of the weak base or K_b for the base.

Example O

Using K_a to calculate pH of a solution of a weak base

For the weak base, B, in aqueous solution:

$$B(aq) + H_2O(\ell) \rightleftharpoons BH^+(aq) + OH^-(aq) \quad \textit{Equation } ①$$

BH^+ is the conjugate acid of B and reacts with water:

$$BH^+(aq) + H_2O(\ell) \rightleftharpoons B(aq) + H_3O^+(aq) \quad \textit{Equation } ②$$

So: $K_a(BH^+) = \dfrac{[H_3O^+][B]}{[BH^+]}$

From Equation ①: $[BH^+] = [OH^-]$.

Also, $[OH^-] = \dfrac{K_w}{[H_3O^+]}$

Assuming [B] = initial concentration of the base = c(B).

Substituting into the K_a expression:

$$K_a(BH^+) = \frac{c(B).[H_3O^+]}{\dfrac{K_w}{[H_3O^+]}}$$

$$= \frac{c(B).[H_3O^+]^2}{K_w}$$

Rearranging: $[H_3O^+]^2 = \dfrac{K_a \times K_w}{c(B)}$

$$[H_3O^+] = \sqrt{\frac{K_a \times K_w}{c(B)}}$$

and $pH = -\log [H_3O^+]$

An alternative method of calculation uses $\mathbf{K_b}$, the **base dissociation constant**, where

$$K_b = \frac{1 \times 10^{-14}}{K_a}$$

Example P

Using K_b to calculate pH of a solution of a weak base

For the reaction of a weak base with water:

$$B(aq) + H_2O(\ell) \rightleftharpoons BH^+(aq) + OH^-(aq)$$

the K_b expression is: $K_b = \dfrac{[BH^+][OH^-]}{[B]}$

From the equation for the reaction: $[BH^+] = [OH^-]$

Substituting into the K_b expression:

$$K_b = \frac{[OH^-]^2}{[B]}$$

If K_b is small, then $[OH^-]$ will be much smaller than [B], ie $[OH^-] <<< [B]$, where [B] = c(B) in the initial solution.

So: $K_b = \dfrac{[OH^-]^2}{c[B]}$

Rearranging: $[OH^-]^2 = K_b \times c(B)$

$$[OH^-] = \sqrt{K_b \times c(B)}$$

$$[H_3O^+] = \frac{K_w}{[OH^-]}$$

and $pH = -\log [H_3O^+]$

Example Q

Calculate the pH of a weak base

Calculate the pH of a 0.400 mol L^{-1} solution of sodium ethanoate, CH_3COONa.

$K_a(CH_3COOH) = 1.74 \times 10^{-5}$ mol L^{-1}.

$$CH_3COONa(s) \xrightarrow{H_2O(\ell)} CH_3COO^-(aq) + Na^+(aq)$$

CH_3COOH is the conjugate acid of the weak base, the ethanoate ion, CH_3COO^-.

$$CH_3COO^-(aq) + H_2O(\ell) \rightleftharpoons CH_3COOH(aq) + OH^-(aq)$$

$$K_a(CH_3COOH) = \frac{[CH_3COO^-][H_3O^+]}{[CH_3COOH]}$$

$$[CH_3COO^-] = 0.400 \text{ mol L}^{-1}$$

$$[CH_3COOH] = [OH^-] = \frac{K_w}{[H_3O^+]}$$

Rearranging and substituting in:

$$[H_3O^+]^2 = \frac{K_w \times K_a}{[CH_3COO^-]}$$

$$[H_3O^+] = \sqrt{\frac{K_w \times K_a}{[CH_3COO^-]}}$$

$$[H_3O^+] = \sqrt{\frac{1 \times 10^{-14} \times 1.74 \times 10^{-5}}{0.400}} = 6.60 \times 10^{-10}$$

Using $pH = -\log [H_3O^+]$: $pH = -\log 6.60 \times 10^{-10} = 9.18$

Unit 12.2 Activity 3C: pH of bases

1. Calculate the pH of the following solutions:
   a. 0.100 mol L^{-1} sodium hydroxide.
   b. 0.500 mol L^{-1} potassium hydroxide.
2. Determine the concentration of hydroxide ions in:
   a. 0.100 mol L^{-1} hydrochloric acid.
   b. 0.20 mol L^{-1} hydrochloric acid.
3. Calculate the pH of the following solutions:
   a. 0.0100 mol L^{-1} ammonia NH_3.
   b. 0.200 mol L^{-1} aminomethane CH_3NH_2.
   c. 0.100 mol L^{-1} solution of sodium ethanoate CH_3COONa.
   d. 0.0500 mol L^{-1} sodium fluoride NaF.
   e. 0.00250 mol L^{-1} sodium methanoate HCOONa.

   pK_a: (CH_3COOH) = 4.76, (HF) = 3.17, (HCOOH) = 3.74
   K_a: (NH_4^+) = 5.75×10^{-10}, ($CH_3NH_3^+$) = 2.78×10^{-11}

Summary of relationships

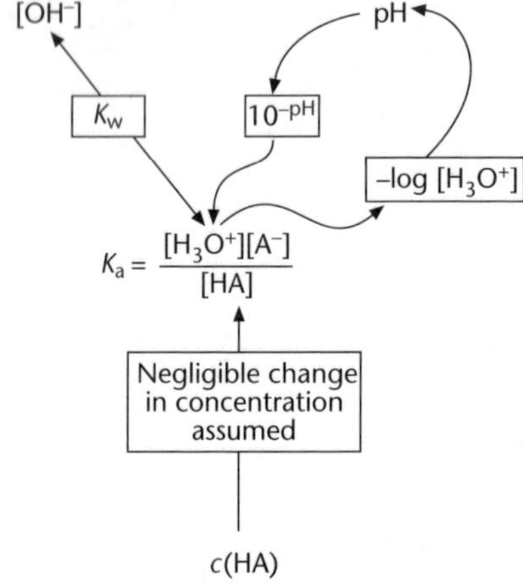

Unit 12.2 Acids, Bases and Salts

Topic 4: Reactions of acids and bases

Topic 4 covers:

- Reactions of acids and bases.
- Salts and water of crystallisation.

Introduction

When the hydrogen present in an acid is replaced by a metal, the product is called a **salt**.

Salts can be formed by reaction of the acids with:

- metals.
- metal oxides.
- metal carbonates and metal hydrogen carbonates.

Acid reactions with metals

Acids will react with an active metal (eg magnesium) to release hydrogen gas. The other product is a salt of the acid, eg:

$Mg(s) + 2HCl(aq) \rightarrow MgCl_2(aq) + H_2(g)$

$Mg(s) + H_2SO_4(aq) \rightarrow MgSO_4(aq) + H_2(g)$

The presence of hydrogen gas can be detected by collecting the gas in a test tube and showing that the gas will burn with a 'pop'.

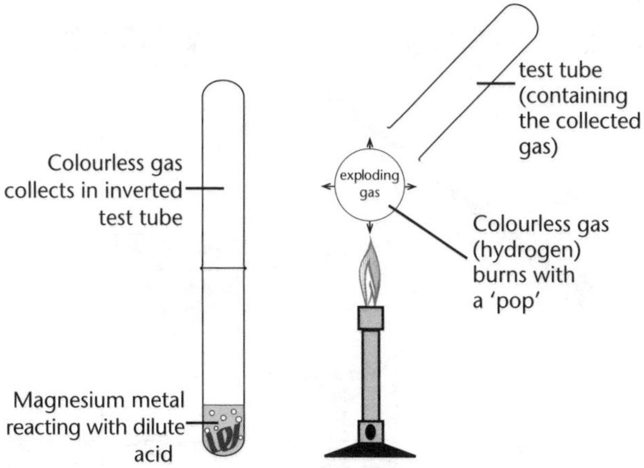

Acids react with metals, and hydrogen gas is formed, which can be tested for by the 'pop' test.

Acid reactions with metal oxides and metal hydroxides

Acids will react with the oxide or hydroxide of a metal to produce a salt and water, eg:

$CuO(s) + H_2SO_4(aq) \rightarrow CuSO_4(aq) + H_2O(\ell)$

$Pb(OH)_2(s) + 2HNO_3(aq) \rightarrow Pb(NO_3)_2(aq) + 2H_2O(\ell)$

In most cases, the solid will react to produce a colourless solution. If the oxide is coloured, the reaction may cause the solution to change colour.

Example A

Copper oxide reacting with dilute sulfuric acid

When black copper(II) oxide reacts with dilute sulfuric acid, a blue solution of copper sulfate will form.

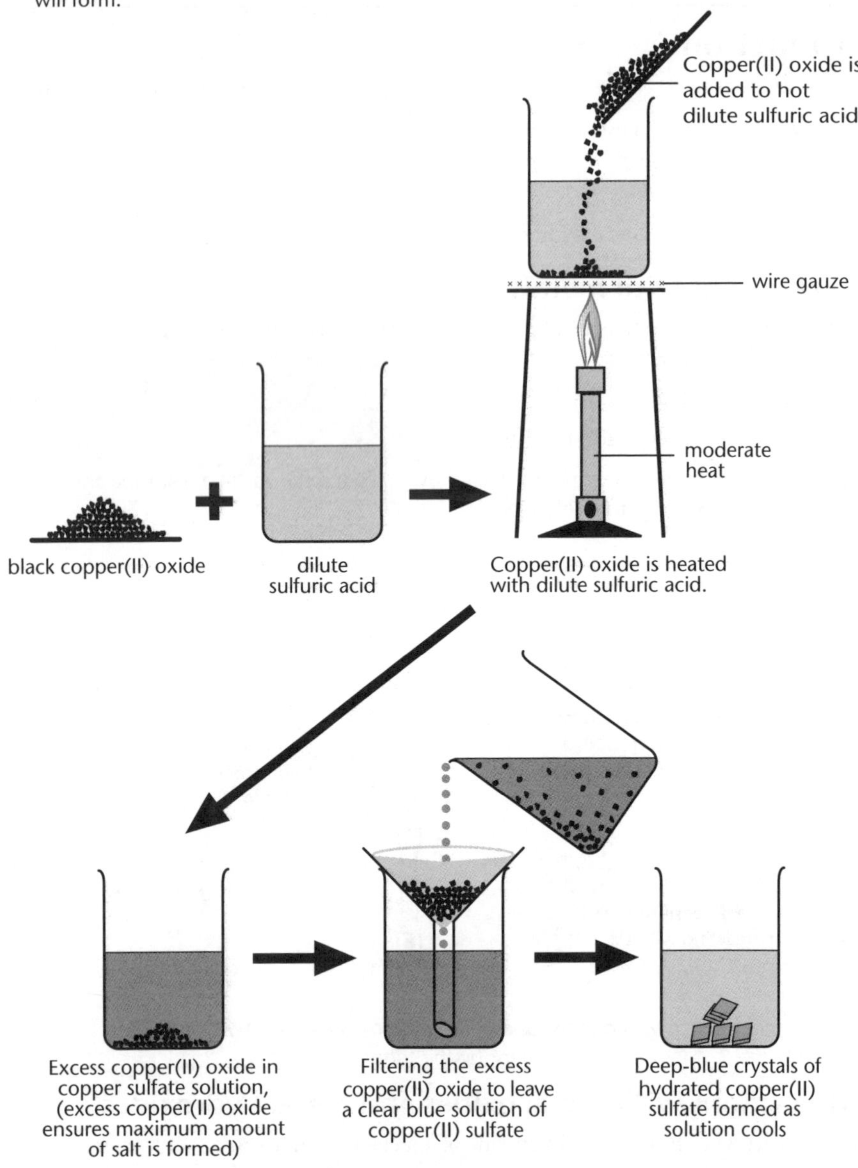

Acids react with a metal oxide to form a salt and water.

Acid reactions with metal carbonates and metal hydrogen carbonates

Acids react with metal carbonates or hydrogen carbonates to produce a salt, water and carbon dioxide, eg:

$CaCO_3(s) + 2HCl(aq) \rightarrow CaCl_2(aq) + H_2O(\ell) + CO_2(g)$

$ZnCO_3(s) + H_2SO_4(aq) \rightarrow ZnSO_4(aq) + H_2O(\ell) + CO_2(g)$

$NaHCO_3(s) + HNO_3(aq) \rightarrow NaNO_3(aq) + H_2O(\ell) + CO_2(g)$

If the gas produced is passed through **limewater** (a solution of calcium hydroxide), the solution will go milky, confirming the presence of carbon dioxide.

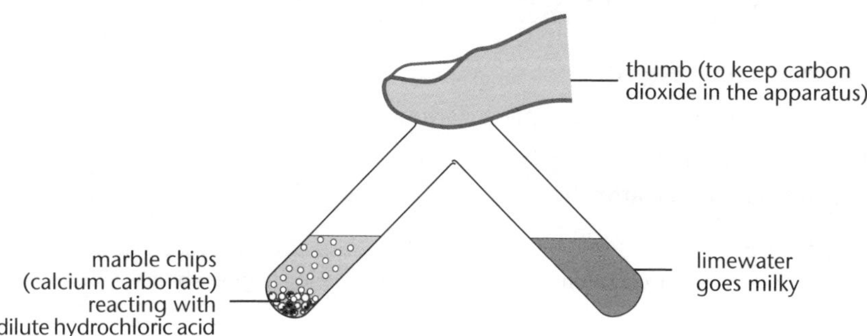

Carbonates and hydrogen carbonates release carbon dioxide when reacting with acids. Carbon dioxide turns limewater milky.

Unit 12.2 Activity 4A: Reaction of acids and bases

1. Describe the observations you would expect to make when:
   a. red litmus paper is added to an acid solution.
   b. blue litmus paper is added to an acid solution.

2. Classify the acidity (high, medium or low) of solutions of the following colours after universal indicator has been added:
   a. Orange.
   b. Red.
   c. Yellow.

3. From the following word list, identify which chemicals and equipment would be needed to produce hydrogen gas and show that it burnt in air with a 'pop'.
   Word list: test tubes; alkali; magnesium ribbon; funnel; glowing splint; burning splint; limewater; dilute acid.

4. State which of the following reactions would fit into the category of:
   base + acid = salt + water.
   a. Copper oxide + sulfuric acid → copper sulfate + water
   b. Magnesium + sulfuric acid → magnesium sulfate + hydrogen
   c. Zinc hydroxide + hydrochloric acid → zinc chloride + water

d. Calcium oxide + water → calcium hydroxide

e. Sodium hydroxide + nitric acid → sodium nitrate + water

5. Name a base and an acid you would require to produce the following salts:

a. Copper nitrate.

b. Zinc sulfate.

c. Magnesium chloride.

6. From the word list below, state which chemicals and equipment you would need to produce carbon dioxide gas and then show that it had been produced.
Word list: test tubes; sodium hydroxide; magnesium ribbon; sodium bicarbonate; glowing splint; burning splint; limewater; dilute acid; beaker; litmus solution.

Reactions of alkalis

Alkali solutions can be recognised in the following ways.

- They will turn red litmus paper blue.
- With **universal indicator**, alkali solutions will indicate a pH greater than 7.

Example B

Universal indicator in alkali solutions produces a colour range of green-blue to purple.

- Alkali solutions will react with an ammonium compound to release ammonia gas. The gas can be identified by its strong smell and by the fact that it will turn red litmus paper blue.

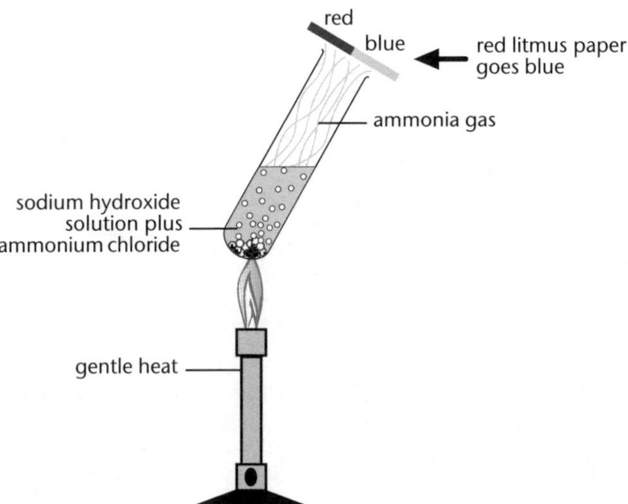

Alkalis release ammonia from ammonium compounds.

- We saw in Topic 2 that alkali solutions can be carefully added to a known volume of acid in a conical flask, which contains a coloured indicator, until the indicator just changes colour – this process is a titration. If the titration were carried out, using the same volumes of solutions, but without the indicator, the solution left in the conical flask is a salt. Crystals of the salt can be obtained from this solution by careful evaporation.

Unit 12.2 Activity 4B: Reactions of alkalis

1. Describe the observations you would expect to make when an alkali is added to:
   a. red litmus solution.
   b. blue litmus solution.

2. What is the pH of a solution whose colour, when added to universal indicator, is:
   a. blue?
   b. green-blue?

3. From the following word list, identify which chemicals and equipment you would need to produce ammonia gas:
   Word list: ammonium chloride; funnel; copper oxide; sodium hydroxide solution; sodium chloride; test tube; Bunsen burner; litmus paper.

4. Explain how you would produce some pure sodium chloride from solutions of sodium hydroxide and hydrochloric acid. Write an equation, in molecular form, for the reaction.

Formulae and equations for reactions of acids and alkalis

Formulae of acids and salts

The formulae for some acids and the salts derived from those acids are given in the table.

Acid	Formula	Name of salt produced	Formula of acid ion in the salt
Hydrochloric	HCl	Chloride	Cl^-
Nitric	HNO_3	Nitrate	SO_4^{2-}
Sulfuric	H_2SO_4	Sulfate	NO_3^-
Ethanoic	CH_3COOH	Ethanoate	CH_3COO^-

Equations for reactions of acids and alkalis

The reactions of acids and bases/alkalis can be written using:

- *General* equations – summarise the reactions that produce a salt from an acid–base reaction, or an acid–metal reaction.
- **Word equations** – state the reagents and products in words. There is no attempt to balance the equation.
- **Formula equations** – state the reagents and products using the correct formulae of the reagents and products. The equations must balance.

The following shows some reactions of acids and bases/alkalis and how they are written as general, word and formula equations.

General equation:	Acid + metal → salt + hydrogen
Word equation:	Hydrochloric acid + magnesium → magnesium chloride + hydrogen
Formula equation:	$2HCl(aq) + Mg(s) \rightarrow MgCl_2(aq) + H_2(g)$
General equation:	Acid + metal oxide → salt + water
Word equation:	Sulfuric acid + copper oxide → copper sulfate + water
Formula equation:	$H_2SO_4(aq) + CuO(s) \rightarrow CuSO_4(aq) + H_2O(\ell)$
General equation:	Acid + metal hydroxide → salt + water
Word equation:	Sulfuric acid + iron(II) hydroxide → iron(II) sulfate + water
Formula equation:	$H_2SO_4(aq) + Fe(OH)_2(s) \rightarrow FeSO_4(aq) + 2H_2O(\ell)$
General equation:	Acid + metal carbonate → salt + water + carbon dioxide
Word equation:	Nitric acid + calcium carbonate → calcium nitrate + water + carbon dioxide
Formula equation:	$2HNO_3(aq) + CaCO_3(s) \rightarrow CaCl_2(aq) + H_2O(\ell) + CO_2(g)$
General equation:	Acid + metal hydrogen carbonate → salt + water + carbon dioxide
Word equation:	Ethanoic acid + sodium hydrogen carbonate → sodium ethanoate +water + carbon dioxide
Formula equation:	$CH_3COOH(aq) + NaHCO_3(s) \rightarrow CH_3COONa(aq) + H_2O(\ell) + CO_2(g)$
General equation:	Acid + alkali → salt + water
Word equation:	Hydrochloric acid + sodium hydroxide → sodium chloride + water
Formula equation:	$HCl(aq) + NaOH\ (aq) \rightarrow NaCl(aq) + H_2O(\ell) + CO_2(g)$

Equations for reactions of acids and bases/alkalis

Formulae of metal oxides, hydroxides, carbonates and hydrogen carbonates

The formulae for some common metal oxides, metal hydroxides, metal carbonates and metal hydrogen carbonates are given in the following table.

Metal oxide and formula	Metal hydroxide and formula	Metal carbonate and formula	Metal hydrogen carbonate and formula
*Sodium oxide Na_2O	*Sodium hydroxide* *$NaOH$*	*Sodium carbonate* *Na_2CO_3*	Sodium hydrogen carbonate (sodium bicarbonate) $NaHCO_3$
Lithium oxide Li_2O	*Lithium hydroxide* *$LiOH$*	*Lithium carbonate* *Li_2CO_3*	# Lithium hydrogen carbonate (lithium bicarbonate) $LiHCO_3$
*Calcium oxide CaO	*Calcium hydroxide* *$Ca(OH)_2$*	Calcium carbonate $CaCO_3$	# Calcium hydrogen carbonate (calcium bicarbonate) $Ca(HCO_3)_2$
Magnesium oxide MgO	Magnesium hydroxide $Mg(OH)_2$	Magnesium carbonate $MgCO_3$	# Magnesium hydrogen carbonate (magnesium bicarbonate) $Mg(HCO_3)_2$
Aluminium oxide Al_2O_3	Aluminium hydroxide $Al(OH)_3$	Does not exist	Does not exist
Zinc oxide ZnO	Zinc hydroxide $Zn(OH)_2$	Zinc carbonate $ZnCO_3$	Does not exist
Iron(II) oxide FeO	Iron(II) hydroxide $Fe(OH)_2$	Iron(II) carbonate $FeCO_3$	Does not exist
Iron(III) oxide Fe_2O_3	Iron(III) hydroxide $Fe(OH)_3$	Does not exist	Does not exist
Lead oxide PbO	Lead hydroxide $Pb(OH)_2$	Lead carbonate $PbCO_3$	Does not exist
Copper(II) oxide CuO	Copper hydroxide $Cu(OH)_2$	Copper(II) carbonate $CuCO_3$	Does not exist
Silver oxide Ag_2O	Does not exist	Silver carbonate Ag_2CO_3	Does not exist

The names and formulae of alkalis are in *italics*. No gold compounds exist.

* Sodium and lithium oxides are extremely reactive and are converted rapidly to their hydroxides and carbonates in contact with the air. Calcium oxide is very reactive and is converted to calcium hydroxide and calcium carbonate in contact with the air.

Lithium, calcium and magnesium hydrogen carbonates exist only in solution.

Unit 12.4 Activity 4C: Word and formula equations

1. Complete the following word equations:

a. Magnesium + hydrochloric acid → magnesium chloride + _____

b. Iron(III) oxide + sulfuric acid → _____ _____ + water

c. Calcium hydroxide + _____ → calcium nitrate + water

d. Zinc carbonate + nitric acid → _____ _____ + water + carbon dioxide

e. Calcium bicarbonate + hydrochloric acid → _____ _____ + water + _____ _____

f. Sodium hydroxide + ethanoic acid → _____ _____ + water

2. a. Write word equations for the reaction between:

i. magnesium and sulfuric acid.

ii. copper hydroxide and hydrochloric acid.

iii. magnesium hydrogen carbonate and sulfuric acid.

b. Write word equations for the formation of:

i. zinc nitrate and water.

ii. sodium ethanoate, water and carbon dioxide.

iii. calcium nitrate, water and carbon dioxide.

3. Complete the following formula equations:

a. $Mg(s) + H_2SO_4(aq) \rightarrow$ _____________ $+ H_2(g)$.

b. $Cu(OH)_2(s) + 2HCl(aq) \rightarrow CuCl_2(aq) +$ _______.

c. $Mg(HCO_3)_2(aq) + H_2SO_4(aq) \rightarrow$ _________ $+ 2H_2O(\ell) + 2CO_2(g)$.

d. _________ + _________ $\rightarrow Zn(NO_3)_2(aq) + 2H_2O(\ell)$.

e. _________ + _________ $\rightarrow CH_3COONa(aq) + H_2O(\ell) + CO_2(g)$.

f. _________ + _________ $\rightarrow Ca(NO_3)_2(aq) + 2H_2O(\ell) + 2CO_2(g)$.

4. Write formula equations for the following reactions:

a. Magnesium added to nitric acid.

b. Solid sodium carbonate added to sulfuric acid.

c. Iron(III) hydroxide added to sulfuric acid.

d. Zinc oxide added to hydrochloric acid.

e. Sodium hydroxide solution added to ethanoic acid.

f. Calcium hydrogen carbonate solution added to hydrochloric acid.

Preparation of salts

There are two practical techniques used to obtain a crystalline sample of a salt.

The first (*see* Example A, page 110) applies to the reaction of an acid with a metal, metal oxide, metal hydroxide, metal carbonate or metal hydrogen carbonate:

- Add excess chemical to the acid to produce the maximum amount of salt.
- Filter excess reagent from the salt solution.
- Cool the salt solution to produce crystals – if the solution is evaporated to dryness, then some crystals will become **dehydrated** and lose their crystallinity.
- Separate crystals from the **mother liquor** (solution remaining after crystals have formed) by **decantating**, wash with distilled water, and then dry the crystals with absorbent (filter) paper.

The second practical technique – the titration – applies to the reaction of an acid with an alkali to produce a solution of the salt. The solution can be evaporated to a small volume, allowed to cool, and then the crystals filtered off, washed, and dried.

Water of crystallisation

The metal in a salt is always present as a positive ion. The positive ion can attract water molecules. As the salt crystallises, the water molecules attached to **positive ions** become part of the crystal. These salts are said to be hydrated and contain **water of crystallisation**.

If the hydrated salt loses its water of crystallisation, it becomes the anhydrous (ie without water) form of the salt.

Example C

Anhydrous copper sulfate can be used to detect water. The anhydrous form of the salt is white but the hydrated form is blue. This colour change is easily seen.

Hydrated salts

Name and formula of hydrated salt	Appearance	Common name of hydrated salt	Name and formula of anhydrous salt	Appearance
Hydrated copper(II) sulfate or copper(II) sulfate pentahydrate $CuSO_4.5H_2O$	Blue crystals	Blue stone or blue vitriol	Anhydrous copper(II) sulfate $CuSO_4$	White powder
Hydrated calcuim sulfate or calcium sulfate dihydrate $CaSO_4.2H_2O$	White crystals	Gypsum	Anhydrous calcium sulfate $CaSO_4$	White powder
Hydrated sodium carbonate or sodium carbonate decahydrate $Na_2CO_3.10H_2O$	White crystals	Washing soda	Anhydrous sodium carbonate Na_2CO_3	White powder

Hydrated salts that lose their water of crystallisation on exposure to air (to become the anhydrous form) are said to **effloresce**.

Example D

Sodium carbonate decahydrate is an efflorescent substance.

Some anhydrous salts never obtain water of crystallisation, eg common salt, NaCl. Other anhydrous salts will absorb so much water that they become a solution. These salts are said to be **deliquescent** – eg anhydrous calcium chloride.

All hydrated salts lose their water of crystallisation on gentle heating.

Unit 12.2 Activity 4D: Preparation of salts, water of crystallisation

1. Describe how you would prepare a sample of zinc sulfate heptahydrate (*hepta* means 'seven') crystals, $ZnSO_4.7H_2O$, starting with solid zinc oxide and dilute sulfuric acid. Assume the usual chemical apparatus is available.
2. Choose a word from the list (alphabetically arranged) below to complete each of the following statements.
   a. Salts are produced from acids by _ _ _ _ _ _ _ _ _ _ _ _ _ _.
   b. 'Mother liquor' is the liquid that remains when _ _ _ _ _ _ _ _ have been formed after a hot salt solution has cooled.
   c. Crystals cannot be dried by _ _ _ _ _ _ _ due to the possibility of water of crystallisation being lost.

   Word list: anhydrous, crystals, heating, hydrated, neutralisation

   d. Some crystals do not contain water of crystallisation. They are said to be _ _ _ _ _ _ _ _ _.
   e. Copper sulfate crystals are described as _ _ _ _ _ _ _ _ because they contain water of crystallisation.
3. Complete the following statement using one or more word(s) from the list below. A particular word may be used more than once.

   Word list: anhydrous, crystal, hydrate, water of crystallisation

   If a __________ contains ______________, it is described as a _________. _________ means that the _________ does not contain _________.
4. From the following list of four formulae, NaCl, $CuSO_4.5H_2O$, $CaSO_4.2H_2O$, $CuSO_4$, identify a salt that:
   a. contains water of crystallisation.
   b. does not contain water of crystallisation.
   c. is anhydrous.
   d. is the anhydrous form of a hydrate.

Unit 12.3 Electrochemistry

Topic 1: Oxidation–reduction reactions

Topic 1 Covers oxidation–reduction reactions:

- Identifying common oxidants and reductants, and the products of their reaction.
- Writing oxidation–reduction equations.
- Determining oxidation numbers.
- Recognising the ability of halogens to act as oxidants in reactions with elements, water or halide ions.

Introduction

Oxidation–reduction reactions involve the transfer of electrons from one substance to another. These reactions are commonly known as **redox** reactions.

Initially, redox reactions were named after the gain of oxygen or loss of hydrogen (oxidation), and the loss of oxygen or gain of hydrogen (reduction). Later it was found that many redox reactions do not involve oxygen or hydrogen; all redox reactions, however, involve the *transfer of electrons*.

- **Oxidation** is the loss of electrons.
- **Reduction** is the gain of electrons.

Oxidation can be:	Reduction can be:
Gain of oxygen	Loss of oxygen
Loss of hydrogen	Gain of hydrogen
Loss of electrons	Gain of electrons

Transfer of oxygen and hydrogen

Redox reactions can sometimes be identified by transfer of oxygen, hydrogen or electrons.

Transfer of oxygen

If a chemical species gains oxygen, it has been oxidised; if oxygen is lost, the chemical species has been reduced.

Example A

Transfer of oxygen

In the reaction $2Fe_2O_3(s) + 3C(s) \rightarrow 4Fe(s) + 3CO_2(g)$, oxygen is transferred from Fe_2O_3 to carbon, C.

$$2Fe_2O_3(s) + 3C(s) \rightarrow 4Fe(s) + 3CO_2(g)$$

oxygen transfer

- Fe_2O_3 is reduced, since it loses oxygen to form Fe.
- C is oxidised, since it gains oxygen to form CO_2.

Transfer of hydrogen

If a chemical species loses hydrogen, then it has been oxidised; if hydrogen is gained, then the chemical species has been reduced.

Example B

Transfer of hydrogen

In the reaction $2H_2S(g) + O_2(g) \rightarrow 2S(s) + 2H_2O(\ell)$, hydrogen is transferred from H_2S to oxygen:

$$2H_2S(g) + O_2(g) \rightarrow 2S(s) + 2H_2O(\ell)$$

hydrogen transfer

- H_2S is oxidised, because hydrogen is lost as H_2S forms S.
- O_2 is reduced, because hydrogen is gained as O_2 forms H_2O.

Oxidation number

In a reaction, changes in **oxidation numbers (ON)** can be used to identify an element that is reduced, an element that is oxidised, and an element that is unchanged.

The oxidation number (ON), sometimes referred to as oxidation state, describes the 'degree' to which an element has been oxidised or reduced. To avoid confusion with the charge on an ion, oxidation numbers are usually written with the sign before the whole number (eg +3 or –1).

Example C

Oxidation numbers of the elements in MnO_4^-

For the permanganate ion, MnO_4^-, the oxidation number of manganese, Mn, is +7 and the oxidation number of oxygen, O, is –2.

ON (Mn) = +7 and ON (O) = –2

Oxidation number and the charge on a polyatomic ion are *not* the same thing and should not be confused.

Oxidation numbers are assigned by a set of rules.

Rule	Examples
1. When atoms exist as elements, they have an oxidation number of zero.	Na, Cl_2, Ne, C and H_2 all have an oxidation number of zero.
2. The oxidation number of a monatomic (one atom) ion is the same as the charge on the ion.	Cu^{2+} has oxidation number +2*, Cl^- has an oxidation number of –1.
3. Hydrogen in compounds has an oxidation number of +1*, except in metal hydrides, where it is –1.	Hydrogen in H_2O, CH_4 and NH_3 has an oxidation number of +1. In NaH, a metal hydride, the oxidation number of hydrogen is –1.
4. Oxygen in compounds has an oxidation number of –2, except in hydrogen peroxide, H_2O_2, where its oxidation number is –1.	Oxygen in MgO, H_2SO_4, H_2O and $KMnO_4$ has an oxidation number of –2. In hydrogen peroxide, H_2O_2, the oxidation number of oxygen is –1.
5. For polyatomic ions (ions containing more than one atom), the sum of the oxidation numbers equals the charge of the ion.	For NH_4^+, the sum of the oxidation numbers is +1. For SO_4^{2-}, the sum of the oxidation numbers is –2.
6. The sum of the oxidation numbers of atoms in a molecule is zero.	The sum of the oxidation numbers for the atoms in each of H_2SO_4, C_4H_{10} and H_2O is zero.

****Note:*** It is important to use the '+' sign when writing positive ONs: use '+1' not '1', '+2' not '2', etc.

Example D

Assigning oxidation numbers

1. Calculate the oxidation number of sulfur in the molecule H_2SO_4.

- **i.** The sum of the oxidation numbers is zero, since H_2SO_4 is a neutral molecule (*Rule 6*), so $2 \times \text{ON (H)} + \text{ON (S)} + 4 \times \text{ON (O)} = 0$.
- **ii.** The oxidation number of hydrogen is +1 (*Rule 3*). Total contribution from hydrogens is +2, since there are two H atoms.
- **iii.** The oxidation number of oxygen is –2 (*Rule 4*). Total contribution from oxygens is $2 \times (-4) = -8$, since there are 4 O atoms.
- **iv.** The oxidation number of sulfur in H_2SO_4 is thus +6, since $(+2) + \text{ON (S)} + (-8) = 0$

Solution:

Atoms present	ON of atom	Total
H (two)	+1	+2
O (four)	–2	–8
S (one)	+6	+6
Sum of ONs = 0		

2. Find the oxidation number of carbon in the polyatomic ion, CO_3^{2-}.

Solution:

- **i.** The charge on the ion CO_3^{2-} is –2, so the sum of the oxidation numbers is –2 (*Rule 5*), so $\text{ON (C)} + 3 \times \text{ON (O)} = -2$.
- **ii.** The oxidation number for oxygen is –2 (*Rule 4*). Total contribution from oxygen is $3 \times (-2) = -6$, since there are three oxygens.
- **iii.** The oxidation number of carbon is +4, since $\text{ON (C)} + (-6) = -2$.

The properties of some ions vary greatly, depending on their oxidation number. The term **oxidation state** is sometimes used to indicate the degree of oxidation of an element.

Example E

Oxidation states of iron ions

The properties of the two oxidation states of iron ions are quite diferent:

Ion	Oxidation state/number	Colour	Reactivity
Fe^{2+}	+2	Pale blue/green	Unstable (easily oxidised to Fe^{3+})
Fe^{3+}	+3	Orange-red/brown	Stable

IUPAC nomenclature

The different oxidation states of elements in ions are identified by using Roman numerals.

Example F

IUPAC nomenclature for compounds of iron

In $FeCl_2$, iron has the +2 oxidation state, and is named iron(II) chloride.

In $FeCl_3$, iron has the +3 oxidation state, and is named iron(III) chloride.

Recognising oxidation and reduction using oxidation numbers

In a redox reaction, the oxidation numbers of some of the elements involved will change. Changes in oxidation number can be used to determine which element is oxidised and which is reduced.

If the oxidation number of an element in a reactant *increases* (ie gets more positive or less negative) during a reaction, the reactant has been *oxidised*.

Example G

Oxidising magnesium

When magnesium metal forms magnesium ions:

Mg	→	Mg^{2+}	+	$2e^-$
ON is 0		ON is +2		
(*Rule 1*)		(*Rule 2*)		

The oxidation number of magnesium increases from 0 to +2, so magnesium metal atoms are said to be oxidised to magnesium ions.

If the oxidation number of a reactant or an atom in that reactant *decreases* (ie gets more negative or less positive) during a reaction, the reactant has been *reduced*.

Example H

Reducing oxygen

When oxygen gas forms oxide ions:

O_2	+	$4e^-$	→	$2O^{2-}$
ON is 0				ON is –2
(Rule 1)				*(Rule 2)*

The oxidation number of oxygen has decreased from 0 to –2, so the atoms in molecules of oxygen gas are said to be reduced to oxide ions.

Oxidation and reduction reactions always occur together, so if one reactant is oxidised, then another reactant must be reduced.

Example I

Magnesium and oxygen

When magnesium burns in oxygen, the magnesium metal atoms are oxidised to magnesium ions and the atoms in molecules of oxygen gas are reduced to oxide ions (Examples G and H):

0		0		+2 –2
2Mg(s)	+	O_2(g)	→	MgO(s)

If any element in a reaction has a change in oxidation number, then the reaction is a redox reaction.

Unit 12.3 Activity 1A: Oxidation numbers

1. What is the oxidation number of:

a. manganese in manganese dioxide, MnO_2. **b.** zinc in zinc metal.
c. zinc in zinc ions, Zn^{2+}. **d.** carbon in carbonate ions, CO_3^{2-}.
e. sulfur in sulfur dioxide, SO_2.

2. What is the oxidation state of chlorine in each of the following?

a. Cl^- **b.** Cl_2 **c.** HOCl **d.** $HClO_4$

3. Give the oxidation number of nitrogen in each of the following.

a. NH_3 **b.** NH_4^+ **c.** N_2 **d.** N_2O **e.** NO
f. HNO_2 **g.** NO_2 **h.** NO_3^-

4. Copy each symbol/formula below and give the oxidation number of each element present.

a. Mg **b.** Cl_2 **c.** H_2O **d.** CaH_2 **e.** Ag^+
f. Fe^{3+} **g.** SO_2 **h.** SO_3 **i.** CH_4 **j.** MgO
k. CO_2 **l.** S_8 **m.** CO **n.** MnO_4^- **o.** CO_3^{2-}
p. H_2O_2 **q.** $S_2O_3^{2-}$ **r.** $Cr_2O_7^{2-}$ **s.** $NaHCO_3$ **t.** $CuCO_3$
u. Na_3PO_4 **v.** H_2SiO_3 **w.** H_3PO_4 **x.** $CuSO_4$ **y.** H_2SO_4

5. For each of the following reactions, state which element is oxidised and which is reduced (if any), and give the oxidation states before and after the reaction.

a. $H_2 + Cl_2 \rightarrow 2HCl$

b. $Zn + Cu^{2+} \rightarrow Zn^{2+} + Cu$

c. $H_2O + CO \rightarrow H_2 + CO_2$

d. $Mg(OH)_2 + 2HCl \rightarrow MgCl_2 + 2H_2O$

e. $2CuS + 3O_2 \rightarrow 2CuO + 2SO_2$

f. $NaCl + AgNO_3 \rightarrow AgCl + NaNO_3$

g. $5Fe^{2+} + MnO_4^- + 8H^+ \rightarrow 5Fe^{3+} + Mn^{2+} + 4H_2$

h. $Fe + \frac{1}{2}O_2 + H_2O \rightarrow Fe^{2+} + 2OH^-$

6. Determine whether each **bolded** chemical species in the following reactions has been *oxidised* or *reduced*. Give reasons for your answer.

a. $\mathbf{Ag^+}(aq) + Cl^-(aq) \rightarrow AgCl(s)$

b. $\mathbf{2Na}(s) + Cl_2(g) \rightarrow 2NaCl(s)$

c. $\mathbf{Co^{2+}}(aq) + Cl_2(g) \rightarrow Co^{3+}(aq) + Cl^-(aq)$

d. $\frac{1}{2}\mathbf{O_2}(g) + H_2(g) \rightarrow H_2O(g)$

e. $\mathbf{Ba}Cl_2(s) \xrightarrow{H_2O} Ba^{2+}(aq) + 2Cl^-(aq)$

f. $Zn(s) + \mathbf{Pb^{2+}}(aq) \rightarrow Zn^{2+}(aq) + Pb(s)$

g. $\mathbf{Hg}(\ell) \rightarrow Hg(g)$

Oxidants and reductants

In redox reactions the reactant that is:

- *oxidised* is called the **reductant** or (**reducing agent**), since it *reduces* the other reactant.
- *reduced* is called the **oxidant** (or **oxidising agent**), since it *oxidises* the other reactant.

Example J

Oxidants and reductants

For $2Mg(s) + O_2(g) \rightarrow 2MgO(s)$

- Magnesium is the reductant because it has been oxidised.
- Oxygen is the oxidant because it has been reduced.

Transfer of electrons

Oxidation, as pointed out already, is loss of electrons, and reduction is gain of electrons.

This can be remembered by using the **mnemonic OIL RIG**:

- **O**xidation **I**s **L**oss of electrons.
- **R**eduction **I**s **G**ain of electrons.

or **LEO GER**:

- **L**oss of **E**lectrons **O**xidation.
- **G**ain of **E**lectrons **R**eduction.

Example K

Transfer of electrons

Zinc metal in copper sulfate solution:

$$Zn(s) + Cu^{2+}(aq) \rightarrow Zn^{2+}(aq) + Cu(s)$$

Zinc metal loses electrons and is oxidised to form zinc ions, $Zn^{2+}(aq)$:

$$Zn \rightarrow Zn^{2+} + 2e^-$$

Copper ions, $Cu^{2+}(aq)$, gain electrons and are reduced to form copper metal, Cu.

$$Cu^{2+} + 2e^- \rightarrow Cu$$

Overall, the electrons lost by the zinc are transferred to the copper ions:

$$Zn(s) + Cu^{2+}(aq) \rightarrow Zn^{2+}(aq) + Cu(s)$$

$2e^-$

electron transfer

Zinc metal is the *reductant* and copper ions are the *oxidant*.

Note: The sulfate ions, SO_4^{2-}, present in the copper sulfate solution, are *not* shown in the equation because they are *not* involved in the reaction. Such 'uninvolved' ions are called **spectator ions**.

Unit 12.3 Activity 1B: Oxidants and reductants

1. For the reaction $2PbO + C \rightarrow 2Pb + CO_2$

a. Identify the oxidant, explaining your reasoning.

b. Identify the reductant, explaining your reasoning.

2. For the reaction $Fe + Cu^{2+} \rightarrow Fe^{2+} + Cu$:

a. Identify the oxidant, explaining your reasoning.

b. Identify the reductant, explaining your reasoning.

3. The reaction of zinc with dilute hydrochloric acid is:

$Zn(s) + 2HCl(aq) \rightarrow ZnCl_2(aq) + H_2(g)$

a. Write an ionic equation for the reaction.

b. Which reactant is oxidised?

c. Which reactant is the oxidant?

d. Which reactant is reduced? To what?

e. Which reactant is the reductant?

f. Describe the direction of the transfer of electrons.

g. Name any spectator ions.

4. Consider the reaction $Mg(s) + Cl_2(g) \rightarrow MgCl_2(s)$.

a. What is oxidised?

b. What is reduced?

c. Which reactant is the oxidant? Why?

d. Describe the direction of the transfer of electrons.

5. Identify the oxidant and the reductant in the following reactions. Give a reason for each of your answers.
   a. $Cu + 2NO_3^- + 4H^+ \rightarrow Cu^{2+} + 2NO_2 + 2H_2O$
   b. $2FeCl_3 + SO_2 + 2H_2O \rightarrow 2FeCl_2 + H_2SO_4 + 2HCl$
   c. $2I^- + H_2O_2 + 2H^+ \rightarrow 2H_2O + I_2$
   d. $Cr_2O_7^{2-} + 3SO_3^{2-} + 8H^+ \rightarrow 2Cr^{3+} + 3SO_4^{2-} + 4H_2O$
   e. $Mg + H_2SO_4 \rightarrow MgSO_4 + H_2$
   f. $C + O_2 \rightarrow CO_2$
   g. $2CuS + 3O_2 \rightarrow 2CuO + 2SO_2$
   h. $CuO + C \rightarrow Cu + CO$

Writing redox equations

To write a balanced equation for a redox reaction, the reaction is broken into two **half-equations** (or **ion–electron equations**).

- The *reduction half-equation* shows how the oxidant is *reduced* to a product.
- The *oxidation half-equation* shows how the reductant is *oxidised* to a product.

Example L

Half-equations

For the reaction of zinc with $CuSO_4$

- The reduction half-equation is:

  $Cu^{2+} + 2e^- \rightarrow Cu$

  The oxidant is Cu^{2+}, and it is reduced since it gains electrons.
- The oxidation half-equation is:

  $Zn \rightarrow Zn^{2+} + 2e^-$

  The reductant is Zn, and it is oxidised since it loses electrons.

Balancing some redox reactions is a relatively simple process.

Example M

Oxidising zinc with hydrogen ions

When zinc metal is placed in hydrochloric acid, the zinc metal dissolves (to form Zn^{2+} ions) and hydrogen gas forms.

The H^+ ions (from HCl) are reduced:

$2H^+ + 2e^- \rightarrow H_2$ [Reduction half-equation]

The Zn metal is oxidised:

$Zn \rightarrow Zn^{2+} + 2e^-$ [Oxidation half-equation]

The overall balanced equation is:

$Zn(s) + 2H^+(aq) \rightarrow Zn^{2+}(aq) + H_2(g)$

$$\left[\begin{array}{l} 2H^+ + \cancel{2e^-} \rightarrow H_2 \\ \quad Zn \rightarrow Zn^{2+} + \cancel{2e^-} \end{array}\right]$$

Chloride ions, Cl^-, from the hydrochloric acid, are spectator ions and are not shown in the reaction.

Balancing half-equations

There are a number of steps involved in balancing half-equations.

Step 1
Identify the species that is oxidised or reduced and the product that is formed.

Step 2
Balance the *atoms* of the element undergoing a change in oxidation number.

Step 3
Balance *oxygen* atoms by adding water to the appropriate side.

Step 4
Balance hydrogen atoms by adding hydrogen ions to the appropriate side.

Step 5
Balance the *charge* by adding electrons to the most positive side.

Example N

Reduction of $Cr_2O_7^{2-}$ to Cr^{3+}
Balance the reduction half-equation
$Cr_2O_7^{2-} \rightarrow Cr^{3+}$

Dichromate ion, $Cr_2O_7^{2-}$ (orange), changes to chromium(III) ion, Cr^{3+} (green):

$$Cr_2O_7^{2-} \rightarrow Cr^{3+}$$

Cr changed from +6 to +3 (reduction):

$$\overset{+6}{Cr_2}O_7^{2-} \rightarrow 2\overset{+3}{Cr^{3+}}$$

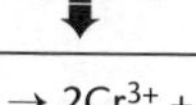

2 Cr atoms

$$Cr_2O_7^{2-} \rightarrow 2Cr^{3+} + 7H_2O$$

7 water molecules needed on RHS, since there are 7O atoms in $Cr_2O_7^{2-}$

$$Cr_2O_7^{2-} + 14H^+ \rightarrow 2Cr^{3+} + 7H_2O$$

$14H^+$ ions needed on LHS, since there are 7 water molecules, ie 14H atoms

$$\underbrace{Cr_2O_7^{2-} + 14H^+}_{} \rightarrow \underbrace{2Cr^{3+} + 7H_2O}_{}$$

charge $-2 + 14 = +12$ $\quad$ $+6 + 0 = +6$
LHS needs to be reduced from +12 to +6; add 6 electrons to LHS:

$$Cr_2O_7^{2-} + 14H^+ + 6e^- \rightarrow 2Cr^{3+} + 7H_2O$$

The last equation is the balanced reduction half-equation for $Cr_2O_7^{2-}$ ions being reduced to Cr^{3+} ions.

Note: Since the dichromate ions react with hydrogen ions, the reduction is occurring in acidic solution and the dichromate solution is said to be **acidified**.

An oxidation half-equation is balanced using the same rules.

Example O

Oxidation of SO_2 to SO_4^{2-}

The oxidation half-equation involves sulfur dioxide forming sulfate ions.

$SO_2 \rightarrow SO_4^{2-}$	[Equation is balanced for sulfur atoms – one atom on each side]
$SO_2 + 2H_2O \rightarrow SO_4^{2-}$	[Balanced for oxygen by adding $2H_2O$ to the LHS]
$SO_2 + 2H_2O \rightarrow SO_4^{2-} + 4H^+$	[$4H^+$ ions added to the RHS – equation is now balanced for hydrogen]
$SO_2 + 2H_2O \rightarrow SO_4^{2-} + 4H^+ + 2e^-$	[$2e^-$ added on the RHS makes both sides neutral – equation is now balanced for charge]

Water molecules and hydrogen ions can appear as reactants *and* products in redox reactions. Electrons appear on the RHS of half-equations for oxidation, the LHS for reduction.

Balancing the overall equation

The oxidation and reduction half-equations are added together to give the *balanced*, **overall equation** for a redox reaction.

Step 1

One or both half-equations may need to be multiplied by an appropriate number, so that the number of electrons lost in oxidation equals the number of electrons gained in reduction.

Then the two half-equations are added together, and the electrons cancelled out.

Step 2

Ions and molecules common to both sides of the equation are cancelled out. (If numbers of ions/molecules are different on both sides of the equation, a reduced number of ions/ molecules will appear on one side of the equation after cancelling.)

Step 3

A final step after writing the overall equation is to double-check that all charges and atoms are balanced.

Example P

Write the balanced equation for sulfur dioxide reacting with acidified dichromate ions.

In this reaction, the sulfur dioxide is oxidised, and this is shown in the oxidation half-equation (Example O); the dichromate ion is reduced, as shown in the reduction half-equation (Example N):

$$SO_2 + 2H_2O \rightarrow SO_4^{2-} + 4H^+ + 2e^-$$

$$Cr_2O_7^{2-} + 14H^+ + 6e^- \rightarrow 2Cr^{3+} + 7H_2O$$

To match the 6 electrons required by the dichromate ion, the oxidation half-equation is multiplied by 3:

$$3SO_2 + 6H_2O \rightarrow 3SO_4^{2-} + 12H^+ + 6e^-$$

The two half-equations are added together:

$$3SO_2 + 6H_2O \rightarrow 3SO_4^{2-} + 12H^+ + 6e^-$$

$$+ \; Cr_2O_7^{2-} + 14H^+ + 6e^- \rightarrow 2Cr^{3+} + 7H_2O$$

$$Cr_2O_7^{2-} + 3SO_2 + 6H_2O + 14H^+ + 6e^- \rightarrow 3SO_4^{2-} + 2Cr^{3+} + 7H_2O + 12H^+ + 6e^-$$

Since the number of electrons is the same on both sides, electrons are cancelled out:

$$Cr_2O_7^{2-} + 3SO_2 + 6H_2O + 14H^+ \rightarrow 3SO_4^{2-} + 2Cr^{3+} + 7H_2O + 12H^+$$

The water and hydrogen ions are cancelled out (ie *Step 2*).

- $6H_2O$ on the LHS and $7H_2O$ on the RHS simplify to one H_2O on the RHS (ie cancel $6H_2O$ on each side).
- $14H^+$ on the LHS and $12H^+$ on the RHS simplify to $2H^+$ on the LHS (ie cancel $12H^+$ on each side).

The overall balanced equation now becomes:

$$Cr_2O_7^{2-}(aq) + 3SO_2(g) + 2H^+(aq) \rightarrow 3SO_4^{2-}(aq) + 2Cr^{3+}(aq) + H_2O(\ell)$$

Checking charges and atoms are balanced (ie *Step 3*).

Charges balanced:

LHS: $Cr_2O_7^{2-} = -2$; $2H^+ = +2$; LHS = $(-2) + (+2) = 0$

RHS: $3SO_4^{2-} = 3 \times -2 = -6$; $2Cr^{3+} = 2 \times +3 = +6$; RHS = $(-6) + (+6) = 0$

Atoms balanced:

LHS: 2Cr from ($Cr_2O_7^{2-}$); 3S (from $3SO_2$); 13O (7O from $Cr_2O_7^{2-}$, 6O from $3SO_2$); 2H (from $2H^+$)

RHS: 3S (from $3SO_4^{2-}$); 13O (12O from $3SO_4^{2-}$, one O from H_2O); 2Cr (from $2Cr^{3+}$); 2H (from H_2O)

LHS [2Cr + 3S + 13O + 2H] = RHS [3S + 13O + 2Cr + 2H]

The balanced, overall equation shows that two hydrogen ions from the acid are a reactant along with the reductant, SO_2, and the oxidant, $Cr_2O_7^{2-}$.

Potassium ions, K^+, from $K_2Cr_2O_7$ are spectator ions, so are not shown in the reaction.

Unit 12.3 Activity 1C: Redox equations

1. Balance the following half-equations and indicate whether they involve oxidation or reduction:
   - **a.** $Na^+ \rightarrow Na$
   - **b.** $H^+ \rightarrow H_2$
   - **c.** $Fe^{2+} \rightarrow Fe^{3+}$
   - **d.** $C \rightarrow Cu^{2+}$
   - **e.** $SO_2 \rightarrow SO_4^{2-}$
   - **f.** $Cr_2O_7^{2-} \rightarrow Cr^{3+}$
   - **g.** $MnO_4^- \rightarrow Mn^{2+}$
   - **h.** $H_2O_2 \rightarrow H_2O$
   - **i.** $Br^- \rightarrow Br_2$
   - **j.** $I_2 \rightarrow I^-$
2. Balance the following equations. In each case, write balanced half-equations and combine the balanced half-equations to give the overall equation.
   - **a.** $Fe^{2+} + MnO_4^- \rightarrow Fe^{3+} + Mn^{2+}$
   - **b.** $Mg + Ag^+ \rightarrow Mg^{2+} + Ag$
   - **c.** $Fe^{2+} + Cr_2O_7^{2-} \rightarrow Fe^{3+} + Cr^{3+}$
   - **d.** $SO_2 + MnO_4^- \rightarrow SO_4^{2-} + Mn^{2+}$
   - **e.** $Cl^- + H_2O_2 \rightarrow Cl_2 + H_2O$
   - **f.** $MnO_4^- + I^- \rightarrow Mn^{2+} + I_2$
   - **g.** $Cr_2O_7^{2-} + H_2S \rightarrow S + Cr^{3+}$
   - **h.** $Zn + VO_2^+ \rightarrow VO^{2+} + Zn^{2+}$
   - **i.** $I^- + IO_3^- \rightarrow I_2$
   - **j.** $Cu + NO_3^- \rightarrow NO_2 + Cu^{2+}$
   - **k.** $H_2S + SO_2 \rightarrow S + H_2O$

Common oxidants and reductants

Oxidants

The following table shows some common oxidants, their properties and their half-equations.

Oxidant	Colour	State	Reduction half-equation
Oxygen, O_2	Colourless	Gas	$O_2(g) + 4e^- \rightarrow 2O^{2-}$
Hydrogen ions, H^+ (ie an acidic solution)	Colourless	Aqueous	$2H^+(aq) + 2e^- \rightarrow H_2(g)$
Chlorine, Cl_2	Pale green	Gas	$Cl_2(g) + 2e^- \rightarrow 2Cl^-$
Potassium permanganate (in an acidic solution, ie H^+/MnO_4^-)	Purple	Aqueous	$MnO_4^-(aq) + 8H^+ + 5e^- \rightarrow Mn^{2+}(aq) + 4H_2O$
Sodium dichromate (in an acidic solution, ie $H^+/Cr_2O_7^{2-}$)	Orange	Aqueous	$Cr_2O_7^{2-}(aq) + 14H^+ + 6e^- \rightarrow 2Cr^{3+}(aq) + 7H_2O$ orange green
Hydrogen peroxide, H_2O_2 (in an acidic solution, ie H^+/H_2O_2)	Colourless	Aqueous	$H_2O_2(aq) + 2H^+ + 2e^- \rightarrow 2H_2O$
Iodine, I_2	Brown	Aqueous	$I_2(aq) + 2e^- \rightarrow 2I^-(aq)$
Iron(III) ions, Fe^{3+}	Orange-red brown	Aqueous	$Fe^{3+}(aq) + e^- \rightarrow Fe^{2+}(aq)$ orange-red brown green

Note: The reduction half-equations (right-hand column) show that oxidants always *gain* electrons and are *reduced* in redox reacions.

Reductants

The following table shows some common reductants, their properties and their half-equations.

Reductant	Colour	State	Oxidation half-equation
Carbon, C	Black	Solid	$C(s) + 2H_2O \rightarrow CO_2(g) + 4H^+(aq) + 4e^-$
Carbon monoxide, CO	Colourless	Gas	$CO(g) + H_2O \rightarrow CO_2(g) + 2H^+(aq) + 2e^-$
Sulfur dioxide, SO_2	White fumes	Gas	$SO_2(g) + 2H_2O \rightarrow SO_4^{2-}(aq) + 4H^+(aq) + 2e^-$
Hydrogen, H_2	Colourless	Gas	$H_2(g) \rightarrow 2H^+(aq) + 2e^-$
Iodide ions, I^-	Colourless	Aqueous	$2I^-(aq) \rightarrow I_2 + 2e^-$ colourless → yellow-brown solution or black solid
Bromide ions, Br^-	Colourless	Aqueous	$2Br^-(aq) \rightarrow Br_2(\ell) + 2e^-$ colourless → orange
Iron(II) ions, Fe^{2+}	Green	Aqueous	$Fe^{2+}(aq) \rightarrow Fe^{3+}(aq) + e^-$ green → orange-red brown
Zinc, Zn	Grey	Solid	$Zn(s) \rightarrow Zn^{2+} + 2e^-$ colourless
Magnesium, Mg	Grey	Solid	$Mg(s) \rightarrow Mg^{2+} + 2e^-$ colourless
Copper, Cu	Pink	Solid	$Cu(s) \rightarrow Cu^{2+} + 2e^-$ usually blue
Iron, Fe	Grey	Solid	$Fe(s) \rightarrow Fe^{2+} + 2e^-$ pale green

Note: The oxidation half-equations (right-hand column) show that reductants always *lose* electrons and are *oxidised* in redox reactions.

Unit 12.3 Activity 1D: Common oxidants and reductants

1. A colourless solution of potassium iodide is added to a purple solution of potassium permanganate acidified with sulfuric acid. The purple colour disappears to leave a light amber solution of iodine, $I_2(aq)$.
   - **a.** Identify both the oxidant and the reductant.
   - **b.** Identify any components of the solutions which are not involved in the reaction.
   - **c.** Write balanced oxidation and reduction half-equations.
   - **d.** Write the overall equation.

2. Write balanced half-equations for:
   - **a.** sulfur dioxide gas reacting as a reductant.
   - **b.** oxygen being reduced to oxide ions.
   - **c.** iron(III) ions as an oxidant.

3. Write balanced half-equations for the following.
   a. Bromide ions acting as reductants.
   b. Hydrogen gas as a reductant.
   c. Iodine, I_2, being reduced.
   d. Hydrogen peroxide, H_2O_2, acting as an oxidant.
4. Write an overall equation showing:
   a. copper(II) ions reacting with magnesium.
   b. sodium being oxidised by chlorine, $Cl_2(g)$.
   c. iodide ions reducing iron(III) ions to iron(II) ions.
   d. sulfur dioxide gas reacting with hydrogen peroxide.
5. Write a balanced equation and describe what would be observed when the following reactions take place.
   a. Sulfur dioxide and bromine water.
   b. A freshly prepared iron(II) sulfate solution is added to an acidified potassium permanganate solution.
6. For each of the following reactions:
   i. write balanced equations.
   ii. describe the expected observations and link these to the species involved.
   a. Sulfur dioxide gas is bubbled into an acidic dichromate solution.
   b. Acidified permanganate solution is added to freshly prepared iron(II) sulfate. (The sulfate is a spectator ion.)
   c. Magnesium metal is added to copper nitrate solution. (The nitrate is a spectator ion.)
   d. Acidified hydrogen peroxide is reacted with bromide ions in solution.
7. Account for the following observations and write balanced equations to support your answers.
   a. A black solid and a colourless solution form when acidified permanganate solution is added to potassium iodide solution.
   b. When copper metal is added to a silver nitrate solution, silver needles form in the copper metal and the solution turns blue.
   c. When acidified hydrogen peroxide is added to aqueous iron(II) sulfate, the colourless solution turns red-brown.
   d. When sulfur dioxide gas, $SO_2(g)$, is bubbled into a solution containing iron(III) ions, $Fe^{3+}(aq)$, the yellow-brown solution loses its colour.

▶ More Online Check the Internet for more information on rusting – the oxidation of iron.

Halogens as oxidants

Halogens can accept electrons from other substances – ie they are all oxidants. The oxidising power decreases as you go down group 17:

$$F_2 > Cl_2 > Br_2 > I_2$$

Strongest oxidant — Weakest oxidant

Halogens reacting with halides

Fluorine, the most powerful oxidant of the halogens, will oxidise the ions of group 17 elements below it (the **halides**) – ie fluorine will oxidise chloride, bromide and iodide ions.

Similarly, chlorine will oxidise bromide and iodide ions; bromine will oxidise iodide ions.

Example Q

Reaction of chlorine and potassium bromide

The following process is used for the extraction of bromine from sea water. Chlorine gas is bubbled through sea water and bromine is liberated.

Full equation: $2KBr(aq) + Cl_2(g) \rightarrow 2KCl(aq) + Br_2(aq)$

Ionic equation: $2Br^-(aq) + Cl_2(g) \rightarrow 2Cl^-(aq) + Br_2(aq)$

Halogens reacting with hydrogen

Halogens oxidise hydrogen to form hydrogen halides.

Hydrogen halides are acidic in water.

Example R

Chlorine and hydrogen

$H_2(g) + Cl_2(g) \rightarrow 2HCl(g)$

Halogens reacting with metals

Halogens oxidise metals to form metal halides.

Example S

Chlorine and metals

$2Na(s) + Cl_2(g) \rightarrow 2NaCl(s)$

$2Al(s) + 3Cl_2(g) \rightarrow 2AlCl_3(s)$

Halogens reacting with water

Fluorine, the strongest oxidant, reacts immediately with water, oxidising oxygen to form oxygen gas and hydrofluoric acid, HF:

$$2F_2(g) + 2H_2O(\ell) \rightarrow 4HF(aq) + O_2(g)$$

Chlorine, when added to water, forms two acids:

$$\underset{}{H_2O(\ell)} + Cl_2(g) \rightarrow \underset{\text{hydrochloric acid}}{HCl(aq)} + \underset{\text{hypochlorous acid}}{HOCl(aq)}$$

The mixture of hydrochloric and hypochlorous acids is commonly known as chlorine water.

Iodine does not react with water.

Unit 12.3 Activity 1E: Halogens as oxidants

1. Write ionic equations for the following *if* a reaction occurs.

a. Aqueous potassium iodide and chlorine.

b. Aqueous silver bromide and iodine.

c. Aqueous silver bromide and chlorine.

d. Aqueous potassium chloride and bromine.

2. a. Write the equations for the reactions between iron (solid) and chlorine and iodine.

b. Identify the more vigorous reaction (ie the one involving the strongest oxidant).

3. Fluorine gas is the strongest oxidant of the halogens. Explain why it is not used in normal laboratory preparations.

4. a. 2 mL of 'Solution A' (KCl, KBr or KI) was put into a test tube and a few drops of 'Solution B' (chlorine, bromine or iodine solution) was added. Complete the table to show the expected observations for the reactions, using the following key:

Colourless – chlorine liberated. Orange – bromine liberated.

Yellow-brown – iodine liberated. No change – no reaction.

Solution B	Solution A		
	Chlorine solution	Bromine solution	Iodine solution
KCl(aq)		No change	i.
KBr(aq)	ii.		Pink
KI(aq)	Yellow-brown	iii.	

b. Write balanced half-equations for any reaction that occurs in **i**, **ii** or **iii**.

Unit 12.3 Electrochemistry
Topic 2: Oxidation–reduction reactions and electrochemistry

Topic 2 covers oxidation–reduction processes:
- Reactions and calculations involving electrochemical cells.
- Cell diagrams.
- Use of standard reduction potentials.
- Predicting whether oxidation–reduction reactions occur.
- Relative strength of oxidants and reductants.
- Knowledge of appearance and state of various oxidants and reductants, and the product to which they are converted in an oxidation–reduction reaction.

Electrochemical cells

A redox reaction is made up of two half-reactions described by corresponding half-equations. In an **electrochemical cell**, the two half-reactions are carried out in separate **half-cells**.

A typical half-cell is made of a metal rod, M, in contact with an aqueous solution of metal ions, M^+, where M^+ and M are the components of a **redox couple** (two species that are the reduction or oxidation product of each other). In this case, the reductant can also act as the **electrode**, connecting the half-cell to the **external circuit**.

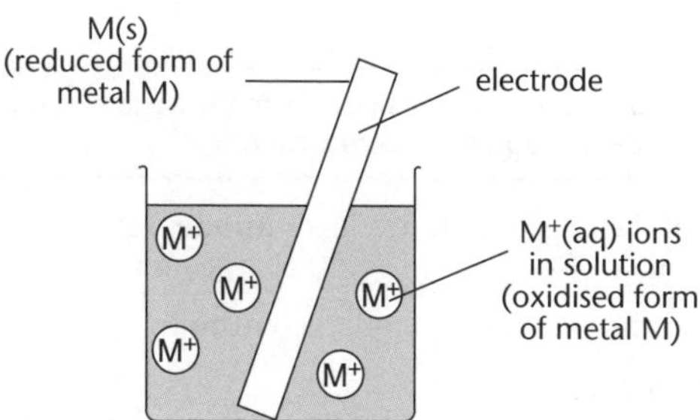

Half-cell for $M^+(aq)/M(s)$

A simple electrochemical cell is made by connecting two half-cells.

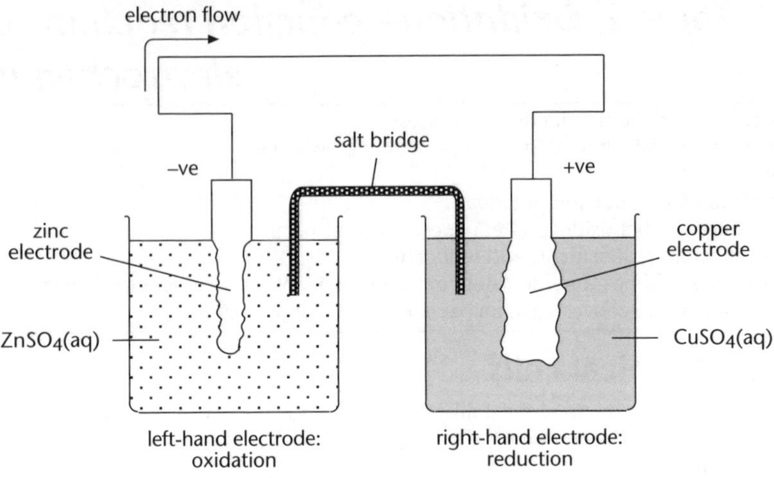

A simple electrochemical cell

As the **oxidation–reduction reaction** proceeds in the electrochemical cell, electrons flow through the **external circuit** (ie the wire connecting the half-cells) from the reductant to the oxidant.

Example A

In the electrochemical cell above, electrons flow from the negative (zinc) electrode to the positive (copper) electrode through the external circuit.

- **Oxidation** occurs at the negative electrode, or **anode**, eg:
  $Zn(s) \rightarrow Zn^{2+}(aq) + 2e^-$
- **Reduction** occurs at the positive electrode, or **cathode**, eg:
  $Cu^{2+}(aq) + 2e^- \rightarrow Cu(s)$

Combining the reactions in the two half-cells gives the overall cell reaction, eg:

$Zn(s) + Cu^{2+}(aq) \rightarrow Zn^{2+}(aq) + Cu(s)$

The **salt bridge** linking the half-cells must be present to complete the circuit. The salt bridge (or a semipermeable membrane) allows a flow of ions so that the charge in each half-cell remains balanced.

Example B

In the electrochemical cell above, the **negative ions** move through the salt bridge into the Zn^{2+}/Zn half-cell to offset the increase in positive charge (due to formation of Zn^{2+}), while positive ions will move into the Cu^{2+}/Cu half-cell where there is a decrease in positive ions (due to reduction of Cu^{2+}).

The salt bridge may be a salt solution in agar jelly in a U-tube or filter paper soaked in the electrolyte solution.

Example C

Solutions commonly used in the salt bridge are potassium chloride or potassium nitrate.

As the cell reaction proceeds, the changes in mass of the reactants and products are related according to the stoichiometry of the balanced equation.

Example D

An electrochemical cell is made up of a Zn^{2+}/Zn half-cell and an Ag^{+}/Ag half-cell.

If the mass deposited at the silver electrode is 0.50 g, calculate the mass lost at the zinc electrode.

From the balanced equation for the cell reaction:

$$2Ag^{+} + Zn \rightarrow 2Ag + Zn^{2+}$$

the mole ratio of Zn lost : Ag gained = 1 : 2.

Since $n(\text{Ag}) = \dfrac{0.50 \text{ g}}{108 \text{ g mol}^{-1}} = 0.004\,63 \text{ mol}$

Then $n(\text{Zn}) = \dfrac{1}{2} \times 0.004\,63 \text{ mol} = 0.002\,315 \text{ mol}$, and

$m(\text{Zn}) = 0.002\,315 \text{ mol} \times 65.4 \text{ g mol}^{-1} = 0.151 \text{ g}$

Cell diagrams

An electrochemical cell can be represented by a **cell diagram**.

Example E

The cell diagram for the electrochemical cell on page 136 is: $Zn(s) \mid Zn^{2+}(aq) \parallel Cu^{2+}(aq) \mid Cu(s)$

For a **standard cell diagram**, the conventions are:

- Two parallel lines || represent the salt bridge.
- The single line | represents a phase boundary, eg between solid and liquid (solution).
- The electrodes through which electrons flow are placed at the start and finish of the cell diagram.
- The anode (half-cell where oxidation occurs) is placed on the left (**LHE** or left-hand electrode).
- The cathode (half-cell where reduction occurs) is placed on the right (**RHE** or right-hand electrode).
- When electrodes are connected, the electrons flow from anode to cathode, which is from left to right in the standard cell diagram.

Other conventions for cell diagrams include:

- Oxidised and reduced species in the same phase (eg in a solution) are separated by a comma in the cell diagram.
- An inert (unreactive) electrode (eg **graphite** or platinum) is placed either first in the cell diagram (for the anode) or last (for the cathode).

Example F

Electrochemical cell with inert electrodes

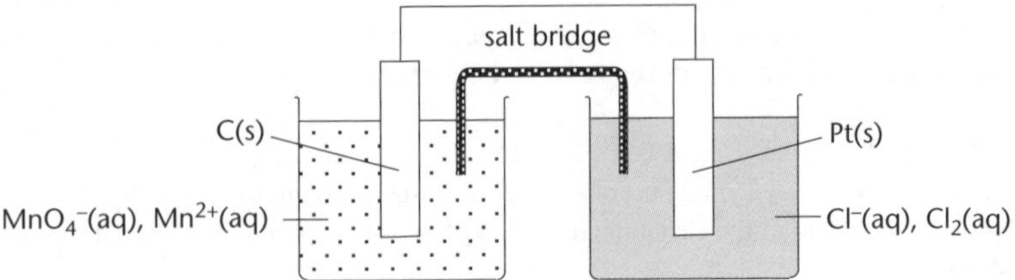

The cell diagram for this cell, which includes a platinum and a graphite, C(s), electrode, is:
Pt(s) I Cl^-(aq), Cl_2(aq) II MnO_4^-(aq), Mn^{2+}(aq) I C(s)

Unit 12.3 Activity 2A: Cell diagrams

1. Give the standard cell diagram for each of the following cells. In each case, the left-hand electrode where oxidation occurs is named first.

a. Zinc electrode in zinc sulfate solution connected to a graphite electrode in a solution of potassium iodate.

b. Iron electrode in iron(II) sulfate connected to a silver electrode in silver nitrate solution.

c. A zinc electrode in zinc sulfate connected to an inert platinum electrode in a solution containing iron(II) and iron(III) ions.

d. A platinum electrode in a solution of potassium iodide connected to a graphite electrode in a solution of sodium dichromate.

Cell potential, $E°_{cell}$

A voltmeter placed in the external circuit of an electrochemical cell will measure the **electromotive force** (**EMF**) or voltage difference between the electrodes. This voltage difference, or **cell potential** ($E°_{cell}$), is the potential of the cell to pull and push electrons through the circuit, and is measured in units of volts, V.

Example G

For the cell Zn(s) I Zn^{2+}(aq) II Cu^{2+}(aq) I Cu(s), the voltmeter would read 1.10 V and the cell diagram would be written as:
Zn(s) I Zn^{2+}(aq) II Cu^{2+}(aq) I Cu(s) $E°_{cell} = +1.10$ V

As the voltage depends on the ion concentrations in each half-cell, the standard cell potential is determined at a concentration of 1.0 mol L^{-1} for all solutions and a temperature of 25°C.

Standard reduction potentials

It is impossible to measure the potential of a single electrode (half-cell) because the circuit is not complete. Thus, the potential of an electrode is measured against a reference, the **standard hydrogen electrode** (**SHE**), whose voltage is set at 0.00 V under the following **standard conditions**:

- All solutions have a concentration of 1 mol L^{-1}.
- All gas pressures are 1 bar or 100 kPa (kilopascals).
- Temperature is 25°C (unless specified otherwise).

The half-cell for the SHE is:

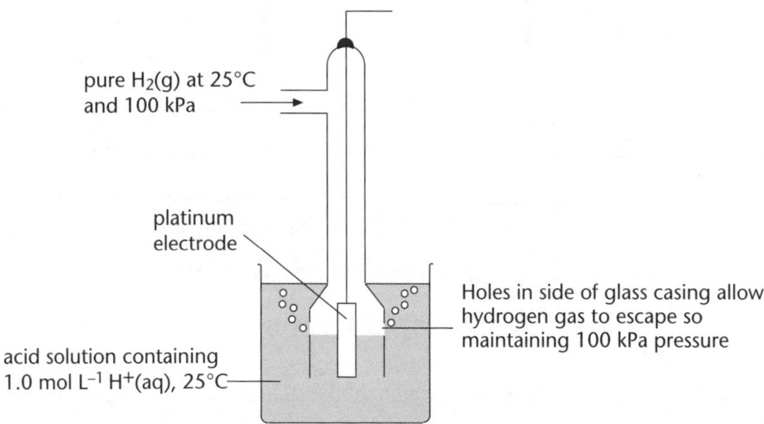

Standard hydrogen electrode (a standard hydrogen half-cell)

The hydrogen half-cell uses a platinum electrode because platinum is unreactive (inert), and it catalyses the reaction that produces hydrogen gas. The reaction in this half-cell is:

$2H^+(aq) \quad + \quad 2e^- \rightarrow H_2(g)$

Measuring the reduction potential of a half-cell

All **standard reduction potentials** (***E*°**), including the standard hydrogen electrode, are measured in volts (V) under standard conditions, for a cell in which the half-reaction occurs as a reduction when paired with the standard hydrogen electrode.

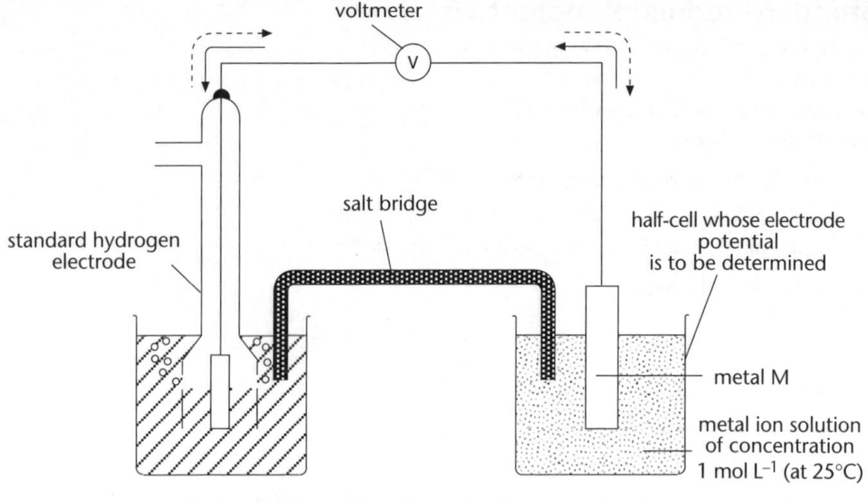

Apparatus to determine the electrode potential of a half-cell

When a half-cell is connected to a standard hydrogen electrode so that the reading on the voltmeter is a positive value the:

- voltmeter reading gives the absolute value of the standard reduction potential of the half-cell.
- *charge* on the half-cell electrode gives the *sign* of the standard reduction potential for that half-cell.

Example H

Reduction potential for Cu^{2+}/Cu

The electrochemical cell made by connecting a standard hydrogen electrode (SHE) to a Cu^{2+}/Cu half-cell is:

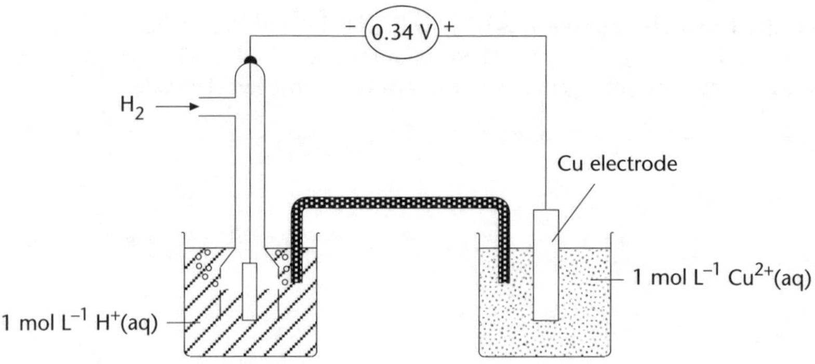

An electrochemical cell using the Cu^{2+}/Cu half-cell and the standard hydrogen electrode (SHE)

Since the Cu electrode is connected to the positive terminal of the voltmeter, the voltmeter reading of 0.34 V means the potential for the Cu^{2+}/Cu half-cell is more positive than H^+/H_2 by 0.34 V, ie:

$E°(Cu^{2+}/Cu) = +0.34$ V

The reactions occurring are:

- At the anode (–) $H_2(g) \rightarrow 2H^+(aq) + 2e^-$
- At the cathode (+) $Cu^{2+}(aq) + 2e^- \rightarrow Cu(s)$
- The overall cell reaction is $Cu^{2+}(aq) + H_2(g) \rightarrow Cu(s) + 2H^+(aq)$

Example I

Reduction potential for Zn^{2+}/Zn

The electrochemical cell made by connecting a Zn^{2+}/Zn half-cell to the SHE is:

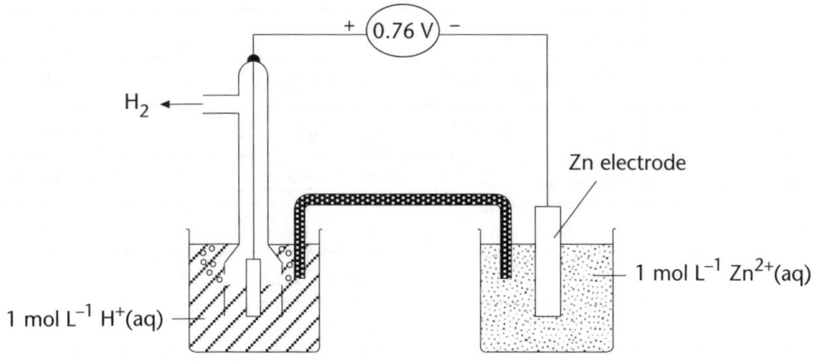

An electrochemical cell using the Zn^{2+}/Zn half-cell and the standard hydrogen electrode (SHE)

Since the Zn electrode is connected to the negative terminal of the voltmeter, the voltmeter reading of 0.76 V means the electrode potential of the Zn^{2+}/Zn half-cell is more negative than the standard hydrogen electrode by 0.76 V, ie:

$E°(Zn^{2+}/Zn) = -0.76$ V

The reactions occurring are:

At the anode (–) $Zn(s) \rightarrow Zn^{2+}(aq) + 2e^-$

At the cathode (+) $2H^+(aq) + 2e^- \rightarrow H_2(g)$

The overall cell reaction is $Zn(s) + 2H^+(aq) \rightarrow Zn^{2+}(aq) + H_2(g)$

Some standard reduction (electrode) potentials are shown in the following table.

Table of standard reduction potentials E°

Redox couple	Ion-electron half-equation		Standard reduction potential, *E*° (V)
F_2/F^-	$F_2 + 2e^-$	$\rightarrow 2F^-$	+2.89
H_2O_2/H_2O	$H_2O_2 + 2H^+ + 2e^-$	$\rightarrow 2H_2O$	+1.76
Au^+/Au	$Au^+ + e^-$	$\rightarrow Au$	+1.69
MnO_4^-/MnO_2	$MnO_4^- + 4H^+ + 3e^-$	$\rightarrow MnO_2 + 2H_2O$	+1.69
OCl^-/Cl_2	$2HClO + 2H^+ + 2e^-$	$\rightarrow Cl_2 + 2H_2O$	+1.63
BrO_3^-/Br_2	$2BrO_3^- + 12H^+ + 10e^-$	$\rightarrow Br_2 + 6H_2O$	+1.51
MnO_4^-/Mn^{2+}	$MnO_4^- + 8H^+ + 5e^-$	$\rightarrow Mn^{2+} + 4H_2O$	+1.51
$2ClO_3^-/Cl_2$	$2ClO_3^- + 12H^+ + 10e^-$	$\rightarrow Cl_2 + 6H_2O$	+1.46
PbO_2/Pb^{2+}	$PbO_2 + 4H^+ + 2e^-$	$\rightarrow Pb^{2+} + 2H_2O$	+1.46
Cl_2/Cl^-	$Cl_2 + 2e^-$	$\rightarrow 2Cl^-$	+1.36
$Cr_2O_7^{2-}/Cr^{3+}$	$Cr_2O_7^{2-} + 14H^+ + 6e^-$	$\rightarrow 2Cr^{3+} + 7H_2O$	+1.36
Au^{3+}/Au	$Au^{3+} + 3e^-$	$\rightarrow Au$	+1.52
MnO_2/Mn^{2+}	$MnO_2 + 4H^+ + 2e^-$	$\rightarrow Mn^{2+} + 2H_2O$	+1.23
O_2/H_2O	$O_2 + 4H^+ + 4e^-$	$\rightarrow 2H_2O$	+1.23*
IO_3^-/I_2	$2IO_3^- + 12H^+ + 10e^-$	$\rightarrow I_2 + 6H_2O$	+1.21
Br_2/Br^-	$Br_2 + 2e^-$	$\rightarrow 2Br^-$	+1.08
NO_3^-/NO_2	$2NO_3^- + 4H^+ + 2e^-$	$\rightarrow 2NO_2 + 2H_2O$	+0.81
Ag^+/Ag	$Ag^+ + e^-$	$\rightarrow Ag$	+0.80
Fe^{3+}/Fe^{2+}	$Fe^{3+} + e^-$	$\rightarrow Fe^{2+}$	+0.77
MnO_4^-/MnO_4^{2-}	$MnO_4^- + e^-$	$\rightarrow MnO_4^{2-}$	+0.56
I_2/I^-	$I_2 + 2e^-$	$\rightarrow 2I^-$	+0.54
Cu^{2+}/Cu	$Cu^{2+} + 2e^-$	$\rightarrow Cu$	+0.34
N_2/NH_3	$N_2 + 6H^+ + 6e^-$	$\rightarrow 2NH_3$	+0.27
S/H_2S	$S + 2H^+ + 2e^-$	$\rightarrow H_2S$	+0.17

* The potential of the oxygen electrode at pH 7.00 is +0.82 V.

Table of standard reduction potentials E° (continued)

Redox couple	Ion-electron half-equation		Standard reduction potential, $E°$ (V)
H^+/H_2	$2H^+ + 2e^-$	$\rightarrow H_2$	0.00
Pb^{2+}/Pb	$Pb^{2+} + 2e^-$	$\rightarrow Pb$	–0.13
Sn^{2+}/Sn	$Sn^{2+} + 2e^-$	$\rightarrow Sn$	–0.14
Fe^{2+}/Fe	$Fe^{2+} + 2e^-$	$\rightarrow Fe$	–0.44
Zn^{2+}/Zn	$Zn^{2+} + 2e^-$	$\rightarrow Zn$	–0.76
Al^{3+}/Al	$Al^{3+} + 3e^-$	$\rightarrow Al$	–1.68
Mg^{2+}/Mg	$Mg^{2+} + 2e^-$	$\rightarrow Mg$	–2.36
Na^+/Na	$Na^+ + e^-$	$\rightarrow Na$	–2.71
Ca^{2+}/Ca	$Ca^{2+} + 2e^-$	$\rightarrow Ca$	–2.87
K^+/K	$K^+ + e^-$	$\rightarrow K$	–2.94
Li^+/Li	$Li^+ + e^-$	$\rightarrow Li$	–3.04

Relative strength of oxidants and reductants

The values of reduction potentials provide information on the strength of species as oxidants or reductants.

Example J

The redox couple F_2/F^- has the most positive reduction potential. This means F_2 is the strongest oxidant, and also the species most readily reduced, while F^- is the weakest reductant.

Metals generally have negative reduction potentials. The more negative the electrode potential, the stronger the metal is as a reductant and the more readily it loses **valence** electrons to form a cation. Metals with:

- negative electrode potentials can reduce aqueous hydrogen ions to hydrogen gas.
- positive electrode potentials do not reduce hydrogen ions and so do not react with dilute acids – they are the unreactive metals, eg Cu, Ag and Au.

Note: Aluminium is a reactive metal but the presence of an oxide coating often makes it appear unreactive.

Unit 12.3 Activity 2B: Using reduction potentials

1. The table gives some standard reduction potentials.

Redox couple	$E°$ (V)	Redox couple	$E°$ (V)
Ca^{2+}/Ca	–2.87	I_2/I^-	+0.54
Mg^{2+}/Mg	–2.36	$HOCl/Cl_2$	+1.63
Fe^{2+}/Fe	–0.44	H_2O_2/H_2O	+1.76

Using the table, identify which of the above is the:

   a. metal that is the strongest reductant.
   b. ion that is the strongest oxidant.
   c. molecule that is the weakest oxidant.
   d. cation that is the strongest oxidant.

2. Identify the strongest oxidant from the following electrochemical data. Give a reason for your choice.
   $E°(Fe^{3+}/Fe^{2+}) = +0.77$ V
   $E°(Fe^{2+}/Fe) = -0.44$ V
   $E°(H_2O_2/H_2O) = +1.76$ V
   $E°(O_2/H_2O_2) = +0.70$ V

Calculating $E°_{cell}$ values

The value of $E°_{cell}$ can be calculated by combining the standard reduction potentials for two half-cells. The half-cell with the most positive (or least negative) electrode potential will:

- contain the strongest oxidant.
- be the cathode (half-cell where reduction occurs).

The $E°_{cell}$ is calculated using:

$E°_{cell} = E°_{cathode} - E°_{anode}$, or

$E°_{cell} = E°_{reduction} - E°_{oxidation}$

This relationship can also be written as:

$E°_{cell} = E°(RHE) - E°(LHE)$

where the terms RHE and LHE refer *only* to the **standard cell diagram** (where reduction occurs in the right-hand electrode).

Example K

The Zn(s) I Zn^{2+}(aq) II Cu^{2+}(aq) I Cu(s) cell

A cell is set up by combining the two half-cells:

$Zn^{2+}(aq) + 2e^- \rightarrow Zn(s)$ $E° = -0.76$ V

$Cu^{2+}(aq) + 2e^- \rightarrow Cu(s)$ $E° = +0.34$ V

- The half-cell with the most positive reduction potential is Cu^{2+}/Cu so this is the half-cell where reduction occurs, ie the cathode. This half-cell must contain the *strongest oxidant*, Cu^{2+}. The reaction will proceed in the direction written.
- Oxidation will occur in the Zn^{2+}/Zn half-cell and the Zn metal will be oxidised. This half-cell must contain the *strongest reductant*, Zn.

- The *overall equation* for the cell reaction is the sum of the two half-cell reactions, ie:
  $Cu^{2+}(aq) + Zn(s) \rightarrow Zn^{2+}(aq) + Cu(s)$
- The voltage (EMF) of the cell is:
  $E°_{cell} = E°_{cathode} - E°_{anode} = +0.34\ V - (-0.76\ V) = +1.10\ V$

Unit 12.3 Activity 2C: Calculating $E°_{cell}$ values

1. a. Use the table of standard reduction potentials on pages 142–3 to calculate the EMF ($E°$cell) for the following cells:

i. $Zn(s) \mid Zn^{2+}(aq) \parallel Sn^{2+}(aq) \mid Sn(s)$

ii. $Cu(s) \mid Cu^{2+}(aq) \parallel MnO_4^-(aq), Mn^{2+}(aq) \mid Pt(s)$

iii. $Fe(s) \mid Fe^{2+}(aq) \parallel Ag^+(aq) \mid Ag(s)$

iv. $Zn(s) \mid Zn^{2+}(aq) \parallel I_2(aq), I^-(aq) \mid C(s)$

b. Write balanced equations for the overall cell reactions occurring in each of the cells in part **a**.

2. A cell is set up as follows:

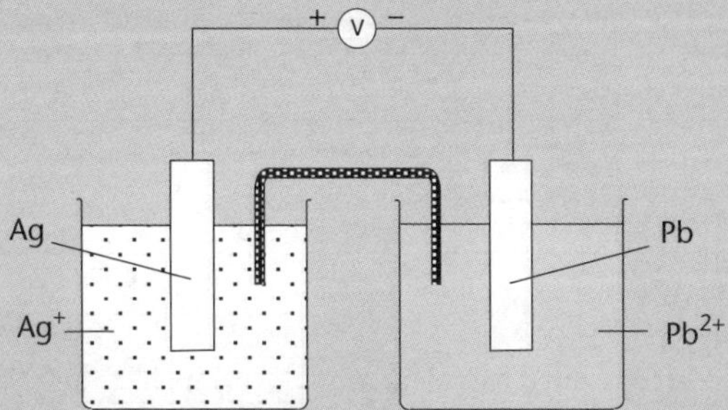

where $E°_{cell} = +0.93$ V and $E°(Pb^{2+}/Pb) = -0.13$ V. Use the information provided to determine the standard potential $E°(Ag^+/Ag)$.

3. An electrochemical cell is set up using two half-cells:

Silver metal in 1.0 mol L^{-1} silver nitrate solution, $E°(Ag^+/Ag) = +0.80$ V.
Nickel metal in 1.0 mol L^{-1} nickel nitrate solution, $E°(Ni^{2+}/Ni) = -0.24$ V.

a. Copy the diagram below and complete it by labelling the solutions and the metals in the boxes provided.

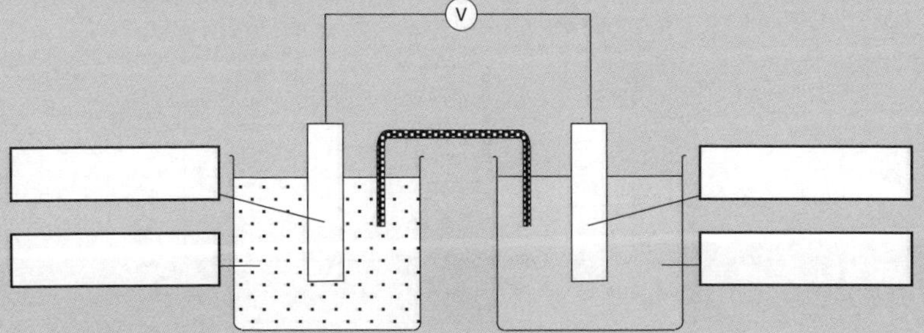

b. Label the diagram to show the direction of flow of charge in both the 'internal' and 'external' circuits.

c. Calculate the expected cell voltage.

4. Three electrochemical cells were constructed by linking the following electrodes:
   - **i.** $E°(Zn^{2+}/Zn) = -0.76$ V and $E°(Ag^{+}/Ag) = +0.80$ V
   - **ii.** $E°(Fe^{3+}/Fe^{2+}) = +0.77$ V and $E°(Sn^{2+}/Sn) = -0.14$ V
   - **iii.** $E°(Cl_2/Cl^{-}) = +1.36$ V and $E°(MnO_4^{-}/Mn^{2+}) = +1.51$ V

   For *each* of the electrochemical cells in **i.–iii.** determine the:
   - **a.** strongest oxidant and reductant.
   - **b.** equations for the reactions in each half-cell.
   - **c.** overall equation for the cell reaction.
   - **d.** $E°_{cell}$.
   - **e.** standard cell diagram.
5. A student set up four half-cells A, B, C and D, each containing a metal (eg A) dipped in a solution containing the metal ion (eg A^{2+}). These half-cells were connected in different combinations to make a series of electrochemical cells. The voltmeter readings and the metal electrode that formed the positive terminal in each case are recorded in the table.

Combination of half-cells	Voltmeter reading (V)	Positive terminal
A + B	+0.29	A
B + C	+0.67	C
B + D	+1.10	D

   - **a.** State the order of reactivity of the four metals. Explain clearly how you made your decision.
   - **b.** If half-cell B is connected to a standard hydrogen electrode (SHE), the reading on the voltmeter is 0.76 V and B is the negative terminal.
     - **i.** Determine the reading on the voltmeter if each of the other half-cells were connected to a SHE.
     - **ii.** In each case, state whether the SHE would be the positive or negative electrode.
6. An electrochemical cell is set up as follows:

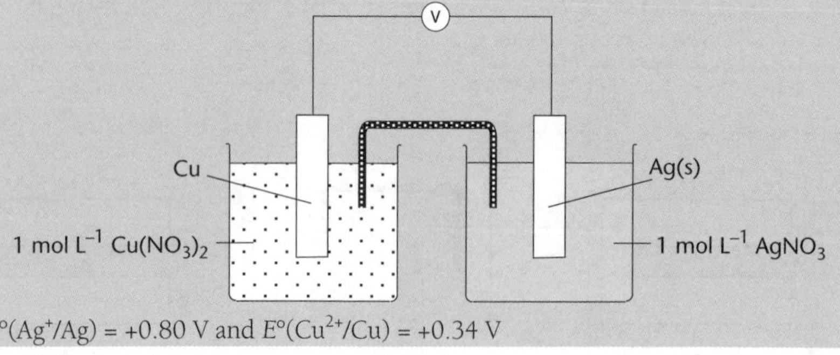

$E°(Ag^{+}/Ag) = +0.80$ V and $E°(Cu^{2+}/Cu) = +0.34$ V

a. Determine the reading on the voltmeter in the above cell.

b. Give the standard cell diagram that represents this cell.

c. Describe the nature and direction of flow of charge in this cell, including the role of the salt bridge.

d. **i.** Write a balanced equation for the reaction occurring.

ii. Describe what would be observed in each half-cell as it discharges (ie the reaction proceeds spontaneously), and explain the observations in terms of the chemical reactions occurring.

e. The cell is allowed to discharge for a period of time during which the mass of the copper electrode changes by 3.20 g. Calculate the mass change that would be expected on the silver electrode.

f. The Ag^+/Ag electrode is replaced with a standard hydrogen electrode, SHE. Discuss how this affects the operation of the cell, including aspects such as the charge on the electrodes, the cell voltage and any changes occurring in the Cu^{2+}/Cu half-cell.

Predicting whether oxidation–reduction reactions occur

Standard reduction potentials can be used to:

- predict whether an oxidation–reduction reaction will occur spontaneously.
- decide whether an ionic compound can exist, ie does not decompose as a result of a spontaneous redox reaction between the two ions present.

For a reaction to be spontaneous, the E°_{cell} value must be positive, where $E^\circ_{cell} = E^\circ_{cathode} - E^\circ_{anode}$.

To determine if a reaction is spontaneous, it is necessary to decide which species must be reduced (forming the cathode half-reaction) and which must be oxidised (the anode half-reaction).

Example L

Determine if magnesium metal will react with lead(II) nitrate solution.

$E^\circ(Mg^{2+}/Mg) = -2.36$ V and $E^\circ(Pb^{2+}/Pb) = -0.13$ V

Solution:

For a reaction to occur, the Mg would have to be oxidised to Mg^{2+} and the Pb^{2+} would have to be reduced according to the equation:

$Mg(s) + Pb^{2+}(aq) \rightarrow Mg^{2+}(aq) + Pb(s)$

For the reaction as written, E°_{cell} is calculated:

$$E^\circ_{cell} = E^\circ_{cathode} - E^\circ_{anode} = E^\circ(Pb^{2+}/Pb) - E^\circ(Mg^{2+}/Mg)$$
$$= -0.13\text{ V} - (-2.36\text{ V})$$
$$= +2.23\text{ V}$$

Since E°_{cell} is positive, the reaction between Mg and Pb^{2+} is spontaneous.

Unit 12.3 Activity 2D: Predicting redox reactions

1. Use the standard reduction potentials on pages 142–3 decide whether the following reactions will occur.
   a. Bromine reacting with silver metal.
   b. Chlorine reacting with a solution of iron(II) nitrate.
   c. A solution of iron(II) nitrate reacting with metallic copper, Cu.
2. Explain why the compound iron(III) chloride can exist whereas iron(III) iodide can not.
3. Show that it is not possible to store a solution of silver nitrate in an iron pot without a spontaneous reaction occurring.
4. Show that hydrochloric acid cannot be used to acidify the oxidation reaction of potassium permanganate with iron(II) sulfate.
5. **a.** Using the standard reduction potentials below, show that a reaction can occur between H_2O_2 and Fe^{2+}.
   $E°(Fe^{3+}/Fe^{2+}) = +0.77$ V
   $E°(Fe^{2+}/Fe) = -0.44$ V
   $E°(H_2O_2/H_2O) = +1.76$ V
   $E°(O_2/H_2O_2) = +0.70$ V
   b. Write a balanced equation for this reaction.
   c. Describe what would be observed as the reaction proceeds.

Limitations to using electrode potentials

There are some limitations to using electrode potentials.

- Electrode potentials used to predict the spontaneity of a reaction give no indication as to the rate of reaction (kinetics). A reaction, despite having a positive $E°_{cell}$ value, may be so slow that there is no observed reaction.

Example M

Consider the cell for the reaction of Cu^{2+} with H_2 gas:
Pt | $H_2(g)$, $2H^+(aq)$ || $Cu^{2+}(aq)$ | Cu(s)
$E°(Cu^{2+}/Cu) = +0.34$ V and $E°(H^+/H_2) = 0.00$ V.
The calculated voltage is +0.34 V, suggesting the reaction would be **spontaneous.** In fact, the reaction is so slow it effectively does not happen.

- Electrode potentials involve equilibrium reactions, which are affected by changes in the concentration of the reactants. Standard potentials are determined when the concentration of the species is 1 mol L^{-1}. Changes in concentration will change the value of the electrode potential and the measured voltage of the cell.

Corrosion

Many metals are oxidised (ie they **corrode**) if left in water containing dissolved oxygen.

Corrosion (rusting) of iron

In iron, corrosion is called **rusting** and the **hydrated** iron(III) oxide formed is **rust**. The initial half-reaction for iron rusting is:

$$Fe(s) \rightleftharpoons Fe^{2+}(aq) + 2e^-$$

Fe and Fe^{2+} are a redox couple since they are interconvertible in redox reactions – Fe is the reduced species, Fe^{2+} is the oxidised species. If electrons are added, the equilibrium in the above reaction shifts to the left – iron is formed and the rusting process is reversed.

Rust prevention

Cathodic protection

A car battery is used to 'negative earth' a car, making the car body a negative electrode, and reducing the effects of rusting. Steel wharves are similarly protected by connecting them to the negative terminal of a power source.

Sacrificial corrosion

Another method of rust prevention is **sacrificial corrosion**. A more reactive metal than iron, eg zinc, is coated onto the iron. The zinc protects the iron metal because:

- The zinc reacts first since Zn is more reactive than Fe.

  $Zn(s) \rightarrow Zn^{2+}(aq) + 2e^-$

- Electrons lost by the zinc move to the iron, protecting the iron from corrosion.

  $Fe^{2+}(aq) + 2e^- \rightarrow Fe(s)$

Zinc-coated iron is called **galvanised** iron. Zinc protects iron through sacrificial corrosion, and also because it forms zinc oxide, which, along with the unreacted zinc below the oxide layer, act as a physical barrier to prevent water and oxygen reaching the iron.

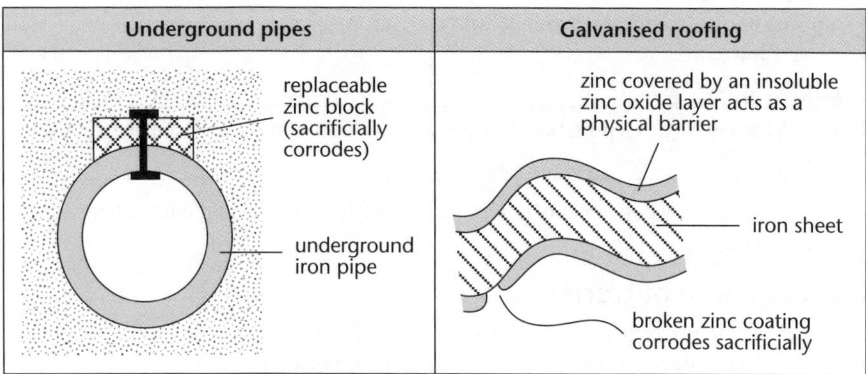

Protecting iron/steel from rusting

Less reactive metals than iron, such as copper, cannot protect iron by sacrificial corrosion, because the iron atoms in iron metal are more reactive than the copper atoms in copper metal. In fact, copper metal in contact with iron will promote rusting of the iron, because the iron will act as a sacrificial metal.

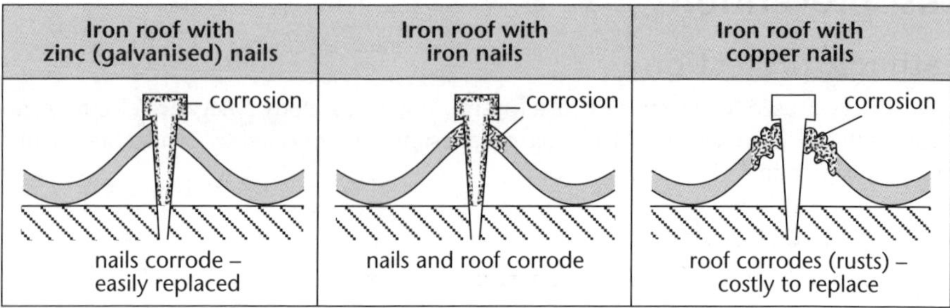

Protection of steel roofing

Batteries

An electrochemical cell may be either:

- an **electrolytic cell** – a cell that uses electrical energy to bring about a chemical reaction in a direction which is opposite to that which proceeds naturally.

  or

- a **galvanic cell** – in which electrons flow through a circuit as a result of a **spontaneous** chemical reaction within the cell. A galvanic cell is commonly referred to as a **battery**.

Example N

The AA battery used in a portable CD player and the lead–acid battery used in a car are galvanic cells.

Galvanic cells are very convenient sources for converting chemical potential energy to electrical energy.

Rechargeable batteries

Because electrode reactions can be reversible, some electrochemical cells and batteries (groups of electrochemical cells in series) can be **recharged**. This is done by supplying them with electrons at a voltage greater than that produced by the cell, which forces the reaction to occur in the direction which is not spontaneous.

Example O

Lead–acid battery

The **lead–acid battery**, which is the common car battery, is rechargeable. It usually consists of six electrochemical cells joined in series, each cell having a Pb and a PbO_2 electrode. When discharging, the overall reaction occurring is:

$PbO_2 + Pb + 2H_2SO_4 \rightarrow 2PbSO_4 + 2H_2O$

The standard cell diagram is:

$Pb(s)\ |\ Pb^{2+}(aq)\ ||\ Pb^{2+}(aq)\ |\ PbO_2(s)$

As the battery **discharges**, water is produced and lead sulfate is precipitated out so that the concentration and density of the sulfuric acid solution decreases.

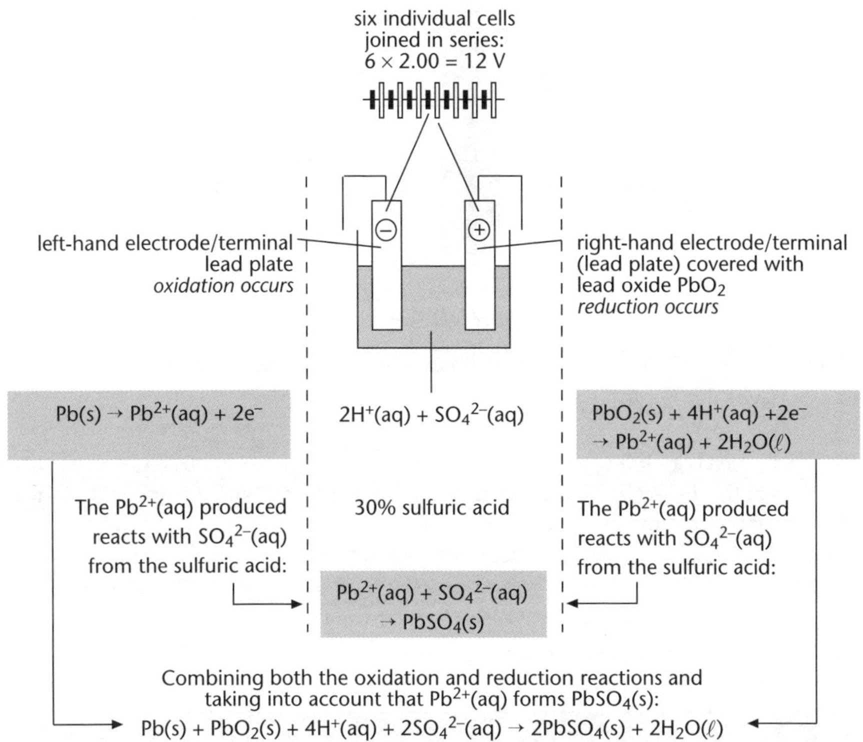

When the battery is being recharged, the reaction at each electrode is simply reversed.

The lead–acid battery

Electrode	Reaction occurring during charging
Negative	$PbSO_4(s) + 2e^- \rightarrow Pb(s) + SO_4^{2-}(aq)$
Positive	$PbSO_4(s) + 2H_2O(\ell) \rightarrow PbO_2(s) + 4H^+(aq) + SO_4^{2-}(aq) + 2e^-$

The overall reaction is reversed, and the concentration and density of the electrolyte will increase:

$2PbSO_4 + 2H_2O \rightarrow Pb + PbO_2 + 2H_2SO_4$

Alkaline dry cell (Leclanché cell)

The dry cell or torch battery was first developed in about 1860. Dry cells have the advantage that they are light, portable and cheap. The main disadvantage is that they are not usually rechargeable.

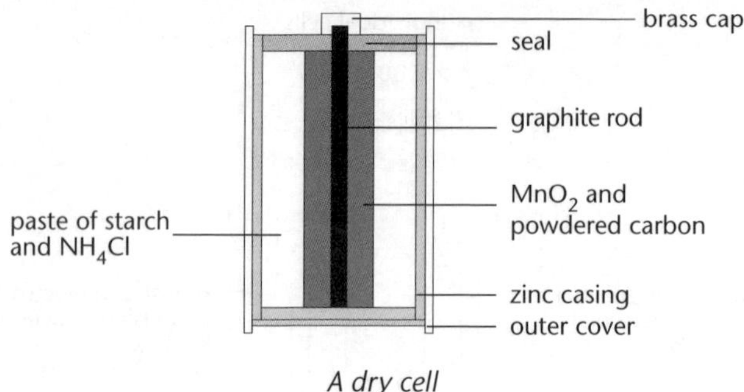

A dry cell

The cathode or positive terminal is the half-cell where reduction occurs, ie:

$2MnO_2 + H_2O + 2e^- \rightarrow Mn_2O_3 + 2OH^-$,

followed by:

$NH_4^+ + OH^- \rightarrow NH_3 + H_2O$

The species in this half-cell are 'contained' in powders or pastes, and a graphite, C(s), rod is used as an inert electrode.

The negative terminal is the anode where oxidation occurs:

$Zn \rightarrow Zn^{2+} + 2e^-$

The zinc metal is the reductant but also acts as the casing for the cell.

The cell diagram for the dry cell is:

$Zn \mid Zn^{2+} \parallel MnO_2, Mn_2O_3 \mid C$

Unit 12.3 Activity 2E: Batteries

1. Mercury–zinc 'button' batteries are used in watches and other small electrical devices. These electrochemical cells have the following construction:

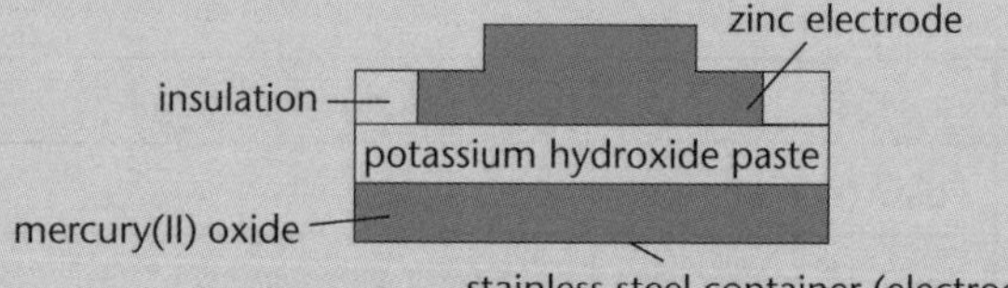

The overall chemical reaction that occurs in such a cell is:

$Zn(s) + H_2O(\ell) + HgO(s) \rightarrow Zn(OH)_2(s) + Hg(\ell)$

a. Write the balanced *oxidation* half-equation.

b. Write the balanced *reduction* half-equation.

c. Identify the *negative* electrode of this type of cell.

d. Identify a major potential chemical hazard which should be allowed for in the safe disposal of a used cell of this type.

2. The standard electrode potentials for the two half-cells in a lead–acid battery are:

$PbO_2(s) + 4H^+(aq) + SO_4^{2-}(aq) + 2e^- \rightarrow PbSO_4(s) + 2H_2O(\ell)$ $E° = +1.69$ V

$PbSO_4(s) + 2e^- \rightarrow SO_4^{2-}(aq) + Pb(s)$ $E° = -0.36$ V

a. Write a balanced equation for the overall cell reaction.

b. Identify the anode and cathode, and the positive and negative terminals of the battery.

c. Calculate the E°_{cell} for this cell.

d. Explain why the density of the solution decreases as the cell discharges.

3. A hydrogen–oxygen fuel cell consists of an electrolyte solution, eg KOH, and two inert electrodes. Hydrogen and oxygen are bubbled into the cell and the following reactions occur:

$H_2(g) + 2OH^-(aq) \rightarrow 2H_2O(\ell) + 2e^-$

$O_2(g) + 2H_2O(\ell) + 4e^- \rightarrow 4OH^-(aq)$

a. Write the overall equation for the cell reaction.

b. Copy the diagram of the cell and label the:

i. anode and cathode.

ii. oxidant and reductant.

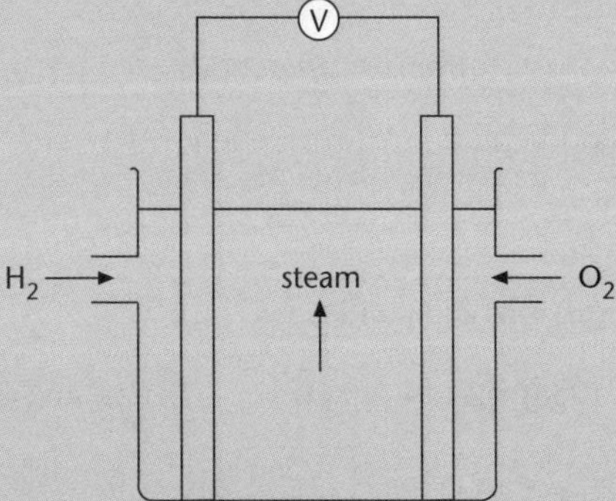

c. The standard cell potential of this fuel cell is +1.23 V. Calculate the standard reduction potential for the conversion of $H_2O(\ell)$ to $H_2(g)$ and $OH^-(aq)$. $E^\circ(O_2/OH^-) = +0.40$ V.

Oxidation–reduction reactions of selected transition metal compounds

The metals in groups 3–12 of the periodic table are referred to as **transition metals**. When ions for compounds of these metals undergo oxidation–reduction reactions, characteristic colour changes occur.

Iron

The iron(II) ion, Fe^{2+}, is easily oxidised to the iron(III) ion, Fe^{3+}:

$$Fe^{2+} \rightarrow Fe^{3+} + e^-$$

Iron(II) solutions or compounds exposed to air are readily oxidised by the air to form iron(III) compounds. Iron(II) solutions or compounds are usually a mixture of iron(II) and iron(III) ions. As oxidation continues, there are noticeable colour changes as iron(III) compounds are formed.

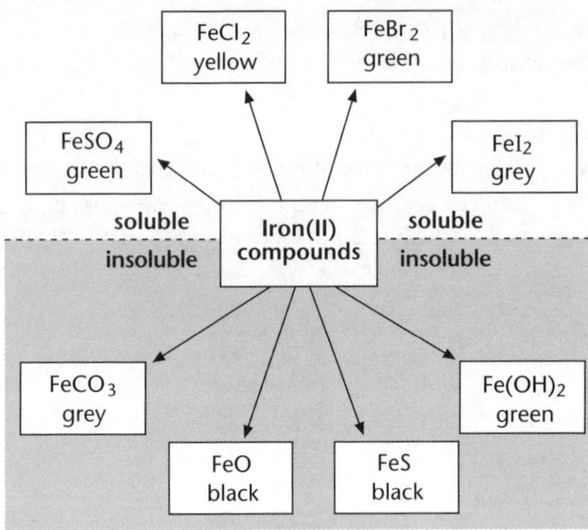

Iron(II) compounds

Iron(III) iodide does not exist because a redox reaction between the iron(III) ions and the iodide ions occurs.

Example P

Reaction between iron(III) ions and iodide ions

$I_2(aq) + 2e^- \rightarrow 2I^-(aq)$ $E° = +0.54$ V

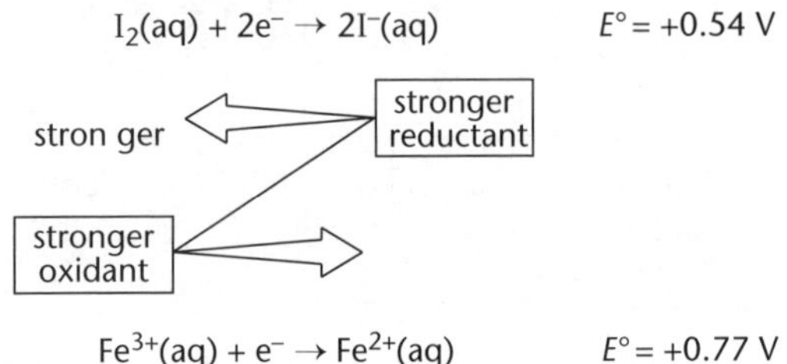

$Fe^{3+}(aq) + e^- \rightarrow Fe^{2+}(aq)$ $E° = +0.77$ V

Z diagram for redox reaction between I_2 and Fe^{3+}

The iron(III) ions are reduced to iron(II) ions and the iodide ions are oxidised to form iodine. The redox reaction is:

$$2Fe^{3+}(aq) + 2I^-(aq) \rightarrow 2Fe^{2+}(aq) + I_2(aq)$$

The iron(II) ions formed in the reaction above react with some of the (unreacted) iodide ions to form a precipitate of iron(II) iodide, FeI_2:

$$Fe^{2+}(aq) + 2I^-(aq) \rightarrow FeI_2(s)$$

The overall reaction between iron(III) and iodide is:

$$2Fe^{3+}(aq) + 6I^-(aq) \rightarrow 2FeI_2(s) + I_2(aq)$$

The $E°_{cell}$ for this spontaneous reaction is:

$E°_{cell} = +0.77 - 0.54 = +0.23$ V

Iron(III) sulfide does not exist for similar reasons.

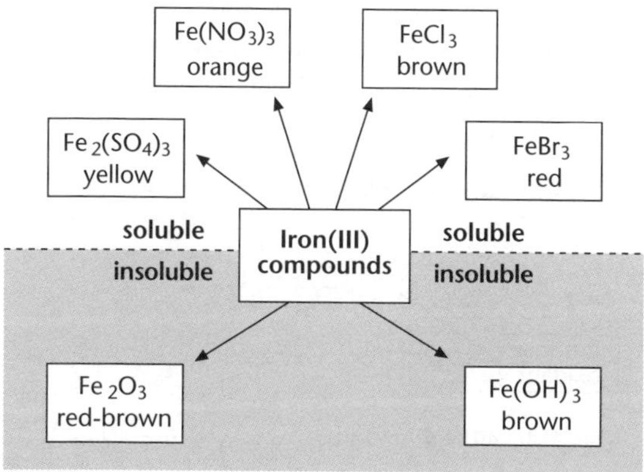

Iron(III) compounds

The test for the presence of Fe^{3+} ion in aqueous solution involves the addition of thiocyanate ion, SCN^-. This results in the formation of the dark-red **complex ion**, $[Fe(H_2O)_5SCN]^{2+}$, often simply written as $FeSCN^{2+}$.

The hydrated aqueous iron(III) ion (hexaaquairon(III) ion) is acidic:

$$H_2O(\ell) + [Fe(H_2O)_6]^{3+}(aq) \rightarrow [Fe(H_2O)_5(OH)]^{2+}(aq) + H_3O^+(aq)$$

Thus, if aqueous sodium carbonate is mixed with aqueous iron(III) ions, the aqueous iron(III) protonates the water forming aqueous hydronium ions, which react with carbonate ions, releasing carbon dioxide gas. The red-brown precipitate formed is $Fe(OH)_3$.

Iron(III) carbonate cannot therefore be formed by **precipitation**.

Unit 12.3 Activity 2F: Iron

1. Explain the following.

- **a.** Grey-green iron(II) hydroxide turns to a red-brown colour when left in moist air.
- **b.** When sodium carbonate is mixed with aqueous iron(III) chloride, bubbles are seen and the precipitate which forms will not effervesce (produce bubbles) with dilute hydrochloric acid.
- **c.** When aqueous iron(III) chloride is added to aqueous potassium iodide, a brown-grey precipitate is seen.

2. An orange solution is found to be acidic. On addition of aqueous sodium hydroxide, an **orange-brown precipitate forms**. When a few drops of potassium thiocyanate solution are added to the orange solution, it **turns a dark-red colour**.

- **a.** Write balanced equations for each of the observed reactions (shown in **bold** above).
- **b.** Write an equation to show why the orange solution is acidic.

Copper

The copper(I) ion is stable only in certain insoluble compounds. The copper(I) ion disproportionates in aqueous solutions. **Disproportionation** is the simultaneous oxidation and reduction of the same element in the same chemical reaction.

Copper(I) ions are both reductants and oxidants.

$$Cu^{2+}(aq) + e^- \rightarrow Cu^+(aq) \qquad E° = +0.16\ V$$

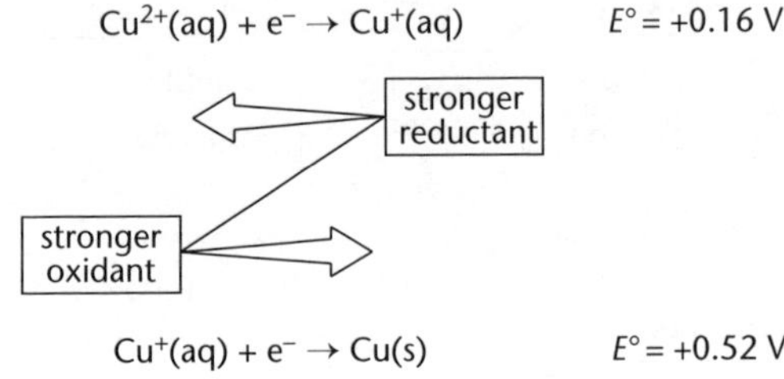

$$Cu^+(aq) + e^- \rightarrow Cu(s) \qquad E° = +0.52\ V$$

'Z diagram' for redox reactions of copper(I) ions

Copper(I) ions are:

- stronger *oxidants* than copper(II) ions.
- stronger *reductants* than copper atoms.

Copper(I) ions therefore reduce and oxidise each other, forming copper and copper(II) ions:

$$2Cu^+(aq) \rightarrow Cu(s) + Cu^{2+}(aq)$$

The $E°_{cell}$ for this spontaneous reaction is:

$$E°_{cell} = +0.52 - 0.16$$
$$= +0.36\ V$$

Thus aqueous copper(I) ions do not exist in solution.

Example Q

The neutralisation reaction between copper(I) oxide and sulfuric acid illustrates the disproportionation of copper(I) compounds:

$$Cu_2O + H_2SO_4 \rightarrow Cu + CuSO_4 + H_2O$$

Ignoring the sulfate ions (which are spectator ions) in the above reaction gives a better indication of the disproportionation reaction that occurs at the same time as the neutralisation reaction:

$$Cu_2O(s) + 2H^+(aq) \rightarrow Cu(s) + Cu^{2+}(aq) + H_2O(\ell)$$

Copper(I) iodide, CuI, is white. Copper(I) oxide, Cu_2O, is red.

Example R

Red Cu_2O is formed in the Benedict's and Fehling's tests for aldehydes.

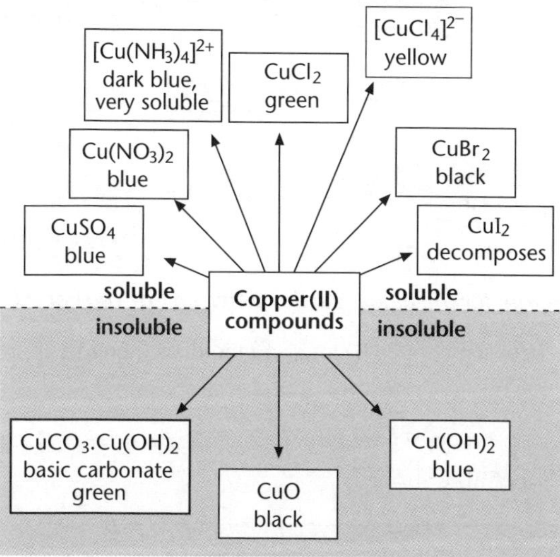

Copper(II) compounds

Copper(II) compounds are easily recognisable by their blue or green solutions.

Copper(II) hydroxide is so insoluble that Cu(II) ions can be precipitated from solution by even a very small amount of hydroxide ions. Adding ammonia to a solution of copper sulfate will precipitate out copper hydroxide:

$$NH_3(aq) + H_2O(\ell) \rightleftharpoons NH_4^+(aq) + OH^-(aq)$$

$$2OH^-(aq) + Cu^{2+}(aq) \rightarrow Cu(OH)_2(s)$$

(from $NH_3(aq)$)

The tetraamminecopper(II) complex ion, $[Cu(NH_3)_4]^{2+}(aq)$, is very soluble. Adding excess ammonia to a precipitate of copper(II) hydroxide formed originally from ammonia will result in the copper(II) hydroxide dissolving to form a solution containing the tetraamminecopper(II) complex ion:

$$Cu(OH)_2(s) + 4NH_3(aq) \rightarrow [Cu(NH_3)_4]^{2+}(aq) + 2OH^-(aq)$$

Insoluble copper(II) compounds can be converted into solutions of aqueous copper(II) ions by reacting them with dilute acid, which gives a characteristic blue or green solution:

$$CuO(s) + H_2SO_4(aq) \rightarrow CuSO_4(aq) + H_2O(\ell)$$

black　　　　　　　　　　blue

Copper(II) iodide does not exist because there is a redox reaction between copper(II) ions and iodide ions.

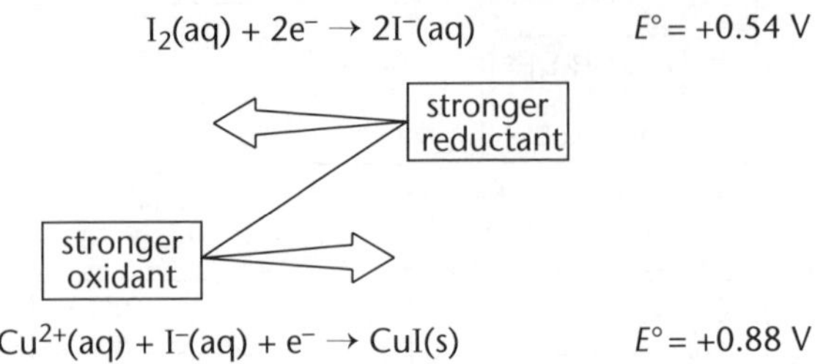

Z diagram for the redox reaction between $Cu^{2+}(aq)$ and $I^-(aq)$

The overall reaction between copper(II) ions and iodide is therefore represented by:

$$2Cu^{2+}(aq) + 4I^-(aq) \rightarrow 2CuI(s) + I_2(aq)$$

For this reaction $E°_{cell} = +0.88 - 0.54 = +0.34$ V

White copper(I) iodide, CuI, is formed.

Unit 12.3 Activity 2G: Copper

1. If the blue precipitate from the reaction of aqueous copper(II) sulfate with aqueous sodium hydroxide is heated strongly, a black solid is produced.
   a. Give the balanced equation for this reaction and name the black solid.
   b. Explain why this reaction is not an oxidation–reduction reaction.
2. Copper metal is used on the domes of various buildings. After several years' exposure to the atmosphere, it turns pale green. Suggest a reason for the colour change.
3. Explain why copper(II) iodide does not exist.

Manganese

Manganese compounds can have oxidation numbers ranging from +2 to +7, although compounds with manganese having +3 and +5 oxidation numbers are rare. Manganese compounds are generally used as oxidants, because, apart from Mn^{2+}, their reduction reactions have positive $E°$ values.

Electrode potentials for reactions of common manganese species

Oxidant	Reductant	$E°$ (V)
Weakest oxidant	*Strongest reductant*	
$Mn^{2+}(aq) + 2e^-$ colourless*	$\rightarrow Mn(s)$ grey	−1.18
In strongly basic solution: $MnO_4^-(aq) + e^-$ purple	$\rightarrow MnO_4^{2-}(aq)$ green	+0.56
$MnO_2(s) + 4H^+(aq) + 2e^-$ black	$\rightarrow Mn^{2+}(aq) + 2H_2O(\ell)$ colourless	+1.23

In acid solution: $MnO_4^-(aq)$ + $8H^+(aq)$ + $5e^-$ purple	→ $Mn^{2+}(aq)$ + $4H_2O(\ell)$ colourless	+1.51
In neutral solution: $MnO_4^-(aq)$ + $4H^+(aq)$ + $3e^-$ purple	→ $MnO_2(s)$ + $2H_2O(\ell)$ black	+1.69
Strongest oxidant	*Weakest reductant*	

*$Mn^{2+}(aq)$ ions in solution are very pale pink although dilute solutions in test tubes will seem colourless.

Unit 12.3 Activity 2H: Manganese

1. Give the formula and colour of:
   - **a.** manganese(II) sulfate.
   - **b.** manganese dioxide.
   - **c.** potassium manganate.
   - **d.** potassium permanganate.
2. Give the oxidation numbers of manganese in the following compounds.
   - **a.** $KMnO_4$
   - **b.** Mn_2O_3
   - **c.** MnO_2
   - **d.** $MnO(OH)$
   - **e.** K_2MnO_4
3. A student forgot to acidify potassium permanganate solution used in a titration. A dark-brown suspension was formed in the flask. Identify the suspension.

Chromium

Chromium compounds can have oxidation numbers ranging from +2 to +6.

Electrode potentials for reaction of common chromium species

Oxidant	Reductant	$E°$ (V)
Weakest oxidant	*Strongest reductant*	
$Cr^{2+}(aq)$ + $2e^-$	→ $Cr(s)$ grey	–0.90
$Cr^{3+}(aq)$ + $3e^-$ blue-green	→ $Cr(s)$ grey	–0.74
$Cr^{3+}(aq)$ + e^- blue-green	→ $Cr^{2+}(aq)$	–0.42
$Cr_2O_7^{2-}(aq)$ + $14H^+(aq)$ + $6e^-$ orange	→ $2Cr^{3+}(aq)$ + $7H_2O(\ell)$ blue-greeen	+1.36
Strongest oxidant	*Weakest reductant*	

As the oxidation number of a transition metal in an oxide increases, the nature of the oxide changes from basic (typical metal oxide), through amphoteric, to acidic:

- Chromium(II) oxide, CrO, a **basic oxide** (black) that dissolves in acid, ie:
  $CrO + 2H_3O^+ \rightarrow Cr^{2+} + 3H_2O$

- Chromium(III) oxide, Cr_2O_3, an amphoteric oxide (green) that dissolves in both acid and base. It forms green $[Cr(OH)_4]^-$ in excess hydroxide.
- Chromium(VI) oxide, CrO_3, an **acidic oxide** (orange) that dissolves in water to give a solution of chromic acid, $H_2Cr_2O_7$ (or $Cr_2O_7^{2-}$ ions).

Chromium(VI) exists as *yellow* chromate (chromate(VI)) ions in basic solutions, and as *orange* dichromate ions in acidic conditions:

$$\underset{\text{yellow}}{2CrO_4^{2-}(aq)} + 2H^+(aq) \rightleftharpoons \underset{\text{orange}}{2Cr_2O_7^{2-}(aq)} + H_2O(\ell)$$

This reaction is not an oxidation–reduction reaction as the oxidation number of chromium is +6 in both CrO_4^{2-} and $Cr_2O_7^{2-}$.

Vanadium

The highest oxidation state of vanadium is +5 and occurs when vanadium metal combines with the highly electronegative element, oxygen.

Example S

Vanadium pentoxide, V_2O_5, is used as a **catalyst** in the production of sulfuric acid.

A solution of vanadium(V) ions can be made by dissolving ammonium vanadate, NH_4VO_3, in sodium hydroxide solution and then adding sulfuric acid. The solution is yellow due to the presence of the VO_2^+ ion in acidic solution. If this yellow solution is shaken with zinc, the colour changes gradually, through green to blue (VO^{2+}), then to the green V^{3+} ions and finally to the violet V^{2+} ions:

$$\underset{\text{yellow}}{VO_2^+(aq)} \rightarrow \underset{\text{blue}}{VO^{2+}(aq)} \rightarrow \underset{\text{green}}{V^{3+}(aq)} \rightarrow \underset{\text{violet}}{V^{2+}(aq)}$$

Unit 12.3 Activity 2I: Chromium and vanadium

1. Give the formula and colour of:

   a. chromium(II) oxide.

   b. chromium(III) oxide.

   c. chromium(VI) oxide.

2. Give the oxidation number of the metal in:

   a. $Cr_2O_7^{2-}$

   b. CrO_4^{2-}

   c. VO_2^+

Unit 12.3 Electrochemistry

Topic 3: Electrolysis

Topic 3 covers electrolysis:

- Describing principles of simple electrolytic cells.

Introduction

Electrolysis means 'to split using electricity'. The process involves **chemical changes** that occur when an electric current is passed through an **electrolyte**. These chemical changes involve *redox* reactions.

An electrolysis reaction occurs in an **electrolytic cell**. An electrolytic cell contains:

- a DC (direct current) power supply.
- an electrolyte – a liquid containing ions that are free to move. An electrolyte can be either a **molten** ionic compound or a solution containing dissolved ions.
- two electrodes – these carry current to and from the electrolyte. The negative electrode is the cathode and the positive electrode is the anode. Electrodes are usually made of inert materials (such as platinum, Pt, or graphite, C) that are *not* reduced or oxidised in electrolysis reactions.

When a current is supplied to an electrolytic cell:

- cations (positive ions) move or migrate towards the cathode (the negative electrode), where reduction occurs.
- anions (negative ions) migrate towards the anode (the positive electrode), where oxidation occurs.

This can be remembered by:

- LEOA – **L**oss of **E**lectrons is **O**xidation at the **A**node
- GERC – **G**ain of **E**lectons is **R**eduction at the **C**athode

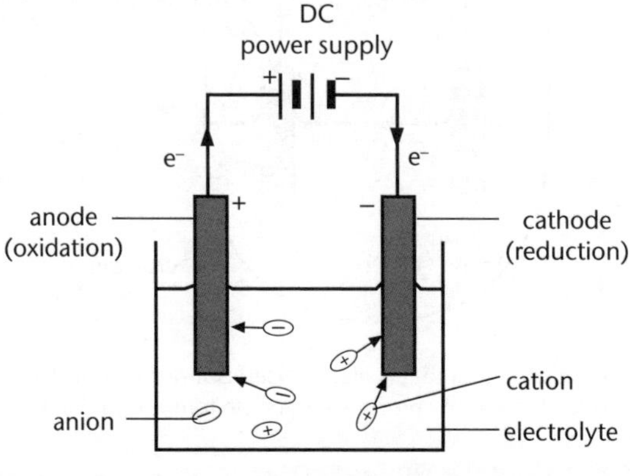

An electrolytic cell

Ions move as they are attracted to the oppositely charged electrode. The specific oxidation and reduction reactions that occur in each cell depend on the ions and other species present in the electrolyte.

Electrolysis of a molten ionic compound

In molten ionic compounds, the species that form the electrolyte are the cations and anions from the ionic compound. The ions are free to move because the ionic compound is in liquid (molten) form.

During electrolysis:

- cations (usually metal ions) collect at the cathode, gain electrons and are reduced to elemental metal.
- anions collect at the anode, give up electrons and are oxidised.

The electrodes used are usually made of graphite because it is inert (does not react), conducts electricity and is relatively cheap.

Example A

Electrolysis of lithium chloride

Lithium chloride, LiCl, is melted then electrolysed using graphite electrodes.
At the cathode, lithium ions, Li^+, are reduced, forming molten lithium metal:

$$Li^+(\ell) + e^- \rightarrow Li(\ell)$$

At the anode, chloride ions are oxidised, to form chlorine gas:

$$2Cl^-(\ell) \rightarrow Cl_2(g) + 2e^-$$

Adding these two half-equations together gives the overall equation for the electrolysis of molten LiCl:

$$2Cl^-(\ell) + 2Li^+(\ell) \rightarrow 2Li(\ell) + Cl_2(g)$$

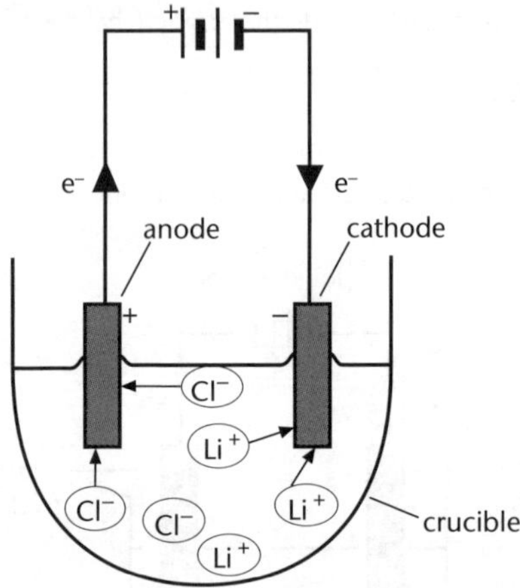

Note: In practice it is not pure LiCl that is electrolysed but rather a mixture of lithium chloride and potassium chloride. Potassium chloride lowers the temperature at which lithium chloride melts.

Electrolysis of aqueous solutions

An aqueous solution contains water and ions that are free to move. When an aqueous solution is electrolysed, some of the species present are oxidised and others are reduced.

The reactions that occur depend on the *ease* with which different ions can be reduced or oxidised compared with water. For metal ions, this relates to their reactivity; ions of metals that are unreactive (eg copper) are more easily reduced than ions of reactive metals (eg sodium).

Example B

Electrolysis of copper chloride solution

In copper chloride solution, the copper(II) ions are reduced at the cathode and a pink-brown layer is observed. At the anode, chloride ions are oxidised to Cl_2 and bubbles of gas are observed:

$2Cl^-(aq) \rightarrow Cl_2(g) + 2e^-$

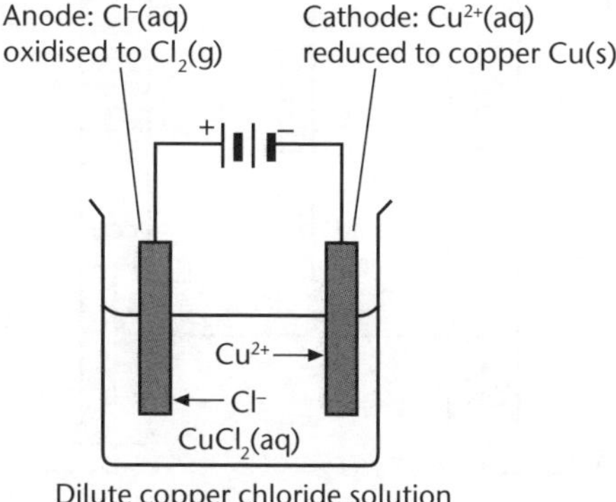

Dilute copper chloride solution

In the electrolysis of dilute aqueous solutions, *water* is often 'split' into its elements rather than the dissolved ions being oxidised or reduced:

$$2H_2O(\ell) \rightarrow O_2(g) + 2H_2(g)$$

Example C

Electrolysis of water

When water is electrolysed, oxygen gas is produced at the anode and hydrogen gas at the cathode.

The anode reaction is:

$2H_2O(\ell) \rightarrow O_2(g) + H^+(aq) + 4H^+ + 4e^-$

The cathode reaction is:

$2H_2O(\ell) + 2e^- \rightarrow H_2(g) + 2OH^-(aq)$

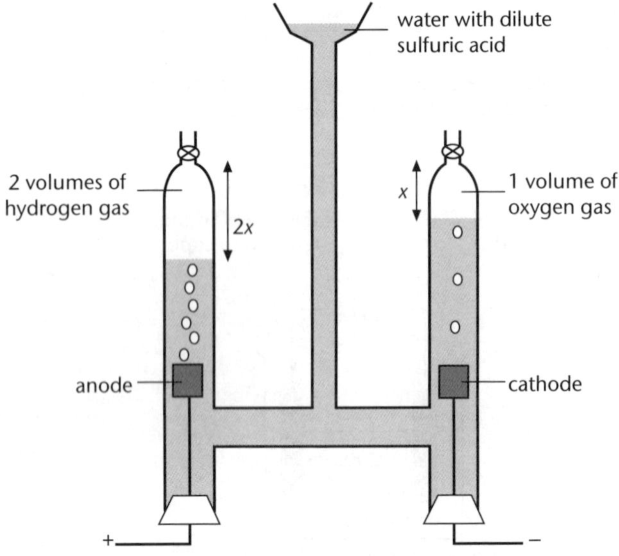

Note: Dilute sulfuric acid is added because pure water does not contain sufficient ions to allow the current to flow.

The *concentration* of ions also affects the reduction and oxidation reactions that occur.

Electrolysis in industry

Most metals in the Earth's crust are found as metal compounds called **ores**. The pure metal is *extracted* from an ore by reducing the ore to the metal.

The cheapest method of extracting a metal is to reduce the ore by heating it with a reductant such as **coke**, a form of carbon. This method of reacting ores with a reducing agent such as carbon (or carbon monoxide) does not work for metals high on the **activity series** (such as aluminium, magnesium and sodium). These metals must be extracted from their ores using electrolysis, a much more costly process compared to reduction using coke. The ore is usually concentrated and purified prior to reduction at the cathode of an electrolytic cell.

Electrolysis is also used to purify metals and to *electroplate* metal objects.

Extraction of aluminium

Aluminium extraction can be broken down into three steps:

- Mining of the aluminium ore.
- Purification of the ore.
- Electrolysis or **smelting** of the ore.

Mining aluminium ore

The ore that contains aluminium is known as **bauxite**. It is a red-coloured clay containing about 20% aluminium oxide, Al_2O_3. The remaining 80% consists of impurities of iron oxide, Fe_2O_3, **silica**, SiO_2, and clay. The aluminium oxide, or **alumina**, is purified in Australia.

Ore purification

The bauxite is first ground and then heated with caustic soda (sodium hydroxide) solution, which dissolves only the aluminium oxide and the silica impurity. The reaction between aluminium oxide and caustic soda is:

$$Al_2O_3(s) + 2NaOH(aq) + 3H_2O(\ell) \rightarrow \underset{\text{sodium aluminate}}{2NaAl(OH)_4(aq)}$$

Iron oxide does *not* dissolve in caustic soda. It forms a sludge, which settles to the bottom and is removed by **filtration**.

The solution containing sodium aluminate and dissolved silica is cooled and small **seed crystals** of aluminium hydroxide are added. Aluminium hydroxide forms, leaving the silica impurities in solution:

$$Na^+(aq) + Al(OH)_4^-(aq) \rightarrow Al(OH)_3(s) + Na^+(aq) + OH^-(aq)$$

The solid aluminium hydroxide is washed and then heated to form 99.5% pure alumina:

$$2Al(OH)_3(s) \xrightarrow{\text{heat}} Al_2O_3(s) + 3H_2O(\ell)$$

Alumina is shipped from Australia to smelters such as that at Tiwai Point, New Zealand.

Tiwai Point is a suitable place for an aluminium smelter because:

- it provides a deep-water harbour for shipping raw materials in and exports out.
- low-cost bulk electricity is available (principally produced from the Manapouri hydroelectric power station).
- it is associated with the developed infrastructure present in Invercargill.

Tiwai Point aluminium smelter in New Zealand

The electrolysis process

Step 1: Alumina is dissolved in **cryolite**, Na_3AlF_6 – this reduces the melting point of alumina from 1200°C (for pure alumina) to 970°C (mixture), which saves considerable amounts of energy. The mixture of alumina and cryolite – the electrolyte – contains the Al^{3+} ions that are reduced to aluminium during electrolysis.

Step 2: Electrolysis takes place in large steel *pots* using carbon electrodes. A lining of carbon at the bottom of the steel pot forms the cathode; carbon anodes are suspended in the molten cryolite mixture.

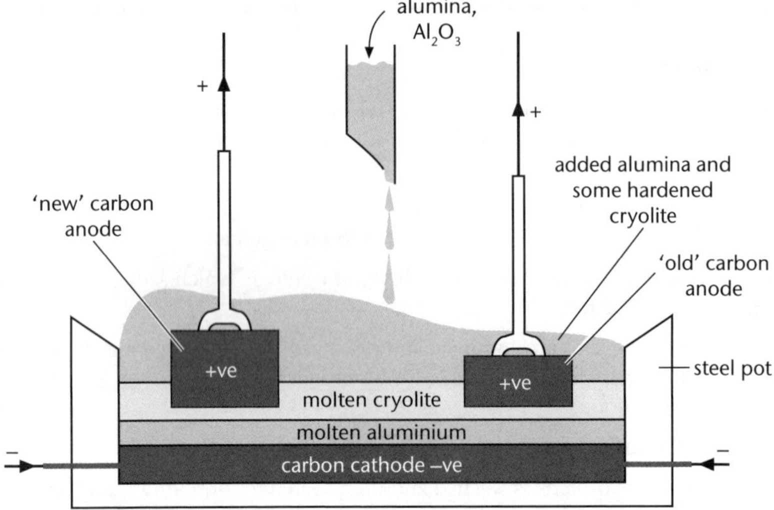

Alumina is added periodically from the hopper as alumina in contact with the molten cryolite dissolves, electrolyses and forms aluminium metal.

Some of the molten cryolite solidifies and forms a 'crust' with some of the added alumina. This crust is broken to get at the molten aluminium that forms.

Both the cathode and the anodes (there are two rows of nine) are made of carbon. The anodes react with oxygen produced from the alumina, so they have to be continually replaced.

Cathode reaction

Aluminium ions are reduced to molten aluminium metal:

$$Al^{3+}(\ell) + 3e^- \rightarrow Al(\ell)$$

The molten aluminium produced sinks to the bottom of the pot where it is removed periodically by siphoning it out.

Anode reaction

At the anode, oxide ions from the alumina are oxidised to form oxygen gas:

$$2O^{2-}(\ell) \rightarrow O_2(g) + 4e^-$$

The passage of an electric current causes oxygen to react with the carbon of the anode to produce carbon dioxide:

$$C(s) + O_2(g) \rightarrow CO_2(g)$$

As the anodes 'wear away' from the reaction above, they are replaced periodically.

Note: Carbon dioxide is the main pollutant produced. Others include carbon monoxide and sulfur dioxide. 'Scrubbers' remove 99% of pollutants prior to waste gases being released.

Electroplating

Electroplating involves the use of electrolysis to coat or *plate* one metal onto another.

Electroplating is used to:

- prevent corrosion of metals, eg tin plated on iron.
- improve appearances of metals, eg silver plating on jewellery and cutlery.

Example D

Chrome plating

Chrome plating involves plating a thin layer of chromium metal onto steel. This makes an object appear shiny and prevents the steel from rusting.

An object to be plated is thoroughly cleaned and then becomes the cathode of the electrolytic cell.

Copper plating

Copper plating involves making the object the cathode in a solution of copper sulfate.

The cathode reaction is:

$Cu^{2+}(aq)$	+	$2e^-$	→	$Cu(s)$
(from copper sulfate)		(supplied by cathode)		(deposits on object)

Dilute sulfuric acid is added to the copper sulfate to help to achieve an even coating of copper.

The anode is a piece of copper, which slowly goes into solution as $Cu^{2+}(aq)$ as the electrolysis proceeds.

The anode reaction is:

$Cu(s) \rightarrow Cu^{2+}(aq) + 2e^-$

(the anode)

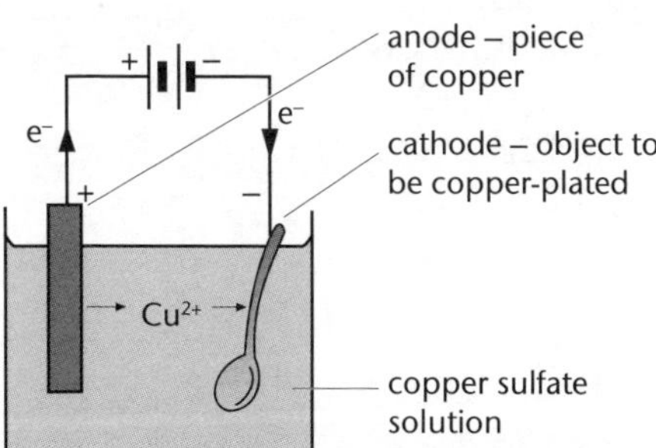

The $Cu^{2+}(aq)$ ions replace those lost from the copper sulfate solution. In effect, the copper from the anode plates onto the object at the cathode.

Unit 12.3 Activity 3A: Electrolysis

1. Where in an electrolytic cell do the following occur?
   a. Oxidation
   b. Reduction
2. For the electrolysis of molten magnesium chloride, write the reactions occurring at the negative electrode and positive electrode.
3. Some molten lead iodide is electrolysed.
   a. Sketch a suitable electrolytic cell for the electrolysis and label the electrodes as cathode, anode, positive and negative.
   b. Write half-equations for the reaction at each electrode.
4. The extraction of aluminium is carried out by the electrolysis of molten aluminium oxide (alumina) dissolved in cryolite (sodium aluminium fluoride).
   a. Write a balanced half-equation for the reaction at the cathode.
   b. Write a balanced half-equation for the reaction at the anode.
   c. Combine the two half-equations from **a.** and **b.** to give a balanced equation for the overall reaction.
5. A piece of filter paper was soaked with dilute sulfuric acid and connected to a DC power supply as shown.

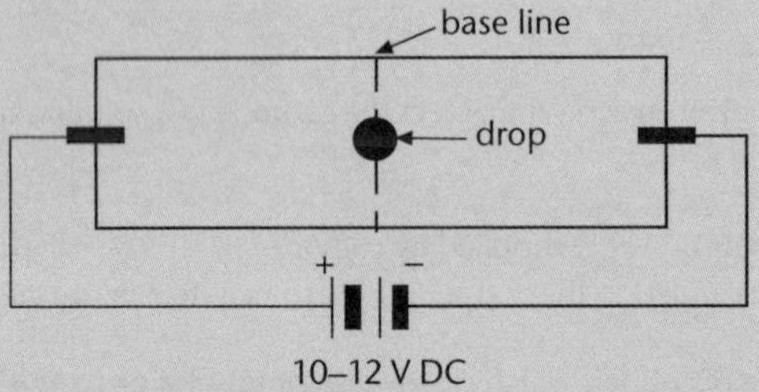

   a. Predict and explain what will happen when one drop of the following solutions is placed on the base line.
      i. Potassium permanganate, $KMnO_4$.
      ii. Copper(II) sulfate, $CuSO_4$.
      iii. Potassium dichromate, $K_2Cr_2O_7$.
      iv. Chromium nitrate, $Cr(NO_3)_3$.
   b. What is the purpose of the dilute sulfuric acid?

6. Sodium metal is produced by the electrolysis of sodium chloride in a *Downs cell*.

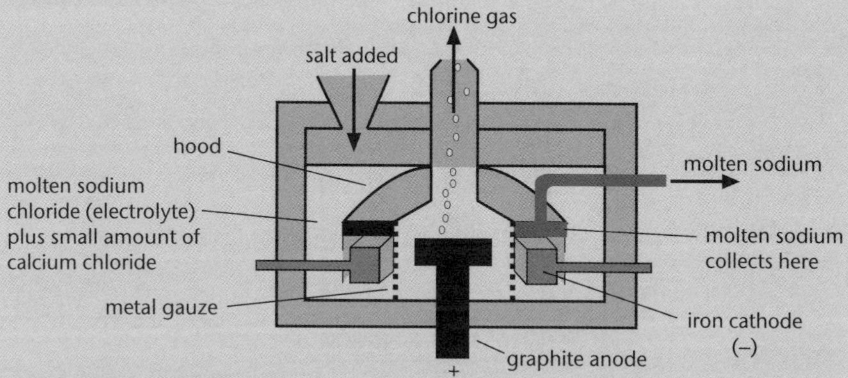

Downs cell

   a. Which ions migrate to the iron cathode?
   b. Give an equation for the reduction process.
   c. Which ions migrate to the graphite anode?
   d. Give an equation for the oxidation process.
   e. Give the overall equation for the electrolysis reaction.

7. The diagram shows the electrolysis of an aqueous solution of copper chloride.

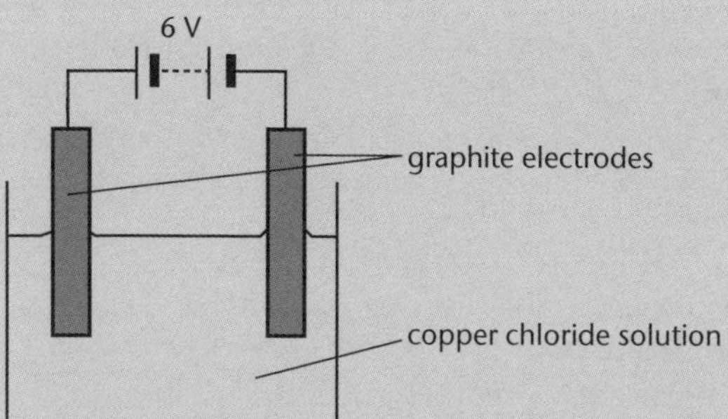

During this electrolysis, a student noticed that copper metal was deposited at the cathode and chlorine gas was given off at the anode.

   a. Write the half-equation for the reaction that occurs at the:
      i. cathode.
      ii. anode.
   b. Write a paragraph discussing how copper is produced at the cathode and chlorine is produced at the anode. Use your knowledge of the principles of electrolysis and refer to the diagram. Use the following key words in your answer: anode, cathode, electrons, anions, cations, negative, positive, oxidation, reduction.

8. Aluminium is a metal that has many uses. It is obtained from aluminium oxide by electrolysis. A simplified diagram of the cell used to produce aluminium is shown:

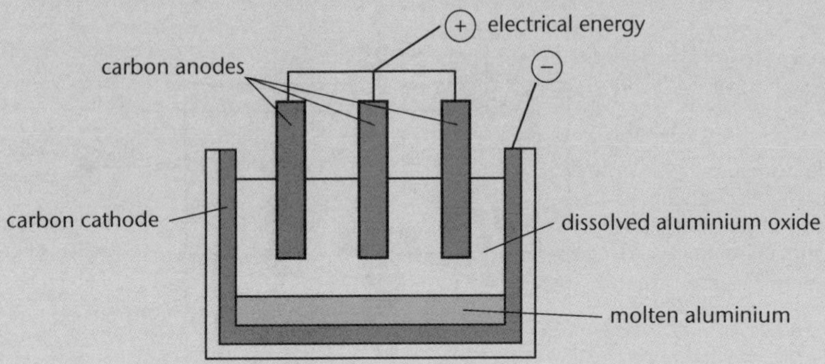

a. Describe the movement of ions when the electric current is applied in the cell.

b. Write a balanced half-equation to show the reaction occurring at the:

i. cathode.

ii. anode.

9. One way to have a pollution-free car is to use hydrogen as the fuel, since the combustion of hydrogen forms water vapour only (along with the energy produced, of course).

Discuss how hydrogen can be made by the electrolysis of water using inert (unreactive) electrodes such as carbon or platinum. In your answer, refer to any other product that may form, the electrodes (anode and cathode), as well as relevant oxidation – reduction half-equations.

10. Copper is extracted from its ore by reduction. For many purposes (eg electrical wiring), copper is required to be of a high state of purity. Such copper is produced by electrolysis.

The anode is made of impure copper and the cathode of pure copper. The electrolyte is copper sulfate solution. During electrolysis, the anode dissolves and pure copper is deposited on the cathode.

a. Copy the diagram then identify the:

i. anode and cathode.

ii. electrode where oxidation takes place.

iii. electrode where reduction takes place.

iv. direction of Cu^{2+} ion movement during electrolysis.

b. Write the balanced half-equations for the processes at the anode and cathode.

Unit 12.4 Carbon Compounds

Topic 1: Properties and uses of carbon

Topic 1 covers the properties and uses of carbon:

- State and appearance at room temperature.
- Allotropes of carbon – structure, physical properties and uses.
- Reaction with oxygen.
- Carbon cycle.
- Carbon as fuel.
- Carbon as an industrial reducing agent.

Properties and reactions of carbon

At room temperature, carbon is a solid. There are three **allotropes** of carbon – **graphite**, **diamond** and **buckminsterfullerene** (buckyball) – which have quite different appearances and properties.

Example A

Graphite has the unusual property for a non-metal of being a very good conductor of electricity; diamond does not conduct electricity.

The two most common allotropes are graphite and diamond. Graphite is the most stable form of carbon.

Properties of carbon

Property	Allotrope of carbon	
	Graphite	Diamond
Physical state at room temperature	Solid; very high melting point	Solid; very high melting point
Appearance	Grey-black; shiny	Clear and dull unless faces cut to reflect light
Hardness	Very soft; used in pencils to leave a trace on paper	Very hard; hardest naturally occurring substance
Electrical conductivity	Very good	Nil
Solubility in water	Insoluble	Insoluble
Reaction with oxygen	Both allotropes: • Produce carbon dioxide when heated in plenty of oxygen: $C(s) + O_2(g) \rightarrow CO_2(g)$ • Produce carbon monoxide when heated in limited supply of oxygen: $2C(s) + O_2(g) \rightarrow 2CO(g)$	

Structures and uses of carbon allotropes

The carbon atom has four electrons available to form covalent bonds.

Diamond

In **diamond**, each carbon atom is covalently bonded to four other carbon atoms. This arrangement is continuous, ie a diamond crystal is an extended arrangement of carbon atoms in three dimensions.

The very strong, rigid, 3-D network of carbon atoms in diamond makes it the hardest naturally occurring substance.

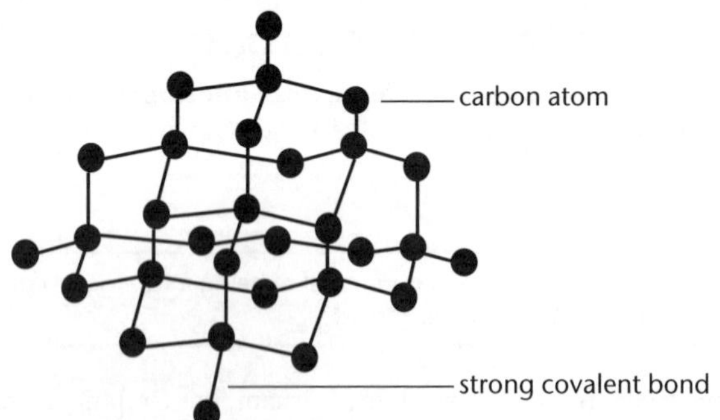

The arrangement of atoms in diamond

Diamond is used:

- For jewellery – the hardness of diamond allows **faces** to be cut into the surface.

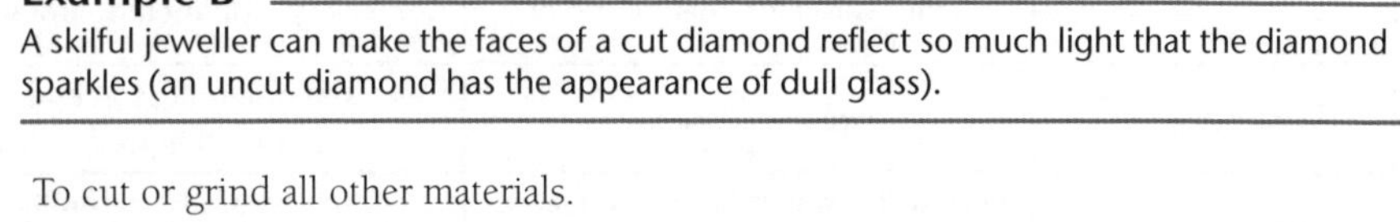

Example B

A skilful jeweller can make the faces of a cut diamond reflect so much light that the diamond sparkles (an uncut diamond has the appearance of dull glass).

- To cut or grind all other materials.

Example C

Diamond is used to cut glass and as the tips of drills used in mining.

Graphite

In graphite, only three electrons per carbon atom are involved in covalent bonding. The fourth electron of each carbon atom bonds very weakly with carbon atoms in layers above and below the layer containing the carbon atom under consideration. The layers can slide over each other (slipperiness) and the loose electron is able to flow through the graphite as an electric charge. Graphite is a good conductor of electricity.

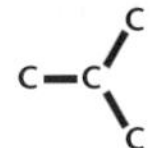

Structure showing one carbon atom joined to three others, with all atoms in the same plane – produces a layer structure

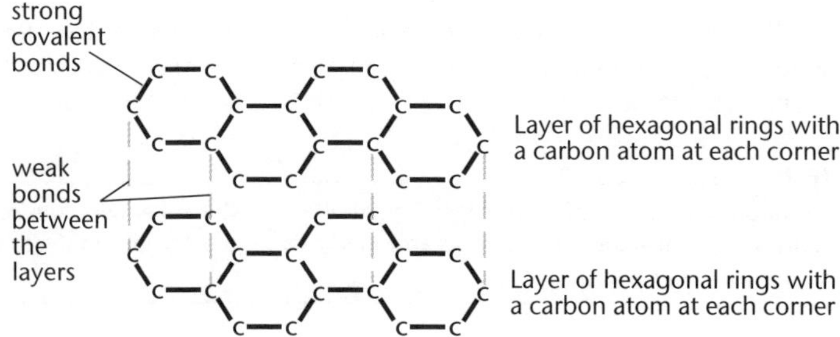

A piece of natural graphite is dark grey-black, it shines and feels very slippery.

The arrangement of carbon atoms in graphite

The structure of graphite explains why graphite:

- Has a slippery texture – the layers can slide over each other. Graphite mixed with clay is used in 'lead' pencils.

Example D

A 7B pencil contains pure graphite and is very soft. A 7H pencil has a large percentage of clay and very little graphite, and is very hard. Between these two extremes, pencils have a range of hardness, with 'HB' in the middle.

- Is a good conductor of electricity – the loose electron is able to flow through the graphite under an electrical potential. Graphite is used as an electrical conductor in electrolysis and electrochemical cells.

Example E

Graphite is used as the electrodes in the commercial cells to produce aluminium metal.

- Has a very high melting point. The multiple covalent bonding in the layers of graphite requires a large amount of energy for the network to be broken.

Example F

The cell temperature to produce aluminium metal is 1000°C but the melting point of graphite used for the electrodes is nearly 4000°C.

- Is stable to most chemicals.

Example G

Magnesium metal is produced commercially by the electrolysis of molten magnesium chloride. The chlorine that is produced at the graphite anode, although very hot and therefore extremely active, does not react with the graphite anode.

Buckminsterfullerene

Buckminsterfullerene has a molecular formula of C_{60} and was discovered in 1985. The molecule is hollow and nearly spherical, with 12 pentagonal (five-sided) faces and 20 hexagonal (six-sided) faces.

Example H

Buckminsterfullerene is named after architect Richard Buckminster Fuller, who designed geodesic domes (light structural frameworks arranged as a set of polygons in the form of a shell).

Because of the likeness to the shape and form of a soccer ball, buckminsterfullerene is commonly referred to as 'soccerene' or 'footballene', or as a 'buckyball'.

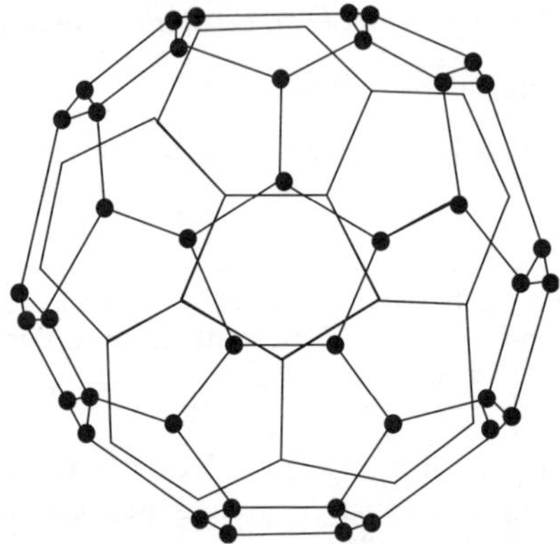

Structure of buckminsterfullerene

Research is developing the use of buckminsterfullerene and related compounds as:

- **Superconductors**.
- Non-metallic magnets.

Unit 12.4 Activity 1A: Physical properties of carbon

1. Explain the meaning of each of the following terms.
   a. Hard. b. Insoluble.
   c. Shiny. d. Melting point.
2. Describe the appearance at room temperature of:
   a. diamond. b. graphite.
3. Explain why:
   a. diamond is very hard.
   b. graphite is very soft and slippery.
4. State three uses of carbon based on the physical properties of carbon. Explain the use in terms of the properties of the material.
5. The terms 'allotrope' and 'isotope' are often confused. Define both these terms and give examples of each, using the element carbon.

Carbon bonding

Carbon atoms can form strong covalent bonds with other carbon atoms. The consequences of this are:

- A chain of carbon atoms can be formed, of very great length and strength, using two of the four possible covalent bonds per carbon atom. Molecules with long chains of carbon atoms form the basic structure of all living species.
- Each carbon atom in the chain has two more bonds it can form with other atoms.
- Many different elements can have atoms attached to the chain, producing a huge variety of large molecules resulting in a colossal variety of species in the living world.

The carbon cycle

Carbon is continually being recycled in the living world – this is known as the **carbon cycle**.

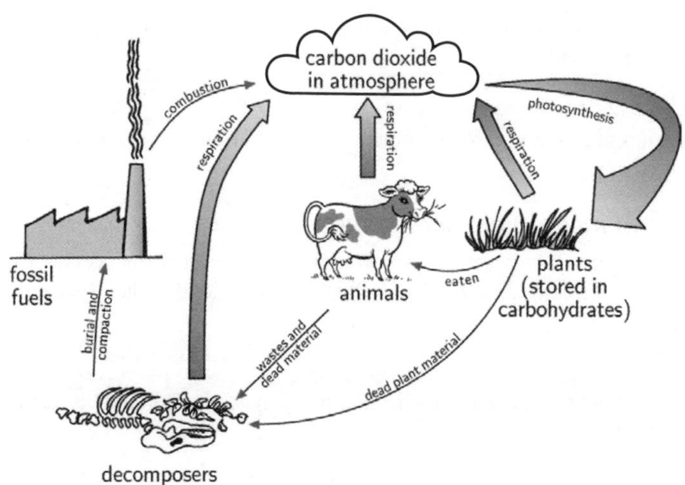

The carbon cycle

There are two main processes involved in the carbon cycle – photosynthesis and respiration.

- **Photosynthesis** requires sunlight and the presence of chlorophyll in the leaves of plants to produce glucose and oxygen from carbon dioxide and water, ie:
  carbon dioxide + water → glucose + oxygen.
- Photosynthesis is an **endothermic** process – it is the *energy-absorbing* (from the sun) process of the living world.
- **Respiration** is an **exothermic** process – it is the *energy-supplying* process for the living world, ie:
  glucose + oxygen → carbon dioxide + water.

Photosynthesis and respiration are the reverse of each other – the two processes work together to keep the amount of carbon dioxide in the atmosphere at about 0.03%.

Uses of carbon

Carbon as a fuel

Carbon is contained in **fuels** such as coal, coke, petroleum and natural gas.

- *Coal* is formed from dead vegetation trapped in swamps, then covered with earth and subjected to intense heat and pressure over millions of years. Current estimates of coal reserves are 1 trillion tonnes (1 000 000 000 000 tonnes), which are predicted to last for approximately 235 years at current usage. Coal can be processed into **coke** – a purer form of carbon from which gases and oils in coal have been removed.
- *Petroleum* and *natural gas* are derived from millions of sea organisms (eg shellfish) that, on dying, fell to the seabed and were covered with silt from rivers. Subjected to intense pressure and heat over millions of years, the products are a thick black liquid and gases that are contained below the Earth's surface by a hard, impervious **cap rock**.
  Petroleum reserves are estimated at 132 300 million tonnes. The world is currently using 3252 million tonnes per year – it is estimated that the petroleum reserves will last 43 years and natural gas reserves will last 66 years.

When fuels containing carbon are burnt, an exothermic chemical reaction occurs, producing carbon dioxide:

$C(s) + O_2(g) \rightarrow CO_2(g)$

Billions of tonnes of coal and coke are burnt every year commercially and domestically to produce heat energy.

Complete and incomplete combustion

If insufficient oxygen is available, the carbon dioxide will react with excess hot carbon to produce poisonous carbon monoxide, $CO(g)$. This reaction is called **incomplete combustion**, ie:

$CO_2(g) + C(s) \rightarrow 2CO(g)$

Some of the heat energy released in the formation of the carbon dioxide is lost in this endothermic reaction.

If more oxygen becomes available, the carbon monoxide formed will burn and carbon dioxide will be produced, again with heat energy being released:

$2CO(g) + O_2(g) \rightarrow 2CO_2(g)$

This reaction is called **complete combustion**.

Example I

The charcoal brazier

In a **charcoal** brazier the combustion of carbon proceeds through various stages, depending on the amount of oxygen available. Starting from the bottom, these stages are:

- formation of carbon dioxide
- conversion of carbon dioxide to carbon monoxide
- conversion of carbon monoxide to carbon dioxide
- release of carbon dioxide to the atmosphere.

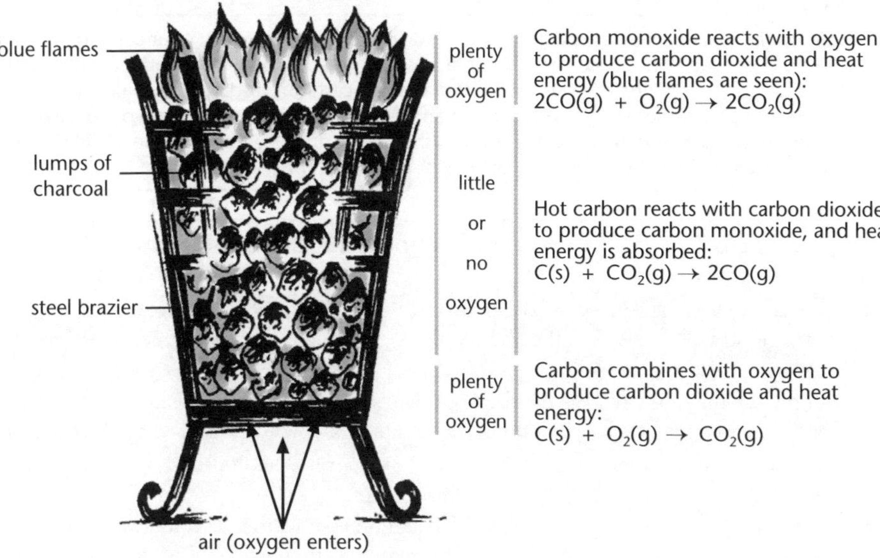

Combustion in a charcoal brazier

Carbon as a reducing agent

Some metals are produced by reduction (the removal of oxygen from a compound) of their oxides. In these reactions, carbon is used as the **reducing agent**.

Example J

Reduction of metal oxides

Copper(II) oxide: $CuO(s) + C(s) \rightarrow Cu(s) + CO(g)$, or
$2CuO(s) + C(s) \rightarrow 2Cu(s) + CO_2(g)$

Lead oxide: $PbO(s) + C(s) \rightarrow Pb(s) + CO(g)$, or
$2PbO(s) + C(s) \rightarrow 2Pb(s) + CO_2(g)$

Iron oxide: $Fe_3O_4(s) + 2C(s) \rightarrow 3Fe(s) + 2CO_2(g)$

Zinc oxide: $2ZnO(s) + C(s) \rightarrow 2Zn(s) + CO_2(g)$

Metal carbides

Carbon can combine with metals, especially iron, to produce carbides.

Example K

Carbon (as coal or coke) is used in the extraction of iron from its oxide. After cooling, the solid product is not pure iron – it contains some iron carbide. If the product contains 2.5–5% carbon, it is a **brittle** but hard material called cast iron. Today, almost all iron produced contains between 0.5% and 2% carbon and is called **steel**. Other types of steel can be produced by mixing different metals with the iron to produce, for example, stainless steel (contains chromium and nickel) and acid-resistant steel (contains silicon).

Unit 12.4 Activity 1B: Carbon in the living world, as a fuel and as a reducing agent

1. For the element carbon, name two sources of the element of commercial value and give some indication of the abundance of the source on planet Earth, eg large, moderate or rare.
2. Supply the missing words to describe the following process (words may be used more than once in the answer – the number of letters in the words is shown by the dashes):
   Coal heated in the absence of air produces _ _ _ _, gases and oils. _ _ _ _ is a purer form of the element _ _ _ _ _ _ than coal.
3. Complete the following equations. If the equation is in words, then use words to complete the equation. If the equation is in symbols, then use symbols to complete the equation. The number of letters in the words or symbols is shown by the dashes.
   a. Carbon + oxygen → _ _ _ _ _ _ _ _ _ _ _ _ _ _ (*two words*).
   b. Carbon dioxide + water → _ _ _ _ _ _ _ + _ _ _ _ _ _ (photosynthesis).
   c. _ZnO(s) + C(s) → _Zn(s)+ _ _ _(g).
4. Explain the meaning of the underlined words in the following passage.
   Carbon can be used as a <u>fuel</u>. It will combine with oxygen <u>exothermically</u> to produce carbon dioxide. <u>Coal</u> is a mineral (naturally occurring substance that is mined) that is mainly carbon. Other minerals of carbon that are fuels are <u>petroleum</u> and natural gas.

Unit 12.4 Carbon Compounds

Topic 2: Properties and reactions of carbon compounds

Topic 2 covers the properties and reactions of carbon compounds:

- Oxides of carbon.
- Properties of carbon dioxide – density, solubility in water, the acidic nature of its aqueous solution, inability to support combustion, reaction with limewater.
- Uses of carbon dioxide related to properties.
- Laboratory preparation of carbon dioxide.
- The impact of carbon and its combustion products on human health and the environment, eg global warming.

Properties of carbon dioxide

Carbon dioxide is a non-poisonous gas that exists in the atmosphere and is vital to the living world.

Colour	Smell	Taste	Density	Solubility in water	Acidity	Combustion
None	None	Experienced when aerated (fizzy) drinks are consumed – tingling effect on tongue as bubbles burst.	Greater than air (1.5 times)	Low	Weak	Does not burn; only supports combustion of fiercely burning materals, eg magnesium.

Laboratory preparation of carbon dioxide

Carbon dioxide can be prepared in the laboratory in two ways:

- By the reaction of any metal carbonate or hydrogen carbonate with any acid.

Example A

The equation for the reaction of calcium carbonate in the form of marble chips with hydrochloric acid is:

$CaCO_3(s) + 2HCl(aq) \rightarrow CaCl_2(aq) + H_2O(\ell) + CO_2(g)$

This is the preferred preparation because the flow of the carbon dioxide gas can be easily controlled by regulating the concentration of the acid. Several jars of carbon dioxide can readily be produced in the apparatus.

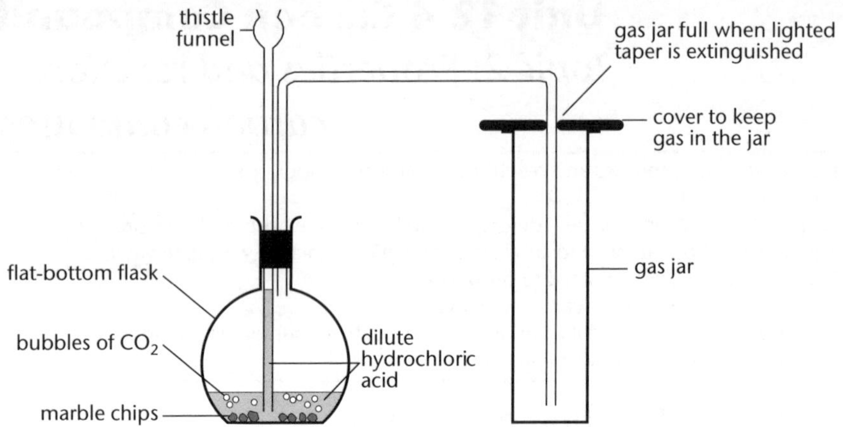

Laboratory preparation of carbon dioxide. The carbon dioxide gas is collected by upward displacement of air.

- By heating any metal carbonate (except sodium and potassium carbonates) or any metal hydrogen carbonate.

Identification of carbon dioxide (limewater test)

Limewater is a **saturated solution** of calcium hydroxide. When carbon dioxide is passed through limewater, calcium carbonate ($CaCO_3$) is precipitated as a white solid. This gives the limewater a milky appearance. No other gas has this effect on limewater and thus this test is used for the identification of carbon dioxide:

$Ca(OH)_2(aq) + CO_2(g) \rightarrow CaCO_3(s) + H_2O(\ell)$

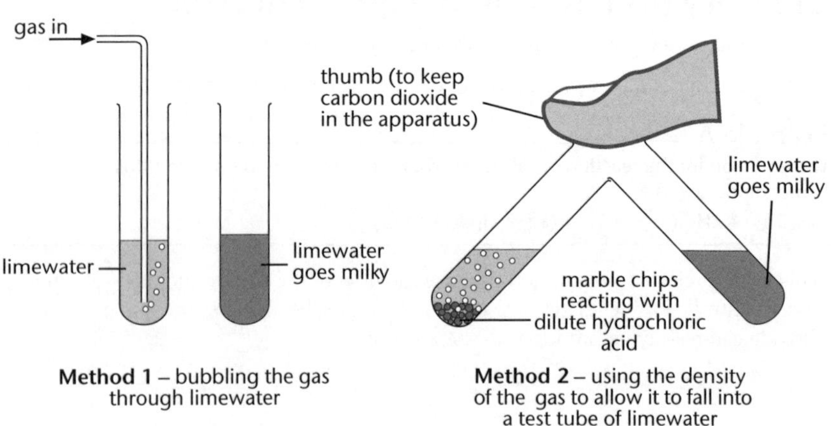

Testing for carbon dioxide by two different methods

A continuous supply of the carbon dioxide gas converts the milkiness (due to insoluble calcium carbonate being formed) into a clear solution (due to the formation of soluble calcium hydrogen carbonate):

$CaCO_3(s) + H_2O(\ell) + CO_2(g) \rightarrow Ca(HCO_3)_2(aq)$

Unit 12.4 Activity 2A: Properties of carbon dioxide

1. State:
   a. three physical properties of carbon dioxide gas.
   b. two chemical properties of carbon dioxide gas.
2. A gas jar is filled with carbon dioxide gas. On top of this jar, and upside down, a gas jar of air is placed.

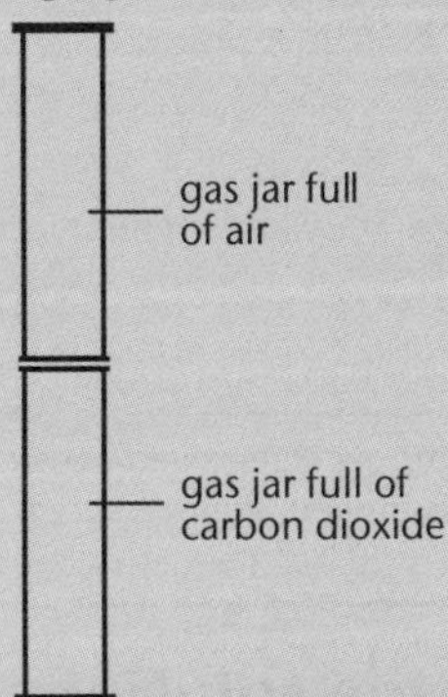

   a. What would be observed on separating the gas jars and adding a few millilitres of limewater to each jar:
      i. a few seconds after the gas jars had been joined?
      ii. one hour after the gas jars had been joined?
   b. Explain the observations.
3. Use the following word list of chemicals to complete parts **a.–d.**
   Word list: copper carbonate, sodium carbonate, sodium hydrogen carbonate, solid tartaric acid.
   a. Choose the chemicals that would react with hydrochloric acid to produce carbon dioxide gas.
   b. Choose the chemical that would react with dilute hydrochloric acid to produce a blue-green solution. Name the product that would make the solution blue-green.
   c. Baking soda is a white commercial product that contains one chemical that, when heated, produces carbon dioxide gas. Identify which chemical listed is sold commercially as baking soda.
   d. Baking powder is a commercial product that is a mixture of baking soda and another dry solid. When water is added to baking powder, carbon dioxide gas is released. Identify which chemical listed is present in baking powder as the other dry solid.

4. a. Design an apparatus to show that when sodium hydrogen carbonate solid is heated, a gas is released that will turn limewater milky.

b. State what would be observed if the gas were passed through the limewater for a long period. Explain this observation.

5. Complete the following equations.

a. Calcium carbonate → calcium oxide + ___________

b. ___________ → sodium carbonate + carbon dioxide + water

c. $MgCO_3(s) \rightarrow$ _____ $+ CO_2(g)$

d. $Na_2CO_3(s) + 2HCl(aq) \rightarrow 2$_____ $+ H_2O(\ell) + CO_2(g)$

e. _________ $+ HCl(aq) \rightarrow NaCl(aq) + H_2O(\ell) + CO_2(g)$

f. $Ca(OH)_2(aq) + CO_2(g) \rightarrow$ ________ + _____

6. If you entered a room full of carbon dioxide, you would die. Explain why.

Solubility of carbon dioxide in water

Carbon dioxide has low **solubility** in water. The solubility of a gas in water is greatly increased by raising the pressure on the gas.

Example B

'Fizzy' drinks (eg lemonade, champagne) **effervesce** when the stopper is removed because the carbon dioxide in the liquid is under pressure. Once the seal is broken, the pressure is released and the gas escapes rapidly to the atmosphere.

Acidic nature of aqueous carbon dioxide solution

An **aqueous solution** is a solute dissolved in the solvent water. Symbols are used to make the following distinctions.

- $CO_2(g)$ represents carbon dioxide gas.
- $CO_2(aq)$ represents carbon dioxide gas dissolved in water.

The two representations are related as shown in the formula equation:

$CO_2(g) + H_2O(\ell) \rightarrow CO_2(aq)$

and by the word equation:

carbon dioxide gas added to *water* produces an *aqueous solution of carbon dioxide*.

The acidity of carbon dioxide gas can be shown by:

- The gas turning a piece of damp blue litmus paper pink.
- Inverting a test tube full of carbon dioxide gas over a beaker of sodium hydroxide solution – the test tube fills rapidly with the solution because the gas reacts rapidly with alkali.

When carbon dioxide gas is in contact with water, two distinct processes occur:

- **Physical dissolving** of carbon dioxide in water, ie:
  $CO_2(g) + H_2O(\ell) \rightarrow CO_2(aq)$
  This process is reversible.

Example C

When bubbles of carbon dioxide gas are being released from a bottle of lemonade, the process is represented by the equation:

$CO_2(aq) \rightarrow CO_2(g) + H_2O(\ell)$

A fizzy drink open to the air for some hours has a quite different taste from a fizzy drink from a freshly opened bottle.

- As the acidic carbon dioxide dissolves in the water, it also reacts with the water, ie:
  $CO_2(g) + H_2O(\ell) \rightarrow H_2CO_3(aq)$
  The product is called carbonic acid.

An aqueous solution of carbon dioxide is acidic.

Uses of carbon dioxide

The major uses for carbon dioxide are:

- To aerate 'fizzy' drinks. The gas has low solubility in water at room temperature and is much more soluble in water under pressure.
- To extinguish fires. Carbon dioxide is a stable compound, is not decomposed by heat and its high density allows it to 'sit' on the fire and exclude air (oxygen). Common fire extinguishers are cylinders of compressed carbon dioxide.
- As a stage effect (**dry ice**). Carbon dioxide **sublimes** – it can go directly from a gas to a solid and from a solid to a gas. Dry ice is solid carbon dioxide formed by compressing carbon dioxide gas in a cylinder – this process is called **sublimation**. Carbon dioxide released from a cylinder is colder than room temperature because the escaping and expanding gas uses energy for the expansion – hence its temperature lowers.

Example D

If a stage scene requires smoke, mist or fog, a cylinder of carbon dioxide is opened on the side of the stage and the gas drifts onto the stage. Because the gas is denser than air, it clings to the floor of the stage, and because it is colder than the air on the stage, water **vapour** in the air condenses, giving the stage the desired appearance.

Unit 12.4 Activity 2B: Solubility in water and uses of carbon dioxide

1. State three uses of carbon dioxide gas. For each use, name the property or properties of carbon dioxide that make the gas useful.

2. State what you would observe if a stoppered test tube containing carbon dioxide gas were turned upside down above a candle flame and the stopper was then removed. Explain the observation.

3. Explain why:

   a. an unopened bottle of lemonade effervesces (bubbles) when the stopper is removed.

   b. a bottle of champagne on the winner's podium erupts when the bottle is shaken.

   c. a can of lemonade that has been open for an hour or so has a 'flat' taste.

4. a. Write the symbol for carbon dioxide gas.
   b. Write the symbol for carbon dioxide gas dissolved in water.
   c. Using the symbols you have provided in **a.** and **b.** of this question, write an equation that illustrates the release of gas from a fizzy drink.
5. Describe one experiment that indicates that carbon dioxide is an acidic gas. Assume a test tube of carbon dioxide gas is available.

Impact of carbon dioxide on the environment

Carbon dioxide dissolves in water – thus rainwater is a dilute solution of carbonic acid that will dissolve calcium carbonate (limestone) to produce soluble calcium hydrogen carbonate:

$CaCO_3(s) + H_2O(\ell) + CO_2(g) \rightarrow Ca(HCO_3)_2(aq)$

Over many years, natural erosion (dissolving) of limestone has created caves, caverns and underground rivers in limestone hills – the reverse process has created **stalagmites** and **stalactites**.

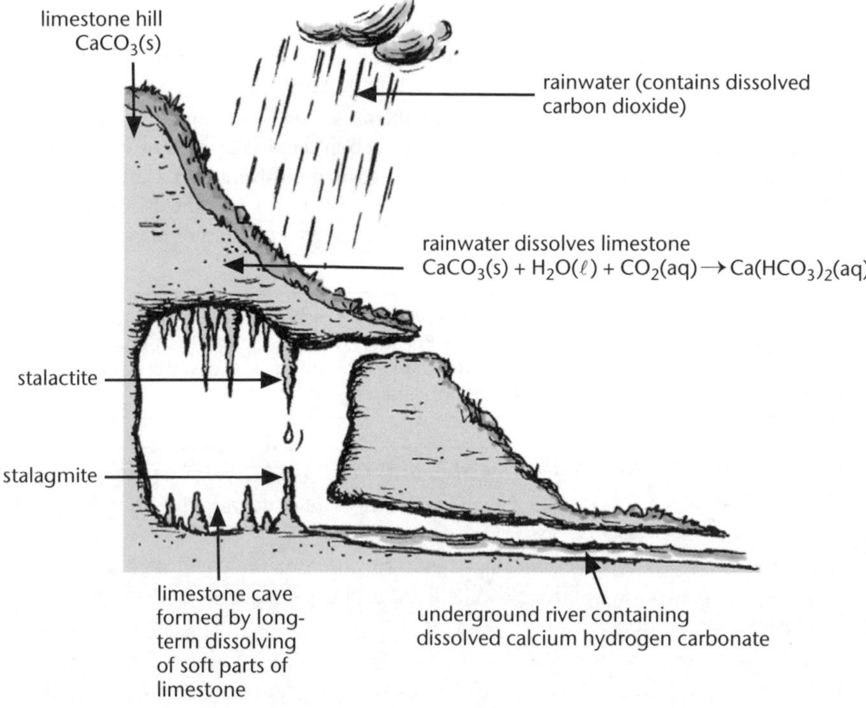

Rainwater evaporates in the warm cave and limestone collects – on the ceiling (stalactite) and on the ground (stalagmite), ie
$Ca(HCO_3)_2(aq) \rightarrow CaCO_3(s) + H_2O(\ell) + CO_2(aq)$

Limestone caves, stalactites and stalagmites, and underground rivers

Marble (a harder form of calcium carbonate than limestone) used for statues and other forms of decoration is also subject to slow attack by natural chemicals, such as carbonic acid, as components of acid rain.

Balance of carbon dioxide in the atmosphere

The amount of carbon dioxide permanently in the atmosphere is about 0.03%. This quantity has been kept constant by the following two processes:

- Photosynthesis by plants removes carbon dioxide from the atmosphere.
- Respiration by plants and animals adds carbon dioxide to the atmosphere.

In addition, all the water on the Earth's surface acts as a huge reservoir for dissolved carbon dioxide.

Human beings are disturbing the balance of carbon dioxide in the atmosphere by:

- burning fossil fuels, which produces carbon dioxide.
- removing forests, to create land for living and cultivation of crops, which reduces the amount of photosynthesis and hinders the removal of carbon dioxide from the atmosphere.

Greenhouse effect and global warming

Carbon dioxide and other gases in the atmosphere trap heat and help maintain Earth's temperatures within a habitable range. This is known as the **greenhouse effect**.

However, the amount of carbon dioxide in the atmosphere is increasing – this contributes to the **enhanced greenhouse effect**. This effect is due to carbon dioxide retaining heat energy better than the other components of the atmosphere – hence the atmosphere warms up (**global warming**) in the same way that a building made of glass (greenhouse) retains heat.

Other greenhouse gases may also contribute to global warming.

Example E

Methane released to the atmosphere by farm animals belching contributes to global warming.

The consequences of global warming are:

- Ice will melt in Earth's polar regions – water levels in the seas will rise and some low-lying islands and shorelines will disappear.
- The higher temperatures above Earth's surface will cause more water to evaporate from seas, lakes and rivers. This water will return to the rivers, lakes and seas of the world as rain, causing flooding and erosion of land.

Pollution from the internal combustion engine

The internal combustion engine uses **hydrocarbons** (eg petrol, diesel oil, kerosene), which burn in air. The engine is very inefficient, so the hydrocarbons do not undergo complete combustion to produce carbon dioxide and water as the only products of combustion. Some fuel escapes unburnt, as well as carbon monoxide and carbon (soot) as products of incomplete combustion.

Example F

Carbon monoxide as a poison to humans

Carbon monoxide is poisonous to the human system because it attaches to haemoglobin at the expense of oxygen:

- When air is inhaled, oxygen attaches to haemoglobin (the red pigment in blood cells) and is carried around the body in the blood. The attachment is loose and easily reversed so that oxygen can be made available at the muscle sites to combine with glucose to produce energy.
- Carbon monoxide also attaches to haemoglobin – more strongly than does oxygen, and irreversibly. Hence, acute poisoning occurs by oxygen denial.

Unit 12.4 Activity 2C: Carbon compounds in the environment

1. **a.** Identify what gases, other than carbon dioxide, are present in the atmosphere and can cause global warming.

b. In five sentences, explain the main features of global warming.

2. Explain why carbon monoxide is a poison to the human system.

3. State the percentage of carbon dioxide in the atmosphere. Explain how this percentage is kept constant.

4. Explain how stalagmites and stalactites are formed in limestone caves. Draw a diagram to help your explanation.

Unit 12.4 Carbon Compounds
Topic 3: Organic chemistry, alkanes and haloalkanes

Topic 3 covers the chemistry of alkanes and haloalkanes:
- Recognising alkanes and haloalkanes.
- Naming or drawing structural formulae of straight or branched chain hydrocarbons (up to eight carbon atoms) and haloalkanes (up to eight carbon atoms).

Introduction

Organic chemistry is the study of the compounds of carbon. All compounds containing carbon (except carbon monoxide, CO, carbon dioxide, CO_2, and the carbonates, CO_3^{2-}) are referred to as **organic compounds**. Organic compounds usually contain hydrogen and sometimes other atoms such as oxygen, nitrogen and chlorine.

Millions of different organic molecules are known, with thousands of new ones identified each year.

Organic molecules are found in:

- all living things – carbon atoms comprise about 20% (by mass) of all animals.
- **fossil fuels** such as coal, petroleum, oil and natural gas.
- common products such as plastics, adhesives, soaps, paper, cosmetics and medicine.

There are several reasons why there are so many organic molecules:

- Carbon atoms form strong covalent bonds with other carbon atoms, giving rise to long *chains* and *rings* of carbon atoms.

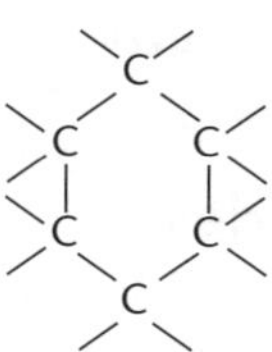

- Carbon atoms readily form single, double and triple covalent bonds, both with carbon atoms and with different atoms.

H—C≡C—H

- The same atoms can join together differently to produce **isomers**.

$$CH_3-CH_2-\underset{\underset{\displaystyle O}{\|}}{C}-H \qquad CH_3-\underset{\underset{\displaystyle O}{\|}}{C}-CH_3$$

The structure of organic compounds is based on a framework of carbon atoms, with hydrogen atoms usually completing the bonding requirements of the carbon atoms.

Hydrocarbons

The simplest organic compounds contain only hydrogen atoms and carbon atoms and are called **hydrocarbons**.

- **Alkanes** contain only C—C single bonds. They are said to be **saturated**, as no more hydrogen atoms can be added to them.
- **Alkenes** have one or more C=C double covalent bonds. They are said to be **unsaturated**, as hydrogen atoms can be added to them.
- **Alkynes** have one or more C≡C triple covalent bonds and are *unsaturated*.
- **Arenes** (or aromatic hydrocarbons) are **cyclic** hydrocarbons with alternating double and single bonds. The simplest arene is **benzene**, C_6H_6, which consists of a six-membered ring of carbon atoms with six attached hydrogen atoms.

Functional groups

Atoms or groups of atoms that influence the chemical behaviour of a compound are called **functional groups**.

Double or triple bonds are examples of functional groups.

Other functional groups include:

- –Cl, –Br, –I found in **haloalkanes**.
- –OH found in **alcohols**.
- $-\underset{\underset{\displaystyle O}{\|}}{C}-OH$ found in **carboxylic acids**.
- $-\underset{\underset{\displaystyle O}{\|}}{C}-O-$ found in **esters**.

Homologous series

A **homologous series** is a family of organic compounds that can be represented by a **general formula** in which each successive member of the series differs by a $-CH_2-$ unit.

Members of a homologous series have similiar methods of preparation and have similiar chemical properties.

Alkanes

Alkanes are an example of a homologous series. The general formula of the acyclic alkanes (those without a ring) is C_nH_{2n+2} or R–H, where **R** represents an **alkyl group**.

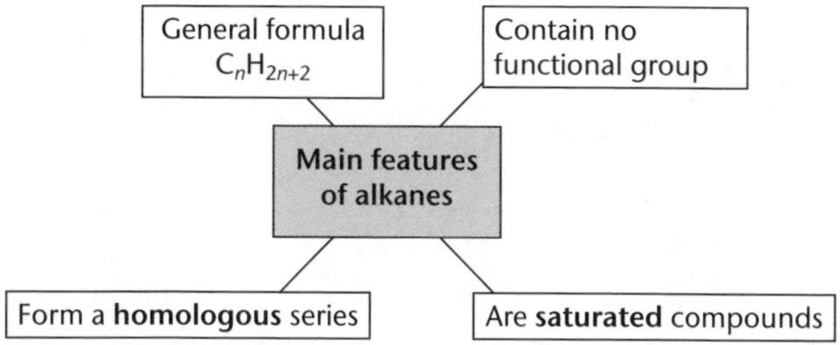

The alkane family builds by lengthening the carbon chain by one $-CH_2-$ group at a time:

- The first member of the alkane family, methane, contains only one carbon atom (n = 1). The number of hydrogen atoms in methane is $2n + 2 = 2 \times 1 + 2 = 4$, so the molecular formula is CH_4.
- The second member of the alkane family contains two carbon atoms (n = 2), so the number of hydrogen atoms is $2 \times 2 + 2 = 6$. This alkane is called ethane, and has the molecular formula C_2H_6.
- The third member of the family has three carbons (n = 3), so the number of hydrogen atoms is $2 \times 3 + 2 = 8$. This alkane is called propane, and has molecular formula C_3H_8.

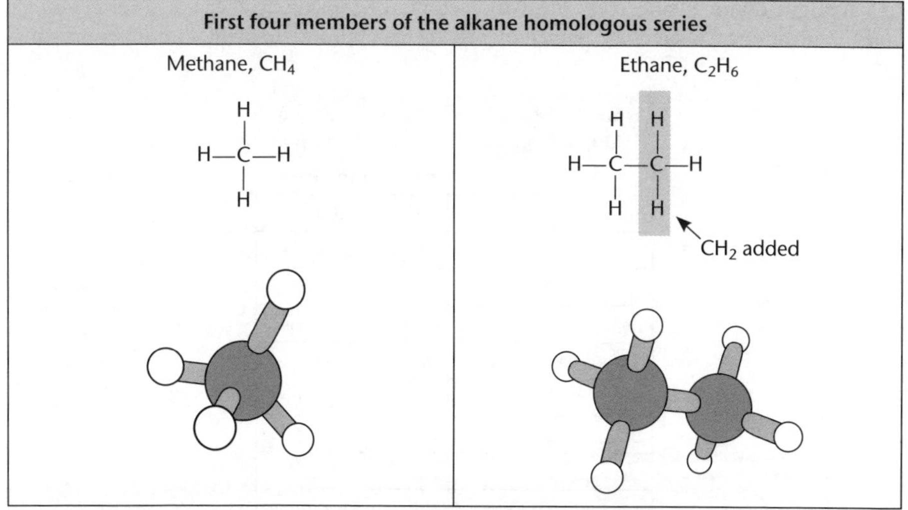

Propane, C_3H_8

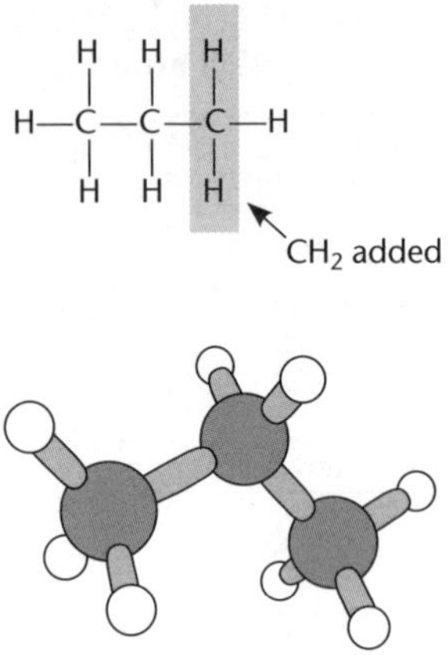

CH_2 added

Butane, C_4H_{10}

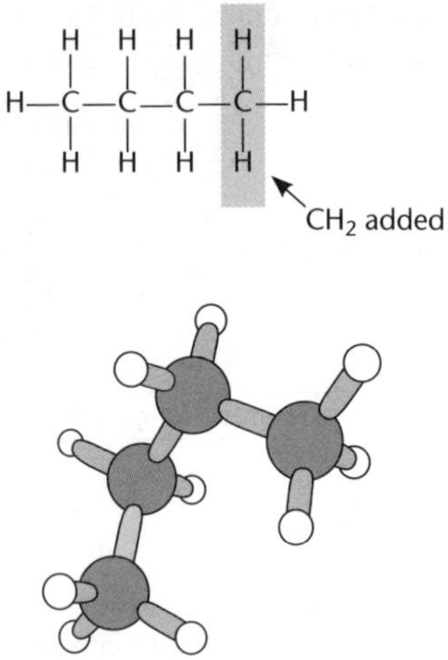

CH_2 added

Naming straight-chain alkanes

The naming of alkanes with four or fewer carbon atoms is based on their historical discovery and is not connected to their structure. Alkanes with more than four carbons are named systematically, ie based on the composition of the alkane. Chemists use the naming system authorised by **IUPAC** (the International Union of Pure and Applied Chemistry) for organic compounds, including alkanes.

In a **straight-chain** alkane, all of the carbon atoms are joined one after the other in a single 'line'. Straight-chain alkanes are named by placing the appropriate prefix in front of the suffix *-ane*:

Number of carbon atoms in chain	Prefix	Alkane name
1	meth-	methane
2	eth-	ethane
3	prop-	propane
4	but-	butane
5	pent-	pentane
6	hex-	hexane
7	hept-	heptane
8	oct-	octane
9	non-	nonane
10	dec-	decane

Example A

Straight-chain alkanes

$CH_3-CH_2-CH_2-CH_3$ is *but*ane because there are four carbons.

C_6H_{14} is *hex*ane because there are six carbon atoms.

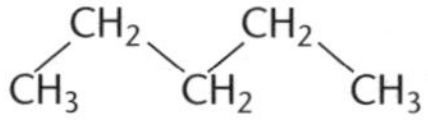

```
   CH2     CH2
  /   \   /   \
CH3    CH2     CH3
```

is *pent*ane because there are five carbon atoms. (Note that the alkane chain is not 'straight' as such. The C atoms are aligned in a zigzag fashion due to the tetrahedral arrangement of bonds around the carbon atom. However, the shape is described as 'straight-chain'.)

Molecular and structural formulae

A **molecular formula** gives information about the type and number of atoms present in a substance. Methane has the molecular formula CH_4. Methane molecules contain one carbon and four hydrogen atoms.

A **structural formula** shows how the atoms are joined together.

Example B

Structural formula of methane, CH_4

Methane has the structural formula shown alongside. The structural formula shows that each hydrogen atom is attached to the carbon atom by a single covalent bond.

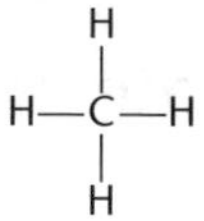

```
    H
    |
H — C — H
    |
    H
```

Usually, a **condensed structural formula** is used to represent organic compounds, especially for larger molecules.

Large alkanes (especially) usually have all the $-CH_2$ groups 'collected together' and written in brackets with a number.

Example C

Octane

Octane has the strutural formula:

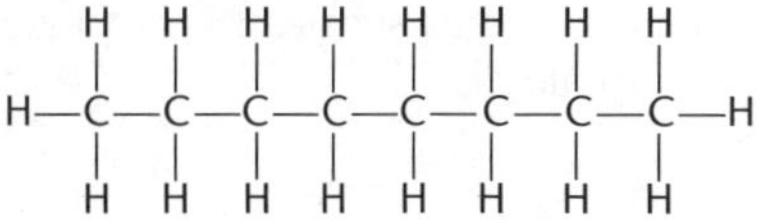

```
    H   H   H   H   H   H   H   H
    |   |   |   |   |   |   |   |
H — C — C — C — C — C — C — C — C — H
    |   |   |   |   |   |   |   |
    H   H   H   H   H   H   H   H
```

A condensed structural formula of octane would be:

$CH_3-CH_2-CH_2-CH_2-CH_2-CH_2-CH_2-CH_3$ or $CH_3CH_2CH_2CH_2CH_2CH_2CH_2CH_3$ or $CH_3(CH_2)_6CH_3$

Alkanes have angles of 109° between the bonds – each C atom has four regions of negative charge, so the geometry is tetrahedral. With several atoms in the chain, a zigzag shape occurs.

In Lewis structures and structural formulae, the molecules are often simplified to give planar views with angles of apparently 90°.

Example D

Bond angles in alkanes

Pentane, C_5H_{12}, has bond angles of 109°, giving a zigzag structure.

```
   H   H   H   H
   |   |   |   |
H—C—C—C—C—H
   |   |   |   |
   H   H   H   H
```

A simplified diagram shows butane as a planar structure.

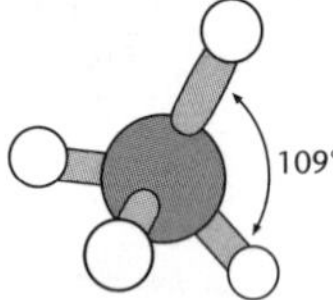

Model of methane showing bond angle of 109°.

Physical state of alkanes

The number of carbon atoms in the alkane chain determines the physical state at room temperature (20°C). As the carbon chain of alkanes gets longer, the alkanes change from gas to liquid to solid.

Example E

The first four members of the alkane series are gases at toom temperature. They are found in natural gas and used as fuel. Alkanes with between 5 and about 17 carbons atoms are liquids at room temperature. They are found in petroleum. Alkane molecules with 18 or more carbon atoms are solids at room temperature – these alkanes can be used for candle wax, and in an unrefined form are used as tar or bitumen for road surfaces.

In the following graph, the melting and boiling points of the first six alkanes are plotted against the number of carbon atoms in the alkane molecule.

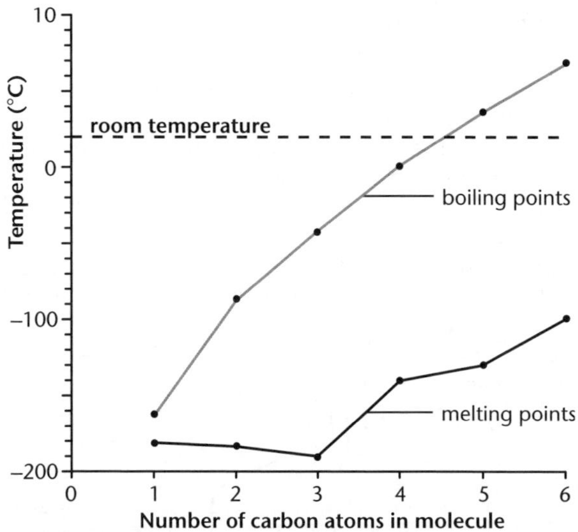

Graph of the melting points and boiling points of the first six alkanes

The graph shows that these six alkanes melt well below room temperature (20°C) and that the first four also boil below this temperature. This means that, at room temperature:

- methane, ethane, propane and butane are gases.
- pentane and hexane are liquids.

Sources of alkanes

Alkanes occur naturally in fossil fuels:

- Natural gas – mainly methane but also contains ethane.
- Petroleum – a mixture of alkanes with up to 30 carbon atoms in the molecule.

Over the last three centuries, by drilling down and breaking through the cap rock, the oil and gas have been released in spectacular black plumes of liquid.

Preparing natural gas for use

Natural gas from the ground is treated for use by passing the gas through potassium carbonate solution to remove carbon dioxide. The potassium carbonate becomes potassium hydrogen carbonate:

$K_2CO_3(aq) + CO_2(g) + H_2O(\ell) \rightarrow 2KHCO_3(aq)$

The potassium carbonate solution is regenerated by passing steam through the potassium hydrogen carbonate solution – the heat supplied by the steam reverses the initial reaction:

$2KHCO_3(aq) \rightarrow K_2CO_3(aq) + CO_2(g) + H_2O(\ell)$

If needed, the carbon dioxide can be dried and used.

Preparing petroleum for use

Petroleum requires more treatment than natural gas because it contains a large number of alkanes (as gases, liquids and solids). **Fractional distillation** – a process of separation of materials based on their different boiling points – separates petroleum into petrol, diesel, etc. In the fractionating column, the materials with the lowest boiling points will rise further than those with higher boiling points.

Example F

Propane and butane are collected at the top of the column as gases. Larger molecule alkanes (of around 20 carbon atoms per molecule) are collected at the bottom, eg diesel, tar, bitumen.

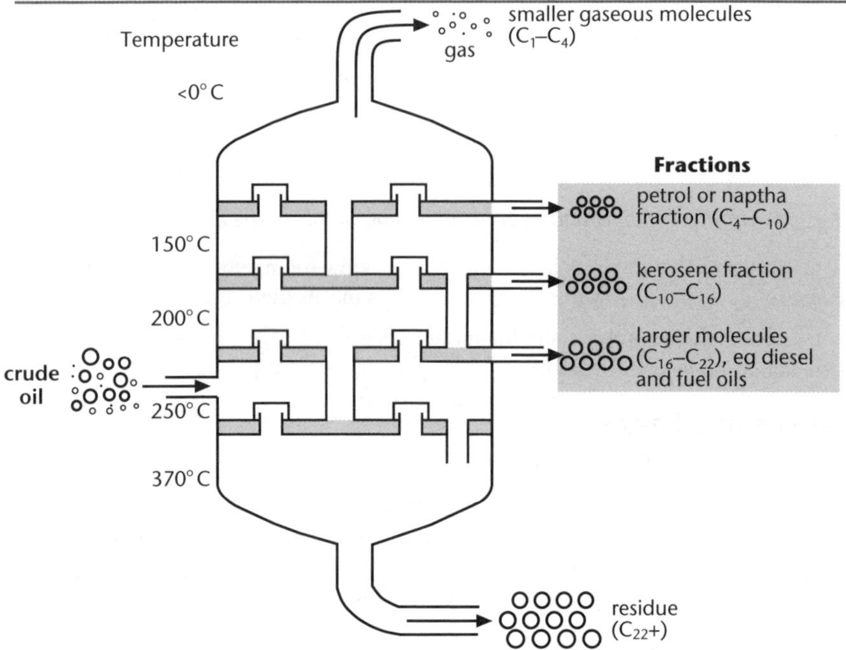

A fractional distillation tower. Distillation separates the hydrocarbons into useful fractions – petrol, jet fuel etc. The distillation tower can be 50 metres high.

Unit 12.1 Activity 3A: Alkanes

1. Write the general formula for an alkane.
2. Using C_2H_6 as an example, explain the meaning of the following terms.
   a. Molecular formula.
   b. Structural formula.
3. Given that the Greek for two is *di*, and for four is *tetra*, suggest alternative names for:
   a. ethane.
   b. butane.

4. a. Methyl propane has a molecular formula of C_4H_{10}. Its structural formula is:

```
        H
        |
      H—C—H
   H    |    H
   |    |    |
 H—C——(C)——C—H
   |    |    |
   H    H    H
```

The compound has a melting point of –159°C and a boiling point of –12°C.

i. Identify the physical state of methyl propane at room temperature (20°C).

ii. State the type of carbon atom that is circled. ***Hint***: The other three carbon atoms are primary.

b. Another alkane is called methyl pentane and has a molecular formula C_6H_{14}. The compound has a melting point of –154°C and a boiling point of 60°C. Identify the physical state of methyl pentane at room temperature.

5. Explain the meaning of the term 'secondary carbon atom' in relation to the structure of alkanes.

6. Name two natural sources of alkanes and indicate the alkanes that are present in the source.

7. a. Write down electron arrangements for carbon and hydrogen atoms.

b. Draw Lewis structures for carbon and hydrogen atoms.

c. Draw Lewis structures for methane and ethane molecules.

8. a. Draw structural formulae for:

i. ethane.

ii. the alkane with three carbon atoms.

iii. the alkane, C_6H_{14}, where the six carbon atoms are in one chain.

b. Give the condensed structural formulae for C_6H_{14}.

9. For the formula C_nH_{2n+2}, write molecular formulae of the alkanes where:

a. $n = 2$ **b.** $n = 4$ **c.** $n = 10$ **d.** $n = 32$

10. Name following the straight-chain compounds.

a. C_5H_{12}

b. C_4H_{10}

c. The alkane with seven carbon atoms.

d. CH_4

e. The alkane with eight carbon atoms.

11. a. Explain the meaning of the following terms.

i. Homologous series.

ii. Saturated hydrocarbon.

b. Give an example for each of the terms in **a.**

Structural isomers

Structural isomers have the same type and number of atoms (ie the same molecular formula), but their atoms are arranged differently.

Structural isomers have different **physical properties** and different names, but they behave similarly in chemical reactions.

Example G

Structural isomers of butane, C_4H_{10}

The alkane C_4H_{10} exists as two structural isomers:

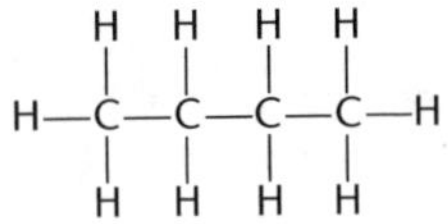

```
     H   H   H   H
     |   |   |   |
 H — C — C — C — C — H
     |   |   |   |
     H   H   H   H
```

$CH_3CH_2CH_2CH_3$

Butane

boiling point, –1°C

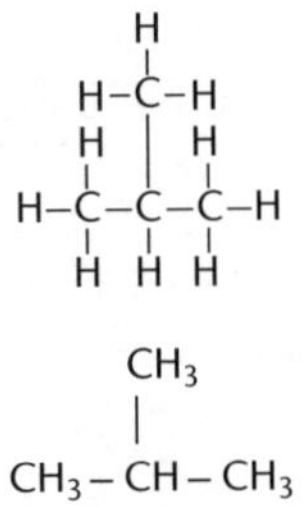

```
         H
         |
       H-C-H
     H   |   H
     |   |   |
   H-C - C - C-H
     |   |   |
     H   H   H
```

```
          CH3
           |
  CH3 – CH – CH3
```

Methylpropane

boiling point, –12°C

Alkane molecules with more than three carbon atoms can have a **branched-chain structure** or a **straight-chain** structure.

- A straight-chain hydrocarbon has all the carbon atoms joined one after another in a chain, eg pentane:
  $CH_3—CH_2—CH_2—CH_2—CH_3$
- A branched hydrocarbon chain has a **parent chain** (the longest chain of carbon atoms) and a **branch chain** (a smaller chain of carbon atoms joined to the parent chain or branch chains), eg methylbutane:

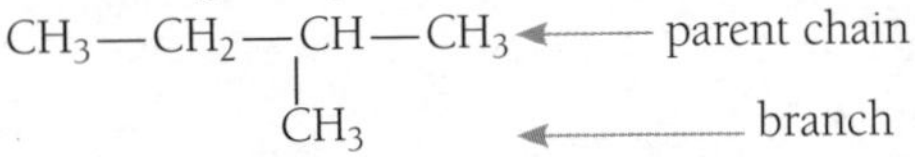

```
CH3—CH2—CH—CH3 ←—— parent chain
         |
        CH3      ←—— branch
```

When drawing branched chain structures, avoid mixing structural and condensed structural formulae:

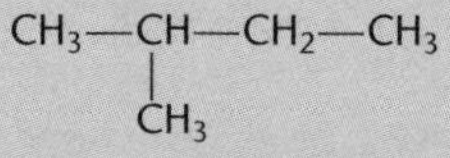

```
CH3—CH—CH2—CH3        or      H   H   H   H
    |                         |   |   |   |
    CH3                   H - C — C — C - C -H
                              |   |   |   |
                              H H-C-H H   H
                                  |
                                  H

              H   H    H   H
              |   |    |   |
        not H—C — C  — C — C—H
              |   |    |   |
              H   CH3  H   H
```

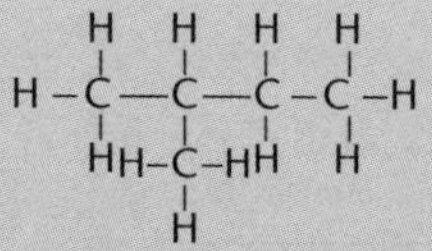

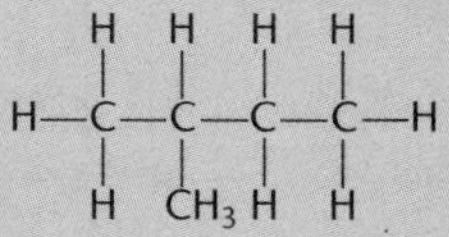

Naming alkanes with branch chains

Branch chains are called **alkyl groups**. Alkyl groups are named after the respective alkane, but wth *-yl* instead of *-ane*.

Alkane	Alkyl group	Alkyl structure
Methane, CH_4	methyl, CH_3–	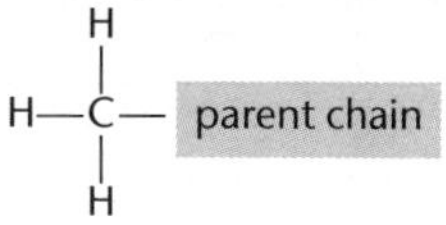 H H—C— parent chain H
Ethane, CH_3CH_3	ethyl, CH_3CH_2–	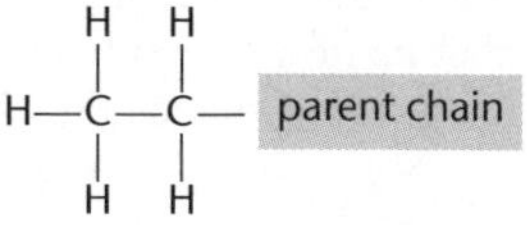 H H H—C—C— parent chain H H
Propane, $CH_3CH_2CH_3$	propyl, $CH_3CH_2CH_2$–	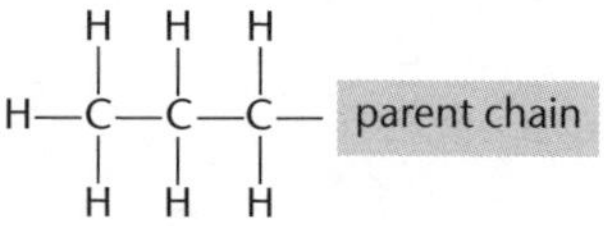 H H H H—C—C—C— parent chain H H H
Other alkyl groups are named accordingly – butyl, pentyl, hexyl, etc.		

Example H

Naming a branched-chain alkane

To name the molecule:

$CH_3—CH_2—CH_2—CH(CH_3)—CH_3$

Steps involved in naming branched-chain alkanes.

Step 1

Identify the parent (longest) chain.
Name with suffix *-ane*.

$CH_3—CH_2—CH_2—CH(CH_3)—CH_3$

Parent chain with 5 carbon atoms – pentane

Step 2

Identify branch chain(s).

$CH_3—CH_2—CH_2—CH(CH_3)—CH_3$

Branch chain → CH_3

Step 3

Count the number of carbons in a branch chain. Name each branch by using the appropriate prefix and adding *-yl*.

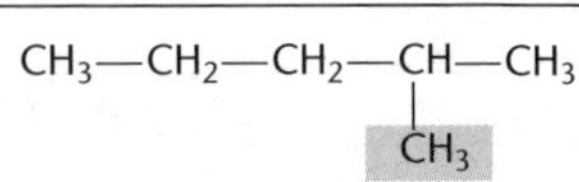

One carbon atom in branch – methyl

Step 4

Identify the position of the branch or branches by numbering from the end of the parent chain which gives the branch the *lowest* number.

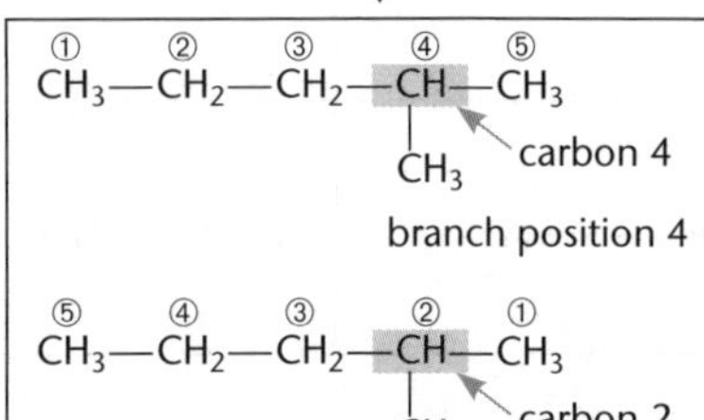

Choose 2 rather than 4.

Step 5

Write full name as:
branch position, 'hyphen', branch name then parent name.

2-Methylpentane

If there are several identical branches on a parent chain, the position of each branch is numbered separately, and the branch names are written together using the following prefixes:

- *di* – if there are two identical branches.
- *tri* – if there are three identical branches.
- *tetra* – if there are four identical branches.

Example I

Two identical branches

The alkane shown is 2,3-dimethylbutane because:

- There are four carbons in the parent chain (hence butane).
- There are two methyl branches (hence dimethyl).
- The branches are on carbon numbers 2 and 3 (hence, 2,3-).

```
         CH3
          |
   CH3 — CH — CH — CH3
               |
              CH3
```

If there are several different branches on a parent chain, they are written in alphabetical order.

Example J

Two different branches

The alkane below is 3-ethyl-3-methylpentane, because 'e' (of ethyl) comes before 'm' (of methyl) in the alphabet.

```
                  CH3
                   |        ← ethyl group on carbon 3
                  CH2
 ①      ②      ③  |    ④       ⑤
CH3 — CH2 — C — CH2 — CH3  ← parent chain is pentane
                   |
                  CH3       ← methyl group on carbon 3
```

If two different groups are attached in equivalent places on the same molecule, then the *lower* number is given to the group which comes *first* alphabetically.

Example K

Different groups at different positions

The following alkane is 3-ethyl-5-methylheptane (rather than 5-ethyl-3-methylheptane), because the lowest number (3) must go with the group which comes first alphabetically (ethyl).

```
 ⑦      ⑥      ⑤      ④      ③      ②      ①
CH3 — CH2 — CH — CH2 — CH — CH2 — CH3
              |             |
             CH3           CH2
                            |
                           CH3
```

If an organic molecule contains a branch which can *only* go in *one* position, then the branch number is omitted.

Example L

Methylpropane

This molecule is namd methylpropane (*not* 2-methylpropane), because the methyl group can *only* be attached to the second carbon.

```
         CH3
          |
   CH3 — CH — CH3
```

Finding all the structural isomers with a given formula

Steps involved with identifying all the different structures possible:

Example M

To find all the isomers with the formula C_6H_{14}:

Step 1

Draw the longest chain carbon skeleton.

C—C—C—C—C—C

Step 2

Remove a 'C' and place it on the 'reduced' skeleton to make different molecules.

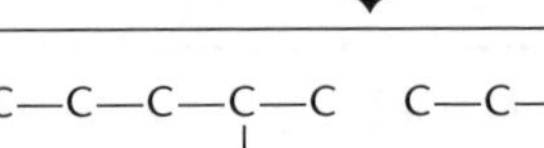

Step 3

Remove a further 'C' and place it on the 'reduced' skeleton to make different molecules.

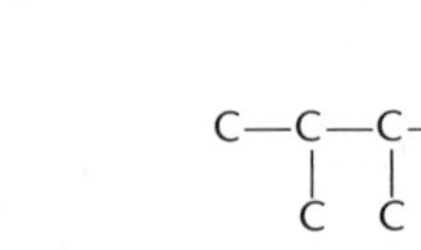

Step 4

Remove a further 'C' and place it on the 'reduced' skeleton to make different molecules.

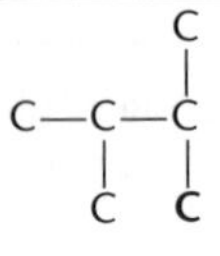

New structure has a 4-carbon chain, which is the same as the isomer created in **Step 3** – ie it is *not* a new isomer.

Step 5

Complete the structures by putting hydrogen atoms in all the bond positions.

$CH_3—CH_2—CH_2—CH_2—CH_2—CH_3$

Hexane

$CH_3—CH_2—CH(CH_3)—CH_2—CH_3$

3-Methylpentane

$CH_3—CH_2—CH_2—CH(CH_3)—CH_3$

2-Methylpentane

$CH_3—CH(CH_3)—CH(CH_3)—CH_3$

2,3-Dimethylbutane

$CH_3—CH_2—C(CH_3)_2—CH_3$

2,2-Dimethylbutane

Cycloalkanes

Alkanes with three or more carbons can form rings or *cyclic* structures, and are called **cycloalkanes**.

Cycloalkanes:

- have the general formula C_nH_{2n}.
- have slightly higher melting and boiling points than the corresponding straight-chain alkanes.
- are named using the prefix *cyclo*.

Example N

Cyclohexane

The molecule is called cyclohexane:

- *cyclo* – in a ring.
- *hexane* – 6 carbon atoms.

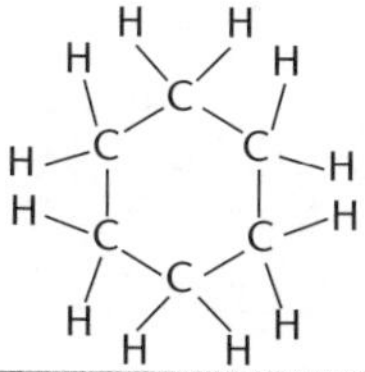

```
      H     H
   H   \   /   H
    \   C     /
     \ / \   /
  H — C     C — H
      |     |
  H — C     C — H
     / \   / \
    H   \ /   H
         C
        / \
       H   H
```

Cycloalkanes are sometimes represented by just the C–C skeleton, eg cyclohexane ⬡.

Cycloalkanes can have alkyl **side chains**, eg $CH_3—CH—CH_2$, methylcyclobutane.

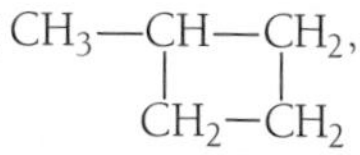

```
CH3—CH—CH2
     |   |
    CH2—CH2
```

Unit 12.4 Activity 3B: Naming alkanes

1. Name each of the following structural formulae:

a.

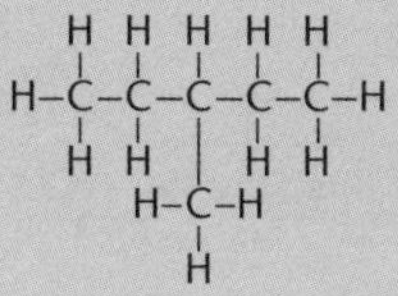

```
  H H H H H
  | | | | |
H-C-C-C-C-C-H
  | |   | |
  H H | H H
    H-C-H
      |
      H
```

b.

```
  H H H H H
  | | | | |
H-C-C-C-C-C-H
  | | |   |
  H H H | H
      H-C-H
        |
      H-C-H
        |
        H
```

c.

```
      H
      |
    H-C-H
      |
    H-C-H H
      |   |
    H-C---C-H
  H   |   |
  |   |   H
H-C---C-H
  |   |
  H H-C-H
      |
    H-C-H
      |
      H
```

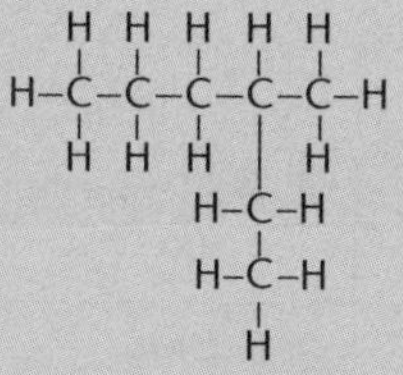

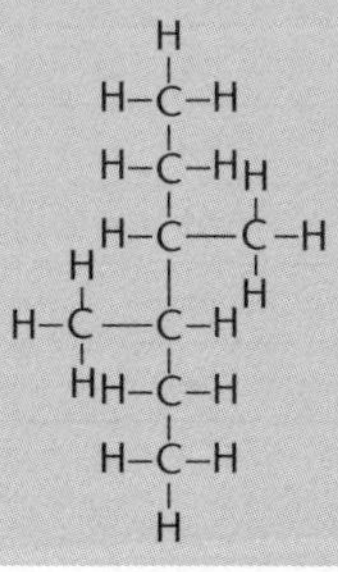

d.

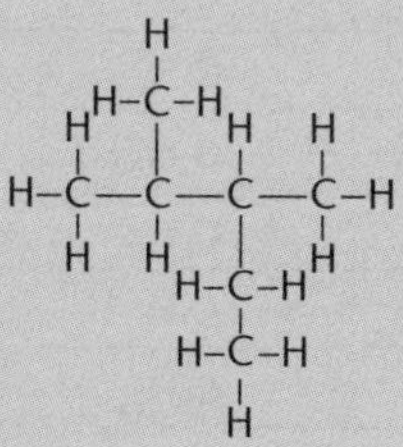

```
      H
      |
    H-C-H
  H   |   H   H
  |   |   |   |
H-C---C---C---C-H
  |   |   |   |
  H   H H-C-H H
          |
        H-C-H
          |
          H
```

2. Draw structural formulae for each of the following hydrocarbons.
   a. Methylpropane
   b. Methylbutane
   c. 3-Methylhexane
   d. 2,2-Dimethylpropane
   e. Octane
   f. 2,2,3-Trimethylbutane
3. Draw a structural formula for 3-ethyl-2-methylpentane.
4. Draw and name all the structural isomers of:
   a. Pentane.
   b. Hexane.
5. Name and draw structures for the *alkanes* with the following molecular formulae:
   a. C_5H_{10}
   b. C_3H_6

Physical properties of alkanes

Alkanes:

- are colourless compounds.
- are *non-polar* and hence are *insoluble* in water.
- have densities less than water (ie < 1 g cm^{-3}), so they float on water.
- dissolve in each other and are often used as solvents and grease removers.
- do not conduct heat or electricity because they do not contain ions or electrons which are free to move.
- show increased melting and boiling points as the length of the carbon chain increases.

Name	Molecular formula	Melting point (°C)	Boiling point (°C)
Methane	CH_4	–183	–162
Ethane	C_2H_6	–183	–89
Propane	C_3H_8	–188	–42
Butane	C_4H_{10}	–138	–1
Pentane	C_5H_{12}	–130	36
Hexane	C_6H_{14}	–95	69
Heptane	C_7H_{16}	–91	98
Octane	C_8H_{18}	–57	126

At room temperature alkanes with:

- up to four carbon atoms are gases.
- between five and 15 carbon atoms are liquids.
- more than 15 carbon atoms are soft, waxy solids.

The smaller liquid alkane molecules are **volatile** – ie they become gaseous with minimal application of heat.

Example O

Boiling points and molar mass

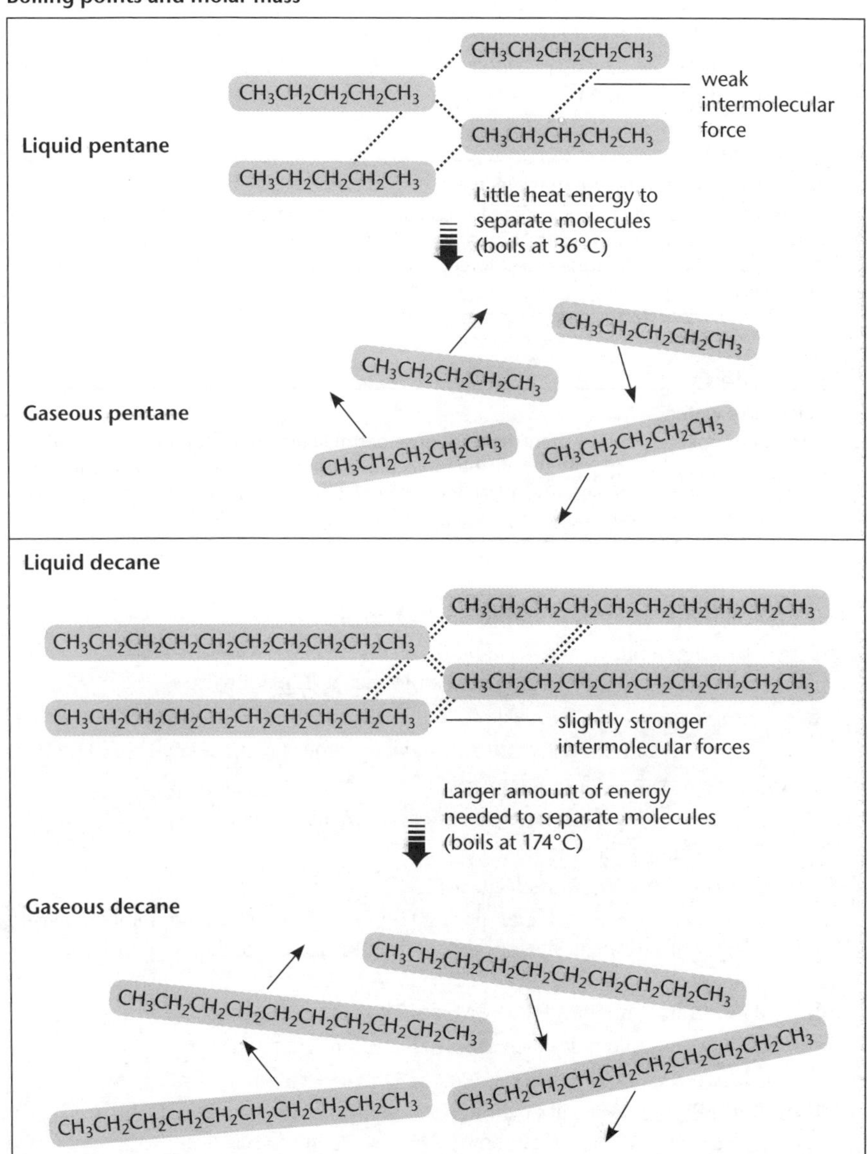

Volatile hydrocarbons burn fiercely and rapidly because plenty of vapour is available to burn.

Solubility of hydrocarbons in water

When a solute dissolves in a solvent, the solute molecules must attract the solvent molecules strongly enough to overcome the forces of attraction between the solute molecules and between the solvent molecules. This does *not* occur with alkanes in water – they are all insoluble in water.

Example P

Forces of attraction between molecules

A large amount of energy is required to separate water molecules, as indicated by the boiling point (100°C) of water. In contrast, pentane, which is a much larger molecule, has a boiling point of 36°C. This indicates weaker forces between pentane molecules than between water molecules.

Example Q

Petrol and water

Petrol does not dissolve in water (the less dense petrol floats on the top of water). Little energy is required to separate the alkane molecules in petrol; greater energy is required to separate water molecules. The attraction between the petrol molecules and the water molecule does not repay the energy of separation of the water molecules.

Unit 12.4 Activity 3C: Physical properties of alkanes

1. Petrol, petroleum oil, candle wax and diesel oil are all mixtures of alkanes. Use this information and the polarity of alkane molecules to explain why:
   a. Petrol does not mix with water, even if the two are shaken vigorously together.
   b. Oil and petrol mix easily to form a homogeneous ('completely mixed') mixture.
   c. Candle wax dissolves in diesel oil.
2. Petrol is a volatile liquid, candle wax is a solid. Which alkane:
   a. Has the highest boiling point?
   b. Consists of the largest molecules?
3. **a.** Using data from the table on page 202, plot a graph of boiling point (°C) against number of carbon atoms for the first six alkanes. Put boiling point on the vertical axis.
   b. Mark room temperature (20°C) on the vertical axis. Name:
      i. The four alkanes which are gases at room temperature.
      ii. Two alkanes which are liquids.
4. Diesel oil spilt in a river will float on top of the water. Explain why the oil and water do not mix. ***Hint:*** Diesel oil is composed of hydrocarbon molecules and, although these molecules are comparatively large, the boiling point of diesel oil is comparatively low.

Chemical properties of alkanes

Combustion

The reaction of alkanes with oxygen is an example of **combustion** (burning). Water is *always* a product, but the other product(s) depend on the amount of oxygen present.

When oxygen is plentiful, **complete combustion** occurs and carbon dioxide, CO_2, is produced. Complete combustion gives a flame that is hot, blue in colour and clean burning (ie produces no soot).

With a *limited* oxygen supply, either carbon monoxide, CO, or carbon (soot) will form (*or* a mixture of CO and soot); this is called **incomplete combustion**. Burning is *incomplete* and the flame may be yellow-coloured. In a *very* limited oxygen supply, some soot always forms.

Example R

Combustion of methane

$CH_4(g) + 2O_2(g) \rightarrow CO_2(g) + 2H_2O(g)$ — complete combustion

$CH_4(g) + \frac{3}{2}O_2(g) \rightarrow CO(g) + 2H_2O(g)$
$CH_4(g) + O_2(g) \rightarrow C(s) + 2H_2O(g)$ } — incomplete combustion

Example S

The Bunsen burner

The methane used as fuel in the Bunsen burner will undergo complete or incomplete combustion depending on the position of the air hole.

When the air hole is open the:

- gas leaving the barrel is a mixture of air and methane – allows complete combustion of the methane.
- outer region of the flame is light blue – this is where complete combustion is occurring.
- inner region of the flame (at the top of the barrel) is deep blue – this is where the pressure of the gas mixture leaving the barrel prevents combustion.

When the air hole is closed:

- there is insufficient air for complete combustion.
- the flame is yellow due to unburnt carbon, which becomes red hot.
- below the yellow flame is a region where the gas is not ignited.
- the heat of the yellow flame makes the air surrounding the flame rise – this creates a current of fresh air that allows complete combustion of the gas at the bottom edges of the flame.

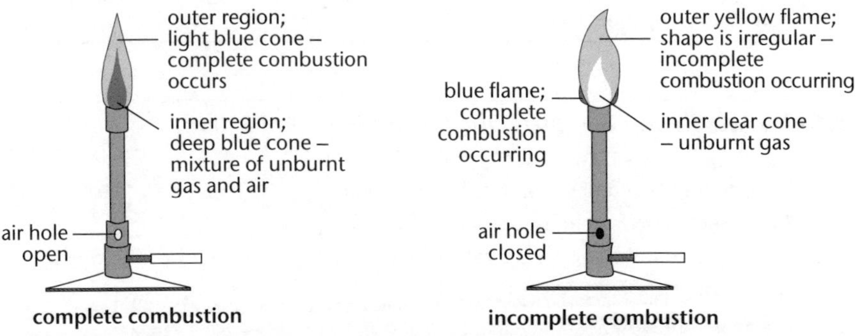

Complete and incomplete combustion in the Bunsen burner

Example T

The bang tin

A 'bang tin' is a metal tin with a hole in the bottom and a tightly fitting lid, also containing a hole in the middle. The bang tin can be used to demonstrate complete and incomplete combustion.

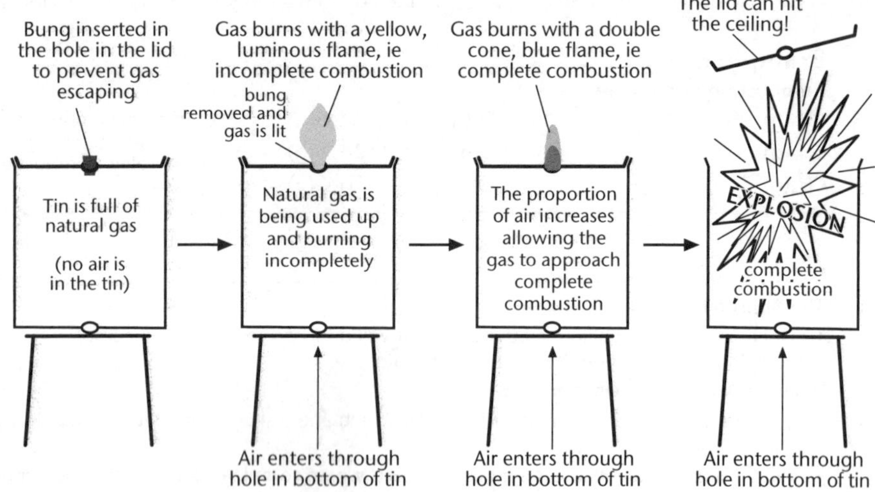

Combustion in the bang tin

When the tin is full of natural gas, there is insufficient air surrounding the gas leaving the tin for complete combustion to occur.

- As the gas is used up, air enters the tin from the hole in the bottom and the mixture in the tin approaches the correct proportions for complete combustion.
- Since the gas and air are intimately mixed, the end result is a perfect mixture to produce an explosion.

The explosion from a tin with dimensions of approximately 20 cm x 20 cm x 20 cm will cause no harm. The lid must be very tightly fitted to get a good bang. Do not attempt this demonstration on a bigger scale.

Unit 12.4 Activity 3D: Complete and incomplete combustion

1. a. State the products for the:

i. complete combustion of a hydrocarbon.

ii. incomplete combustion of a hydrocarbon

b. State the colour flame that is seen when natural gas is burnt:

i. completely.

ii. incompletely.

2. The diagram shows a burning candle.

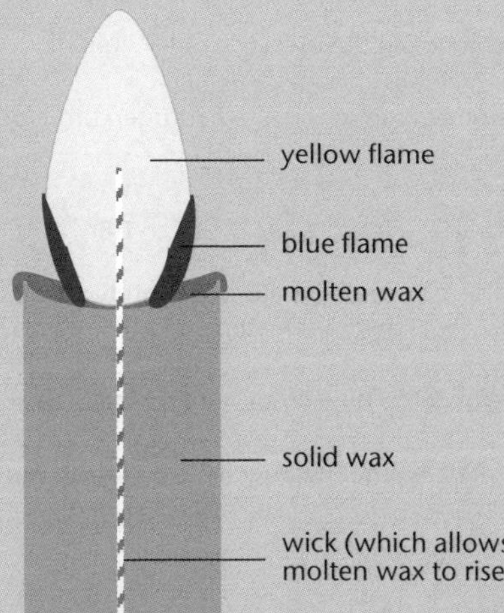

   a. Explain the purpose of the wick.
   b. Explain what is burning.
   c. Explain what is happening in the blue flame.
   d. Explain what is happening in the yellow flame.

3. The equations show the three different processes for the combustion of hexane:

   $C_6H_{14}(\ell) + 9\frac{1}{2}O_2(g) \rightarrow 6CO_2(g) + 7H_2O(g)$

   $C_6H_{14}(\ell) + 6\frac{1}{2}O_2(g) \rightarrow 6CO(g) + 7H_2O(g)$

   $C_6H_{14}(\ell) + 3\frac{1}{2}O_2(g) \rightarrow 6C(s) + 7H_2O(g)$

   Use these equations to explain the meaning of:
   a. Complete combustion.
   b. Incomplete combustion.
   c. The term 'soot' and how soot is formed.

4. Explain why water in a beaker will take longer to boil using a yellow Bunsen flame than a blue Bunsen flame.

Substitution

Although alkanes are very good fuels, reacting well with oxygen in combustion reactions, they are generally unreactive with most other substances.

Under special conditions, alkanes will undergo **substitution** reactions with members of the halogen family (eg chlorine, Cl_2, and bromine, Br_2).

Substitution reactions occur when an atom (or a group of atoms) on a hydrocarbon chain is replaced by another atom (or group of atoms).

Substitution reactions of alkanes involve:

- Loss of a hydrogen atom – involves breaking a C–H bond.
- Replacement of the lost hydrogen atom by a chlorine or bromine atom.
- The presence of a catalyst – usually ultraviolet (UV) light from sunlight.

Example U

Substitution

Methane undergoes substitution reactions with chlorine in the presence of UV light or bromine in the presence of light.

methane + bromine ⟶ bromomethane + hydrogen bromide

$$\begin{array}{c} H \\ | \\ H-C-H \\ | \\ H \end{array} + Br_2 \xrightarrow{\text{light}} \begin{array}{c} H \\ | \\ H-C-Br \\ | \\ H \end{array} + HBr$$

bromine atom substituted for a hydrogen atom

The production of HBr can be detected by a piece of damp blue litmus paper held at the mouth of the test tube; it turns pink.

The bromomethane can react further, because there are other hydrogen atoms which can be substituted.

$$\begin{array}{c} H \\ | \\ H-C-Br \\ | \\ H \end{array} + Br_2 \xrightarrow{\text{light}} \begin{array}{c} Br \\ | \\ H-C-Br \\ | \\ H \end{array} + HBr$$

Dibromomethane

Tribromomethane, $\begin{array}{c} Br \\ | \\ H-C-Br \\ | \\ Br \end{array}$, and tetrabromomethane $\begin{array}{c} Br \\ | \\ Br-C-Br \\ | \\ Br \end{array}$, will also be produced.

As bromine is used in substitution reactions, the brown colour of bromine disappears over time (slow reaction) – the bromine is 'used up'. The bromine is said to have been *decolorised*.

When halogen atoms such as bromine and chlorine are substituted into alkanes, they form **haloalkanes**.

Haloalkanes

Haloalkanes are named in the same way as alkanes, with a prefix for the halogen atom. The position on the parent chain and the names in alphabetical order are used if there is more than one type of halogen present.

Atom	Name used in haloalkane
Bromine	bromo
Chlorine	chloro
Fluorine	fluoro
Iodine	iodo

Example V

Naming haloalkanes

$$CH_3—CH(Cl)—CH(Br)—CH_3$$

is named 2-bromo-3-chlorobutane, because:

- there are 4 carbons in the straight chain (butane).
- bromine comes before chlorine alphabetically, so the bromo is on the lower number carbon (2) than the chloro (which is on carbon 3).

Classifying haloalkanes

Haloalkanes are classified as **primary** (**1°**), **secondary** (**2°**) or **tertiary** (**3°**) according to the carbon atom to which the halogen is attached.

Example W

Primary, secondary and tertiary haloalkanes

Chloroethane, CH_3CH_2Cl, is a *primary* haloalkane because the chlorine atom is attached to a carbon atom which itself is only attached to *one* carbon atom.

2-Chloropropane, $CH_3CHClCH_3$, is a *secondary* haloalkane because the chlorine atom is attached to a carbon atom which is itself attached to *two* carbon atoms.

2-Chloromethylpropane, $CH_3C(CH_3)ClCH_3$, is a *tertiary* haloalkane because the chlorine atom is attached to a carbon atom which is itself attached to *three* carbon atoms.

Unit 12.4 Activity 3E: Haloalkanes

1. $CH_3—CH_2—CH(CH_3)—CH_3 + Br_2 \xrightarrow{\text{catalyst}} CH_3—CH(Br)—CH(CH_3)—CH_3 + HBr$

a. Name the reactants and products.

b. What type of reaction is this?

c. What catalyst is needed?

d. What would be observed as the reaction proceeds?

2. a. Write a balanced equation for bromine reacting with methane, and name the products of the reaction.

b. How could you tell if the reaction was proceeding?

3. a. Name the following haloalkanes:

i. $CH_3CH_2CH_2Br$

ii. CH_3CBr_3

iii. 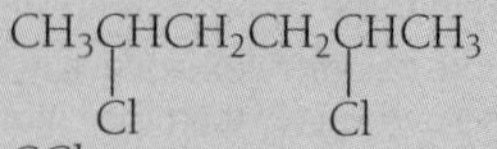
```
CH3CHCH2CH2CHCH3
   |        |
   Cl       Cl
```

iv. CCl_4

b. Classify **i.** and **ii.** in **a.** as primary, secondary or tertiary haloalkanes.

4. For each of the following haloalkanes:

i. Write the name.

ii. Classify as primary, secondary or tertiary.

a. $BrCH_2CH_2CH_2Br$

b. 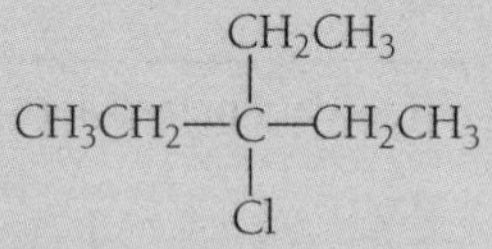
```
           CH2CH3
           |
CH3CH2—C—CH2CH3
           |
           Cl
```

5. Draw structural formulae for the following haloalkanes:

a. 1,2-dibromoethane

b. 2-chloropropane

c. 1,2,3-trichloro-2-methylbutane

6. 1-Bromobutane, 2-bromobutane and 2-methyl-2-bromopropane are isomers of C_4H_9Br.

a. Draw structural formulae for these isomers, and classify each as a primary, secondary or tertiary haloalkane.

b. There is a fourth isomer of C_4H_9Br.

i. Write its name and give its structural formula.

ii. Classify it as a primary, secondary or tertiary haloalkane.

Unit 12.4 Carbon Compounds
Topic 4: Alkenes, polymers and alkynes

Topic 4 covers alkenes, polymers and alkynes:

- Recognising reactions of alkenes (includes asymmetric alkenes and polymerisation) and alkynes.
- Naming or drawing the structural formulae of straight- or branched-chain hydrocarbons (containing up to 8 carbon atoms).

Alkenes

Alkenes are unsaturated hydrocarbons that contain a carbon–carbon double bond (C=C). Double bonds require four electrons. The general formula of the alkenes is C_nH_{2n}.

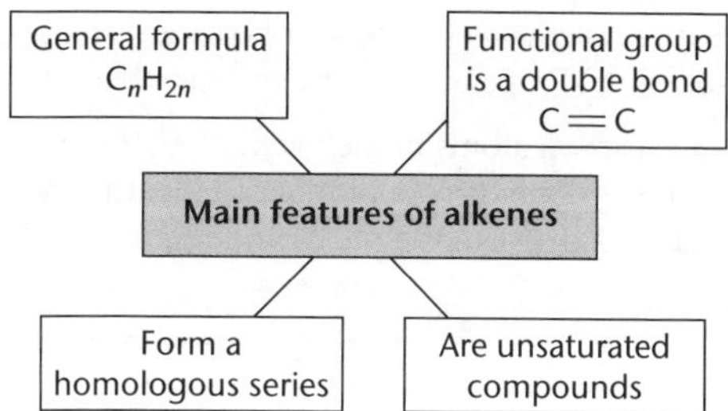

Chemical reactions occur at the alkene functional group (C=C).

In alkenes, two carbon atoms are joined by a double bond. Each of these two carbons atoms is bonded to two other atoms. All six atoms lie in a plane, with angles of 120° between the bonds.

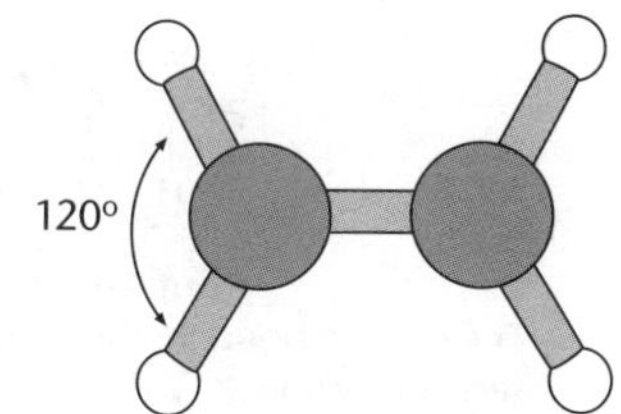

The first two alkenes are **ethene** and propene. (Since *two* carbon atoms are required for a double bond, *methene* does not exist.)

Name	Molecular formula	Structural formula
Ethene	C_2H_4	$H_2C{=}CH_2$ (H and H on each C)
Propene	C_3H_6	$(H)(CH_3)C{=}C(H)(H)$

The structural formula of ethene could be written as:

```
H   H
|   |
C = C
|   |
H   H
```

Because the bond angle between the double bond and a single bond to a H atom is 120°

(120° between C=C and C–H), ethene is often drawn as:

```
H         H
  \     /
   C = C
  /     \
H         H
```

Naming alkenes

Straight-chain alkenes use the same prefix as alkanes but with the suffix *-ene*, as in alk*ene*. The first member of the series is ethene. (Since a carbon–carbon double bond must be present, *methene* does not exist.) Propene is the second member.

If there are more than three carbon atoms in the chain, then the position of the double bond must be specified. This is done using a number to identify the first carbon in the double bond. The number goes between the prefix and the -ene suffix.

Example A

Butene

Butene, C_4H_8, has four carbon atoms. This gives two possible positions for the double bond, giving two different molecules (ie two different isomers) with different names.

```
H  ①   ②   H ③  H ④
  \         |    |
   C = C  — C  — C — H
  /    |    |    |
H      H    H    H
```

But-1-ene

The double bond lies between the first and second carbons.

```
    H               H
    |①   ②    ③    |④
H — C — C  = C  — C — H
    |   |     |     |
    H   H     H     H
```

But-2-ene

The double bond lies between the second and third carbons.

For branched-chain alkenes the:

- parent chain is the chain that contains the double bond.
- position of the double bond must be identified by numbering the chain from whichever end gives the double bond the *lowest* number.

Example B

A branched alkene

To name the molecule:

$$CH_3—\underset{\displaystyle CH_3}{\underset{|}{C}}=CH—CH_3$$

Steps involved in naming branched-chain alkenes.

Step 1

Identify and name the parent chain.

$$CH_3—\underset{\displaystyle CH_3}{\underset{|}{C}}=CH—CH_3$$

Butene (4 C atoms)

Step 2

Identify the position of the double bond.

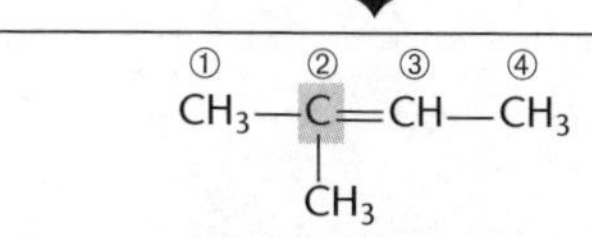

Double bond between carbons 2 and 3; but-2-ene

Step 3

Identify any branch chains and their positions.

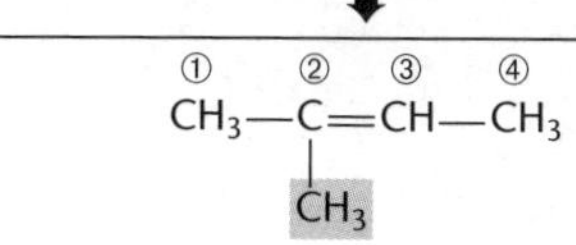

Methyl branch chain on carbon 2; 2-methyl

Step 4

Write full name as:

branch position, 'hyphen', branch name, parent name, 'hyphen', double bond position.

2-Methylbut-2-ene

Cycloalkenes

Cycloalkenes have the general formula C_nH_{2n-2}. Cycloalkenes are named similarly to cycloalkanes.

Example C

Cyclohexene

The molecule is called cyclohexene:

- *cyclo* – in a ring.
- *hex* – 6 carbon atoms.
- *ene* – double bond.

Cycloalkenes are sometimes represented by just the carbon skeleton, eg cyclohexene .

Unit 12.4 Activity 4A: alkenes

1. Write the general formula for an alkene.
2. Explain the difference between the two formulae for the ethene molecule: C_2H_4 and

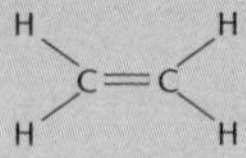

$$\begin{array}{ccccc} H & & & & H \\ & \diagdown & & \diagup & \\ & & C{=}C & & \\ & \diagup & & \diagdown & \\ H & & & & H \end{array}$$

3. **a.** Suggest a name for a compound with molecular formula C_4H_8.
   b. Draw a structural formula for the molecule C_4H_8. **Note:** There are three possible structures depending upon where the double bond is placed – any one structure is sufficient.
4. Using the molecule C_3H_6 as an example, explain what is meant by:
   a. empirical formula.
   b. molecular formula.
   c. structural formula.
5. Write down the molecular formula for the alkene with:
   a. 4 carbon atoms.
   b. 12 carbon atoms.
   c. 5 carbon atoms.
6. Give names for the following alkenes:

   a.

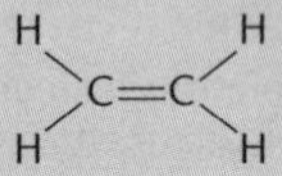

$$\begin{array}{ccccc} H & & & & H \\ & \diagdown & & \diagup & \\ & & C{=}C & & \\ & \diagup & & \diagdown & \\ H & & & & H \end{array}$$

   b. $CH_3—CH{=}CH_2$

   c.

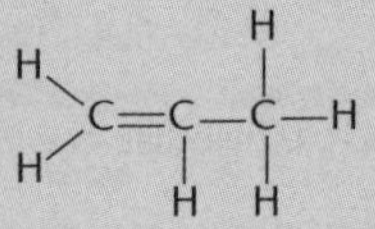

$$\begin{array}{ccccccccc} & & & & & & H & & \\ H & & & & & & | & & \\ & \diagdown & & & & & & & \\ & & C & = & C & - & C & - & H \\ & \diagup & & & | & & | & & \\ H & & & & H & & H & & \end{array}$$

   d.

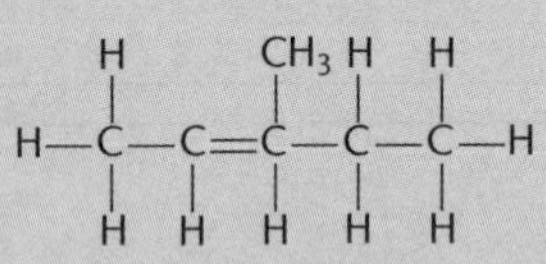

$$\begin{array}{ccccccccccccc} & & H & & & & CH_3 & & H & & H & & \\ & & | & & & & | & & | & & | & & \\ H & - & C & - & C & = & C & - & C & - & C & - & H \\ & & | & & | & & | & & | & & | & & \\ & & H & & H & & H & & H & & H & & \end{array}$$

   e.

$$\begin{array}{ccccccc} & & & & CH_3 & & \\ & & & & | & & \\ CH_3 & - & C & = & C & - & CH_3 \\ & & | & & & & \\ & & CH_2 & & & & \\ & & | & & & & \\ & & CH_3 & & & & \end{array}$$

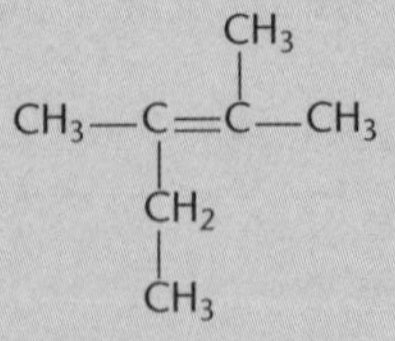

7. Draw structures for:
   a. pent-2-ene
   b. 2-methylhex-3-ene
   c. methylpropene
   d. 2,3-dimethylbut-2-ene
8. Draw and name all possible alkene isomers of C_5H_{10}.

Geometric isomers

Geometric isomers exist for alkenes that have different groups attached to both of the carbon atoms that form the double bond.

- ***cis* isomers** – the same groups are on the same side of the double bond.
- ***trans* isomers** – the same groups are on opposite sides of the double bond.

Geometric isomerism exists because free rotation around the carbon–carbon double bond cannot occur, fixing the position in space of the groups attached to the carbon atoms of the double bond.

Example D

***cis* and *trans* isomers of but-2-ene**

But-2-ene is the simplest alkene to have geometric isomers.

$$\begin{array}{ccc} CH_3 & & CH_3 \\ & C{=}C & \\ H & & H \end{array} \qquad\qquad \begin{array}{ccc} CH_3 & & H \\ & C{=}C & \\ H & & CH_3 \end{array}$$

***cis*-But-2-ene**
CH_3 groups on same side of double bond.

***trans*-But-2-ene**
CH_3 groups on opposite sides of double bond.

Geometric isomers have some different **physical properties** (eg the boiling point of *cis*-but-2-ene is 3.7°C, and that of *trans*-but-2-ene is 0.88°C) but usually have similar chemical properties.

If alkenes have the same groups (or atoms) attached to one of the two carbon atoms of the double bond, they will *not* exhibit geometric isomerism.

Example E

But-1-ene

Geometric isomerism does *not* occur for but-1-ene, because one of the C atoms in the double bond (C number 1) has two identical atoms or groups of atoms (H atoms) attached.

$$\begin{array}{ccc} H & & H \\ & C{=}C & \\ H & & CH_2CH_3 \end{array}$$

But-1-ene

Unit 12.4 Activity 4B: Isomers of alkenes

1. Draw the following alkenes, and identify which one(s) exist as *cis/trans* isomers.
  a. 2-Methylbut-1-ene
  b. Pent-1-ene
  c. Hex-3-ene
  d. 2-Methylpent-2-ene

2. Draw and name four alkene isomers of C_4H_8. Identify the types of isomerism involved.

3. Consider the following two compounds.
  a. $(CH_3)_2C{=}CHCH_2CH_3$ **b.** $CH_3CH{=}CHCH_2CH_3$
    i. Name them.
    ii. Give their molecular formula.
    iii. Draw the two geometric isomers if geometric isomerism exists for the compound.

4. Draw the structural formulae of the following molecules. If the molecule exhibits geometrical isomerism, draw and label both the *cis* and *trans* isomers for the molecule.
   a. 2,3-Dichlorobut-2-ene
   b. 1,1-Dibromopropene
5. Explain why geometrical isomerism is only exhibited by some alkenes and is never found in alkanes.
6. **a.** A compound of molecular formula $C_2H_2Cl_2$ can exist as *cis* and *trans* isomers. Draw and name these isomers.
   b. There is another isomer of $C_2H_2Cl_2$ that cannot exist as *cis* and *trans* isomers. Draw the structural formula for this compound, and name it.
   c. Explain why the molecule drawn in **b.** cannot exist as *cis* and *trans* isomers while the one drawn in **a.** can.

Physical properties of alkenes

Alkenes:

- are non-polar molecules and hence are insoluble in water and in other polar solvents.
- are good solvents for other non-polar molecules.
- show an increase in melting and boiling points as the molar mass (or chain length) increases.
- have densities < 1 g cm^{-3}, so float on water.

At room temperature, ethene, propene and butene are gases. Alkenes of higher molar mass are liquids. The double bond in alkenes makes them much more reactive than alkanes.

Chemical properties of alkenes

Combustion

The combustion reactions of alkenes are much the same as those for alkanes. Like alkanes, alkenes react with oxygen to form water and either carbon dioxide, carbon monoxide or carbon (soot), depending on the amount of available oxygen. Alkenes tend not to burn as cleanly as alkanes and often produce a yellow, sooty flame.

Example F

Combustion of an alkene

Equations illustrating the complete and incomplete conbustion of ethene are:

$C_2H_4(g) + 3O_2(g) \rightarrow 2CO_2(g) + 2H_2O_2(g)$ – complete combustion

$C_2H_4(g) + 2O_2(g) \rightarrow 2CO(g) + 2H_2O_4(g)$ – incomplete combustion

$C_2H_4(g) + O_2(g) \rightarrow 2C(s) + 2H_2O_4(g)$ – incomplete combustion

Addition reactions

Since alkenes are unsaturated, atoms can be added to the molecules.

Reactions at the double bond are **addition reactions**. These involve breaking the double bond to produce a single carbon–carbon bond and two new single covalent bonds:

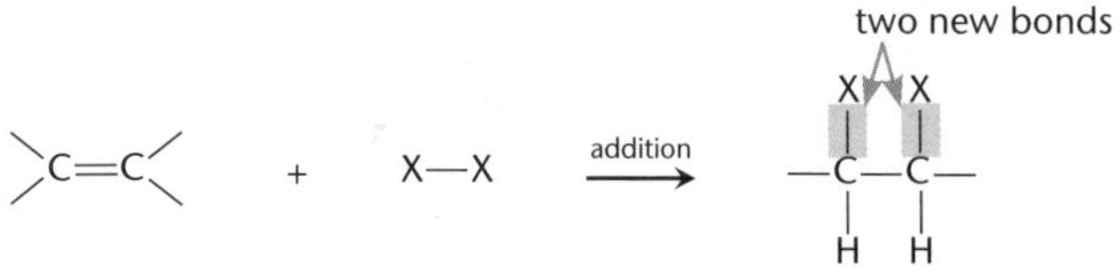

$$>C=C< \quad + \quad X{-}X \xrightarrow{\text{addition}} {-}\underset{H}{\overset{X}{C}}{-}\underset{H}{\overset{X}{C}}{-}$$

two new bonds

The resulting compound from an addition reaction will be saturated.

Addition reactions include *hydrogenation*, *halogenation* and *hydration*.

Hydrogenation

Hydrogenation is an addition reaction where a hydrogen molecule (two hydrogen atoms) 'adds across the double bond' of the alkene to form an alkane. A catalyst is used, usually either nickel at a temperature of 150°C, or platinum at room temperature.

Example G

Hydrogenation of ethene

Addition of hydrogen to ethene produces ethane:

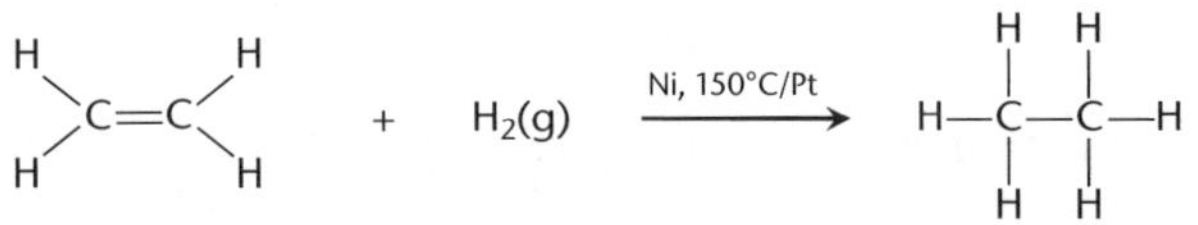

$$H_2C=CH_2 \quad + \quad H_2(g) \xrightarrow{\text{Ni, 150°C/Pt}} H{-}\underset{H}{\overset{H}{C}}{-}\underset{H}{\overset{H}{C}}{-}H$$

Halogenation

Halogenation is an addition reaction where a halogen molecule (two halogen atoms), usually chlorine or bromine, 'adds across the double bond'.

Example H

Halogenation of ethene

Ethene reacts with bromine to form 1,2-dibromoethane:

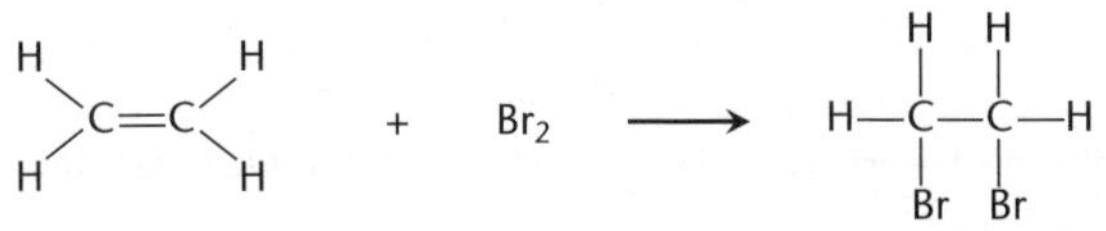

$$H_2C=CH_2 \quad + \quad Br_2 \longrightarrow H{-}\underset{Br}{\overset{H}{C}}{-}\underset{Br}{\overset{H}{C}}{-}H$$

Ethene **1,2-Dibromoethane**

Bromine is commonly used to test for **unsaturation**, since bromine loses its orange-brown colour as it reacts with alkenes. This loss of colour can be used to determine the **rate** of reaction and as a test to distinguish alkenes from alkanes. Alkenes react readily without requiring a catalyst, whereas alkanes require sunlight and react slowly.

Chlorine reacts more rapidly than bromine in addition reactions.

Example I

Chlorination of propene

Propene reacts with chlorine to form 1,2-dichloropropane in a **chlorination** reaction.

$$CH_3CH{=}CH_2 + Cl_2 \longrightarrow CH_3{-}CHCl{-}CH_2Cl$$

Propene **1,2-Dichloropropane**

An addition reaction also occurs between alkenes and the hydrogen halides HCl and HBr.

Example J

Ethene reacting with hydrogen chloride

Ethene reacts with hydrogen chloride to form chloroethane:

$$CH_2{=}CH_2 + HCl \longrightarrow CH_2Cl{-}CH_3$$

Ethene **Chloroethane**

Hydration

Hydration is an addition reaction in which an alkene reacts with steam, H_2O, to form an alcohol.

Example K

Hydration of ethene

Ethene reacts with steam at 330°C and at a pressure of 60 atmospheres in the presence of an acid catalyst to form ethanol:

$$CH_2{=}CH_2 + H_2O(g) \xrightarrow{\text{catalyst}} CH_3{-}CH_2{-}OH$$

Ethene **Ethanol**

In this reaction, the water molecule, H–OH, adds –H to one 'C' of the double bond, and –OH to the other.

This process is used for the industrial preparation of ethanol.

This reaction also takes place when concentrated sulfuric acid is added to ethene and the mixture is then diluted with water.

Addition to unsymmetrical alkenes

An **unsymmetrical alkene** is one that has different groups attached to at least one of the two carbon atoms of the double bond.

The addition of HCl or H–OH (water) to an unsymmetrical alkene produces two possible products. One product (the major product) is produced to a greater extent than the other (the **minor product**).

The rule for determining which is the major product, known as **Markovnikov's rule**, states that:

'The hydrogen atom from HCl or H–OH will bond to the carbon atom in the double bond that has the most hydrogen atoms bonded to it.'

Example L

Addition of HCl to an unsymmetrical alkene (propene)

Propene is an unsymmetrical alkene, because one of the two carbon atoms of the double covalent bond has different groups (CH_3 and H) attached to it.

$$\begin{array}{ccc} CH_3 & & H \\ & C{=}C & \\ H & & H \end{array}$$

Propene, $CH_3CH{=}CH_2$

Addition of HCl to propene produces two products:

$$\underset{\text{carbon 2} \quad \text{carbon 1}}{CH_3CH{=}CH_2} + HCl \longrightarrow \underset{\substack{\textbf{2-Chloropropane} \\ \text{(major product)}}}{CH_3CHClCH_3} + \underset{\substack{\textbf{1-Chloropropane} \\ \text{(minor product)}}}{CH_3CH_2CH_2Cl}$$

2-chloropropane is the major product, because the H atom from the HCl molecule primarily goes to carbon atom number 1 of the propene molecules. Carbon atom number 1 of the double bond has more H atoms (2) than does carbon atom number 2 (only has 1 H atom).

Oxidation reactions

Alkenes undergo an *oxidation reaction* when they react with acidified potassium permanganate (H^+/MnO_4^-). The purple-coloured permanganate decolorises as Mn^{2+} ions form.

$$\underset{\textbf{Ethene}}{CH_2{=}CH_2} \xrightarrow{H^+/MnO_4^-} \underset{\textbf{Ethane-1,2-diol}}{\begin{array}{cc} CH_2 - & CH_2 \\ | & | \\ OH & OH \end{array}}$$

The importance of the reaction is the rapid colour change produced between acidified potassium permanganate and unsaturated molecules. It is used as a test for the presence of double bonds, since alkanes do *not* react with acidified potassium permanganate.

Unit 12.4 Activity 4C: Reactions of alkenes

1. Ethene reacts with steam to produce ethanol according to the equation:

$$CH_2{=}CH_2 + H_2O(g) \xrightarrow[60\text{ atm}]{330°C} CH_3CH_2OH$$

   What is the name given to this type of reaction?
2. Describe what would be observed when ethene is bubbled through bromine water, writing an equation for the reaction.
3. Write balanced equations for the hydrogenation of the following.
   a. Ethene
   b. But-2-ene
   c. 2-Methylbut-2-ene

4. Name and draw the main organic product formed when propene is reacted with hydrogen bromide.
5. Name and draw the main organic product formed when 2-methylpent-2-ene is reacted with concentrated sulfuric acid and water.
6. Use condensed structural formulae to write balanced equations for the following reactions and name the product(s):
   a. Propene reacting with hydrogen over a platinum catalyst.
   b. Propene reacting with chlorine gas.
   c. Bromine, Br_2, reacting with ethene.
   d. Ethene reacting with hydrogen bromide.
   e. Bromine reacting with propene.
   f. Steam reacting with ethene in the presence of a catalyst.
7. Describe two tests you could use to distinguish ethene from ethane.
8. Fats are organic molecules which are often labelled 'saturated' or 'unsaturated'.
   a. What is meant by the term 'unsaturated'?
   b. **i.** Describe a test, using bromine dissolved in a non-polar solvent, that could be carried out in the laboratory to compare the degree of unsaturation of two fats.
   ii. Explain why it is necessary to dissolve the bromine in a non-polar solvent.
9. Complete the following equation by placing a symbol for an atom or group of atoms in each box, and adding a letter or number for each dash.

```
CH3        H                            □   H
   \      /                             |   |
    C = C          +   Br2   ——→    □ — C — C — □
   /      \                             |   |
 H          H                           H   H
Propene, C3H6                   1,2-Dibromo_ _ _ _ _ _ _ _, C_H_Br_
```

Polymers

Polymers are large molecules (**macromolecules**) made of many small **repeating units** known as monomers. Some natural polymers exist, such as starch, wool and DNA. **Synthetic** polymers, made since about 1907, consume over 80% of the organic chemicals used in industry. Most synthetic polymers cannot be broken down by bacteria, so they are said to be **non-biodegradeable**.

Organic polymers form when small organic molecules called **monomers** join together by forming covalent bonds between the monomer molecules. This process is called **polymerisation**:

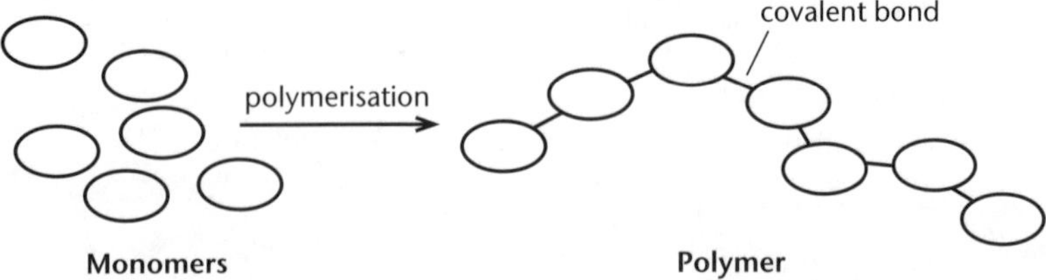

Even the smallest polymer molecule consists of at least 500 monomer units. Chains of monomers can be tangled and held together by weak **intermolecular forces of attraction**:

intermolecular forces of attraction
covalent bond
monomers
polymer

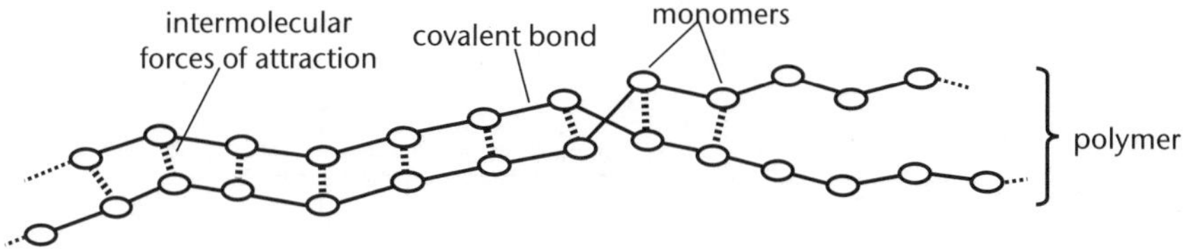

As a polymer chain length grows:

- chains become more tangled.
- intermolecular forces increase in number.
- the melting point becomes higher.
- the polymer becomes stronger.

Alkenes are monomers for a number of common and widely used polymers. The reaction involved is called **addition polymerisation**.

Polymers of this type tend to have properties that make it easy to process the material into sheet and foil form, as well as thread. Since these materials are easily moulded, they are described as **plastics**.

Polyethene (polythene)

Two ethene monomers join together, in the presence of a catalyst, high temperatures and pressure.

Polymerisation of ethene monomers:

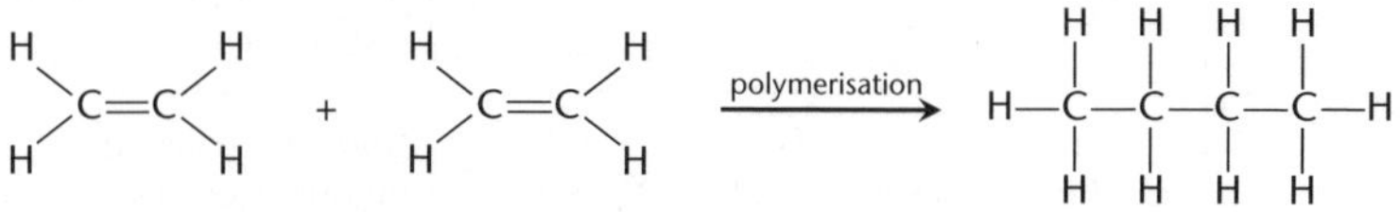

$$H_2C{=}CH_2 + H_2C{=}CH_2 \xrightarrow{\text{polymerisation}} H{-}CH_2{-}CH_2{-}CH_2{-}CH_2{-}H$$

Two ethene monomers

This process is repeated; thousands of monomers join together:

$$n\,(H_2C{=}CH_2) \longrightarrow \left[\!-CH_2{-}CH_2-\!\right]_n$$

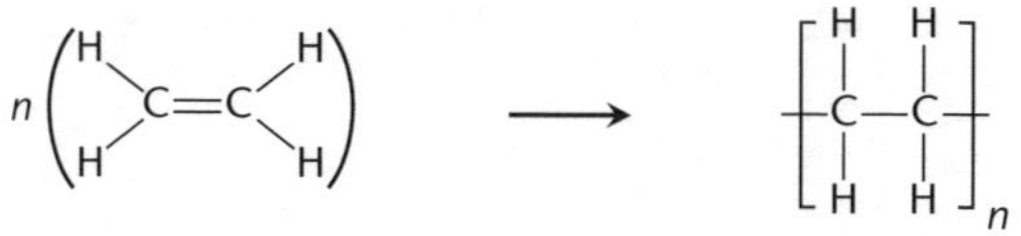

or $\quad n(CH_2{=}CH_2) \longrightarrow {+}\!\left[CH_2{-}CH_2\right]\!{+}_n$

n, the number of monomer units, is very large

Polyvinyl chloride (PVC)

Chloroethene, $CH_2{=}CHCl$, commonly known as vinyl chloride, is the monomer polymerised to make PVC, polyvinyl chloride.

Vinyl chloride monomers are polymerised to produce PVC:

$$n\left(\begin{matrix} H & & & & H \\ & \diagdown & & \diagup & \\ & & C{=}C & & \\ & \diagup & & \diagdown & \\ H & & & & Cl \end{matrix}\right) \xrightarrow{\text{polymerisation}} \left[\begin{matrix} H & & H \\ | & & | \\ C & - & C \\ | & & | \\ H & & Cl \end{matrix}\right]_n$$

Vinyl chloride **PVC**

Properties of some common addition polymers and their uses

Polymer	Monomer		Uses
Polyethene	Ethene	$H_2C{=}CH_2$ (H, H / C=C / H, H)	Bottles, carrier bags, toys, buckets, tubing for carrying cold water
Polyvinyl chloride (PVC)	Vinyl chloride (chloroethene)	$ClHC{=}CH_2$ (H, H / C=C / Cl, H)	Downpipes and guttering, wiring and cable insulation, lino, artificial leather (eg car seats), protective clothing, ground covers for tents
Polypropylene	Propene	$H_2C{=}CH(CH_3)$ (H, CH_3 / C=C / H, H)	Thermal clothing, storage containers, rope, chairs, packaging film
Polystyrene	Styrene	$H_2C{=}CH(C_6H_5)$ (H, H / C=C / H, C_6H_5)	Bean-bag filler, yoghurt pots, expanded form in insulation
Teflon	Tetrafluoroethene	$F_2C{=}CF_2$ (F, F / C=C / F, F)	Non-stick cooking utensik, Gore-Tex, coating underneath skis

Unit 12.4 Activity 4D: Polymers

1. **a.** Explain what is meant by the term 'polymerisation'.

   b. Draw the monomers from which the following polymers are made.

   i. $---CH_2{-}CCl_2{-}CH_2{-}CCl_2----$

   ii. $---CH_2{-}CH(CH_2CH_3){-}CH_2{-}CH(CH_2CH_3){-}CH_2----$

2. Draw the monomer that will form the following polymers.

   a. $---C(CH_3)(H){-}C(H)(H){-}C(CH_3)(H){-}C(H)(H)----$

b. 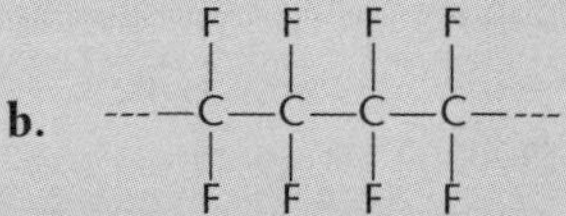

3. Draw a section of the polymer chain formed when the following molecules undergo addition polymerisation.
   a. $CH_2{=}CHCl$
   b. $CH_3CH{=}CHCH_3$
   c. $CH_2{=}CH{-}CN$

4. Match the following terms with their explanations.

a. Polymer	**i.** Two atoms joined together in a molecule using four electrons.
b. Alkene	**ii.** A simple molecule that can be polymerised.
c. Monomer	**iii.** A material that can be easily shaped.
d. Polymerisation	**iv.** A hydrocarbon that contains one double bond.
e. Double bond	**v.** The process of making a polymer from a monomer.
f. Plastic	**vi.** The product of a polymerisation process.

5. **a.** State one use for:
   i. polythene.
   ii. polypropylene.
   b. State what material is used for:
   i. yoghurt cups.
   ii. rainwear.
   iii. non-stick frying pans.

Alkynes

Alkynes are hydrocarbons which contain a carbon–carbon triple bond ($C{\equiv}C$). They have the general formula C_nH_{2n-2}.

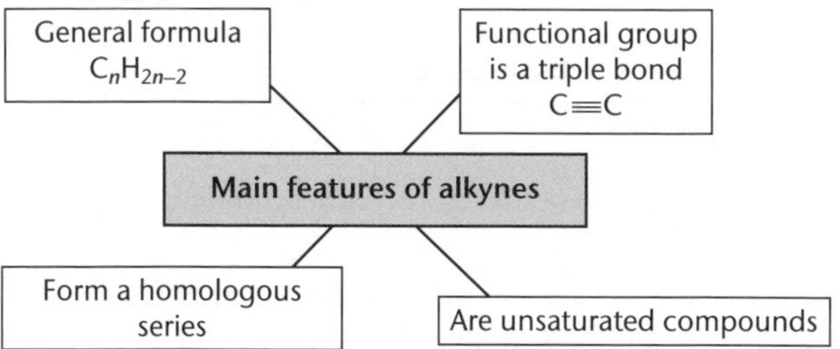

The carbon atoms involved in the triple-bond functional group are each bonded to one other atom in a linear arrangement.

–C≡C–
180°

The first two members of the alkyne family are ethyne and propyne.

Name	Molecular formula	Structural formula
Ethyne (commonly called acetylene)	C_2H_2	H—C≡C—H
Propyne	C_3H_4	H H—C—C≡C—H H

Naming alkynes

Straight-chain alkynes are named by using the same prefixes as for alkanes and alkenes but with the suffix *–yne*, as in alk*yne*. The alkyne with two carbon atoms is ethyne, C_2H_2, and the alkyne with three carbon atoms is propyne, C_3H_4.

Where necessary, the position of the triple bond must be noted (in the same way as the double bond for alkenes).

In branched-chain alkynes, the longest chain that contains the triple bond is the parent chain. The rest of the molecule is named in the same way as for alkanes and alkenes. Branches are numbered in the same way as for alkenes.

Example M

Naming branched alkynes

The structure $CH_3C{\equiv}CCH(CH_3)CH_3$ is named as follows:

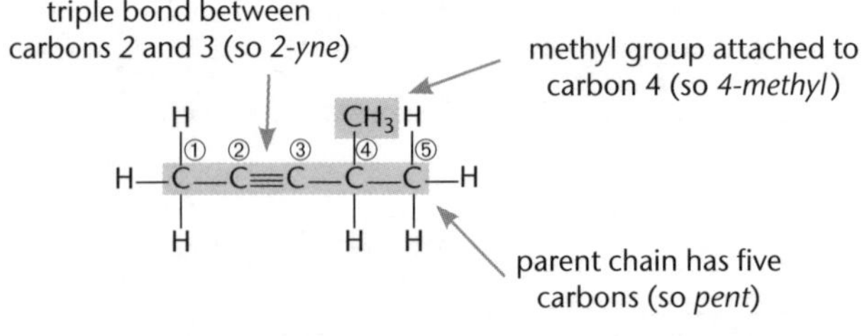

4-Methylpent-2-yne

Reactions of ethyne

Addition reactions occur when alkynes react with either bromine or hydrogen. The reaction occurs in two steps.

- Step 1: One reacting molecule adds to the alkyne. The triple bond becomes a double bond and a substituted alkene is produced.
- Step 2: Another reacting molecule adds to the substituted alkene, producing a saturated product. The double bond becomes a single bond.

Example N

Hydrogenation of ethyne

Step 1: An addition reaction converts the triple bond to a double bond.

$$\underset{\textbf{Ethyne}}{H-C\equiv C-H} + H_2 \xrightarrow{Ni/150^\circ C} \underset{\textbf{Ethene}}{CH_2=CH_2}$$

Step 2: A further addition reaction converts the double bond to a single bond.

$$\underset{\textbf{Ethene}}{CH_2=CH_2} + H_2 \xrightarrow{Ni/150^\circ C} \underset{\textbf{Ethane}}{CH_3-CH_3}$$

Hydrogenation of ethyne usually results in a mixture of gases – ethene, ethane and unreacted ethyne.

Example O

Bromination of ethyne

Ethyne undergoes **bromination** across the triple bond.

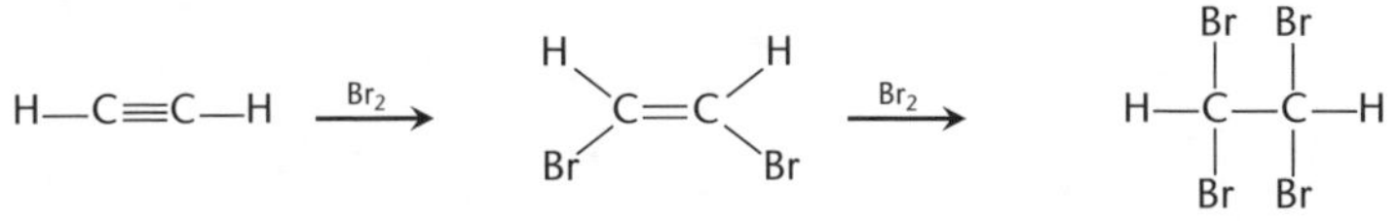

$$H-C\equiv C-H \xrightarrow{Br_2} (H)(Br)C=C(H)(Br) \xrightarrow{Br_2} H-CBr_2-CBr_2-H$$

Ethyne **1,2-Dibromoethene** **1,1,2,2-Tetrabromoethane**

Unit 12.4 Activity 4E: Alkynes

1. Write down the molecular formulae for the alkynes with:

a. 4 carbon atoms.

b. 10 carbon atoms.

c. 12 carbon atoms.

2. Name the following alkynes.

a. $H-C\equiv C-H$

b. $CH_3-C\equiv C-H$

c. 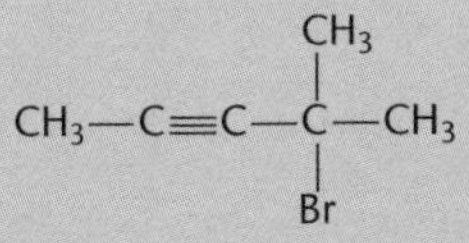 $CH_3-C\equiv C-C(CH_3)(Br)-CH_3$

3. a. Describe what you would observe when ethyne is bubbled into a solution of bromine water.

b. Draw a structural formula for, and name, the organic products for the reaction in **a.**

Unit 12.4 Carbon Compounds

Topic 5: Alcohols

Topic 5 covers alcohols:

- Naming or drawing the structural formulae of alcohols containing up to 8 carbon atoms.
- Classifying alcohols as primary, secondary or tertiary.
- Describing the reactions of alcohols.

Alcohols are organic compounds that contain a hydroxyl (–OH) functional group. Their systematic (IUPAC) name is **alkanol**, and their general formula is $C_nH_{2n+1}OH$.

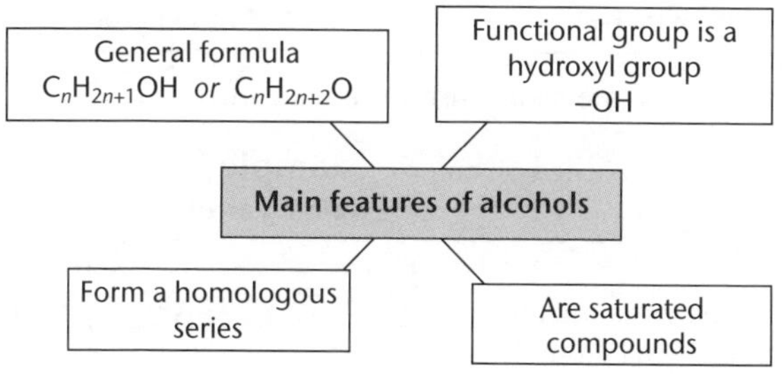

The carbon atom to which the –OH (hydroxyl) functional group is attached has three other atoms bonded in a *tetrahedral* arrangement.

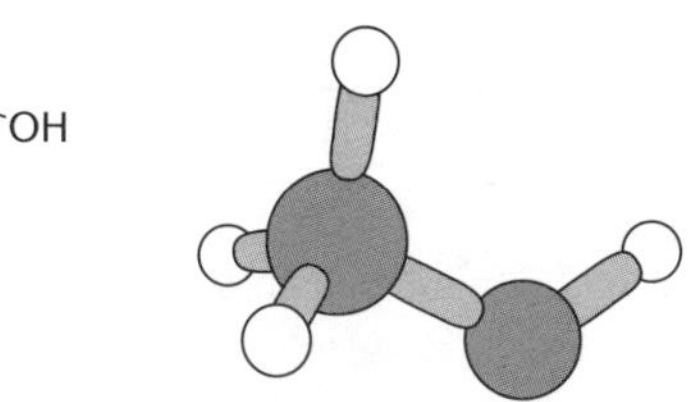

The bonding around the O atom is also tetrahedral, as there are 4 electron clouds around the O; 2 bonding and 2 non-bonding – this results in a 'bent' shape around O, with angle 109°.

Naming alcohols

Alcohols are named by replacing the final -e in the corresponding alkan*e* with -*ol*, giving alkan*ol*.

The first two alcohols are methanol and ethanol.

Name	Molecular formula	Structural formula
Methanol	CH_4O or CH_3OH	H | H—C—O—H | H
Ethanol	C_2H_6O or C_2H_5OH or CH_3CH_2OH	H H | | H—C—C—O—H | | H H

The position of the hydroxyl group must be indicated in the names for alcohols with three or more carbon atoms.

Example A

Propan-1-ol and propan-2-ol

The hydroxyl group in the alcohol with three carbon atoms can either be on the middle carbon or on the end/first carbon.

$$H-\overset{H}{\underset{H}{C}}{}^{③}-\overset{H}{\underset{H}{C}}{}^{②}-\overset{H}{\underset{H}{C}}{}^{①}-\boxed{OH}$$

Propan-1-ol

$$H-\overset{H}{\underset{H}{C}}{}^{③}-\overset{H}{\underset{\boxed{OH}}{C}}{}^{②}-\overset{H}{\underset{H}{C}}{}^{①}-H$$

Propan-2-ol

Alcohols with branch chains are named using the same method as for hydrocarbons.

Steps involved in naming branched-chain alcohols.

Step 1
Identify the parent chain. This is the longest chain with the –OH group attached.

Step 2
Identify the branch chain(s).

Step 3
Number the chain so that the –OH group has the lowest possible number. Give the position of the branch(es).

Step 4
Write the full name as:
branch position(s), 'hyphen', branch name(s), parent name, 'hyphen', OH position, 'hyphen', ol.

Example B

Naming alcohols

$$\boxed{CH_3-\overset{CH_3}{CH}-\underset{OH}{CH}-CH_3}$$

Parent chain has 4 carbon atoms – butanol

$$CH_3-\overset{\boxed{CH_3}\ \text{methyl}}{CH}-\underset{OH}{CH}-CH_3$$

Branch chain – methyl

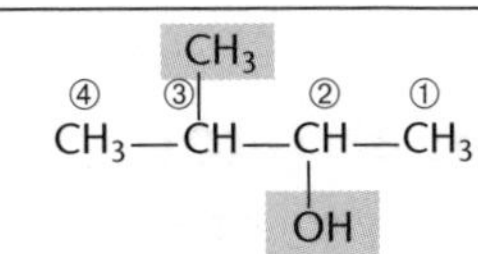

Branch chain on carbon 3, –OH on carbon 2

3-Methylbutan-2-ol

The alcohol is *not* named 2-methylbutan-3-ol, since the functional goup (–OH) must be given the *lowest* possible number.

Alcohols with more than one –OH functional group are called **polyalcohols**. The -ol suffix becomes -*di*ol to signify two hydroxyl groups, and -*tri*ol to signify three hydroxyl groups.

Example C

Polyalcohols

Ethane-1,2-diol is used as an **antifreeze** in car radiators.

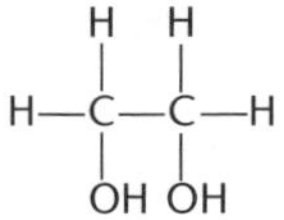

```
    H   H
    |   |
H — C — C — H
    |   |
    OH  OH
```

Propane-1,2,3-triol, also known as **glycerol** or glycerine, is a **viscous** liquid used in hand lotions, the confectionery industry and photography.

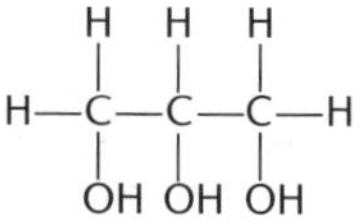

```
    H   H   H
    |   |   |
H — C — C — C — H
    |   |   |
    OH  OH  OH
```

Unit 12.4 Activity 5A: Naming alcohols

1. Name each of the following alcohols:

a. $CH_3—CH_2—OH$

b.

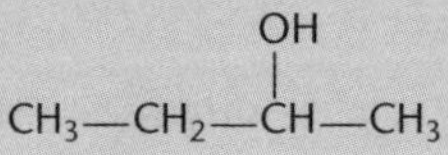

```
              OH
              |
CH3 — CH2 — CH — CH3
```

c.

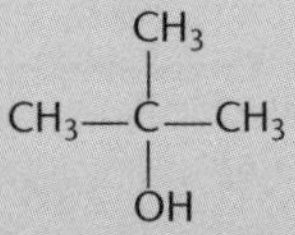

```
       CH3
       |
CH3 — C — CH3
       |
       OH
```

d. $CH_3—CH_2—CH_2—CH_2—OH$

e.

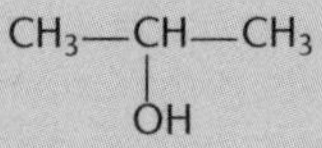

```
CH3 — CH — CH3
       |
       OH
```

f. $CH_3—CH_2—CH_2—OH$

g.

```
              CH3
              |
CH3 — CH2 — C — CH2 — OH
              |
              CH3
```

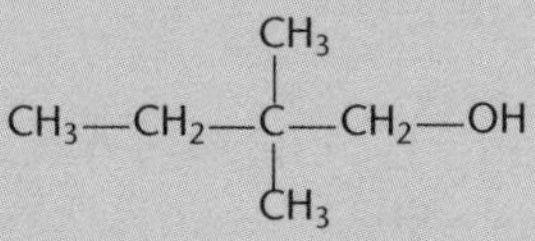

2. Draw structural formulae for the following.

a. Methylpropan-1-ol

b. 3-Methylbutan-2-ol

c. Methanol

d. 3,4-Dimethylpentan-2-ol

3. Two common substances are sorbitol and cholesterol:

- Sorbitol, $CH_2OH(CHOH)_4CH_2OH$; artificial sweetener in toothpaste and breath freshener.
- Cholesterol, $C_{27}H_{45}OH$; found in animal fat.

a. What feature of their structural formulae shows that they are alcohols?

b. What feature of their names shows that they are alcohols?

Isomers

Propan-1-ol and propan-2-ol (see Example A) are structural isomers. Both have molecular formula C_3H_8O (or C_3H_7OH), but the placement of the hydroxyl group differs.

Classifying alcohols

Alcohols are classified as **primary** (**1°**), **secondary** (**2°**) or **tertiary** (**3°**), depending on the number of carbon atoms attached to the carbon atom to which the –OH group is attached.

Example D

Primary, secondary and tertiary alcohols

Propan-1-ol, $CH_3CH_2CH_2OH$, is a primary alcohol, since the –OH group is attached to a carbon atom which itself is attached to only one carbon atom (ie the –OH group is at the end of the chain):

H
CH_3CH_2—C—OH
H

Propan-2-ol, $CH_3CH(OH)CH_3$, is a secondary alcohol, since the –OH group is attached to a carbon atom which itself is attached to two carbon atoms:

H
CH_3—C—CH_3
OH

Methylpropan-2-ol, $CH_3C(CH_3)OHCH_3$, is a tertiary alcohol, since the –OH group is attached to a carbon atom which itself is attached to three carbon atoms:

CH_3
CH_3—C—CH_3
OH

Physical properties of alcohols

Alcohols are colourless compounds.

Small alcohol molecules ($\leq$ 4 carbons atoms) are soluble in water. As the non-polar carbon chain becomes longer, the alcohol molecules become less soluble. Solubility is due to the –OH group. The –OH group is polar due to the high electronegativity of the oxygen atom. In larger alcohol molecules, the non-polar hydrocarbon chain has the greatest influence on solubility.

Example E

Physical properties of methanol and butan-1-ol

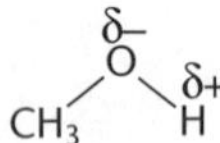

Methanol

Soluble (polar); volatile; boiling point 64°C.

CH_3 CH_2 CH_2 CH_2 O(δ–) H(δ+)

Butan-1-ol

Less soluble (longer hydrocarbon chain); not very volatile; boiling point 118°C.

The melting and boiling points of alcohols are much higher than those of the corresponding alkanes, because **hydrogen bonds** can form between alcohol molecules. These hydrogen bonds are stronger than the forces between alkane molecules.

The boiling points of the alcohols increase with the increasing size of the molecules. This is because there are stronger intermolecular forces present.

Straight-chain alcohols tend to have higher boiling points than branched-chain alcohols.

Alcohols with fewer than seven carbon atoms are liquids at room temperature. The smaller (ie lower molar mass) alcohols are more volatile than the larger (ie higher molar mass) ones.

Name	Boiling point (°C)
Methanol	64
Ethanol	78
Propan-1-ol	98
Butan-1-ol	118
Methylpropanol	108

Unit 12.4 Activity 5B: Isomers and physical properties of alcohols

1. Draw and name the structural isomers for the alcohols with formula $C_4H_{10}O$.

2. The table below sets out the molar masses (*M*) and boiling points of a number of straight-chain alcohols.

Name	*M* (g mol^{-1})	Boiling point (°C)
Methanol	32	64
Ethanol	46	78
Butan-1-ol	74	118
Pentan-1-ol	88	138

a. Plot a graph of molar mass (on the horizontal axis) against boiling point.

b. From the graph, estimate the boiling point of propan-1-ol, C_3H_8O.

c. Another straight-chain alcohol with the molecular formula $C_4H_{10}O$ has a boiling point of 82.6°C. Suggest a structure for a molecule of this compound and give its name.

d. Explain the trend in the boiling points of the four alcohols in the table, with reference to intermolecular bonding.

3. Classify the following as primary, secondary or tertiary alcohols.

a. $CH_3CH_2CH(OH)CH_3$

```
CH3CH2CHCH3
       |
       OH
```

b. $(CH_3)_3CCH_2OH$

```
       CH3
        |
CH3—C—CH2OH
        |
       CH3
```

4. Name and classify each of the following as primary, secondary or tertiary alcohols.
   a. $CH_3CH_2CH_2OH$
   b. $CH_3CHOHCH_2CH_2CH_3$
   c. $CH_3COH(CH_3)CH_3$
   d. $CH_3CH(CH_3)CH_2OH$
   e. $CH_3CH(CH_3)CHOHCH_3$
5. a. Classify each of the following as a primary, secondary or tertiary alcohol:
      i. Ethanol
      ii. Butan-2-ol
      iii. Pentan-3-ol
      iv. 2-Methylbutan-2-ol
      v. Pentan-1-ol
      vi. 3-Methylbutan-2-ol
      vii. 2-Methylpropan-2-ol
      viii. 3-Methylpentan-3-ol
   b. Which of the above are isomers?
6. Two structural isomers of $C_4H_{10}O$ are primary alcohols. Draw their structural formulae and name them.

Chemical properties of alcohols

Combustion

Alcohols burn in air, and, like hydrocarbons, the products of these reactions depend on how much oxygen is available.

Oxidation

Most alcohols can be oxidised by an oxidant such as acidified potassium dichromate, $Cr_2O_7^{2-}/H^+$, or acidified potassium permanganate, MnO_4^-/H^+. Primary alcohols (ie the hydroxyl functional group is on the end of the parent chain) are oxidised to form carboxylic acids.

Example F

Oxidising ethanol

Ethanol, CH_3CH_2OH, is oxidised by acidified potassium dichromate to form ethanoic acid, CH_3COOH (also called **acetic acid**). During the reaction, orange dichromate ions, $Cr_2O_7^{2-}$, are reduced to green chromium(III) ions, Cr^{3+}.

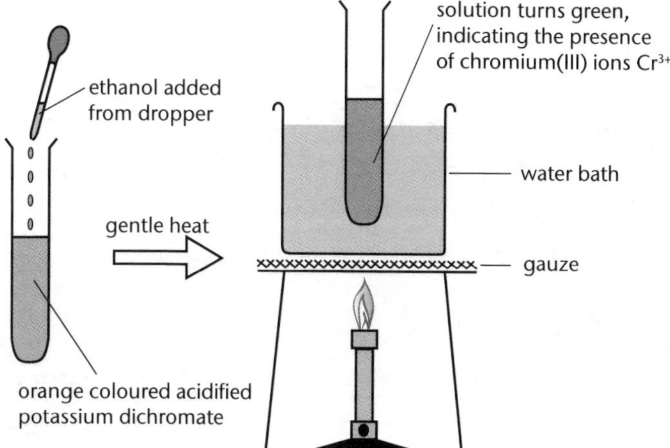

The half-equation for the oxidation of ethanol is:

$$\underset{\text{ethanol}}{CH_3CH_2OH(aq)} + H_2O(\ell) \rightarrow \underset{\text{ethanoic acid}}{CH_3COOH(aq)} + 4H^+(aq) + 4e^-$$

The half-equation for the reduction of dichromate ions is:

$$\underset{\text{orange}}{Cr_2O_7^{2-}(aq)} + 14H^+(aq) + 6e^- \rightarrow \underset{\text{green}}{2Cr^{3+}(aq)} + 7H_2O(\ell)$$

The overall balanced redox equation is:

$$3CH_3CH_2OH(\ell) + 2Cr_2O_7^{2-}(aq) + 16H^+(aq) \rightarrow 3CH_3COOH(aq) + 4Cr^{3+}(aq) + 11H_2O(\ell)$$

Similar oxidation reactions occur for other alcohols: methanol oxidises to methanoic acid, propan-1-ol oxidises to propanoic acid, butan-1-ol oxidises to butanoic acid, etc.

Dilute solutions of ethanol can also be oxidised by *Acetobacter* bacteria to form ethanoic acid. This is how vinegar is produced commercially.

Concentrated ethanol kills bacteria; high-percentage ('high proof') alcoholic drinks don't 'go off' when uncorked/exposed to air.

Elimination

Removing a water molecule from an alcohol produces an alkene. The reaction occurs when an alcohol is heated with concentrated sulfuric acid.

The removal of two atoms or groups of atoms is known as an **elimination reaction**; the –OH group and a hydrogen atom from a neighbouring atom are removed.

Example G

Dehydration (elimination) of ethanol

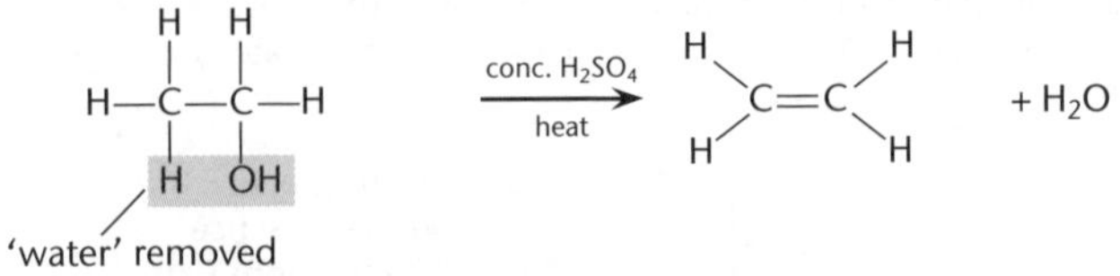

$$H_3C\text{–}CH_2OH \xrightarrow[\text{heat}]{\text{conc. } H_2SO_4} H_2C{=}CH_2 + H_2O$$

'water' removed

Unit 12.4 Activity 5C: Reactions of alcohols

1. Write an equation for the combustion of the following with excess oxygen.
   - **a.** Methanol.
   - **b.** Ethanol.
2. Give an observation for the reaction occurring when propan-1-ol is warmed with dilute acidified:
   - **a.** potassium permanganate.
   - **b.** potassium dichromate.
3. Draw structural formulae for the organic products of the following reactions.
   - **a.** $CH_3CH_2CH_2CH_2OH \xrightarrow{Cr_2O_7^{2-}/H^+}$
   - **b.** $CH_3CH_2CH_2OH \xrightarrow[\text{heat}]{\text{conc. } H_2SO_4}$
   - **c.** $CH_3CH_2CH(OH)CH_2CH_3 \xrightarrow[\text{heat}]{\text{conc. } H_2SO_4}$
   - **d.** $CH_3OH \xrightarrow{MnO_4^-/H^+}$

Unit 12.4 Activity 5D: Alcohols – multiple choice

1. The name of the alcohol with three carbons in the chain and a hydroxyl group on the second one is:
   - **A.** propan-1-ol.
   - **B.** butan-2-ol.
   - **C.** propan-2-ol.
   - **D.** methanol.
2. Name the alcohol $CH_3CH_2CH(OH)CH_2CH_3$.
   - **A.** Pentan-2-ol.
   - **B.** Butan-2-ol.
   - **C.** Hexanol.
   - **D.** Pentan-3-ol.
3. The relatively high boiling and melting points of alcohols compared to alkanes is due to:
   - **A.** covalent bonding.
   - **B.** ionic bonding.
   - **C.** hydrogen bonding
   - **D.** none of the above.
4. The product of the reaction of butanol with acidified potassium dichromate is:
   - **A.** butene.
   - **B.** butanoic acid.
   - **C.** CO_2
   - **D.** butane-1,2-diol.
5. The product of the reaction of heptan-1-ol with concentrated sulfuric acid is:
   - **A.** heptene.
   - **B.** heptanoic acid.
   - **C.** CO_2
   - **D.** heptane-1,2-diol.

Unit 12.4 Carbon Compounds

Topic 6: Carboxylic acids, esters and soaps

Topic 6 covers:

- Describing reactions of carboxylic acids and esters.
- Naming or drawing the structural formulae of carboxylic acids and esters containing up to eight carbon atoms.

Carboxylic acids

Carboxylic acids, or **organic acids**, all contain the —C(=O)—OH (or –COOH) functional group.

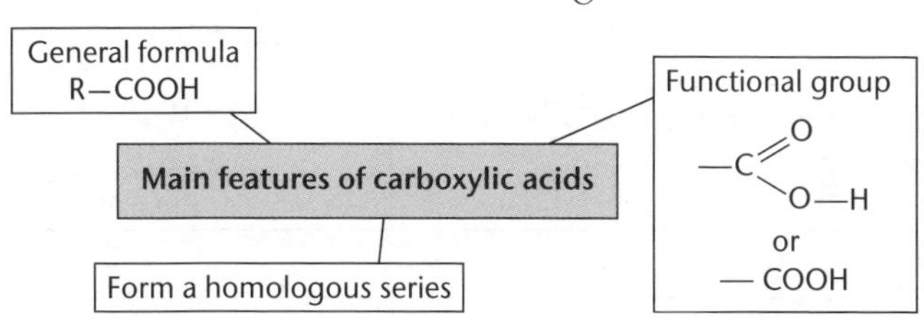

In the general formula, R–COOH, R represents a hydrogen atom or a hydrocarbon chain (or *alkyl* group). Different hydrocarbon chains give rise to the different members of the carboxylic acid family:

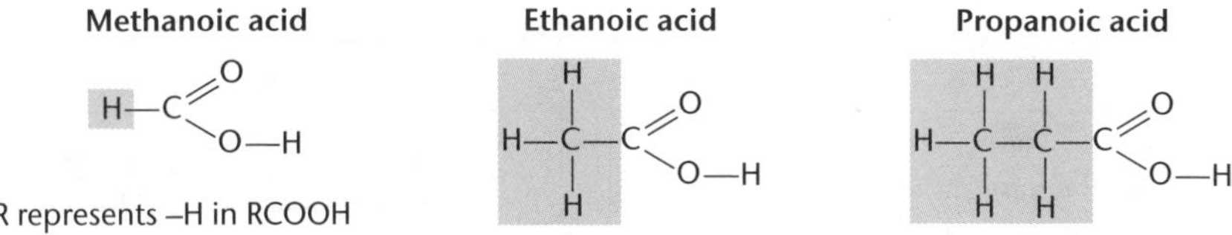

R represents the $-CH_3$ group in RCOOH

R represents the $-C_2H_5$ group in RCOOH

Naming carboxylic acids

The straight-chain carboxylic acids are named by using the corresponding alkane name and replacing the ending *-e* with *-oic acid*.

Name	Molecular formula	Lewis structure	Structural formula
Methanoic acid (formic acid)	HCOOH	H—C(=O:)—Ö—H (with lone pairs on both O atoms)	H—C(=O)—O—H
Ethanoic acid (acetic acid)	CH_3COOH	H—CH₂... : H—C(H)(H)—C(=O:)—Ö—H (with lone pairs on both O atoms)	H—C(H)(H)—C(=O)—O—H

A branched-chain carboxylic acid is named in the same way as other branched-chain molecules.

Example A

Naming a branched-chain carboxylic acid

To name the molecule:

```
        CH3
        |
CH3—C—CH2—C—OH
        |        ||
        H        O
```

Steps involved in naming branched-chain carboxylic acids.

Step 1

Identify the parent chain. This is the longest carbon chain with the COOH functional group attached to the end. The C of COOH is always carbon 1.

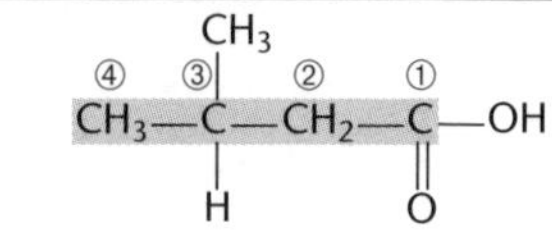

4-carbon atom parent chain – butanoic acid

Step 2

Identify the branch chain.

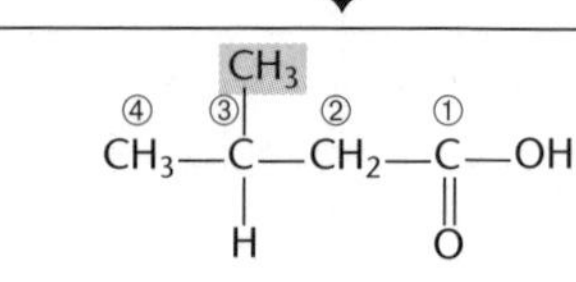

Branch chain – methyl

Step 3

Identify the position of the branch(es).

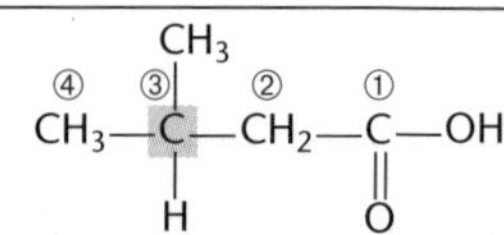

Methyl group at carbon atom 3

Step 4

Write the full name as:

branch position, 'hyphen', branch name and parent name, acid.

3-methylbutanoic acid

The parent chain could equally be numbered:

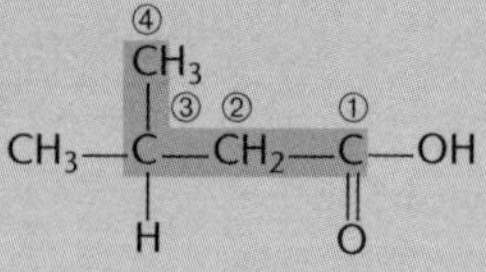

This would still be named 3-methylbutanoic acid.

Unit 12.4 Activity 6A: Carboxylic acids – structure and naming

1. a. Draw the structural formula for propanoic acid, $C_3H_6O_2$.

b. Explain why the carboxylic acids are an homologous series, using ethanoic acid and propanoic acid as examples.

2. Draw structural formulae for the following molecules.

a. Butanoic acid **b.** 2-Methylpentanoic acid **c.** 2-Ethylbutanoic acid

3. Name the following carboxylic acid molecules:

a.
```
HO—C—CH3
   ||
   O
```

b.
```
            CH3
HO          |
  \C——C——CH3
 O//    |
        CH3
```

c.
```
        CH3
        |
CH3—C—CH2—CH2—C—OH
        |              ||
        CH3            O
```

d.
```
CH3—CH——C—OH
       |     ||
       CH2CH3 O
```

Physical properties of carboxylic acids

Small (ie low molar mass) carboxylic acids, such as methanoic acid and ethanoic acid, are soluble in water, because the –COOH group is polar due to the high electronegativity of the two oxygen atoms. Like alcohols, the solubility of acids decreases as the non-polar carbon chain length increases.

Carboxylic acids are colourless liquids or white solids at room temperature.

Short-chain carboxylic acids have strong, often unpleasant, odours (eg smell of rancid butter is caused by butanoic acid). Higher molar mass (and therefore lower volatility) carboxylic acids are less 'smelly'.

Carboxylic acids have higher melting points and boiling points than comparable hydrocarbons and alcohols, because of their increased molar mass and their ability to form hydrogen bonds to each other.

```
      O·······H—O
     /           \
R—C               C—R
     \\          //
      O—H·······O
              \
```

H bond (a relatively strong intermolecular bond)

Chemical properties of carboxylic acids

Acid reactions

Carboxylic acids have typical acid properties, but they are weak acids – only about one acid molecule in every 100 will ionise to form $H_3O^+(aq)$.

Example B

Ethanoic acid – a weak acid

Ethanoic acid, CH_3COOH, will *partially* ionise in water to form hydronium ions, H_3O^+, and ethanoate ions, CH_3COO^-:

$$\underset{\text{ethanoic acid}}{CH_3COOH(aq)} + H_2O(\ell) \rightleftharpoons \underset{\text{ethanoate ion}}{CH_3COO^-(aq)} + H_3O^+(aq)$$

The ions formed from carboxylic acids are named by replacing the *–ic acid* by *–ate*; ethanoic *acid* forms ethano*ate* ions, methano*ic acid* forms methano*ate* ions.

Since they are acidic, solutions of carboxylic acids will:

- Turn blue litmus red.
- Conduct electricity – because sufficient numbers of ions are present and free to move.
- React slowly with metals like magnesium to form hydrogen gas.
  $2CH_3COOH(aq) + Mg(s) \rightarrow Mg^{2+}(aq) + 2CH_3COO^-(aq) + H_2(g)$
- React with bases like sodium hydroxide solution in a *neutralisation* reaction, forming water.
  $CH_3COOH(aq) + OH^-(aq) \rightarrow CH_3COO^-(aq) + H_2O(\ell)$
- React with carbonates and hydrogen carbonates to form water and carbon dioxide gas.
  $2CH_3COOH(aq) + CO_3^{2-}(aq) \rightarrow 2CH_3COO^-(aq) + H_2O(\ell) + CO_2(g)$
  $CH_3COOH(aq) + HCO_3^-(aq) \rightarrow CH_3COO^-(aq) + H_2O(\ell) + CO_2(g)$

Carboxylic acid reactions with metals, bases and carbonates produce organic salts. These salts appear as crystalline solids when the aqueous solution is evaporated.

Example C

Forming organic salts

Magnesium metal reacts with ethanoic acid. The magnesium ions, Mg^{2+}, formed during the reaction, and the ethanoate ions, CH_3COO^-, present in solution, combine to form magnesium ethanoate, an organic salt.

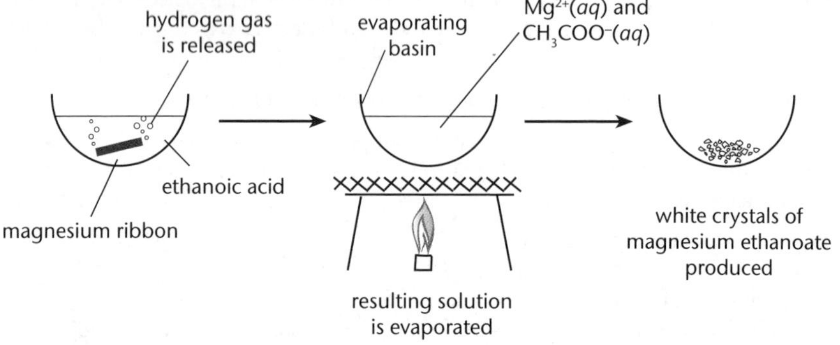

The reaction can be written:

magnesium + ethanoic acid → magnesium ethanoate + hydrogen gas

$Mg(s) + 2CH_3COOH(aq) \rightarrow (CH_3COO)_2Mg(aq) + H_2(g)$

Example D

Neutralising organic acids

Methanoic acid reacts with the base, sodium hydroxide. The solution produced can be evaporated to give salt crystals of sodium methanoate:

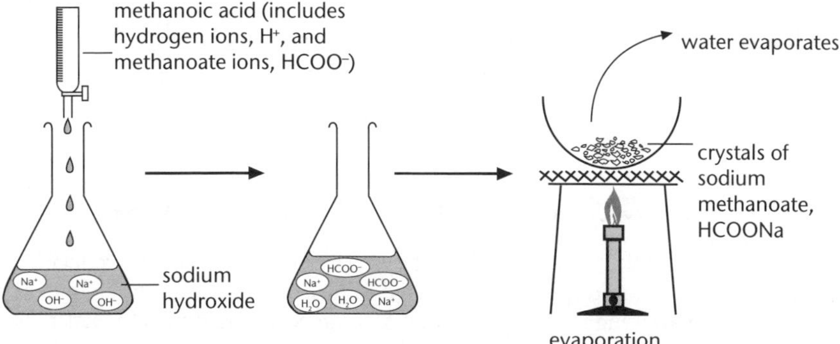

$$HCOOH(aq) + NaOH(aq) \rightarrow HCOONa(aq) + H_2O(\ell)$$

Unit 12.4 Activity 6B: Properties of carboxylic acids

1. Name the salt formed when propanoic acid reacts with sodium hydroxide solution.
2. Write a balanced equation showing:
   a. methanoic acid dissolving in water.
   b. zinc metal reacting with a solution of ethanoic acid.
3. Select, with reasons, the reaction(s) that show carboxylic acids acting as acids:
   a. $HCOOH + NaHCO_3 \rightarrow HCOONa + H_2O + CO_2$
   b. $KOH + C_3H_7COOH \rightarrow C_3H_7COOK + H_2O$
   c. $C_2H_5OH + C_2H_5COOH \rightarrow C_2H_5COOC_2H_5 + H_2O$
   d. $2CH_3COOH + Mg \rightarrow (CH_3COO)_2Mg + H_2$
4. Write an equation to show ethanoic acid reacting with sodium hydroxide, and name the salt formed.

Esters

Esters are organic compounds with the formula $R—\underset{\underset{O}{\|}}{C}—O—R'$, where R and R′ are alkyl groups.

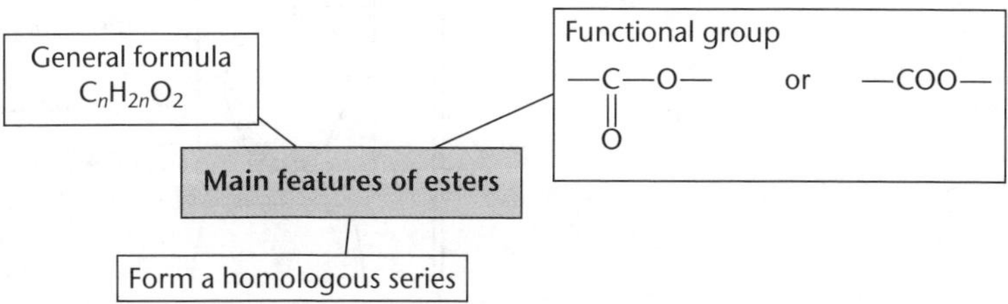

Most esters are oily, volatile liquids, only very slightly soluble in water because they are only slightly polar.

Esterification

Carboxylic acids react with alcohols to form esters. This process, called **esterification,** is an example of a **condensation reaction**, since the –OH part of an acid group and the –H part of a hydroxyl (alcohol) group combine together to make water, H_2O, as a product.

Using R and R′ to represent the different alkyl groups on the carboxylic acid and the alcohol respectively, the general reaction occurring can be written:

$$\underset{\text{Carboxylic acid}}{R{-}\overset{\displaystyle O}{\overset{\|}{C}}{-}OH} + \underset{\text{Alcohol}}{R'{-}OH} \rightleftharpoons \underset{\text{Ester}}{R{-}\overset{\displaystyle O}{\overset{\|}{C}}{-}O{-}R'} + \underset{\text{Water}}{H_2O}$$

The reaction is reversible (shown by $\rightleftharpoons$), as the ester can react with water, producing the acid and alcohol again. Addition of concentrated sulfuric acid increases the amount of ester formed by preventing the backward reaction from occurring; conc H_2SO_4 acts as a dehydrating agent, removing water from the reaction mixture.

Example E

Esterification

The esterification reaction between propanoic acid and ethanol with sulfuric acid forms the ester ethyl propanoate:

$$\underset{\text{Propanoic acid}}{CH_3CH_2COOH} + \underset{\text{Ethanol}}{CH_3CH_2OH} \xrightleftharpoons{\text{conc. } H_2SO_4} \underset{\text{Ethyl propanoate}}{CH_3CH_2\overset{\overset{\displaystyle O}{\|}}{C}OCH_2CH_3} + \underset{\text{Water}}{H_2O}$$

In esterification, the yield can be improved if the reagents are heated together in a process known as **refluxing**. The reaction mixture boils and the product and some of the reagents vaporise and move up the condenser. The gases liquefy when they reach the colder parts of the condenser and run back down into the flask. This process provides maximum product formation, as any unreacted reagent that **vaporises** is returned to the reaction flask. Unreacted acid is removed by adding sodium carbonate solution after refluxing is finished.

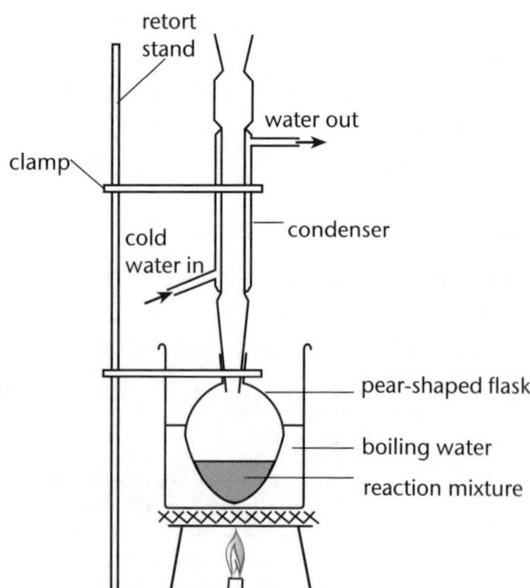

When refluxing, a water bath is used rather than heating directly with a Bunsen burner, as the Bunsen flame could cause the volatile reagents and products to ignite.

Naming esters

The naming system for esters involves identifying the carboxylic acid and alcohol from which the ester was made.

Example F

Naming esters

To name the molecule:

$CH_3—CH_2—C(=O)—O—CH_3$

Steps involved in naming esters.

Step	Example
Step 1 'Split' the molecule in two about the –O–.	$CH_3—CH_2—C(=O) \vdots O—CH_3$
Step 2 Identify the alcohol as the part with –O–. Name as for branch chain.	alcohol group: $O—CH_3$ — methyl group
Step 3 Identify the acid as the part with —C(=O)—. Name as for alkane, but replacing *-ane* with *-oate*.	acid group: $CH_3—CH_2—C(=O)$ — propanoate group
Step 4 Place the alcohol derived part (*–yl* part) before the acid part (*–oate* part).	Methyl propanoate

The alcohol and acid 'parts' of the name are kept separate – not 'run together' as in most organic compound names. Thus:

- $CH_3—C(=O) \vdots O—CH_3$ is called methyl ethanoate. ($CH_3—C(=O)$ from ethanoic acid; CH_3 from methanol)
- $H—C(=O) \vdots O—C_3H_7$ is called propyl methanoate.
- $C_2H_5—O \vdots C(=O)—C_4H_9$ is called ethyl pentanoate. (Note that the molecule shown has the alcohol part first.)

Hydrolysis of esters

Hydrolysis of esters is the 'reverse' of esterification and results in the ester molecule being split apart.

- In the presence of an acid catalyst, a carboxylic acid and an alcohol will be produced.
- In basic conditions, an alcohol and a carboxylate ion, $RCOO^-$, will be produced, resulting in the salt of the acid being formed.

Example G

Hydrolysis of ethyl ethanoate

In acid conditions: $CH_3COOCH_2CH_3 + H_2O \xrightleftharpoons{H^+} CH_3COOH + CH_3CH_2OH$

In basic conditions: $CH_3COOCH_2CH_3 + OH^- \rightarrow CH_3COO^- + CH_3CH_2OH$

or $CH_3COOCH_2CH_3 + NaOH \rightarrow CH_3COONa + CH_3CH_2OH$

Sources and uses of esters

Many esters occur naturally. They often have pleasant fruity smells and are found in flowers and fruit, eg ethyl methanoate is found in raspberries, ethyl butanoate is found in pineapples, methyl butanoate is found in apples.

Synthetic esters are used as flavourings, fragrances and in the biological control of insect pests. Synthetic **polymeric esters**, such as terylene, are very important in the fabric industry.

Some insects produce esters as sex attractants.

Triglycerides are naturally occurring esters found in fats and edible oils.

Unit 12.4 Activity 6C: Esters

1. Draw structural formulae for the organic products of the following reactions:

a. $H_3C-\underset{\underset{O}{\|}}{C}-OH + CH_3CH_2CH_2OH \rightarrow$

b. $CH_3CH_3\underset{\underset{O}{\|}}{C}OH + CH_3CH_2OH \rightarrow$

c. $H\underset{\underset{O}{\|}}{C}OH + CH_3CH_2CH_2CH_2OH \rightarrow$

d. $CH_3CH_2OH + CH_3\underset{\underset{O}{\|}}{C}OH \rightarrow$

2. Write equations and name the products for:

a. Methanoic acid reacting with methanol in the presence of concentrated sulfuric acid.

b. The reaction between methanol and propanoic acid in the presence of concentrated sulfuric acid.

3. Draw structural formulae for each of the following esters.

a. Methyl butanoate.

b. Propyl propanoate.

c. Pentyl methanoate.

d. Butyl ethanoate.

e. Ethyl methanoate.

f. Ethyl butanoate.

4. Name the following esters and the alcohol and carboxylic acid that could be used to make each of them.

a. 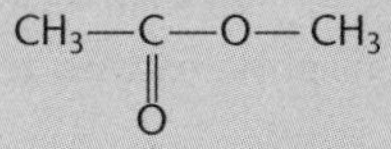 $CH_3-\underset{\underset{O}{\|}}{C}-O-CH_3$

b. 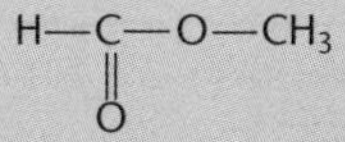 $H-\underset{\underset{O}{\|}}{C}-O-CH_3$

c. $CH_3COOCH_2CH_2CH_3$

d. 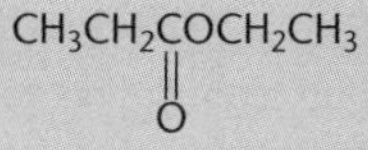 $CH_3CH_2\underset{\underset{O}{\|}}{C}OCH_2CH_3$

e. 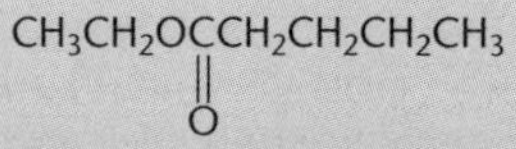 $CH_3CH_2O\underset{\underset{O}{\|}}{C}CH_2CH_2CH_2CH_3$

f. 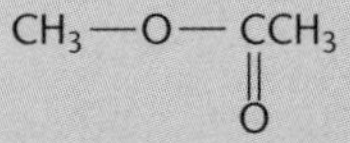 $CH_3-O-\underset{\underset{O}{\|}}{C}CH_3$

5. a. Describe a characteristic property of esters of low molar mass.

b. Draw and name two different esters containing three carbon atoms.

c. Name the organic acid and alcohol used to prepare methyl pentanoate.

d. Briefly describe the laboratory procedure required to obtain a sample of ethyl pentanoate from the reaction mixture in **c**.

6. What would be observed when propan-1-ol is warmed with methanoic acid in the presence of concentrated sulfuric acid?

7. Complete the equation for the following hydrolysis reactions:

a. 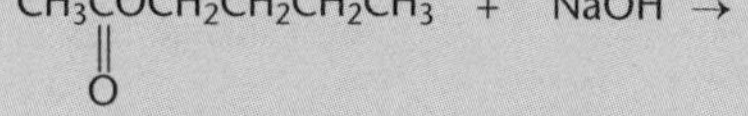 $CH_3\underset{\underset{O}{\|}}{C}OCH_2CH_2CH_2CH_3 \;+\; NaOH \rightarrow$

b. $CH_3CH_2\underset{\underset{O}{\|}}{C}OCH_3 \;+\; H_2O \xrightarrow{H^+}$

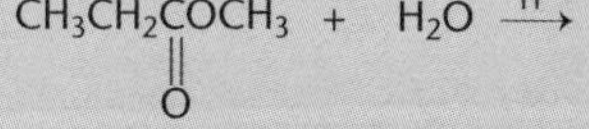

c. $CH_3CH_2CH_2O\underset{\underset{O}{\|}}{C}H \;+\; H_2O \xrightarrow{H^+}$

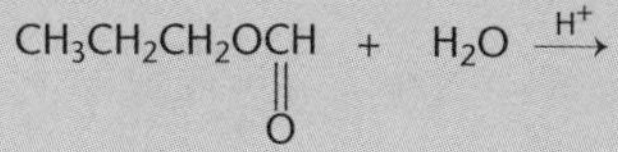

d. 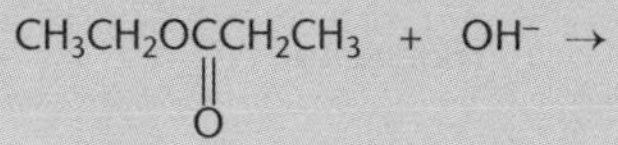 $CH_3CH_2O\underset{\underset{O}{\|}}{C}CH_2CH_3 \;+\; OH^- \rightarrow$

Fats and oils

Fats and edible **oils** are esters of fatty acids.

- Fats are soft solids at room temperature. They are saturated compounds and are found in animals.
- Oils are liquids at room temperature. They are unsaturated compounds and come from plant sources such as seeds, nuts and berries.

Fatty acids are carboxylic acids with *long* hydrocarbon chains containing even numbers of carbon atoms (usually between 12 and 20 carbon atoms).

Fatty acids can be represented, like other carboxylic acids, as RCOOH, where R stands for the long chain. Fatty acids can be saturated, monounsaturated or **polyunsaturated**.

Example H

Stearic acid

Octadecanoic acid or **stearic acid,** $CH_3(CH_2)_{16}COOH$, is a fatty acid:

```
CH3  CH2  CH2  CH2  CH2  CH2  CH2  CH2  CH2   OH
   \ /  \ /  \ /  \ /  \ /  \ /  \ /  \ /  \ /
   CH2  CH2  CH2  CH2  CH2  CH2  CH2  CH2   C
                                            ||
                                            O
```

Stearic acid is a **saturated fatty acid** because the long hydrocarbon chain has no double bonds between any carbon atoms.

Oleic acid

9-Octadecenoic acid or **oleic acid,** $CH_3(CH_2)_7CH{=}CH(CH_2)_7COOH$, is an **unsaturated fatty acid** because it contains a double bond between carbons 9 and 10. Oleic acid is monounsaturated:

```
   17   15   13   11            8    6    4    2
   CH2  CH2  CH2  CH2          CH2  CH2  CH2  CH2  OH
  /  \ /  \ /  \ /   \        /  \ /  \ /  \ /  \ /
CH3  CH2  CH2  CH2     C == C    CH2  CH2  CH2   C
18   16   14   12   H/ 10  9 \H  7    5    3     ||
                                                 O
```

Some fatty acids from plant sources have more than one double bond. These are described as **polyunsaturated fatty acids**.

Fats and oils are **triesters** of the alcohol glycerol, propane-1,2,3-triol. They are sometimes known as triglycerides, as *three* fatty acids are joined to the *glycerol* molecule through ester linkages, as the –OH groups react with the –COOH groups of fatty acids.

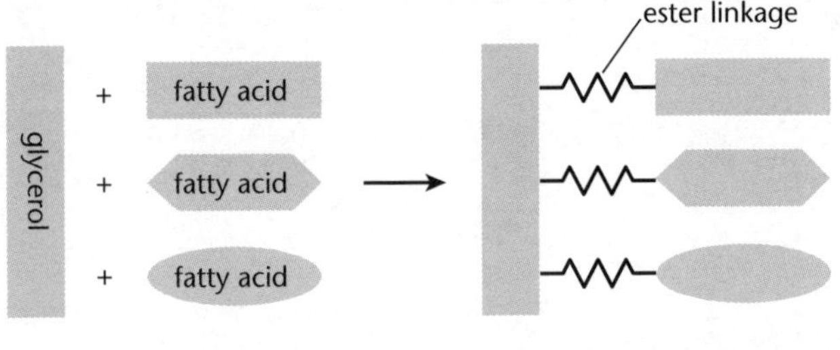

Formation of triglycerides

Example I

Glycerol tristearate is a triglyceride formed from the esterification reaction between glycerol and three stearic acid molecules:

$$\begin{array}{l} H_2C{-}OH \\ \quad | \\ HC{-}OH \\ \quad | \\ H_2C{-}OH \end{array} + 3CH_3(CH_2)_{16}COOH \rightarrow \begin{array}{l} H{-}\overset{H}{\overset{|}{C}}{-}O{-}\underset{O}{\underset{\|}{C}}{-}(CH_2)_{16}CH_3 \\ H{-}C{-}O{-}\underset{O}{\underset{\|}{C}}{-}(CH_2)_{16}CH_3 \\ H{-}\underset{H}{\underset{|}{C}}{-}O{-}\underset{O}{\underset{\|}{C}}{-}(CH_2)_{16}CH_3 \end{array} + 3H_2O$$

From glycerol, with its three alcohol groups | From three stearic acid molecules

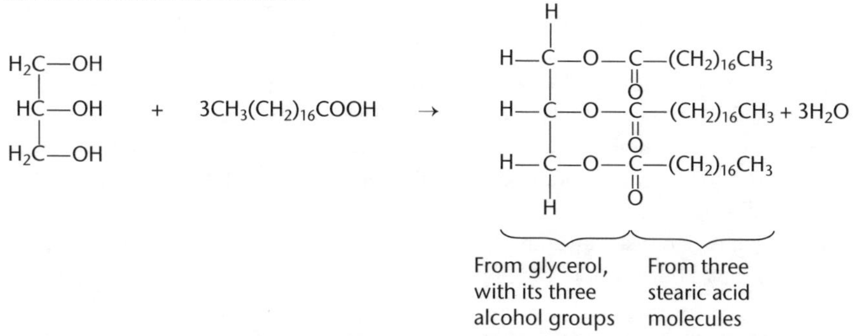

Glycerol + stearic acid → glycerol tristearate

Soap

Soap is formed when sodium hydroxide is heated with triglyceride molecules to form sodium salts. Hydroxide ions break ester linkages to form fatty acid anions and glycerol (as a by-product). The reaction is known as **saponification** and is the *reverse* of esterification:

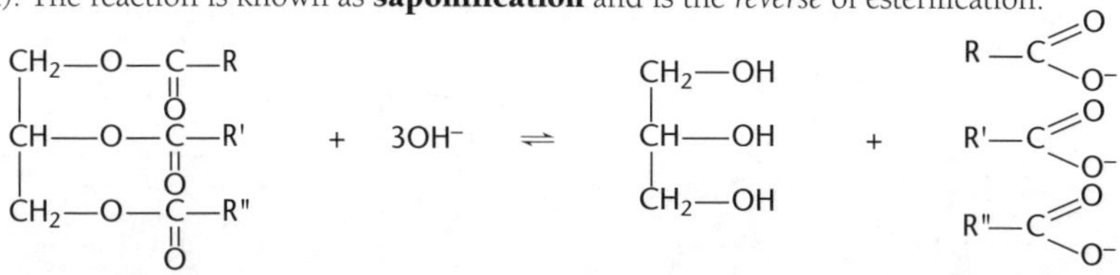

$$\begin{array}{l} CH_2{-}O{-}\underset{O}{\underset{\|}{C}}{-}R \\ CH{-}O{-}\underset{O}{\underset{\|}{C}}{-}R' \\ CH_2{-}O{-}\underset{O}{\underset{\|}{C}}{-}R'' \end{array} + 3OH^- \rightleftharpoons \begin{array}{l} CH_2{-}OH \\ CH{-}OH \\ CH_2{-}OH \end{array} + \begin{array}{l} R{-}C({=}O)O^- \\ R'{-}C({=}O)O^- \\ R''{-}C({=}O)O^- \end{array}$$

Triglyceride **Glycerol** **Fatty acid anions**

As the mixture cools, the sodium *cations* from sodium hydroxide and the fatty acid *anions* from the triglyceride combine to form soap. The hydrocarbon chains (R, R' and R") determine the type of soap produced.

Example J

Making sodium stearate

In beef tallow, ester linkages form from (three) stearic acid molecules in combination with a glycerol molecule; a stearate ester forms. When the stearate ester reacts with sodium hydroxide solution, the soap formed is made of sodium stearate, $CH_3(CH_2)_{16}COONa$:

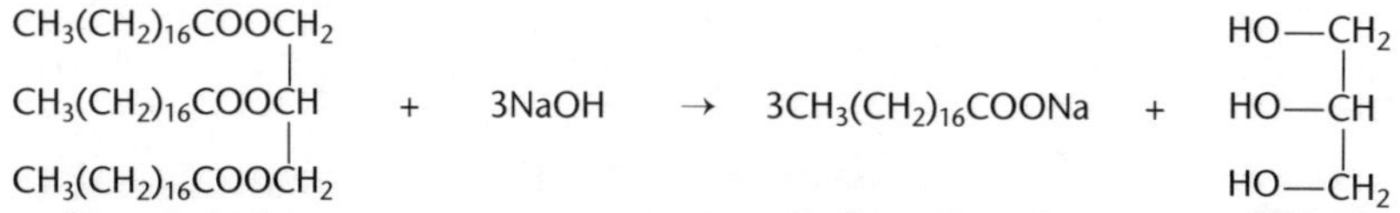

$$\begin{array}{l} CH_3(CH_2)_{16}COOCH_2 \\ CH_3(CH_2)_{16}COOCH \\ CH_3(CH_2)_{16}COOCH_2 \end{array} + 3NaOH \rightarrow 3CH_3(CH_2)_{16}COONa + \begin{array}{l} HO{-}CH_2 \\ HO{-}CH \\ HO{-}CH_2 \end{array}$$

Stearate ester **Sodium stearate** **Glycerol**

If potassium hydroxide is used instead of sodium hydroxide, soft liquid (or flowing) soaps are produced.

Unit 12.4 Activity 6D: Fatty acids and soaps

1. Give three differences between fats and oils.
2. Write a balanced equation for the reaction of oleic acid, 9-octadecenoic acid, with sodium hydroxide solution.
3. A particular triglyceride is formed from glycerol and three stearic acid molecules.
   - **a.** Name the organic functional group in a triglyceride.
   - **b.** Draw a structural formula for the triglyceride.
   - **c.** Write a balanced equation showing the triglyceride reacting with sodium hydroxide.
   - **d.** What type of reaction is this?
4. The molecule shown is found in an animal fat.

$$\begin{array}{l} \overset{\displaystyle O}{\|} \\ CH_2O\overset{}{C}(CH_2)_{14}CH_3 \\ |\quad \overset{\displaystyle O}{\|} \\ CHOC(CH_2)_{14}CH_3 \\ |\quad \overset{\displaystyle O}{\|} \\ CH_2OC(CH_2)_{14}CH_3 \end{array}$$

   - **a.** What functional group is found in this molecule?
   - **b.** **i.** When animal fats are reacted with NaOH and heated, the sodium salt of the fatty acid is formed. This salt is a soap.
     Write the formula of the sodium salt formed from the fat shown.
     ii. The other molecule formed in the manufacture of soap is an alcohol. Draw the structure of this alcohol molecule.
   - **c.** Explain how the structure of the molecules found in plant oils will differ from those found in animal fats.

Unit 12.4 Carbon Compounds

Topic 7: Methanol, ethanol and ethanoic acid

Topic 7 covers methanol, ethanol and ethanoic acid:

- Properties of methanol, ethanol and ethanoic acid – solubility in water, melting and boiling points.
- Complete and incomplete combustion (including balanced equations) of methanol, ethanol and ethanoic acid.
- Production of ethanol by fermentation.
- Formation of ethanoic acid from ethanol.

Methanol production

Methanol is made from natural gas in a two-stage process:

- In the first stage, the natural gas is mixed with steam at 800–850°C with a nickel catalyst to produce **synthesis gas**, a mixture of carbon monoxide and hydrogen:
  $CH_4(g) + H_2O(g) \rightarrow CO(g) + 3H_2(g)$
- In the second stage, the synthesis gas is compressed to 100 atmospheres (10 MPa), heated to 200–300°C, and passed over a copper catalyst to produce methanol:
  $CO(g) + 2H_2(g) \rightarrow CH_3OH(g)$

Ethanol production

Ethanol – commonly called 'alcohol' – is made by a natural process of **fermentation**, ie:

- Wine is produced by fermenting the **glucose** found in fruit juice (eg grape juice); the flavour of the wine depends on the juice used.
- The brewing industry produces ethanol by fermenting starch found in grains (eg barley, rice, maize) in a three-stage process:

Fermentation of starch and glucose to produce ethanol

Stage of fermentation	Equations illustrating stages of the fermentation process
1	$2(C_6H_{10}O_5)_n(aq) + nH_2O(\ell) \xrightarrow[\text{amylase}]{\text{enzyme}} nC_{12}H_{22}O_{11}(aq)$ starch + water → maltose
2	$C_{12}H_{22}O_{11}(aq) + H_2O(\ell) \xrightarrow[\text{maltase}]{\text{enzyme}} 2C_6H_{12}O_6(aq)$ maltose + water → glucose
3	$2C_6H_{12}O_6(aq) \xrightarrow[\text{zymase}]{\text{enzyme}} 2C_2H_5OH(aq) + 2CO_2(g)$ glucose → ethanol + carbon dioxide

Ethanol from starch requires stages 1 to 3; ethanol from glucose requires stage 3 only.

The final product in all cases of fermentation is a dilute solution (up to 15%) of ethanol in water – above this concentration, the ethanol poisons the **enzymes** used in the process.

Example A

Beer

Beer is the fermented liquor solution of cereal grains, most commonly barley and wheat. It is coloured and flavoured by the addition of hops (flavour only) and malt (flavour and colour) during the fermentation process. (Hops are dried flowers from a vine and malt is obtained by allowing grain to germinate.)

Higher concentrations of ethanol can be achieved by distilling the liquor to produce 'spirits'.

Example B

Spirits include brandy (distilled wine), whisky (the flavour is dependent upon the water used in fermentation) and gin (the distilled liquor is flavoured with the juice from juniper berries).

Example C

Alcohol use and humans

Socially, much alcohol (ethanol) is consumed in a wide variety of beverages. Regarded by many as a stimulant, alcohol is a depressant to the human nervous system. In moderation, it has no lasting deleterious effects on the human system – in excess it is a poison.

Unit 12.4 Activity 7A: Methanol and ethanol – structure of molecules and commercial production

1. a. Write the general formula for the alcohol series.
   b. Write the molecular formula of methanol.
   c. Draw the structural formula of ethanol.
2. CH_4 and CH_4O are molecular formulae of two carbon compounds.
   a. Draw the structural formula of each compound.
   b. Name each compound.
   c. State the valency of each element in each compound.
3. Outline how:
   a. methanol is obtained commercially from natural gas.
   b. ethanol is obtained from starch.

Physical properties of methanol and ethanol

Methanol and ethanol are colourless liquids at room temperature.

Melting points and boiling points

Alcohols have higher melting points and boiling points than the corresponding alkanes because the forces between alcohol molecules are stronger than those between alkane molecules, and more energy is required to break those forces.

Comparison of the melting and boiling points of alkanes and alcohols

Name	Melting point (°C)	Boiling point (°C)
Methane	–182	–164
Methanol	–94	65
Ethane	–183	–88
Ethanol	–117	79

The boiling points of methanol and ethanol are both below 100°C. For this reason, these liquids are described as volatile (easily vaporised).

The stronger forces between alcohol molecules are due to the alcohol molecule being polar, ie there is a charge difference in the alcohol molecules – the oxygen atom attracts electrons in covalent bond formation more strongly than carbon and hydrogen atoms do.

Example D

Alcohols as polar molecules

The stream of alcohol from a burette is deflected by a positively charged rod because the negative end of the alcohol molecule is attracted to the positively charged rod.

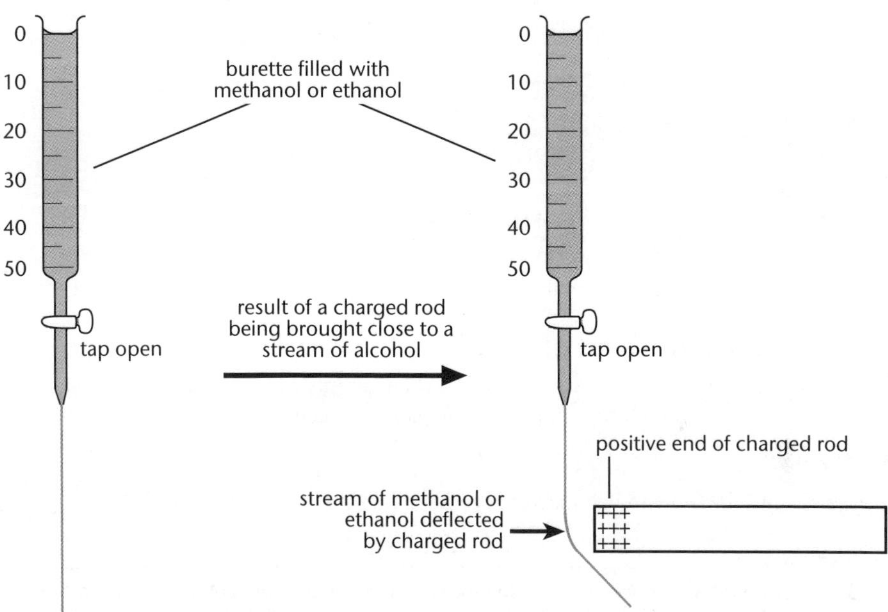

Polar nature of alcohol molecules. The negative ends of the alcohol molecules are attracted towards the positive end of the charged rod.

Solubility in water

Methanol and ethanol are soluble in water in all proportions because the water and alcohol molecules are strongly attracted to each other. These attractive forces are stronger than the forces that exist between the water molecules and than those between the alcohol molecules.

Chemical properties of methanol and ethanol

Combustion

Methanol and ethanol burn in air. The products of combustion depend on the amount of oxygen available.

- The equations for the *complete* combustion of methanol (CH_3OH) and ethanol (C_2H_5OH) are:
  $CH_3OH(\ell) + 1\frac{1}{2}O_2(g) \rightarrow CO_2(g) + 2H_2O(\ell)$
  $C_2H_5OH(\ell) + 3O_2(g) \rightarrow 2CO_2(g) + 3H_2O(\ell)$
  The flame for complete combustion is blue.
- If insufficient oxygen is available, *incomplete* combustion will occur, forming the products carbon monoxide and/or carbon instead of carbon dioxide.
  For methanol:
  $CH_3OH(\ell) + O_2(g) \rightarrow CO(g) + 2H_2O(\ell)$ *or*
  $CH_3OH(\ell) + \frac{1}{2}O_2(g) \rightarrow C(s) + 2H_2O(\ell)$.
  For ethanol:
  $C_2H_5OH(\ell) + O_2(g) \rightarrow 2C(s) + 3H_2O(\ell)$ *or*
  $C_2H_5OH(\ell) + 2O_2(g) \rightarrow 2CO(g) + 3H_2O(\ell)$.
  The flame for incomplete combustion is yellow due to the presence of red-hot carbon.

Uses of methanol and ethanol

Both methanol and ethanol are excellent fuels when complete combustion occurs. The low volatility of the alcohols allows complete combustion to occur readily.

Methanol

Methanol is used mainly to manufacture synthetic petrol from natural gas.

Example E

Synthetic petrol is produced from methanol. Octane-91 petrol contains a large proportion of synthetic petrol and is marketed as lead free. The realisation that natural gas supplies are running low has accelerated the search for other sources of natural gas.

Other uses for methanol are:

- in the production of other carbon chemicals.
- as fuel in high-performance motor cars.
- as a component of methylated spirits.

Example F

Methylated spirits ('meths')

Methylated spirits – commonly known as 'meths' – is a purple liquid readily available in hardware shops. It consists of:

- 90% rectified spirit – ethanol produced by fermentation and then distilled until the concentration of ethanol is 96% (it is impossible to get a higher concentration of ethanol by distillation of the liquor).
- 9.5% methanol – added as a deterrent to human consumption (methanol is a much greater poison than ethanol).

- small quantities of pyridine – pyridine has an unpleasant smell (and taste) and is used as a further deterrent to human consumption.
- purple dye – added as a visual warning that the liquid is not water.

Although methylated spirits contains a small quantity of water, it is a very good fuel that is used:

- in restaurants to keep to keep food hot at a buffet meal.
- as a portable source of fuel, eg for camping.

Ethanol

Ethanol is the 'alcohol' in all alcoholic beverages. Other uses of ethanol are as a:

- chemical for the manufacture of other carbon chemicals.
- fuel.
- solvent, eg for quick-drying inks.

Unit 12.4 Activity 7B: Properties of alcohols

1. **a.** If an alcohol has a melting point below room temperature (20°C) and a boiling point above room temperature, state what its physical state at room temperature is.
   b. If an alcohol is a solid at room temperature (20°C), state what must be true about the values of its melting point and boiling point.
2. Propanol has a melting point of –126°C and a boiling point of 98°C. Butanol has a melting point of –89°C and a boiling point of 117°C.
   a. State the physical state of these alcohols at room temperature.
   b. State which of these two alcohols has the stronger forces between its molecules and explain your choice.
3. State which of the following liquids, in the form of a stream from a tap or burette, would be deflected by a charged rod. Explain your choices.
   a. Water.
   b. Ethanol.
   c. Hexane.
   d. Petrol (a mixture of alkanes).
4. Complete the following equations (do not change any formulae or numbers in front of formulae).
   a. Ethanol + oxygen → carbon monoxide + ________
   b. $C_2H_5OH(\ell)$ + __$O_2(g)$ → $2CO_2(g)$ +$3H_2O(\ell)$
   c. $C_2H_5OH(\ell) + O_2(g) \rightarrow$ _____ + _____
   d. $CH_3OH(\ell)$ + ______ → $CO(g)$ + _____
5. Christmas puddings can be set alight by pouring warm brandy over the pudding and applying a match.
   a. State what is burning.
   b. Explain why it is difficult to see the flame unless the room is darkened.
6. State two uses for:
   a. methanol.
   b. ethanol.
7. Name one component of methylated spirits other than methanol and ethanol and explain why it is present.
8. State two reasons why ethanol is a component of some quick-drying inks.

Ethanoic acid

The most common use for ethanoic acid is in vinegar, which is a 3–6% solution of ethanoic acid.

Example G

Making vinegar

To make vinegar, wine (for white vinegar) or beer (for malt vinegar) is poured through casks drilled with air holes and filled with beech shavings on which bacteria adhere. The bacteria, *Mycocarderma aceti*, use oxygen to convert the alcohol present into ethanoic acid – the liquor goes 'sour'. ***Note:*** It is the malt present in beer that gives malt vinegar its characteristic brown colour.

Fortified wines (eg sherry and port) and spirits do not go sour because their high alcohol content destroys the bacteria. Alcohol in the laboratory does not go sour because there is no food source available for the bacteria.

The formation of ethanoic acid from ethanol can be summarised in the equation:

$$C_2H_5OH(\ell) + O_2(g) \rightarrow CH_3COOH(s) + H_2O(\ell)$$

Ethanoic acid

Name	Molecular formula	Structural formula
Ethanoic acid (acetic acid)	$C_2H_4O_2$ or CH_3COOH	H on C (above); H—C—C—O—H; H on C (below), =O on second C

The structural formula of ethanoic acid is often shortened to CH_3COOH.

Properties of ethanoic acid

Ethanoic acid:

- is a colourless solid.
- is soluble in water.
- is a weak acid. A weak acid (eg ethanoic acid) has a lower acidity than a strong acid (eg hydrochloric acid) even when they are in the same concentration.
- has a sour taste due to its acidity (in dilute solution, the taste is characteristic of vinegar).

Unit 12.4 Activity 7C: Ethanoic acid – manufacture and properties

1. State the approximate percentage of ethanoic acid in vinegar.

2. State the difference between white vinegar and malt vinegar.

3. Outline how wine or beer can be converted into vinegar.

4. a. Write the molecular formula of ethanoic acid.

b. Draw the structural formula of ethanoic acid.

c. Name the acid whose formula is CH_3CH_2COOH. ***Hint:*** Count the number of carbon atoms in the molecule.

Unit 12.4 Carbon Compounds
Topic 8: Aldehydes and ketones

Topic 8 covers aldehydes and ketones:

- Structures and systematic naming of aldehydes and ketones using IUPAC conventions.
- Oxidation reactions of aldehydes using the reagents MnO_4^-/H^+, $Cr_2O_7^{2-}/H^+$, Tollens' reagent, Fehling's solution and Benedict's solution.

Introduction

Aldehydes and **ketones** are known as **carbonyl** compounds because they contain the carbonyl group, C=O. The carbon-to-oxygen double bond is polar:

$$\overset{\delta+}{>C}=\overset{\delta-}{\ddot{O}}:$$

Names and structures

Aldehydes

The systematic name for aldehydes is **alkanals**, but the name *aldehyde* is widely used. Aldehyde is derived from *al*cohol *dehyd*rogenated.

The aldehydes are a homologous series with the carbonyl group, C=O, on the *end* carbon atom along with a hydrogen atom, H, forming a CHO functional group. Aldehydes have the general formula $C_nH_{2n+1}CHO$ or R–CHO:

R(H)$\overset{\delta+}{C}=\overset{\delta-}{\ddot{O}}:$ or R–CHO

Aldehydes are named according to the following conventions:

- The *-e* of the alkane is changed to *-al*.
- Methanal, ethanal and propanal are the first members of the aldehydes with one, two and three carbon atoms respectively.
- Any alkyl groups on the parent chain are indicated by numbering the carbon atoms from the carbon of the CHO group.
- The prefixes *di-* and *tri-* are used to indicate more than one alkyl group.
- The common name for methanal, H–C(=O)–H or HCHO, is **formaldehyde**.
- The common name for ethanal, CH_3C(=O)–H or CH_3CHO, is **acetaldehyde**.

Example A

Name $CH_3CH_2CH(CH_3)CHO$

This is a branched-chain aldehyde. Its structural formula is:

```
      H    H    H         O
      |    |    |        //
H  -  C4 - C3 - C2 - C1
      |    |    |        \
      H    H    CH3       H
```

It is named as follows:

- There are four carbon atoms in the longest continuous chain containing the –CHO group, hence *butanal*.
- There is a methyl group, CH_3, attached to the second atom from the –CHO carbon, giving *2-methyl*.

The name of $CH_3CH_2CH(CH_3)CHO$ is 2-methylbutanal.

Ketones

The systematic name for ketones is **alkanones**. Many ketones are used as solvents. The functional group C=O of a ketone is *not* an end carbon. This means that ketones always have two alkyl groups attached to the carbon atom of the carbonyl group, ie RCOR′.

```
R    δ+   δ–
 \
  C = O:
 /
R′
```

Ketones are named according to the following conventions:

- The *-e* of the alkane is changed to *-one*.
- Propanone, CH_3CCH_3 (with =O on the central C), and butanone, $CH_3CCH_2CH_3$ (with =O on the second C), are the first two members of the ketones. Propanone is commonly known as **acetone**.
- For ketones with five or more carbon atoms, a number is used to identify the position of the carbon of the carbonyl group.

Example B

$CH_3COCH_2CH_2CH_3$ is pentan-2-one, $CH_3CH_2COCH_2CH_3$ is pentan-3-one.

- Any alkyl groups on the parent chain are indicated by numbering.
- The prefixes *di-* and *tri-* are used to indicate more than one alkyl group.

Example C

Naming ketones, eg $CH_3COCH(CH_3)CH_2CH_3$

Inspection shows that the molecule below is a branched-chain ketone:

```
      H         CH3   H    H
      |         |     |    |
H  -  C1 - C2 - C3 -  C4 - C5  - H
      |    ||   |     |    |
      H    O    H     H    H
```

- The parent chain is five carbons long, thus *pentan*.
- The carbonyl group C=O is present, thus pentan*one*.

- The carbon forming the functional group is the 2nd carbon (C^2), thus pentan-2-one.
- The third carbon (C^3) has a methyl group attached, hence *3-methyl.*

The name of $CH_3COCH(CH_3)CH_2CH_3$ is therefore 3-methylpentan-2-one.

Unit 12.4 Activity 8A: Names and structures

1. Draw the formulae indicated in brackets for each of the following compounds.
   a. Methanal (structural).
   b. Hexan-2-one (structural).
   c. 2-Methylpentanal (molecular and structural).
   d. 3-Methylpentan-2-one (molecular and structural).
2. Name $CH_3CH_2CH_2{-}\underset{\underset{O}{\|}}{C}{-}CH_3$.
3. Two structural isomers of C_4H_8O are aldehydes and one is a ketone. For each of these three compounds:
   a. write the IUPAC name.
   b. give the structural formula.

Bonding, properties and uses

Intermolecular bonding results from relatively strong intermolecular forces due to the polarity of the carbonyl group in aldehyde and ketone molecules. There is no hydrogen bonding between aldehyde and ketone molecules, as there is no H atom attached to an oxygen atom.

Melting and boiling points

The melting and boiling points of the aldehydes and ketones increase with increasing molar mass. At room temperature methanal is a gas. Most ketones and aldehydes are volatile liquids with definite odours:

- Aldehydes tend to have unpleasant, pungent smells.
- Ketones tend to have more pleasant, sweeter smells.

Solubility

The oxygen of the carbonyl group has a lone pair of electrons, which means the aldehyde or ketone molecules form hydrogen bonds with the hydrogen atoms of water molecules:

- C_1–C_3 aldehydes are soluble, C_4–C_6 aldehydes are slightly soluble.
- C_3 and C_4 ketones are soluble, C_5 ketones are only slightly soluble.

Aldehydes and ketones tend to be soluble in non-polar solvents due to their non-polar alkyl groups.

Uses

- **Formalin**, a solution of about 40% methanal and 60% water, is used as a preservative and disinfectant.
- Propanone (acetone) is a major industrial solvent.

Unit 12.4 Activity 8B: Bonding, properties and uses

1. Explain why aldehydes have higher melting and boiling points than alkanes of similar molar mass.

2. a. Explain why the solubility of aldehydes in water decreases as the molar mass of the aldehyde increases.

b. Discuss whether you expect the solubility of ketones to decrease in the same way.

3. Discuss, with reasons, the differences in the boiling points (bp) of: ethanol (M = 46 g mol^{-1}), bp = 78°C, and ethanal (M = 44 g mol^{-1}), bp = 21°C.

Preparation of carbonyl compounds

Preparation of aldehydes

Aldehydes are prepared from the oxidation of primary alcohols. In practice, the aldehydes are themselves often then oxidised to a carboxylic acid.

Example D

Oxidation of ethanol

$$CH_3CH_2OH \xrightarrow{Cr_2O_7^{2-}/H^+} CH_3CHO \xrightarrow{Cr_2O_7^{2-}/H^+} CH_3COOH$$

ethanol — ethanal — ethanoic acid

To prevent the oxidation of ethanol to ethanoic acid, the ethanal is removed from the reaction vessel as it is formed. This is done by adding the mixture of ethanol and aqueous dichromate dropwise to a hot solution of sulfuric acid.

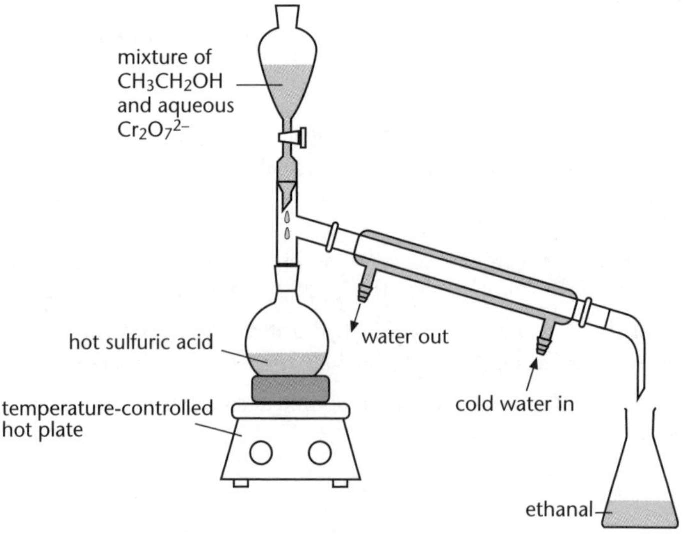

Preparation of ethanal from ethanol

Depending on their boiling point, other aldehydes can be prepared in a similar manner.

Preparation of ketones

Ketones are prepared by oxidising secondary alcohols.

Example E

Preparation of pentan-2-one

To prepare pentan-2-one, oxidise pentan-2-ol:

$$\underset{\textbf{Pentan-2-ol}}{CH_3CH_2CH_2\underset{\displaystyle OH}{\underset{|}{C}}HCH_3} \xrightarrow[\text{warm}]{Cr_2O_7^{2-}/H^+} \underset{\textbf{Pentan-2-one}}{CH_3CH_2CH_2\underset{\displaystyle O}{\underset{\|}{C}}CH_3}$$

This reaction is slow. To ensure all the alcohol is oxidised, the reaction is carried out under **reflux**. This involves heating the mixture in a flask with a condenser attached. Although the volatile materials evaporate, they are condensed and returned to the reacting solution.

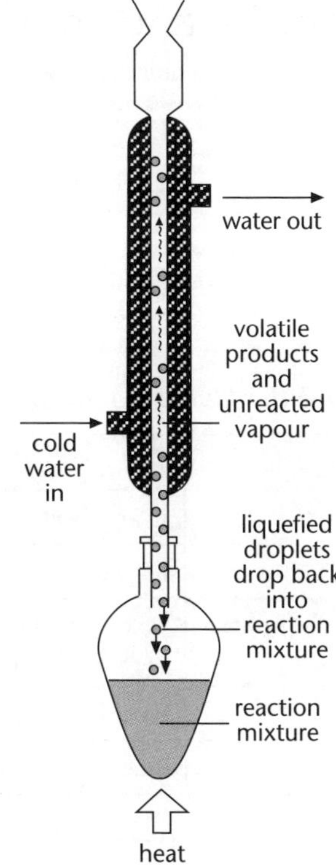

Reflux apparatus for oxidising pentan-2-ol under reflux to produce pentan-2-one

Unit 12.4 Activity 8C: Preparing aldehydes and ketones

1. Consider the laboratory preparation of propanal from propan-1-ol.
  a. Identify the most likely impurity in the product.
  b. Describe how the products are separated.
  c. Explain why there is no similar impurity during the production of propanone from propan-2-ol.

2. Explain why ethanal is easily separated from an original reaction mixture of ethanol and acidified potassium dichromate.

3. Give the structural formulae and names of the alcohols used to prepare:
  a. CH_3CH_2CHO
  b. $CH_3CH_2COCH_3$
  c. $CH_3CH(CH_3)CHO$

Distinguishing aldehydes and ketones

The basis of tests used to distinguish between aldehydes and ketones is that aldehydes are easily oxidised to carboxylic acids whereas ketones are not.

Tests for aldehydes

Aldehydes are easily oxidised to carboxylic acids. As the aldehyde is oxdised, it reduces the other reactant. The test for aldehydes is based on the fact that aldehydes are oxidised by weak oxidants, eg Tollens' reagent, Fehling's solution and Benedict's solution, but primary and secondary alcohols are not. Stronger oxidants (eg dichromate and permanganate) oxidise aldehydes, as well as primary and secondary alcohols.

Colour changes associated with reduction of oxidising agents by aldehydes

Oxidising agent	Colour change
Potassium dichromate (acidified)	Orange solution to green
Potassium permanganate (acidified)	Purple solution to colourless
Tollens' reagent – ammoniacal silver nitrate	Colourless solution forms silver precipitate or silver mirror
Fehling's solution	Blue solution forms orange-red precipitate
Benedict's solution	Blue solution forms orange-red precipitate

Tollens' reagent contains the diamminesilver(I) complex ion, $[Ag(NH_3)_2]^+$, made up by:

- Placing 2–3 mL of aqueous silver nitrate, $AgNO_3$, in a *clean* test tube.
- Adding one drop of NaOH(aq) so that a precipitate of silver oxide, Ag_2O, forms.
- Adding just enough dilute ammonia, $NH_3(aq)$, to redissolve the precipitate, forming $[Ag(NH_3)_2]^+(aq)$.

Aldehydes reduce the silver ions, Ag^+, to metallic silver, Ag, which is often seen as a *silver mirror* on the inside of the test tube.

Example F

Ethanal with Tollens' reagent

$$\underset{\text{ethanal}}{CH_3CHO(aq)} + \underset{\text{Tollens' reagent}}{2[Ag(NH_3)_2]^+(aq)} + H_2O(\ell) + 2H^+(aq) \rightarrow \underset{\text{ethanoic acid}}{CH_3COOH(aq)} + \underset{\text{silver mirror}}{2Ag(s)} + 4NH_4^+(aq)$$

Ketones do not react with Tollens' reagent, so no change is observed.

Benedict's solution contains aqueous copper(II) ions, Cu^{2+} (which give the reagent a blue colour), in basic conditions. To prevent the Cu^{2+} from precipitating out as $Cu(OH)_2$, the Cu^{2+} is complexed with citrate ions.

When Benedict's solution is added to an aldehyde and warmed, an orange-red **precipitate** (ppt) of copper(I) oxide, Cu_2O, forms.

Example G

Ethanal with Benedict's solution

$$\underset{\text{ethanal}}{CH_3CHO(aq)} + \underset{\text{Benedict's solution}}{4OH^-(aq) + 2Cu^{2+}(aq)} \rightarrow \underset{\text{ethanoic acid}}{CH_3COOH(aq)} + \underset{\text{orange-red precipitate}}{Cu_2O(s)} + 2H_2O(\ell)$$

Fehling's solution is another reagent with copper(II) ions in a basic solution. It behaves exactly the same as Benedict's solution does.

Ketones do not react with Benedict's solution or Fehling's solution because a ketone cannot reduce metal ions.

Reaction summary for ethanal

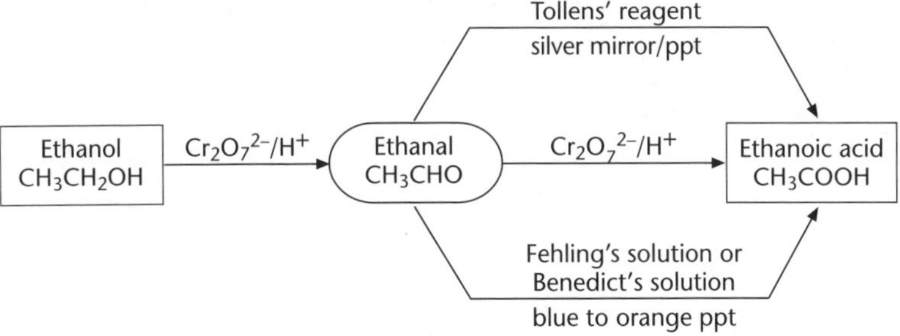

Unit 12.4 Activity 8D: Distinguishing aldehydes and ketones

1. a. When an aldehyde reacts with Fehling's solution, does the aldehyde undergo oxidation, reduction, substitution, addition, hydrolysis or elimination?

b. State the original colour of Fehling's solution and identify the species responsible for this.

c. Identify the precipitate formed in the Fehling's test, and its colour.

2. Aqueous silver nitrate (a few mL) is placed in a clean test tube and 1 drop of NaOH is added. The precipitate that forms is dissolved by adding just enough dilute ammonia.

a. Name the reagent that has been prepared above.

b. Describe what is observed when this reagent is warmed with a few drops of each of the following.

i. Butanone.

ii. Butan-1-ol.

iii. Butanal.

3. a. Write half-equations and an overall balanced equation for the reaction of:

i. propanal with Tollens' reagent.

ii. ethanal with Fehling's solution.

iii. methylpropanal with acidified potassium dichromate.

b. For each of the reactions **i.–iii.,** state what would be observed and link these observations to the species involved.

Unit 12.4 Carbon Compounds

Topic 9: Organic reactions summary

Topic 9 summarises the organic reactions introduced in Unit 12.4:

- Describing structures and reactions of organic compounds containing the functional groups haloalkane, alcohol, alkene, alkyne, ester, carboxylic acid, aldehyde and ketone.

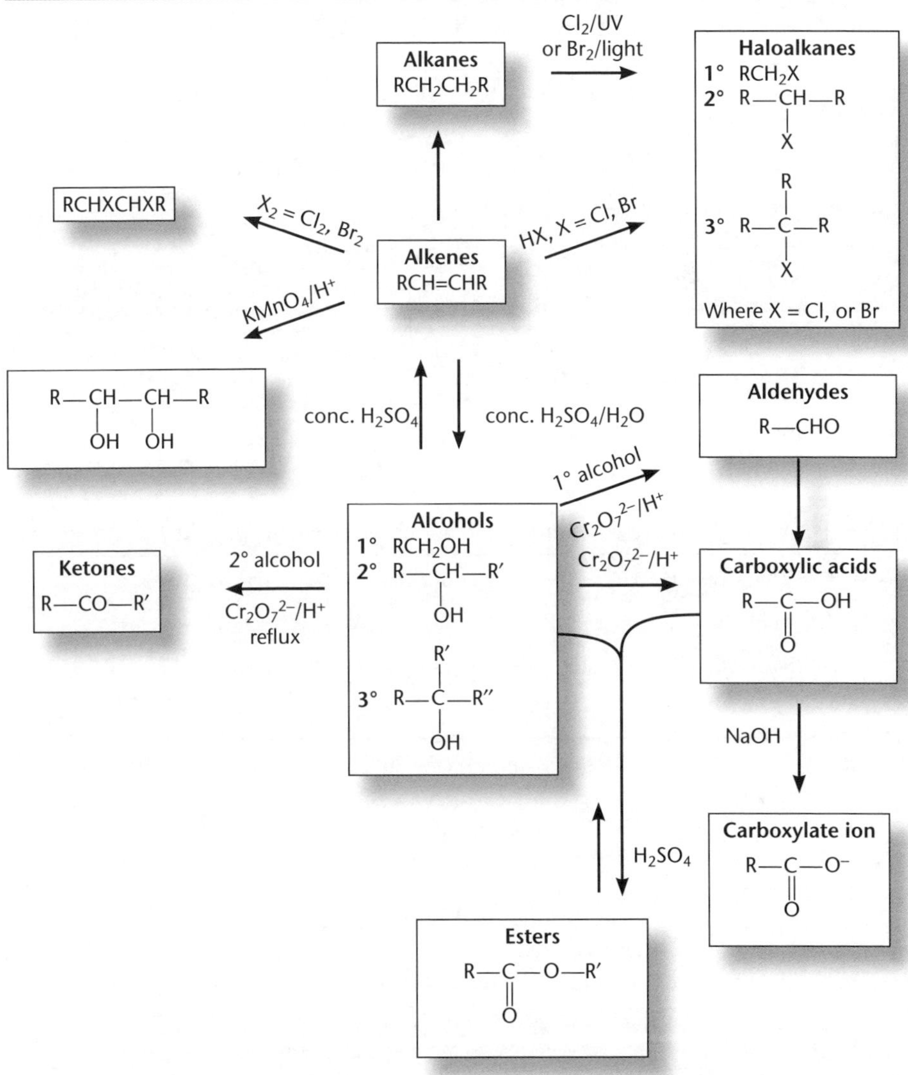

Unit 12.4 Activity 9A: Organic reactions summary

1. Identify the functional groups **A–E** on the following steroid molecule:

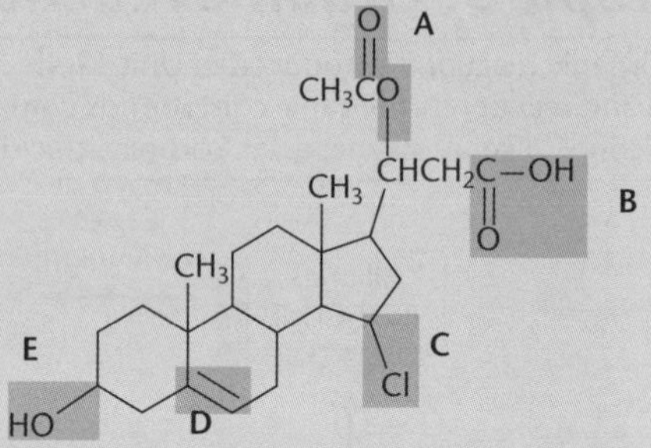

$$\text{Steroid skeleton with: } \overset{\text{A}}{CH_3\overset{\overset{\displaystyle O}{\|}}{C}O}\text{–},\ CH_3,\ CHCH_2\underset{\underset{\displaystyle O}{\|}}{C}\text{–}OH\ (\text{B}),\ CH_3,\ Cl\ (\text{C}),\ C{=}C\ (\text{D}),\ HO\ (\text{E})$$

2. For each of the following molecules:

a. identify the functional group.

b. write the name.

i. $CH_3CH_2CH{=}CH_2$ **ii.** $CH_3CH_2\underset{\underset{O}{\|}}{C}\text{—}OH$ **iii.** $CH_3C{\equiv}CCH_2CH_3$

iv. $CH_3\text{—}\underset{\underset{OH}{|}}{\overset{\overset{CH_3}{|}}{C}}\text{—}CH_3$ **v.** $CH_3CH_2CH_2\underset{\underset{O}{\|}}{C}\text{—}OCH_2CH_3$

3. Refer to the following compounds to answer questions **a.–g.** below.

A	$CH_3CH_2CH_2CH_2OH$	B	$CH_3CH{=}CHCH_3$	C	$H\underset{\underset{O}{\|}}{C}OCH_2CH_2CH_3$
D	$CH_3CH_2CH_2CH_3$	E	$CH_3CH_2CH_2\underset{\underset{O}{\|}}{C}OH$	F	$CH_3\underset{\underset{CH_3}{|}}{C}{=}CH_2$

a. What type of alcohol is compound A – primary, secondary or tertiary?

b. State the type of reaction that will change compound A to compound E.

c. Describe what would be observed if compound E is added to sodium carbonate.

d. Contrast the reactions of compound B and compound D with bromine. Include relevant observations and equations, and identify the type of reaction that occurs in each case.

e. Explain what is meant by the term 'structural isomer'. Use examples from A–F to illustrate your answer.

f. Compound B has two geometric isomers (*cis* and *trans*). Draw and name these *two* isomers.

g. Compounds **A**, **C** and **E** are all colourless liquids. Describe tests that would identify *each* of these compounds. Use the physical *and* chemical properties of these compounds to select suitable tests. Include any observations that would be made when carrying out these tests.

4. a. Draw the structural formulae of the organic products in the following reactions.

i. $CH_3CH(CH_3)CH_2CH_2OH \xrightarrow{\text{conc. } H_2SO_4}$

ii. $CH_3CH_2COOH + CH_3CH_2OH \xrightarrow[\text{and heat}]{\text{conc. } H_2SO_4}$

iii. $CH_3{-}CH(CH_3){-}CH_2{-}OH \xrightarrow{MnO_4^- / H^+}$

b. Name another reagent that can be used for reaction **iii**. Describe what would be observed when the reaction is carried out.

5. An organic reaction is shown by the equation below:

$$\begin{array}{l} CH_2{-}\underset{A}{\underbrace{C({=}O){-}O}}{-}CH_2CH{=}CHCH_2CH_3 \\ CH{-}C({=}O){-}O{-}(CH_2)_{16}CH_3 \\ CH_2{-}C({=}O){-}O{-}(CH_2)_{16}CH_3 \end{array} + 3H_2O \longrightarrow 2CH_3(CH_2)_{16}COOH + CH_3CH_2\underset{B}{\underbrace{CH{=}CH}}CH_2\underset{C}{\underbrace{COOH}} + \begin{array}{l} CH_2OH \\ CHOH \\ CH_2OH \end{array}$$

Compound Z

a. Name the functional groups A, B and C.

b. Identify the part of a molecule in the equation which is regarded as *unsaturated*.

c. Give the common name for compounds similar to compound Z.

6. a. Give the structural formula of the organic products for the following reactions.

i. $CH_3CH_2CH_2CH_3 + Cl_2 \xrightarrow{\text{ultraviolet light}}$

ii. $CH_3{-}CH(OH){-}CH_3 \xrightarrow{\text{conc. } H_2SO_4}$

iii. $CH_3CH_2CH{=}CH_2 \xrightarrow{KMnO_4}$

iv. $CH_3CH_2CH_2OH \xrightarrow{Cr_2O_7^{2-} / H^+}$

v. $CH_3CH_2CH_2{-}C({=}O){-}OH + CH_3CH_2OH \xrightarrow{H_2SO_4}$

b. Classify each of the reactions above as addition, elimination, substitution, oxidation or esterification.

7. Compounds X, Y, Z, W, V and V′ are involved in a series of reactions, as shown by the scheme below. In this scheme, only the organic reaction products are indicated. Molecular formulae for compounds X and W are given. Additional information about the compounds is given in the box on the right of the reaction scheme.

   Identify the compounds V, V′, W, X, Y and Z. Write a structural formula and name for *each* compound. (***Hint***: Begin with compound X, which is a straight-chain molecule.)

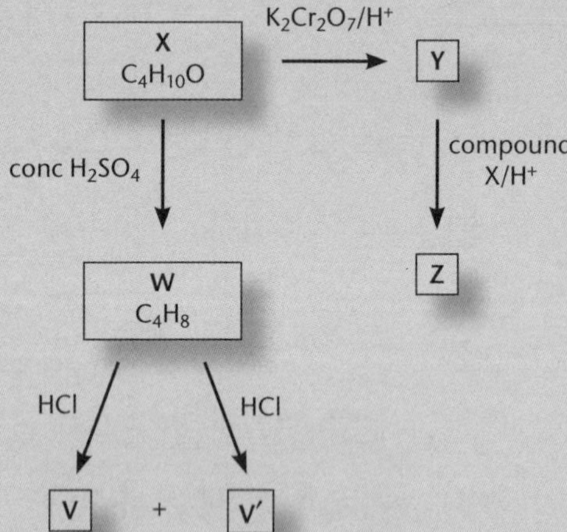

Additional information
Compound Y turns blue litmus red.
Compound Z has a characteristic pleasant, fruity smell. It is formed by the reaction between compounds X and Y in the presence of sulfuric acid.
Compound W is unsaturated.
Compounds V and V′ are structural isomers.

8. The flow diagram below shows a series of organic reactions.

   Draw structural formulae and write IUPAC (systematic) names for the compounds A, B, C and D, and identify reagents E and F.

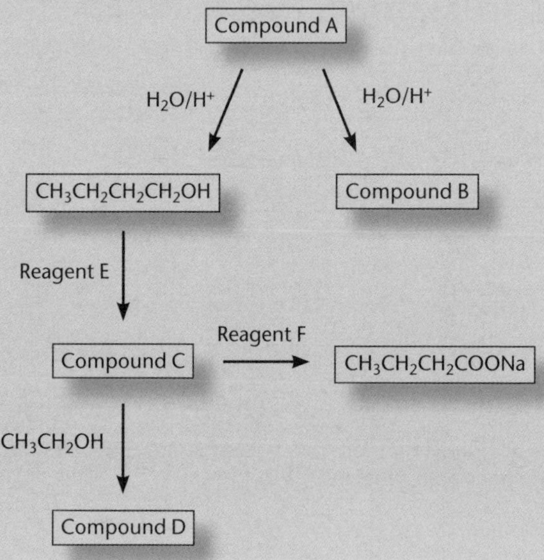

9. Complete the following equations by drawing the structural formulae of the organic products.

a. $CH_3CH_2CH{=}CH_2 \xrightarrow{H_2/Pt}$

b. $CH_3CH(CH_3)CH_2OH \xrightarrow[\text{heat}]{\text{conc. } H_2SO_4}$

c. $CH_3CH_2CH_2CH_2OH + HCOOH \xrightarrow[\text{heat}]{\text{conc. } H_2SO_4}$

d. $CH_3CH(CH_3)CH_2CH_2OH \xrightarrow{Cr_2O_7^{2-}/H^+}$

e. $CH_3CH{=}CHCH_3 \xrightarrow{MnO_4^-/H}$

10. Consider the following pairs of compounds.

a. Cyclohexene and propyl ethanoate.

b. Ethanoic acid and 1-chlorobutane.

c. Butan-1-ol and hexane.

d. Cyclohexane and cyclohexene.

Identify tests to distinguish between the pairs of compounds.

i. Describe the test to be carried out, with expected observations.

ii. Clearly explain how the test results can be used to distinguish between the molecules in each pair of compounds, and why the test used is a suitable one.

11. An unsaturated compound, A, C_4H_8, reacts with water under acidic conditions to form two new products, B and C. Product B reacts with acidified potassium permanganate solution to form compound D. Compound D reacts with a solution of sodium carbonate, producing bubbles of gas. Compound C reacts with D in the presence of sulfuric acid, and compound E is formed. Compound E has a characteristic smell.

Write the structural formula for compounds A–E, and write names for compounds A–D.

12. A compound, F, $C_6H_{12}O_2$, in acidic solution, reacts to form two new organic compounds G and H. Compound G reacts with acidified dichromate solution to form compound H. Compound H reacts with sodium hydroxide to form a salt. Compound G, on reaction with concentrated H_2SO_4, produces an unsaturated compound, I. On reaction of compound I with dilute sulfuric acid, a new compound, J, is formed as a major product, along with small amounts of compound G. Compound J is an isomer of compound G.

Write the structural formula and name for each compound F, G, H, I and J.

13. Describe tests that would distinguish between the following pairs of compounds. Explain how the test results can be used to distinguish between the molecules in each pair of compounds and why the test used is a suitable one.

a. Ethanol and chloroethane.

b. Butanoic acid and hexane.

14. Use the following reaction scheme to answer the questions that follow.

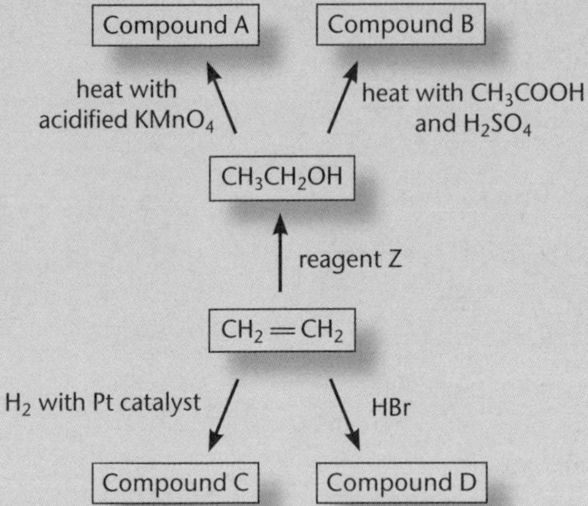

a. Write the name of reagent Z.

b. For compounds A, B, C and D, write the class of compound to which they belong, the structural formula, and their name.

c. If propene had been reacted with HBr instead of ethane, two possible products could be formed.

Use structural formulae to write an equation for this reaction and indicate which of the two products is most likely to be formed.

15. Complete the following equations:

a. Methane + steam → carbon monoxide + __________

b. $CO(g) + 2H_2(g) \rightarrow$ __________

c. Starch + ________ $\xrightarrow{\text{amylase}}$ maltose

d. $C_{12}H_{22}O_{11}(aq) + H_2O(\ell) \xrightarrow{\text{maltose}}$ __________

16. If petrol, a mixture of alkanes in the range C_5 to C_8, is added to water, two layers are obtained, with the less dense petrol floating on the surface of the water. Explain why a bottle of gin, which is a mixture of ethanol and water, forms only one layer.

17. State the products for the complete combustion of methanol.

18. Two bottles, one containing sulfuric acid and the other containing ethanoic acid, have the same concentration.

a. State whether the acidities of the two solutions are the same or different.

b. Explain your choice.

19. Explain why fortified wines and spirits do not go sour when they are exposed to air.

20. The flow diagram below shows a series of organic reactions.

Draw structural formulae and write IUPAC (systematic) names for the compounds A, B, C and D. Identify reagent E.

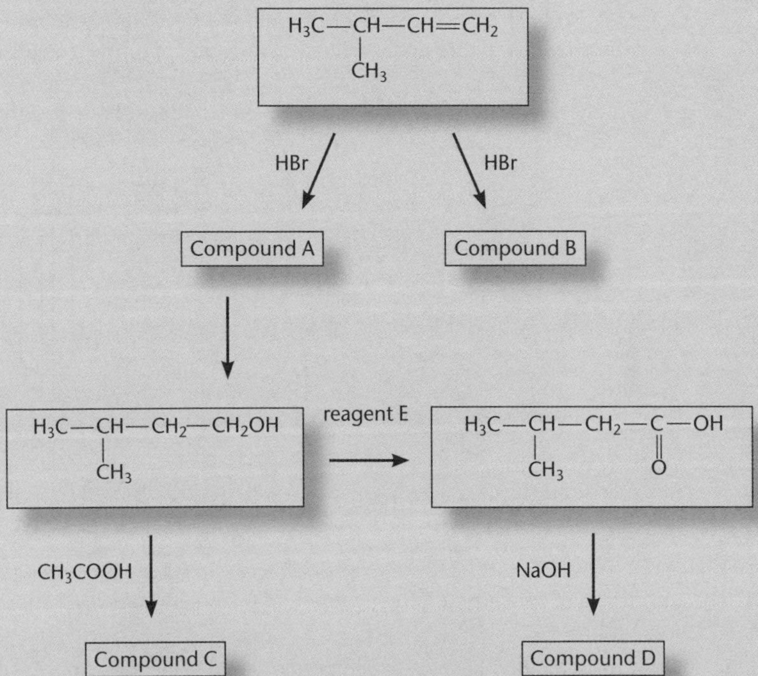

21. Three bottles of colourless liquids are known to contain 1-pentene, ethyl ethanoate and propanoic acid.

Describe how the three liquids could be identified using only bromine water.

22. Draw structural formulae for the reagents that could be used to make the following compounds. State any conditions or catalysts that are needed.

a. A + B → $CH_3CH_2CH_2CH_2Br$ + HBr

b. C + D → CH_3CHCH_3 (with Cl on the central C)

c. E + F $\xrightarrow{H^+}$ $CH_3COCH_2CH_2CH_3$ (with =O on the C of the CO group)

23. The following sequence shows two steps in a series of organic reactions:

$$\underset{A}{CH_3CH_2CH_2OH} \rightarrow B \rightarrow \underset{C}{CH_3CH_2COOH}$$

a. Compound A can be directly converted to compound C by adding an oxidising agent and heating the mixture under reflux. Identify an appropriate oxidising agent.

b. Select one of the diagrams (**i.–iii.**) below and use it to explain how the process of reflux works and why the reaction is carried out this way.

i.

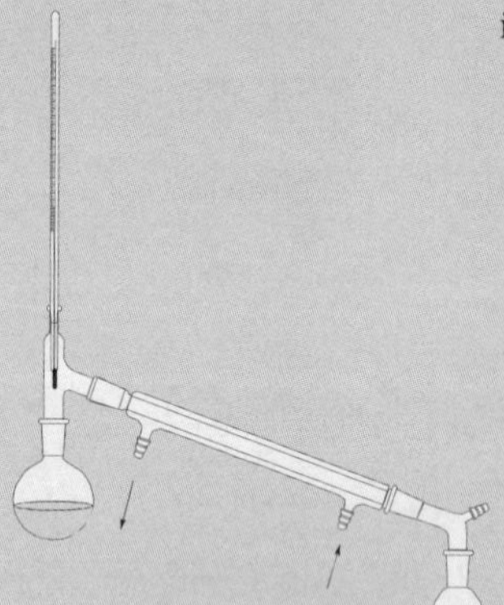

ii.

iii.

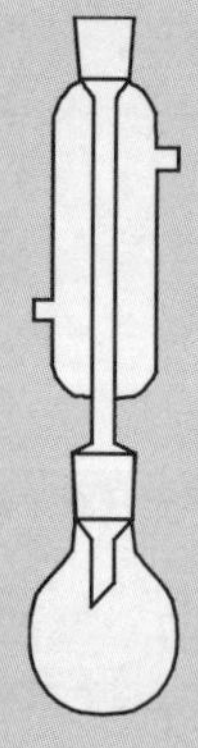

c. The conversion of compound A into compound C initially involves the formation of compound B.

- **i.** Draw the structural formula of compound B and name it.
- **ii.** Describe how the conditions in **a.** could be modified to produce compound B, rather than compound C, as the major product. Give reasons for these modifications.

Unit 12.5 Natural Resources and Chemical Industries in Papua New Guinea

Topic 1: Extraction of oil and gas

Topic 1 covers the extraction of oil and gas in Papua New Guinea, by discussing:

- The carbon cycle.
- Petroleum refining.
- Natural gas and liquefied natural gas.

Introduction

Papua New Guinea is not only a consumer; it is also an active participant in developing and using the principles of chemistry. Papua New Guinea has long been known as region with rich oil and gas deposits. Large-scale exploration began after the 1970s Arab oil embargo, at which time Papua New Guinea was still a colony of Australia.

This exploration established the Hides Anticlines, an area in the Southern Highlands, as the richest area in oil, with the main discoveries taking place within this tectonic region.

All of Papua New Guinea's existing oil and gas fields are located in the region of the Papuan Fold Belt. 'Fold belt' is a geological term that refers to a region of the Earth's surface where the rocks have been subjected to severe folding and thrusting.

The major oil and gas reserves in Papua New Guinea are shown on the following map.

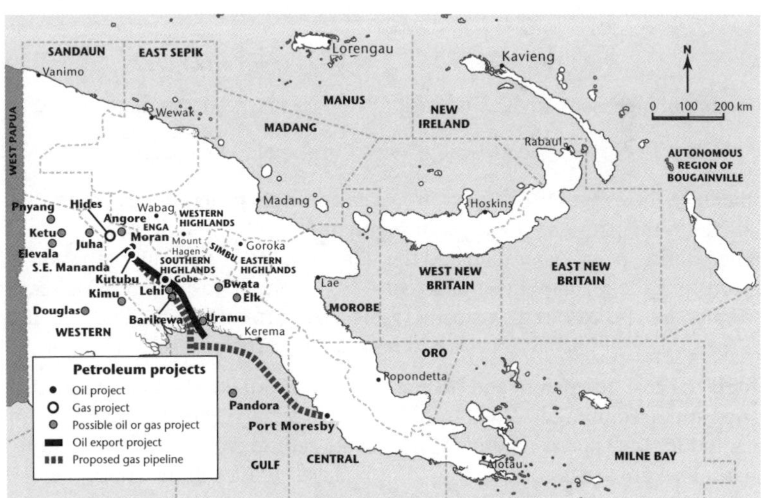

Petroleum projects in PNG

The carbon cycle

Carbon dioxide makes up about 0.04% of air. In normal circumstances, this value is kept constant by the cycling of CO_2 through nature by the processes of photosynthesis and respiration in the **carbon cycle**, shown in the following flow diagram.

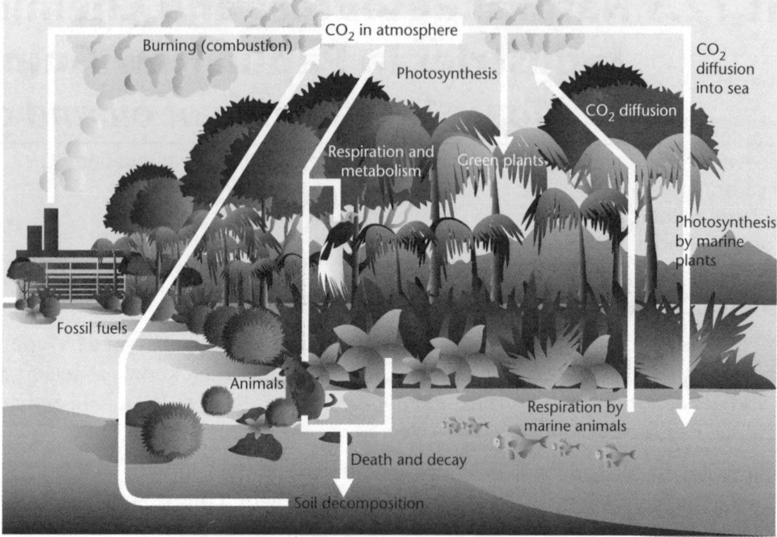

Basic carbon cycle flow diagram

The main pathway of the carbon cycle is the:

- conversion by plants of CO_2 gas in the atmosphere into glucose.

$$6CO_2(g) + 6H_2O \xrightarrow[\text{chlorophyll}]{\text{sunlight}} C_6H_{12}O_2(aq) + 6O_2(g)$$

- conversion of glucose in organisms back to CO_2 gas in the atmosphere.

$$C_6H_{12}O_2(aq) + 6O_2(g) \rightarrow 6CO_2(g) + 6H_2O + \text{energy}$$

Plants take in about 25% of the CO_2 gas from the atmosphere and photosynthesise it into carbohydrates in the presence of the green **pigment** chlorphyll. Some of the carbohydrate is consumed during plant respiration and the rest is used to build plant tissue and growth. Animals consume the carbohydrates and return CO_2 to the atmosphere during respiration. Carbohydrates are also oxidised by small organisms, returning CO_2 gas to the atmosphere. Through the process of photosynthesis and respiration, carbon is cycled in nature.

Fossil fuels are coal, petroleum and natural gas. These formed when ancient life forms were buried and decomposed. When fossil fuels are burnt to generate electricity, to run industries and for transportation, they also generate carbon dioxide and release it to the atmosphere. This process is contributing to rising levels of CO_2 in the atmosphere.

Unit 12.5 Activity 1A: The carbon cycle

1. Write the general word equation and chemical equation for animal respiration.
2. Write the general word equation and chemical equation for photosynthesis.
3. Research and explain the chemical composition of chlorophyll as an important substance in photosynthesis.
4. Explain the importance of oceans in the carbon cycle.

Coal

Coal is a black or brownish-black rock composed of a mixture of carbon, hydrocarbons, nitrogen, sulfur and water. It formed from plants that lived hundreds of millions of years ago. Coal is the most abundant of the fossil fuels. Some countries have large deposits of coal.

Example A

Coal

Large coal mines are found in areas such as the Hunter Valley in New South Wales, Australia. There are no coal mines in Papua New Guinea.

Coal has many uses. In some countries coal is used directly for heating. The main use of coal has been for generating electricity. Coal is also used to purify iron in a blast furnace, where it removes oxygen from iron oxide (Fe_2O_3) via a metal reduction process.

Example B

Coal and iron extraction

Coal reacts with oxygen to produce carbon monoxide gas, which reduces iron oxide in the blast furnace in the following series of reactions:

$$C(s) + O_2(g) \rightarrow CO_2(g)$$

$$C(s) + CO_2(g) \rightarrow 2CO(g)$$

$$Fe_2O_3(s) + 3CO(g) \rightarrow 2Fe(s) + 3CO_2(g)$$

Unit 12.5 Activity 1B: Coal

1. Describe the physical appearance of coal.
2. Explain the uses of coal.
3. Write the general reaction equation for the reduction of a metal compound, MX, by a reducing agent, R.
4. Research and explain the reduction of iron oxide in a blast furnace.

Oil (petroleum)

Oil formed from the remains of animals and plants that lived millions of years ago. Over time, the remains became covered by layers and layers of mud. The heat and pressure from these layers caused the animal and plant remains to turn into crude oil.

Petroleum literally means 'rock oil'. It is a complex mixture of hydrocarbons and other chemicals. The hydrocarbons are mostly alkanes, cycloalkanes and aromatic hydrocarbons. Their relative proportions vary depending on where and how the pertoleum was formed. The composition of crude oil varies from one deposit to another.

Generally, the elemental composition of crude oil is carbon (83–7%), hydrogen (11–15%), sulfur (1–6%), nitrogen (0.1–2%), oxygen (0.05–1.5%) and metals (<0.1%).

The amount of sulfur in crude oil also varies from one deposit to another. Crude oil with a high sulfur content is referred to as 'sour' crude oil. Crude oil with a low sulfur content is referred to as 'sweet' crude oil.

Crude oil is the source of most of our liquid fuels, and petroleum oil is used in the forms of petrol and diesel that power most vehicles. Other useful crude oil products include propane, butane, fuel oil, lubrication oil, greases, paraffin wax and bitumen. Petroleum products have a wide range of uses, including in the manufacture of polymers and fibres, adhesives and sealants, cosmetics, fragrances and food additives, paints and packaging.

Example C

Arabian heavy crude oil and light crude oil have the same components but in different proportions.

Example D

Sweet and sour

Australian crude oil is sour crude – it has a high sulfur content. Papua New Guinean crude oil is sweet crude – it has a very low sulfur content.

Unit 12.5 Activity 1C: Oil (petroleum)

1. Write the chemical formulae for propane and butane.
2. Find out where the following products are used.
   a. Fuel oil
   b. Paraffin wax
   c. Bitumen
   d. Kerosene
3. Research and explain the difference between kerosene and jet fuel.

Oil refining

Although petroleum refineries are generally similar, operations can vary from one refinery to another because of the type of crude oil and the type of products required. The following operations are carried out in many refineries.

Desalting is usually the first process in crude oil refining. Salt, **minerals** and other water-soluble impurities are removed from the crude oil by washing it with water. The salt dissolves into the water. After desalting, the remaining salt content is usually measured in PTB (pounds of salt per thousand barrels of crude oil).

Desalting is necessary to protect refinery equipment against plugging, corrosion and catalytic poisoning.

Example E

Desalting

Calcium, sodium and magnesium chlorides are the most common salts in crude oil. They are removed by desalting.

Petroleum can be separated and purified into the different hydrocarbons in two ways: non-chemically and chemically.

Non-chemical separation methods

Separation methods that don't involve chemical reactions include solvent extraction, absorption, distillation and crystallisation.

Solvent extraction

Solvent extraction removes impurities that would affect the performance of the final petroleum product. It removes heavy aromatic compounds from lubricating oils. It also removes impurites that would contribute to corrosion and poisoning of catalysts in subsequent processes at the refinery.

Solvent extraction and solvent dewaxing usually remove these impurities at intermediate refining stages or just before the product is sent to storage.

The specific processes and solvents used depend on the type of petroleum being treated, the contaminants present and the finished product requirements.

Example F

Extraction solvents

The most widely used extraction solvents are phenol, furfural (an aromatic aldehyde) and cresol. Furfural is used to remove coloured compounds containing sulfur and oxygen. Other solvents include liquid sulfur dioxide, nitrobenzene and 2,2′-dichloroethyl ether. **Ethylene** glycol is used to extract the aromatic compounds benzene, toluene (methylbenzene) and xylene (1,4-dimethylbenzene) (also known as the BTX fraction).

Absorption

Molecular sieves are materials with microscopic holes that block large molecules and let smaller molecules pass through into the sieve. The larger molecules pass by with the solvent and the smaller molecules are trapped (adsorbed) in the sieve until desorption occurs. The holes are measured in Ångstroms (1 Å = 10^{-10} m) or nanometres (1 nm = 10^{-9} m). The petroleum industry uses molecular sieves consisting of **porous** aluminosilicates, which adsorb linear hydrocarbons in preference to branched ones.

Example G

Molecular sieves in petroleum refining

During petrol refining, moleculer sieves with 3–10 Å pores are used to separate mixtures of the C_4 compounds 2-methylpropene (isobutene), 1-butene, 2-butene and butane. 2-Methylpropene is cleanly separated as it is too bulky to be absorbed.

Distillation

The next step in the refining of oil is **fractional distillation.** This is carried out in tall steel towers up to 50 metres high. Crude oil is heated in a furnace to about 600°C and then fed into the bottom of a fractionation tower.

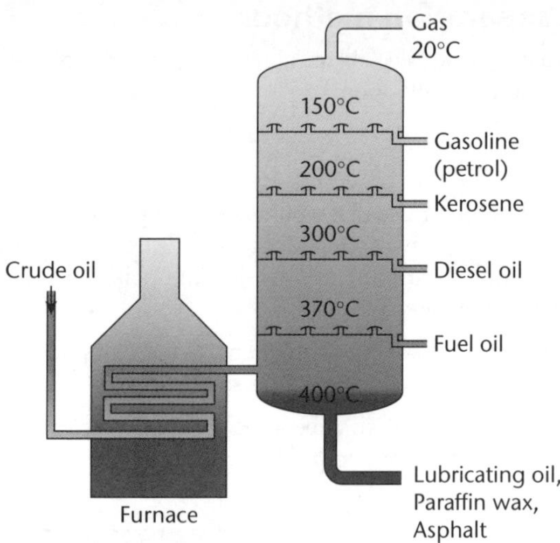

Fractional distillation column at a petroleum refinery

The boiling points of organic compounds increases with molecular mass, so distillation separates crude oil into fractions of different size and boiling points, from gases to heavy oils, waxes and asphalt. The chemicals that result from the first fractional distillation are called **straight-run fractions**. They are listed in the following table.

Straight-run fractions from petroleum distillation

Fraction	Boiling point (°C)	Uses	Sent to
Gases (C_1–C_4 compounds)	<20	Natural gas	Gas processing
Light naphtha (C_5–C_8 compounds)	20–150	Jet fuel	Petroleum blending
Heavy naphtha (C_9–C_{12} compounds)	150–200	Feedstock for fuel, chemicals	Catalytic reforming
Kerosene (C_9–C_{16} compounds)	175–275	Jet fuel	Hydro-treating
Gas oil (C_{15}–C_{25} compounds)	200–400	Diesel, cracking to olefins	Distillate fuel blending
Lubricating oil	>350	Lubrication and catalytic cracking to lighter fractions	Fluid catalytic cracking
Heavy fuel oil	>350	Boiler fuel and catalytic cracking for lighter fractions	Fluid catalytic cracking
Asphalt		Paving, coating and construction purposes	Coking

The C_1–C_4 alkanes (methane (CH_4) to butane (C_4H_{10})) are gases, which are sent to the gas processing unit. The C_5–C_8 alkanes (pentane (C_5H_{12}) to octane (C_8H_{18})) are refined into petroleum. The C_9–C_{16} alkanes (nonane (C_9H_{20}) to hexadecane ($C_{16}H_{34}$)) are refined into diesel fuel and kerosene. Higher alkanes are refined into fuel oil and lubricating oil. At the heavier end of the range is wax and asphalt, which can be converted to more valuable products.

Crystallisation

Waxes are removed from lubricating oil fractions by crystallisation. A solvent such as a benzene/butan-2-one mixture is added to the oil and the solution is then cooled to about –20°C. The wax precipitates out as crystals and can be removed.

Unit 12.5 Activity 1D: Processes in oil refining

1. Explain how waxes are removed from petroleum by crystallisation.
2. Research petroleum refining further to explain the separation of fractions by linearity and aromaticity.
3. Name the natural gases in the C_1–C_4 range.

Chemical separation methods

Straight-run fractions from petroleum distillation are further refined to produce fuels for high-compression engines. The refining converts hydrocarbons greater than C_{10} or smaller than C_4 into hydrocarbons in the petrol range (C_4–C_{10}).

The aim is to produce compounds with a high **octane rating**. The octane rating (or octane number) of a fuel is a measure of its performance. The higher the octane rating, the more it can be compressed before it spontaneously ignites, which causes damaging engine 'knocking'. The rating is based on a scale on which isooctane (2,2,4-trimethylpentane) is 100 (minimal knock) and heptane is 0.

Example H

Octane rating 92

Petrol with an octane rating of 92 has the same knock as a mixture of 92% isooctane and 8% heptane. This does not mean that the petrol contains these hydrocarbons in these proportions. It simply means that it has the same resistance to spontaneous ignition as the described mixture.

Some fuels are even more knock-resistant than isooctane, so the scale has been extended to allow for octane numbers higher than 100.

Example I

Octane ratings >100

Petrol with the same knocking characteristics as a mixture of 90% isooctane and 10% racing fuels, avgas, liquefied petroleum gas (LPG) and alcohol fuels, such as methanol, may have octane ratings of 110 or significantly higher.

An octane rating is not an indicator of the energy content of a fuel. It is only a measure of the fuel's tendency to burn in a controlled manner, rather than exploding in an uncontrolled manner.

Lead in the form of tetraethyl lead was once a common fuel additive because it reduced knocking, but since the 1970s its use in fuels has been progressively phased out worldwide, because of lead's negative effects on the environment and health.

Other chemicals that are added to fuel to increase its performance include aromatic hydrocarbons such as toluene, ethers, alcohol and isooctane.

Thermal cracking

Thermal cracking is the use of high temperatures(400–900°C) to convert alkanes to alkenes (ie form double bonds). At the cracking temperature, a C–C bond breaks and two free radicals form. Radicals are atoms or groups of atoms that contain one or more unpaired electrons (represented by a single dot in the formula). They are very reactive.

Example J

Thermal cracking

In thermal cracking, octane forms ethene or propene and a shorter-chain radical:

$$CH_3(CH_2)_6CH_3 \rightarrow CH_3(CH_2)_3CH_2^{\bullet} + {}^{\bullet}CH_2CH_2CH_3$$

$$CH_3(CH_2)_3CH_2^{\bullet} \rightarrow \underset{\text{ethene}}{H_2C{=}CH_2} + \underset{\text{shorter radical}}{{}^{\bullet}CH_2CH_2CH_3}$$

$$CH_3(CH_2)_6CH_3 \rightarrow \underset{\text{propene}}{H_2C{=}CHCH_3} + \underset{\text{shorter radical}}{CH_3(CH_2)_3CH_2^{\bullet}}$$

Fluid catalytic cracking

Fluid catalytic cracking is a process that uses heat and catalysts such as zeolite to convert long-chain alkanes to aromatics and branched-chain molecules.

Example K

Fluid catalytic cracking

Gas oil (C_{15}–C_{25}) breaks up to form 2-methyl-1-butene and 2-methyl-2-butene.

Hydrocracking and hydrotreating

Hydrocracking is a chemical process that converts long-chain alkanes in crude petroleum into shorter-chain alkanes, alkenes and aromatic compounds and then hydrogenates them. Catalysts such as palladium on zeolite are used in the hydrocracking process.

The result is a relatively large amount of isobutane (methylpropane), which is used in later alkylation and isomerisation reactions (to produce high-quality jet fuel).

Hydrotreating is a catalytic hydrogenation process that removes contaminants such as oxygen, sulfur, nitrogen, metals and unsaturated hydrocarbons from liquid petroleum fractions. Hydrotreating catalysts are usually transition metals, such as cobalt and molybdenum. The contaminants are converted to ammonia, hydrogen sulfide and water.

Polymerisation

In this process, two short-chain alkenes, such as ethene, propene or butene, combine to form a longer-chain alkene, at high temperatures and pressures and using a catalyst of phosphoric acid (H_3PO_4) on clay. The process is terminated at the dimer or trimer stage and the products are blended with petrol or added to the petroleum feedstock.

Example L

Polymerisation

Isobutene combines with propene to yield a C_8 or C_{12} alkene mixture for detergent manufacture.

Alkylation

Reactions of short-chain alkenes are useful for producing hydrocarbons in the petrol range with more branched products. Catalysts such as sulfuric acid and hydrofluoric acid are used. The process is very useful in the petrol refinery. The alkylate product consists of a mixture of heptane and octane isomers and so can be added to petrol.

Example M

Alklyation

Alklyation of isobutene and isobutane produces branched products such as 2,2,4-trimethylpentane and the 2,3,4- and 2,3,3-trimethyl isomers.

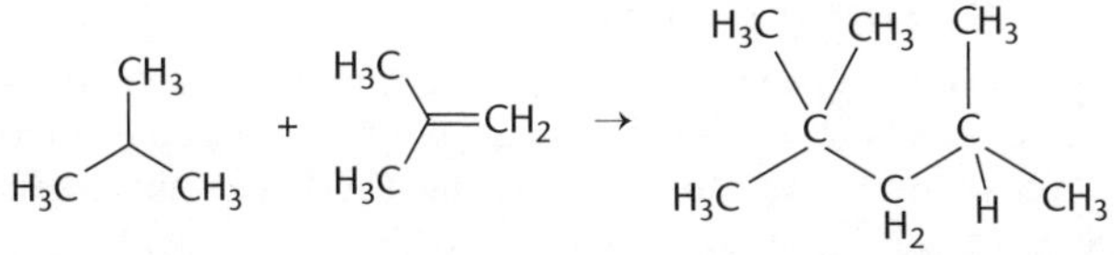

Isobutane (2-methylpropane) + **Propene** → **2,2,4-Trimethylpentane**

Catalytic reforming

Catalytic reforming is the opposite process to cracking and is the most widely used method for petroleum production. The process converts low-octane compounds to high-octane products, called reformates, and hydrogen. This is achieved by rearranging some hydrocarbons and breaking other hydrocarbons into smaller molecules. The catalysts are Pt or Pt–Re on alumina and the process occurs at high temperatures and pressures.

Example N

Catalytic reforming

Catalytic reforming of octane forms 2,5-dimethylhexane. Methylcyclohexane rearranges to form toluene and 3 moles of hydrogen.

Catalytic reforming of the hydrocarbon $C_{15}H_{32}$ breaks it down into ethene, propene and octane.

The less valuable product of the reaction can be converted to other, more commercially valuable hydrocarbons.

Example O

Metathesis

Propene breaks up to form ethene and butene. The reaction is applicable to many alkenes and is called *metathesis*.

$2CH_2{=}CHCH_3 \rightarrow CH_2{=}CH_2 + CH_3CH{=}CHCH_3$

Propene Ethene Butene

Unit 12.5 Activity 1E: Conversion methods

1. Explain how catalytic reforming works.
2. Define octane rating and explain how it affects motor vehicles.
3. Draw structures for:
   a. isobutene.
   b. isobutane.
   c. 2,2,4-trimethylpentane.
4. Write the chemical reaction for the catalytic reforming of:
   a. methylcyclohexane to toluene and 3 moles of hydrogen.
   b. octane to 2,5-dimethylhexane.

Petroleum refining in PNG

The ten top oil-producing countries are Saudi Arabia (producing 12% or the world's oil), Russia (11%), United States (9%), Iran (5%), China (4.5%), Canada (4%), United Arab Emirates (3%), Mexico (3.6%), Kuwait (3%) and Iraq (3%). Australia produces 0.70% of the world's oil and Papua New Guinea produces 0.04%.

The refining of oil in Papua New Guinea began in 2004. The InterOil Napa Napa Oil Refinery is the only oil refinery in Papua New Guinea. InterOil gets its crude oil from Kutubu in Southern Highlands Province and some from overseas. The refinery has the capacity to produce 36 000 barrels per day.

Napa Napa Oil Refinery

Petroleum products such as diesel, kerosene, petrol and liquefied petroleum gas (LPG) are used domestically. Products such as naphtha are not used in Papua New Guinea and are exported to China and Australia.

The following table compares the crude compositions of products from the Napa Napa Oil Refinery and Australian Mutineer oil.

Comparative yields of products made from local Kutubu crude oil and Australian Mutineer crude oil

Stream	Volume (%)	
	PNG	Australia
Fuel gas	0.4	0
Propane (LPG)	0.6	0
Butane (LPG)	3.9	0
Light virgin naphtha	11.6	5
Mixed naphtha	11.4	13
Unleaded petrol	10.7	8.9
Kerosene	7.7	7.4
Diesel	38.0	57.8
Low-sulfur waxy residue	15.7	7.8

Unit 12.5 Activity 1F: Petroleum in PNG

1. Like all oil from PNG, Kutubu crude oil is referred to as 'sweet' crude. Explain why this is.
2. The Napa Napa Oil Refinery imports crude oil from Australia and Asia. Use the information from the table to justify this practice.
3. From the information in the table, should Mutineer Oil Refinery import crude oil from PNG? Explain.

Natural gas and liquefied natural gas

Natural gas

Natural gas formed from the decaying remains of ancient plants and animals. Natural gas is found deep underground, often near oil deposits. It is extracted by using drilled wells.

Natural gas can be considered the most environmentally friendly fossil fuel, for two reasons:

- It has the lowest CO_2 emissions per unit of energy.
- It is suitable for use in combined-cycle power stations, which use waste heat from the power station to make steam to generate additonal electricity.

The major constituent of natural gas is methane, $CH_4(g)$. It also contains other hydrocarbons, CO_2, nitrogen and hydrogen sulfide.

Liquefied natural gas

Liquefied natural gas (LNG) is produced relatively cheaply both worldwide and in Papua New Guinea. LNG is produced when natural gas is cooled below its boiling point to –160°C through a process known as liquefaction. Liquefaction removes or reduces the levels of hydrocarbons, water, carbon dioxide, oxygen and some sulfur compounds. This increases the methane concentration to 80–90% and is known as 'sweetening'. Liquefaction also reduces the density to 1/600th of that of the original gas, making transport more economically viable.

Liquefied natural gas is:

- clear, colourless, odourless, non-corrosive and non-toxic.
- less than half as dense as water and floats on water.
- not explosive in its liquid state.

LNG does not burn by itself. It needs to be in vapour form in an enclosed space and mixed with air. It is only combustible at concentrations of 5–15% by volume of natural gas in air.

LNG is cleaner burning than diesel fuel. It is particulary suited to heavy-duty vehicles because of its high storage density and low nitrogen oxides (NO_x) and particulate emissions.

Example P

LNG

LNG is used by garbage trucks, local delivery (eg grocery) trucks and transit buses. Vehicles that are classified as Class 8 (above about 15 000 kg, gross vehicle weight) typically use LNG. LNG is also used in gas stoves and heaters.

Unit 12.5 Activity 1G: Liquefied natural gas

1. What is LNG?
2. Draw the structure of methane.
3. What is the difference between natural gas and liquefied natural gas?
4. Under what conditions does liquefied natural gas become explosive?

Gas refining in PNG

The countries with the largest natural gas reserves are Algeria, Australia, Brunei, Indonesia, Libya, Malaysia, Nigeria, Oman, Qatar, Trinidad and Tabago, and Papua New Guinea.

The ExxonMobil and Oil Search Port Moresby LNG plant is the only gas refinery in Papua New Guinea. The consortium gets its gas from the Hides gas fields at Kutubu in Southern Highlands Province.

There are five steps involved in getting natural gas into homes and businesses in the cities from gas fields in remote locations.

1. Natural gas is pumped from below ground at the Hides gas fields to the surface. The gas is transported from the site through pipelines to Australia and Port Moresby.
2. The gas is liquefied in Port Moresby to produce a stable liquid ready for shipping.
3. *Transportation* to China occurs in double-hulled ships specifically designed to handle the low temperature required for LNG. These carriers are insulated to limit the amount of LNG that evaporates. The LNG is off-loaded as a liquid and pumped from the jetty to storage tanks at the terminal. The LNG remains at –160°C throughout.
4. When LNG is received at the terminals, it is transferred to insulated storage tanks specifically built to hold LNG. Tanks can be above or below ground and keep the liquid at low temperature to avoid evaporation. Full containment systems use two tanks – an inner tank that contains the LNG, and an outer tank to contain any leaks from the inner tank. The tanks are well insulated to keep the natural gas at –160°C to avoid evaporation.
5. *Re-gasification* of the LNG occurs when it is pumped out of the tanks and warmed so that it returns to its natural gaseous state. The natural gas is then pumped into the natural gas transmission system to cities and towns.

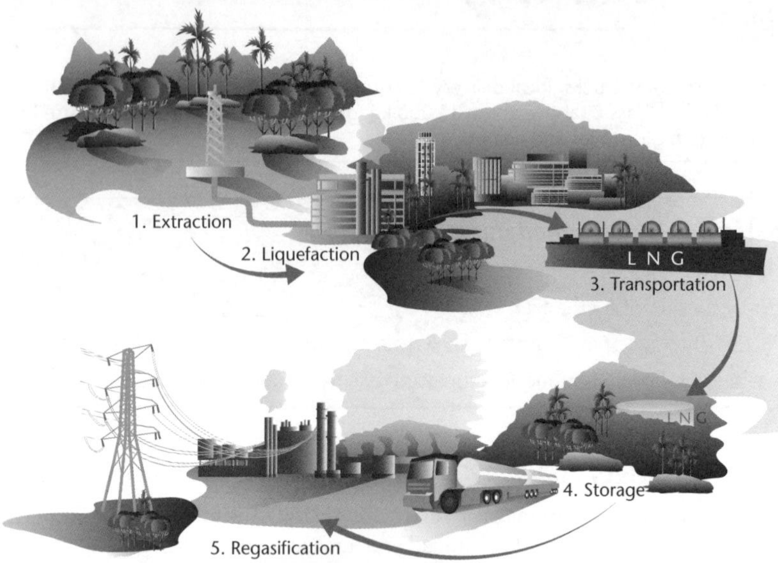

Processes in LNG in Papua New Guinea

Unit 12.5 Activity 1H: LNG refining

1. How is LNG transported?
2. How is LNG stored?
3. Is LNG flammable or explosive?
4. What are the benefits of LNG in transportation applications?
5. What vehicles or niche markets use LNG?

Unit 12.5 Natural Resources and Chemical Industries in Papua New Guinea

Topic 2: Polymers, polymerisation and production of rubber

Topic 2 covers the production of polymers in Papua New Guinea:

- Polymers and polymerisation.
- Production of plastics.
- Production of rubber in Asia and PNG.

Polymers

Polymers from natural and synthetic sources are used frequently in Papua New Guniea. A **polymer** is a large organic molecule or **macromolecule** made of many small **repeating units** called **monomers** joined together (see Unit 12.4 Topic 4).

Naturally occurring polymers include silk and cellulose. Synthetic polymers include polypropylene and **Teflon**.

The properties of polymer depend on the:

- type of bonding between the monomers.
- chemical nature of these repeating units.
- presence of bonding between the polymer chains – these are called cross-linkages.

There are two main categories of polymers, which depend on their methods of preparation:

- **Addition polymers** are products of addition reactions.
- **Condensation polymers** are products of condensation reactions.

Addition polymerisation

Monomers used to make addition polymers all contain double bonds. The bonds that form the polymer chain involve only the carbon atoms from the double bond of the monomer molecules. The polymer chain is the only product formed. No molecule is lost as the result of polymerisation.

Example A

Polyethene

Polyethene (polythene) is formed by the polymerisation of ethene in an addition reaction, ie:

$n(H_2C{=}CH_2) \rightarrow -[H_2C{-}CH_2]_n-$

where n is a very large number.

Common addition polymers include **PVC** (polyvinyl chloride or polychloroethene) and Teflon.

Properties of monomers and their corresponding addition polymers

Property	Monomer	Polymer
Saturation (double bonds)	Has double bond	No double bond
Molecular weight	Low	High
Boiling point	Relatively low	Relatively high
Reactivity	High	Low
Reaction with bromine water	Reacts – turns brown bromine water colourless	Does not react

Unit 12.5 Activity 2A: Addition polymerisation

1. Saran is a useful polymer with the structure
$\ldots CH_2–CCl_2–CH_2–CCl_2–CH_2–CCl_2\ldots$
Give the name and structure of the monomer from which it is made.

2. Identify the functional group that all monomers require to form addition polymers.

3. Draw a section of the polymer chain formed by the following monomers.

a. $CH_2{=}CH–CN$

b. $CH_2{=}CH(CH_3)$

c. $F_2C{=}CF_2$

d. $CH_2{=}CHCl$

Condensation polymerisation

Condensation polymerisation involves the elimination of a small molecule (such as water) as monomers join together.

Example B

Condensation polymers

Polyesters, polyamides and proteins are all condensation polymers.

Polyesters

When carboxylic acid reacts with an alcohol, an ester is formed and water is eliminated or 'condenses' out:

$RCOOH + R'OH \rightarrow RCOOR' + H_2O$

acid alcohol ester

This is an esterification reaction. In the reaction the acid and alcohol have just one functional group each – an acid group or a hydroxy group.

Polyesters are produced when the esterification reaction is repeated many times. To form a polyester, each monomer must have two functional groups. In this way, the monomers can join end to end. Usually one monomer has two hydroxyl (OH) groups at each end – it is a diol. The other monomer usually has two carboxylic acid (COOH) groups at each end – it is a **dicarboxylic acid**. It could also have or two acid chloride (COCl) groups at each end – it is a diacid chloride.

Example C

Terylene

Terylene is the polyester formed from ethane-1,2-diol and benzene-1,4-dicarboxylic acid.

$HO–CH_2–CH_2–OH + HOOC–C_6H_4–COOH \rightarrow HO–CH_2–CH_2–OOC–C_6H_4–COOH + H_2O$

After one reaction the product molecule still has two functional groups, which continue to react, forming repeating units $–[O–CH_2–CH_2–OOC–C_6H_4–CO]_n–$.

Polyesters can also be formed from single momers containing both functional groups.

Example D

Polylactic acid

Polylactic acid is a biodegradable polyester formed from the monomer lactic acid (2-hydroxypropanoic acid).

$HO–C(CH_3)(H)–COOH \xrightarrow{-H_2O} HO–CH(CH_3)–CO–[O–CH(CH_3)–CO]_n–O–CH(CH_3)–COOH$

Lactic acid **Polylactic acid**

Unit 12.5 Activity 2B: Condensation polymerisation and polyesters

1. Describe a condensation reaction.
2. Describe an elimination reaction.
3. Water is one substance that is eliminated during a condensation reaction. Can you think of another?
4. Draw a section of the polymer chain formed by the condensation polymerisation of 4-hydroxybutanoic acid.

Polyamides

A carboxylic acid group (COOH) and an amino group (NH_2) can react to form an **amide** or **peptide link** (–CONH–). The formation of many amide links produces a **polyamide**.

Nylon was one of the first useful fibres made. Nylons are formed by condensation polymerisation of either a dicarboxylic acid or a diacid chloride with a diamine (containing two NH_2 groups)

Example E

Nylon 6,6

Nylon 6,6 is a synthetic polymer used in clothing. It is produced from 1,6-diaminohexane and 1,6-hexanedioic acid (dissolved in cyclohexane, C_6H_{12}):

$H_2N–(CH_2)_6–NH_2 + HOOC–(CH_2)_4–COOH \rightarrow H_2N–(CH_2)_6–HN–OC–(CH_2)_4–COOH + H_2O$

After one reaction, the product molecule still has two functional groups, which continue to react and form more repeating units $–[HN–(CH_2)_6–HN–OC–(CH_2)_4–CO]_n–$.

Unit 12.5 Activity 2C: Polyamides

Nylon 5,5 is made from the monomers 1,5-pentanedioic acid and 1,5-diaminopentane.

a. Draw the structural formula of each monomer.

b. Give the formula of one repeating unit of the polymer.

c. Give the formula of the polymer.

Polypeptides

Polypeptides are polyamides made from **amino acid** monomer units (NH_2–CRH–COOH). R groups are present as **side chains**.

Amino acids contain both basic (the **amine**) and acidic groups. Therefore, they are amphiprotic – they can accept or donate protons. Depending on the pH of the solution they are dissolved in, they can exist as a positively charged ion, a negatively charged ion or a neutral ion in which the amine group has deprotonated the carboxylic acid. This neutral molecule that has both positive and negative charges is called a **zwitterion**.

$NH_2\text{–CRH–COOH} \rightleftharpoons NH_3^+\text{–CRH–COO}^-$

The molecule formed when two amino acids join together in a condensation polymerisation reaction is called a **dipeptide**. Many amino acids joined together are known as a **polypeptide**.

Example F

Polypeptide formation

$NH_2\text{–CRH–COOH} + NH_2\text{–CR'H–COOH} \rightarrow NH_2\text{–CRH–CO–NH–CR'H–COOH} + H_2O$

Amino acid **Amino acid** **Dipeptide**

Proteins

When a polypeptide has a specific function, it is known as a **protein**. Proteins are made up of different combinations of 20 amino acid molecules and are fundamental components of all living organisms. Proteins form hormones, enzymes, muscle tissue and antibodies. Humans can only synthesise some amino acids – the remaining ones, called **essential amino acids**, must be obtained from our food.

Example G

Proteins

Insulin is a hormone and a soluble protein. **Collagen** is an insoluble protein that gives strength and flexibility to tendons. All enzymes are proteins.

Protein structure

There are several levels to the structure of protein.

- The **primary structure** is the basic amino acid sequence consisting of up to 100 amino acids.
- The **secondary structure** is the three-dimensional arrangement caused by hydrogen bonding between oxygens of C=O groups and hydrogens of NH_2 groups.

- The **tertiary structure** is the three-dimensional folding resulting from interactions between R groups.

The overall shapes of proteins are very complex.

Hydrolysis of **peptides** and proteins produces the various amino acids from which the proteins are made.

Unit 12.5 Activity 2D: Proteins

1. Describe the relationship between proteins, amino acids, polypeptides and dipeptides.
2. Draw the structural formulae for the two possible dipeptides formed from aminoethanoic acid (glycine, H_2NCH_2COOH) and 2-aminopropanoic acid (alanine, $CH_3CH(NH_2)COOH$).
3. Explain why it is impossible to draw a repeating unit for a protein.
4. Explain why the structure of 2-aminopropanoic acid is more correctly drawn as a zwitterion ($H_3N^+–CH(CH_3)–COO^-$).

Plastics

A **plastic** is any of a wide range of synthetic or semisynthetic organic solids that can be moulded. The term 'plastic' is derived from Greek word *plasticos*, meaning 'to mould'.

Plastics are typically organic polymers of high molecular mass, but they often contain other substances. They are usually synthetic, most commonly derived from petrochemicals, but many are partially natural.

Plastics are relatively cheap and easy to make, have a wide range of uses and are impervious to water. This means they can be used in a wide range of products, from paper clips to spaceships. Plastics have displaced many traditional materials, such as wood, stone, horn and bone, leather, paper, metal, glass and ceramic, in most of their former uses.

Production of plastics

The production of plastic takes place ove four stages:

1. Acquiring the raw material or monomer.
2. Synthesising the basic polymer.
3. Compounding the polymer into a material that can be used for fabrication.
4. Moulding or shaping the plastic into its final form.

Acquiring raw materials or monomers

Plant resins were used to produce the first plastics.

Example H

Early plastics

Cellulose (from cotton), furfural (from oat hulls), oils (from seeds) and various starch derivatives were used to make the first plastics. Early plastics included polystyrene and polyvinyl chloride (PVC). Sources were natural plastic materials such as chewing gum and chemically modified natural materials such as rubber and nitrocellulose.

Today, most plastics are produced from petrochemicals, which are cheap and widely available. However, as supplies of oil will eventually be depleted, other sources of raw materials are being explored, such as coal gasification.

Synthesising the basic polymer

The first step in plastic manufacturing is polymerisation, by either addition or condensation reactions.

Compounding the polymer

Most plastics contain other organic or inorganic additives to achieve certain characteristics. Additives include:

- **antioxidants** to protect against degradation by **ozone** (O_3) or oxygen.
- ultraviolet stabilisers to protect against weathering.
- plasticisers to increase the polymer's flexibility.
- lubricants to reduce friction problems.
- pigments to add colour.
- flame retardants to lower the flammability of the plastic.

The amount of additives ranges from 0% for polymers used to wrap foods to more than 50% for polymers used in electronic applications.

Example I

Plastic additives

Glass or carbon fibres are added to some plastics, such as Kevlar, to increase their strength and stability.

Gas is added to polystyrene to form foamed polystyrene, such as Styrofoam™, which is used in disposable coffee cups.

Moulding the plastic

Common types of moulding include injection, extrusion and compression moulding.

In injection moulding, melted plastic is injected into a cold mould, where it sets to the required shape.

In extrusion moulding, an extruder forces softened plastic through a shaped die. The plastic can emerge in almost any form, including a circular rod or tube, and a wide, flat sheet. The driving force is supplied by a screw, which provides constant pressure.

Compression moulding uses pressure to force the plastic into a certain shape. Half of a two-piece mould is filled with plastic and then the two halves of the mould are brought together and the plastic is melted under high pressure.

Classification of plastics

Plastics are usually classified by:

- the chemical structure of the polymer's backbone and side chains, eg acrylics, polyesters, silicones, polyurethanes and halogenated plastics.
- how they are synthesised, eg condensation, polyaddition or cross-linking.

- qualities that are relevant for manufacturing or product design.
- various physical properties, such as density, tensile strength, glass transition temperature, and resistance to various chemical products.

There are two types of plastics: thermoplastics and thermosetting plastics.

Thermoplastics can be moulded again and again. Their chemical composition does not change when heated.

Example J

Thermoplastics

Polyethylene, polypropylene, polystyrene, polyvinyl chloride and polytetrafluoroethylene (PTFE) are thermoplastics.

Thermosetting plastics can only melt and take shape once. Once they have solidified, they stay solid. In the thermosetting process, an irreversible chemical reaction occurs. Most thermosetting plastics are shaped by compression moulding.

Example K

Thermosetting plastics

Rubber is a thermosetting plastic. Before heating with sulfur (a process called vulcanisation), the polyisoprene is a tacky, slightly runny material, but after vulcanisation the product is rigid and non-tacky. Other thermosetting plastics are polyurethanes, fibreglass and Bakelite.

Biodegradability

If a plastic is **biodegradable**, then it breaks down (degrade) upon exposure to sunlight (ultraviolet radiation), water or moisture, bacteria, enzymes and wind abrasion. Attack by rodents, pests or insects can also be regarded as a type of biodegradation or environmental degradation.

Uses of plastics

The following tables list some uses of plastics.

Common uses of plastics

Plastic	Uses
Polypropylene	Thermal clothing, storage containers, rope, chairs, packaging film
Polyester	Fibres, textiles
Polyethylene terephthalate (PET)	Carbonated drink bottles, peanut butter jars, plastic film, microwavable packaging
Polyethene	Bottles, carrier bags, toys, buckets, tubing for carrying cold water
High-density polyethylene (HDPE)	Detergent bottles, milk jugs, moulded plastic cases
Polyvinyl chloride (PVC)	Downpipes and guttering, wiring and cable insulation, lino, artificial leather (eg car seats), protective clothing, ground covers for tents, shower curtains, window frames
Polyvinylidene chloride (PVDC) (Saran)	Food packaging

Common uses of plastics (continued)

Plastic	Uses
Polycarbonate	Compact discs, spectacles, riot shields, security windows, traffic lights, lenses
Polycarbonate/ acrylonitrile butadiene styrene (PC/ABS)	Car parts, mobile phones

Special purpose plastics

Plastic	Characterisitcs	Uses
Plastarch material	Biodegradable, heat-resistant, thermoplastic composed of modified corn starch	Food packaging, personal care items, carrier bags, window insulation
Polyetheretherketone (PEEK)	Strong, chemical- and heat-resistant, biocompatible thermoplastic	Medical implants, aerospace mouldings. One of the most expensive commercial polymers
Poly(methyl methacrylate) (PMMA)	Transparent, high molecular weight thermoplastic	Hard contact lenses, glazing (eg as Perspex, Oroglas or Plexiglas), fluorescent light diffusers, rear light covers for vehicles
Polytetrafluoroethylene (PTFE) (Teflon)	Heat-resistant, non-reactive thermoplastic	Non-stick cooking utensils, plumber's tape and water slides, Gore-Tex, coating underneath skis

Unit 12.5 Activity 2E: Plastics

1. Define the term 'plastic'.
2. Describe the process and steps in plastic manufacture.
3. Differentiate between extrusion moulding and compression moulding.

Rubber

Rubber is a natural polymer that is produced by the rubber tree (*Hevea brasiliensis*). It is an elastomer (an elastic hydrocarbon polymer) found in the latex, a milky colloidal suspension found in the sap of the tree. Latex is collected by cutting a thin strip of bark from the tree and allowing the latex to flow into a collecting vessel over a period of hours.

Latex is generally processed into a latex concentrate or coagulated (at low pH) to produce more specialist block rubbers.

The monomer unit of natural rubber is isoprene (2-methyl-1,2-butadiene):

$$CH_2{=}C(CH_3){-}CH{=}CH_2 \rightarrow -[CH_2C(CH_3){=}CH(CH_2)]_n-$$

Isoprene **Polyisoprene (natural rubber)**

Synthetic rubber is made from monomers derived from petroleum. One of the most common synthetic rubbers is the styrene–butadiene rubbers.

$C_6H_5CH{=}CH_2 + CH_2CH{=}CH{=}CH_2 \rightarrow -[CHC_6H_5CH_2{-}CH_2CH{=}CH{-}CH_2]_n-$

Styrene **1,2-Butadiene**

Rubber production in Asia and Papua New Guinea

Rubber was introduced to Asia from South America in the 19th century. Asia supplies around 92% of the world's natural rubber, raw material for the production of thousands of everyday articles.

Indonesia, Thailand and Malaysia lead the world in natural rubber production. Other Asia–Pacific producers include China, India, Vietnam and Sri Lanka.

China is the largest world producer of synthetic rubber. Other major producers are Japan, South Korea and Taiwan. Thailand, Malaysia, India, Indonesia and Iran have lower production levels of synthetic rubber.

Asia is also a big consumer of rubber. It has a large number of vehicles, which require around 70% of the world's tyre production. Most of the multinational companies that make these tyres have bases in Asia.

Asia also dominates the production of other rubber goods such as gloves, condoms and medical rubber products.

Example L

Rubber

Rubber is used in industrial and engineering products, footwear products, latex-based products, latex foam and fibre foam, gloves and condoms.

The rubber tree was introduced to Papua New Guinea in the early 1900s, but the rubber industry has not developed to the same extent as in other countries in the region. By world standards, rubber production in Papua New Guniea is very low, as seen in the following table. There is the potential for a healthy synthetic rubber industry to take advantage of Papua New Guinea's thriving petroleum industry.

Major natural rubber producers (×1000 tonnes)

Country	2002	2003	2004
Brazil	96	97	98
Guatemala	40	41	41
Other Latin America	23	23	23
Cameroon	58	62	62
Cote d'Ivoire	120	126	126
Ghana	12	12	12
Liberia	109	110	110
Nigeria	45	46	46
DR of Congo	8	8	8
Other Africa	11	11	11
Bangladesh	5	5	5
Cambodia	47	47	47
China	468	477	477
India	640	690	690
Indonesia	1630	1761	1810
Malaysia	589	678	681
Myanmar	30	30	30
Papua New Guinea	4	4	5
Philippines	74	78	78
Sri Lanka	91	93	93
Thailand	2615	2873	2931
Vietnam	373	384	383
Total world	**7270**	**7880**	**7928**

Unit 12.5 Activity 2F: Rubber

1. From where was natural rubber introduced to Asia?
2. Draw the structures for the monomer and polymer of natural rubber.
3. Which three Asian countries lead the world in the production of natural rubber?
4. Which Asian country leads the world in the production of synthetic rubber?
5. Refer to the table to identify the country showed the greatest increase in natural rubber production between 2002 and 2004.
6. Explain the low amount of natural rubber production in PNG between 2002 and 2004.
7. What raw material is available in PNG for synthetic rubber products?

Unit 12.5 Natural Resources and Chemical Industries in Papua New Guinea

Topic 3: Metallic ores and mining in PNG

Topic 3 covers metallic ores and mining in Papua New Guinea by discussing:

- Extraction of copper, gold, nickel, silver and cobalt.
- Environmental consequences of mining.

A metallic **ore** is a naturally occurring **mineral** from which a metal can be extracted. Papua New Guinea has a number of valuable ores that contain gold, silver, copper, nickel and cobalt.

So far, ores have been extracted from land-based deposits. However, larger ore concentrations of gold and copper have been discovered in the seabed of the Bismarck Sea. The PNG Government and mining companies are exploring this area with a view to extracting metallic ores from the ocean floor.

The following map shows the locations of current mining projects in PNG.

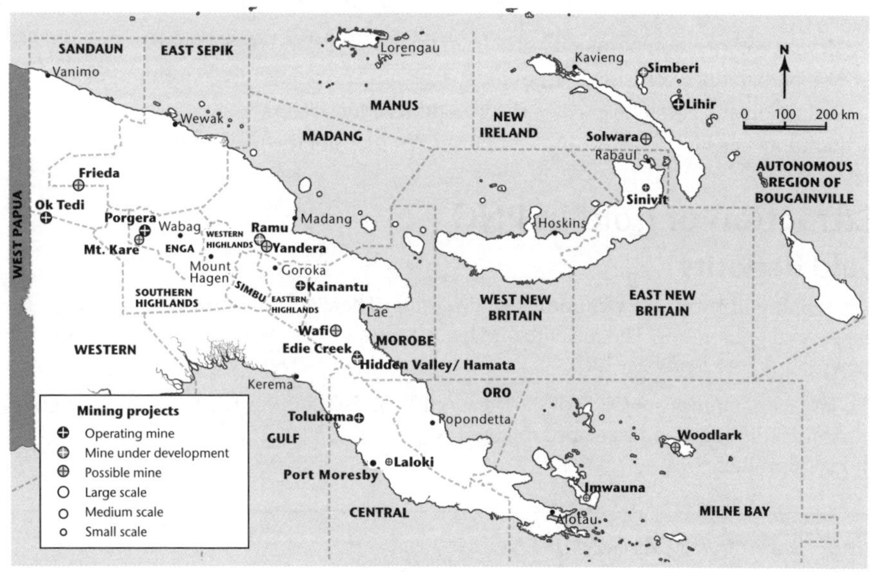

Mining projects in PNG

Metal extraction

The process of extracting and purifying metals from their ores for commercial use is called **extractive metallurgy**.

The principal processes of general metallurgy are pyrometallurgy, hydrometallurgy and electrometallurgy.

- Pyrometallurgy uses high temperatures to carry out reduction and smelting reactions.
- Hydrometallurgy uses aqueous solutions and inorganic solvents to achieve reactions (leaching chemistry).
- Electrometallurgy uses electricity to extract and refine metals in aqueous solution.

Example A

Metal extractions

The reduction of Fe_2O_3 to Fe in a blast furnace is a pyrometallurgical process.
Acid leaching of zinc oxide to obtain zinc metal is a hydrometallurgical process.
Purifying alumina in cryolite (AlF_3 + 3NaF) by electrorefining is an electrometallurgical process.

There are four general stages in the extraction of metals from their ores:

1. Mining – excavation of mineral ores from the ground (land or sea floor).
2. Ore preparation – crushing or grinding, screening or sizing and concentration.
3. Reduction and extraction – uses the general reaction $M^{n+} \rightarrow M^0$.
4. Refining – ensures high quality and purity (99.99%) of the recovered metal.

Extraction of gold in PNG

Gold deposits

Most mining in Papua New Guinea is gold mining. There are a number of gold deposits and several gold mines. The metallic ore deposits and mines in Papua New Guinea are shown on the map on page 293.

The large-scale mines are Ok Tedi, Porgera and Lihir gold mines. The medium-scale gold mines are Tolukuma and Kainantu mines. There are also a few small ones that may develop into larger mines.

Unit 12.5 Activity 3A: Gold in PNG

1. Use a map to find the locations of gold ore deposits and mines in PNG.
2. With the help of other sources, select a gold mine and discuss the socio-economic impacts of the gold-mining activities in the immediate locality.

Properties and chemistry of gold

Gold (Au) is one of the oldest metals known and used extensively. It occurs naturally as an attractive yellow-coloured metal. The melting point of gold is 1064°C and its boiling point is 2807°C.

Gold:

- is **ductile** and corrosion resistant. **Alloys** containing 50% or more Au are also corrosion resistant.
- has a high density (19.32 g/cm^3) and electrical and **thermal conductivity** (conducts heat well).
- does not react with oxygen, sulfur or selenium at any temperature.
- reacts with halogens in the presence of water.
- reacts with various oxidising agents at room temperature provided a good **ligand** is present to lower the redox potential below that of water.
- dissolves in aqua regia – a highly corrosive 3:1 mixture of concentrated hydrochloric acid (HCl) and nitric acid (HNO_3).

Example B

Gold and aqua regia

On reaction with aqua regia, gold forms a chlorine complex according to the following reaction:

$Au(s) + 6H^+(aq) + 3NO_3^-(aq) + 4Cl^-(aq) \rightarrow AuCl_4^-(aq) + 3NO_2(g) + 3H_2O(\ell)$

The gold is recovered by electrolysis or solvent extraction.

- dissolves in alkaline cyanide in the presence of air.

Applications of gold

Gold has a long history of use in jewellery, coins and as a means of currency, and as a store of wealth (gold bullion). Today, gold is used in medicine and dentistry, electronics, catalytic converters in cars and trucks, and nanotechnology. It has many applications in engineering and space exploration, where its resistance to corrosion and ability to reflect heat are very useful.

Unit 12.5 Activity 3B: Properties and chemistry of gold

1. Conduct some research to verify the statement: 'Gold is one of the oldest metals known and used extensively'.
2. What properties of gold make it attractive for use as a means of exchange or currency?
3. Explain why deposits of gold in the Panguna copper mine are found under deposits of copper.
4. Explain why gold has not been a traditional means of exchange in Papua New Guinea.
5. Research the medical applications of gold.
6. Name the following gold complexes and determine the oxidation number of gold in each.
   a. $AuCl_4^-(aq)$
   b. $[Au(CN)_2]^-(aq)$

Extraction and processing of gold

Gold is extracted by a mixture of pyrometallurgy, hydrometallurgy and electrometallurgy. The method of mining depends on the gold deposits, whether they are lode or vein deposits or placer deposits.

Example C

Gold-mining methods

Lode or vein deposits occur within a fissure in a rock. Lode or vein gold deposits are mined by blasting or drilling.

A placer deposit is an accumulation of minerals on the surface of the Earth, usually as river sediment. Placer deposits of gold are minded by hydraulic, dredging or power shovelling.

Ore preparation involves crushing and pulverising the gold ore to a fine powder.

Extraction of the gold may occur by one or more of the following processes.

- **Flotation** – air is pumped through a slurry of water and ore. The gold minerals are carried to the surface in the air bubbles and skimmed off to be refined.
- Amalgamation – a mercury–gold **amalgam** is formed, from which the gold is liberated by heating.
- Cyanidation – a cyanide solution (NaCN or KCN) reacts with gold in air to form a cyanide–gold complex, which gets reduced by zinc to gold metal.

Example D

Cyanidation of gold

Gold reacts with alkaline cyanide in the presence of air according to the following reaction:

$4Au(s) + 8CN^-(aq) + O_2(g) \rightarrow 4[Au(CN)_2]^-(aq) + 4OH^-(aq)$

The Au^+ in $[Au(CN)_2]^-$ is reduced to Au^0 with powdered zinc in the following redox reaction:

$2[Au(CN)_2]^-(aq) + Zn(s) \rightarrow [Zn(CN)_4]^{2-}(aq) + 2Au(s)$

- Carbon-in-pulp – a cyanide solution (NaCN or KCN) reacts with gold in air to form a cyanide–gold complex, which gets reduced by carbon to gold metal.

Smelting purifies the gold further. The gold is heated with a substance called a flux. The flux bonds with the contaminant and floats on top of the melted gold. The gold is cooled and allowed to harden in moulds while the contaminated flux (slag) is removed as a solid waste.

Electrochemical refining is required to obtain high-purity gold (99.99%). If even higher purity gold (99.999%) is required, then electrolytic refining is used. Anodes are made from 99.5% Au and cathodes are made of pure Au. During electrolysis, oxidation takes place at the anode, resulting in Au^+ ions. At the cathode 99.999% pure Au is formed.

Unit 12.5 Activity 3C: Extraction and processing of gold

1. What is the difference between hydrometallurgy and pyrometallurgy of gold?
2. Explain the difference between recovery of gold by the cyanidation and carbon-in-pulp methods.
3. Describe the principle of froth flotation of gold minerals.
4. Why would the amalgamation method be dangerous for small-scale gold miners in PNG?

Extraction of copper in PNG

Copper deposits

Papua New Guinea has a number of copper deposits and several copper mines. The largest copper mines are the Panguna, Ok Tedi and Frieda copper mines.

The Panguna copper mine, developed by Bougainville Copper Limited (BCL), was one of the first major investments in Papua New Guinea, and was at one stage one of the world's largest copper and gold mines. However, work on it has been limited due to civil unrest in the province.

Copper is very reactive and occurs in two classes of ores:

- sulfide ores: $CuFeS_2$ and CuS_2.
- oxidised ores: CuO, $Cu_2(OH)_2CO_3$ and $Cu_3(OH)_2(CO_3)_2$.

Unit 12.5 Activity 3D: Copper in PNG

1. Use a map to find the locations of copper ore deposits and mines in PNG.
2. With the help of other sources, discuss the socio-economic impact of the Panguna copper mine on Bougainville and on PNG.
3. Name the following copper compounds.
   a. $CuFeS_2(s)$
   b. $CuO(s)$
4. Draw possible structures of the following compounds.
   a. $Cu_2(OH)_2CO_3$
   b. $Cu_3(OH)_2(CO_3)_2$

Properties and chemistry of copper

Copper (Cu) plays a vital role in shaping our lives. People have been using copper since 9000 BCE. One of the reasons copper is so important is that it can be combined with other metals to make alloys, such as brass and bronze. These alloys are harder, stronger and more resistant to corrosion than pure copper. Copper is an attractive reddish colour and has many desirable properties. The melting point of copper is 1083°C and its boiling point is 2567°C.

Copper:

- is tough, non-magnetic, ductile and resistant to corrosion.
- has a very high **electrical conductivity**, second only to that of silver.
- has a high **thermal conductivity**.
- is antibacterial.
- is easy to alloy and can be easily joined to other metals.
- is recyclable.
- can be used in catalysts.

Applications of copper

Copper has many applications. Copper is used in:

- electrical wiring. Copper wires allow electric current to flow without much loss of energy. This is why copper wires are used in mains cables in houses and under ground. (Overhead cables tend to be made of aluminium because it is lighter.)
- electromagnets in locks, scrapyard cranes and electric bells.
- motors in pumps, domestic appliances (washing machines, dishwashers, refrigerators, vacuum cleaners), computers (disc drives, fans), entertainment systems (CD and DVD players) and cars (starter motors, windscreen wipers, electric windows).
- dynamos in bicycles and power stations.
- transformers in mains adaptors, electricity substations and power stations.

Unit 12.5 Activity 3E: Applications of copper

1. What properties of copper make it useful for everyday use?

2. Explain why deposits of copper in the Ok Tedi copper mine are found on top of the gold deposits.

Extraction and processing of copper

Copper is mostly mined in open-cut mines. Copper extraction requires a mixture of pyrometallurgy, hydrometallurgy and electrometallurgy.

Ore preparation involves crushing and pulverising the copper ore to a fine powder.

Extraction of the copper occurs by:

- flotation (in a similar way to the extraction of gold).
- roasting, which uses pyrometallurgy (in which the material is roasted in a furnace) to convert sulfides to oxides, thus separating the iron and sulfur. The reaction is:
  $4CuFeS_2(s) + 3O_2(g) \rightarrow 2CuS_2(s) + 2FeO(s) + 2FeS(s) + 2SO_2(g)$
- **smelting**, which involves heating the copper in a blast furnace with coke and **silica** to convert the iron to iron silicates ($FeSiO_3$). The remaining copper-rich component is called matte (CuS_2 and FeS).

The matte undergoes a bessemerisation process – air is blown through the molten matte, which converts it to copper. The reactions are:

$2CuS_2(s) + 3O_2(g) \rightarrow 2Cu_2O(s) + 2SO_2(g)$

$2Cu_2O(s) + CuS_2(s) \rightarrow 6Cu(s) + SO_2(g)$

$2FeS(s) + 3O_2(g) \rightarrow 2FeO(s) + 2SO_2(g)$

During bessemerisation, SO_2 bubbles escape, leaving the material with a blistered appearance. For this reason, the product of the bessemerisation process is called blister copper. Blister copper contains 97–8% copper.

Electrorefining purifies the copper further. The anodes cast from blister copper are placed in an aqueous solution of 3–4% $CuSO_4$ and 10–15% H_2SO_4. Cathodes are made of sheets of high-purity copper. An electric potential of 0.2–0.4 volts is applied. Pure copper is formed at the cathode. The reactions are:

- At anodes: $Cu(s) \rightarrow Cu^{2+}(aq) + 2e^-$
- At cathodes: $Cu^{2+}(aq) + 2e^- \rightarrow Cu(s)$

Non-copper materials in copper anodes, such as traces of gold, silver, selenium and tellurium, settle to the bottom of the electrochemical cell. This is called anode mud and is a saleable product. Arsenic and zinc, which were present in the blister copper, remain in solution.

Unit 12.5 Activity 3F: Extraction and processing of copper

1. Why is it necessary for copper ore to undergo roasting?
2. Explain the bessemerisation process.
3. What is blister copper?
4. Smelting and refining of copper are not done in PNG. Find out why.

Extraction of nickel and cobalt in PNG

Nickel and cobalt deposits

Deposits of nickel ore are located in Canada, Australia, Cuba and Indonesia and a small amount is found in Papua New Guinea.

Cobalt ore deposits are found in Zaire, Morocco and Canada and a small amount is found in Papua New Guinea. Cobalt is found in silver, nickel, copper and iron ores. It is usually mined as a **by-product** of copper or nickel. The ore mined usually contains only 0.1% of cobalt.

Papua New Guinea has a large deposit of nickel and cobalt in Madang Province where it is mined at the recently opened Ramu project. This is a joint venture between the PNG Government and Metallurgical Corporation of China.

It is expected that the annual output will be nearly 33 million tonnes of nickel and 3200 tonnes of cobalt over 20 years. The Kurumbakar nickel and cobalt resource has an estimated reserve of 143 million tonnes of 1.01% nickel and 0.1% cobalt. This will support a very long-life operation.

Unit 12.5 Activity 3G: Nickel and cobalt in PNG

1. Use a map to find the locations of Ramu project nickel and cobalt deposits in PNG.
2. Explain the make-up of the ownership of Ramu NiCo.

Properties and chemistry of nickel

Nickel (Ni) is a silvery white transition metal that can be polished. It is a member of the iron–cobalt group of metals. The melting point of nickel is 1453°C and its boiling point is 2732°C. Nickel's specific gravity is 8.902 (25°C) and it has a **valence** of 0, 1, 2 or 3.

Some nickel compounds (nickel carbonyl, nickel sulfide) are considered to be highly toxic or carcinogenic.

Nickel:

- is hard, ductile, **malleable** and **ferromagnetic** (can be magnetised).
- has a fair electrical and **thermal conductivity**.
- does not react with water or aqueous sodium hydroxide under normal conditions or air under ambient conditions.
- reacts readily with air when finely divided.

Example E

$2Ni(s) + O_2(g) \rightarrow 2NiO(s)$

- reacts with halogens, but only reacts slowly with F_2. This makes nickel an important metal for storage containers of fluorine.

Example F

$Ni(s) + Cl_2(g) \rightarrow NiCl_2(s)$ (yellow)
$Ni(s) + Br_2(g) \rightarrow NiBr_2(s)$ (yellow)
$Ni(s) + I_2(g) \rightarrow NiI_2(s)$ (black)

- dissolves slowly in dilute sulfuric acid to form solutions containing the aquated Ni^{II} ion together with hydrogen gas, H_2. In practice, the Ni^{II} is present as the **complex ion** $[Ni(OH_2)_6]^{2+}$.

Example G

$Ni(s) + H_2SO_4(aq) \rightarrow Ni^{2+}(aq) + SO_4^{2-}(aq) + H_2(g)$
$Ni(s) + H_2SO_4(aq) \rightarrow [Ni(OH_2)_6]^{2+}(aq) + SO_4^{2-}(aq) + H_2(g)$

Applications of nickel

The main application of nickel is in alloys. It will alloy with most metals to form very useful corrosion- and heat-resistant alloys.

Example H

Nickel alloys

Alloys of nickel include stainless steel, superalloys in jet engine turbine blades, copper–nickel alloy tubing in desalination plants and shape-memory alloys.

Nickel is also used in coins (eg the toea), armour plating, ceramics, magnets and batteries. Nickel is added to glass to give it a green colour. Nickel plating provides a protective coating to other metals. Finely divided nickel is used as a catalyst for hydrogenating vegetable oils.

Some enzymes contain nickel. Urease is found in plants such as jack beans and soy beans. It also occurs in some animals and microorganisms. This enzyme catalyses the hydrolysis of urea into carbon dioxide and ammonia; its active site contains nickel. Hydrogenases catalyse the hydrolysis or reduction of a chemical by hydrogen. Some of these enzymes (the NiFe-hydrogenases, which also contain iron–sulfur clusters) contain nickel.

Currently, 46% of global nickel production is used for making nickel steels, 34% is used in non-ferrous alloys and superalloys, 14% is used in electroplating and 6% is used elsewhere.

Unit 12.5 Activity 3H: Properties and chemistry of nickel

1. What properties of nickel make it useful for everyday use? Write the electronic configuration for nickel.
2. Name the following nickel compounds.
   a. $Ni(OH)_2(s)$
   b. $[Ni(OH_2)_6]^{2+}$
3. Draw possible structures of the following compounds.
   a. $Ni(OH)_2(s)$
   b. $[Ni(OH_2)_6]^{2+}$
4. Conduct some research to find out what a shape-memory alloy is.

Properties and chemistry of cobalt

Cobalt (Co) is a silver-grey transition metal and is similar to nickel in its properties. The melting point of cobalt is 1495°C and its boiling point is 2870°C. Its magnetic properties are similar to those of iron.

Cobalt:

- is **brittle**, hard and lustrous.
- is **ferromagnetic**.
- does not react with water and is stable in air.
- reacts slowly with acids.

Example I

Cobalt sulfides (cobaltite, Co_3S_4) and arsenides (CoAsS) (smaltite $CoAs_2$) are important sources of cobalt.

Applications of cobalt

Cobalt salts have been used for many centuries to colour glass and porcelain a beautiful deep blue colour.

The artificially produced isotope cobalt-60 is used as a source of gamma (γ) rays. Its high-energy radiation is useful for sterilisation in medicine and of foods.

Example J

The nuclear reaction for producing cobalt-60 from cobalt-59 by bombardment with neutrons is:

$^{59}Co + n \rightarrow {}^{60}Co + \gamma\text{-rays (7.492 MeV)}$

Cobalt is an essential element for all multicellular organisms. It is an active component of the coenzymes called cobalamins.

Example K

Marmite is a source of vitamin B12, a compound containing cobalt.

Cobalt is used as a component of magnetic alloys. Its strong resistance to high temperatures also makes it ideal for use in surface coatings, cutting tools, high-speed steel and diamond tools.

Cobalt is used in the production of gas turbines and jet engines, rechargeable batteries, **drying agents** for paints, pigments and steel-belted radial tyres. Because of its hardness and resistance to oxidation, cobalt metal is also used in electroplating, a process in which an electric field causes metal ions to coat an electrode.

Cobalt is fed to sheep to prevent disease and improve the quality of their wool.

Unit 12.5 Activity 31: Properties and chemistry of cobalt

1. What properties of cobalt make it useful for everyday use?

2. Name the following cobalt compounds.

a. $Co(OH)_3$

b. $CoAs_2$

3. Draw possible structures of the following compounds. You may need to undertake further research or ask your teacher to help you.

a. Co_3S_4

b. Vitamin B12

Extraction and processing of nickel

Conventional roasting and reduction of nickel ores yield a metal that is more than 75% pure.

Example L

In many stainless steel applications, 75% pure nickel can be used without further purification, depending on the composition of the impurities.

Hydrometallurgy – in which nickel sulfide ores undergo flotation and then smelting – produces much purer nickel.

Solvent extraction is used to separate the cobalt and nickel, with the final nickel concentration being greater than 99%. Solvent extraction separates compounds on the basis of their solubility in two immiscible liquids, such as an organic solvent and water. Hence, the nickel is extracted from one liquid into another liquid phase.

For higher purity, nickel oxides are further reduced by the Mond process, which increases the nickel concentrate to greater than 99.99% purity in two stages.

1. Nickel is reacted with carbon monoxide at 40–80°C in the presence of a sulfur catalyst to form nickel tetracarbonyl. The reaction is exothermic and reversible, so a low temperature and high pressure are maintained to ensure higher yield of the nickel tetracarbonyl, which boils at 42°C.

$Ni(s) + CO \rightleftharpoons Ni(CO)_4 \qquad \Delta H = -163 \text{ kJ mol}^{-1}$

2. The nickel tetracarbonyl is decomposed via the reverse endothermic reaction at 230°C to create fine nickel powder. The resultant carbon monoxide is re-circulated and re-used in the first stage of the process.

$Ni(CO)_4 \rightleftharpoons Ni(s) + CO \qquad \Delta H = +163 \text{ kJ mol}^{-1}$

Extraction and processing of cobalt

Industrially, cobalt is normally produced as a **by-product** from the production of copper, nickel and lead.

Many ores contain cobalt but not many are of economic importance. Ore is mined by blasting the earth with explosives or digging into rock with shovels and picks to extract the ore from the rock. Miners crush the ore in primary cone crushers, using wet magnetic separators to separate the copper or nickel, then reduce it to mesh in a ball mill, which is a mineral grinder.

Cobalt is then extracted in a variety of ways.

- Solvent extraction extracts the cobalt from the nickel or copper.
- Smelting, using heat from a blast furnace and a metal reducing agent such as carbon, changes the ore's oxidised state. The carbon removes oxygen and leaves the 99.9% pure metal cobalt.
- Roasting forms a mixture of metals and metal oxides. Treatment with sulfuric acid leaves metallic copper as a residue and dissolves out iron, cobalt and nickel as the sulfates. Iron is obtained by precipitation with lime (CaO).

  Cobalt is produced as the hydroxide by precipitation with sodium **hypochlorite** (NaOCl):

  $2Co^{2+}(aq) + NaOCl(aq) + 4OH^-(aq) + H_2O \rightarrow 2Co(OH)_3(s) + NaCl(aq)$

  The cobalt trihydroxide, $Co(OH)_3$, is heated to form the oxide and then reduced with carbon (as charcoal) to form cobalt metal:

  $2Co(OH)_3 \xrightarrow{heat} Co_2O_3 + 3H_2O$

  $2Co_2O_3 + 3C \rightarrow Co + 3CO_2$

Unit 12.5 Activity 3J: Extraction of nickel and cobalt

1. Explain the differences between the extraction processes of nickel and cobalt.
2. Explain the Mond process used in the extraction of nickel.

Extraction and production of silver in PNG

Silver deposits

Currently in Papua New Guinea, silver is mined at Hidden Valley in Morobe Province. This mine started in 2011 and is an open pit, gold and silver mine. The ore is transported to Australia for processing. Silver is also mined at Porgera (Enga Province) and Lihir (New Ireland Province). Other silver-mining projects being explored include the Mt Kare project, on the border of Enga Province and the Southern Highlands. A silver mine on Misima Island, eastern Papua New Guinea has recently been closed and is being rehabilitated.

Properties and chemistry of silver

Silver (Ag) is an attractive and valuable transition metal. The melting point of silver is 961°C and its boiling point is 2162°C.

Silver:

- is very soft, lustrous, **malleable** and ductile.
- has the highest thermal and electrical conductivity of any metal, making it extremely useful in electronic equipment.
- is antibacterial and non-toxic.
- does not react with air or water.
- reacts readily with sulfur, which causes silver to **tarnish** easily.
- reacts with sulfuric acid (H_2SO_4) and nitric acid (HNO_3).

Applications of silver

Silver has a long history of use in coins and as a means of currency, and as a store of wealth (silver bullion). Silver is used in jewellery and has become an important metal in electronic, industrial and medical applications.

Silver nitrate is an important silver salt. It is a grey solid that is an antiseptic (inhibits growth of microorganisms) and is also used as a reagent in analytical chemistry. Silver nitrate is formed by dissolving silver in nitric acid. On heating to about 320°C, $AgNO_3$ loses oxygen and forms silver nitrite. At higher temperatures, it decomposes and forms silver.

Silver nitrate is a precursor to silver halides, which are used in photography.

Unit 12.5 Activity 3K: Chemistry and applications of silver

1. Write the equation for the decomposition of $AgNO_3$.
2. Write the equation for the reaction between solutions of HCl and $AgNO_3$.
3. Devise a test for confirming the presence of silver ions in a solution.
4. Find out and explain how silver halides (eg AgBr) are used in photography.
5. Research the medical applications of silver.

Extraction and processing of silver

Most silver is mined as a co-product of lead (in galena), copper (in chalcopyrite), zinc (in sphalerite) and gold. It is only rarely found as pure silver nuggets. The minerals most commonly mined in primary silver operations include argentite (Ag_2S), chlorargyrite (AgCl) and pyragyrite (Ag_3SbS_3).

Hydrometallurgy

Silver ore is crushed and ground into a powder. Froth flotation then separates out the silver-bearing ore from waste rock particles. The froth is skimmed off and the metal sulfide is collected. This metal sulfide contains about 0.1% silver as well as the primary metal (lead, copper, zinc or gold). The sulfide is then smelted and drossed (further purified) to increase the silver concentration to about 0.2%.

Pyrometallurgy and electrometallurgy

Once the primary metal (such as gold, copper or zinc) has been extracted, the waste will contain trace amounts of silver. These wastes are treated with chemicals that react with the silver, which is then extracted by electrolysis and roasted.

To extract the silver from the roasted ore, impurites are first removed by heating the ore with a flux.

There are two types of flux – acidic and basic. Acidic fluxes such as silica (SiO_2) remove basic impurities (gangue) such as lime (CaO) or MgO. Basic fluxes include lime (CaO) and magnesium oxide (MgO) and they are used to remove acidic gangue such as SiO_2. Flux combines with the impurities to form slag – molten mixtures of calcium and magnesium silicates – and is removed.

The silver is obtained from the roasted ore by chemical or electrolytic reduction.

Unit 12.5 Activity 3L: Extraction and processing of silver

1. Find out the chemical formulae for chalcopyrite and sphalerite.
2. Write the chemical formulae for argentite and pyragyrite.
3. What is slag refining?

Ocean mining in PNG

The extraction of metals from the sea floor around Papua New Guinea has been proposed for the Bismarck Sea and Solomon Sea.

Canadian mining company Nautilus Minerals Inc.'s Solwara 1 project plans to mine gold and copper deposits associated with deep-sea hydrothermal vents at a depth of 1500 metres in the Bismarck Sea.

The total area of sea floor mined will be 0.11 km^2. It is planned that the mining operation will be conducted in two phases, which may make it easy to stop operations if problems arise. The initial mining phase will be about 30 months and ore will be extracted at a rate of 5900 tonnes per day.

The deposits contain 2.2 million tonnes of ore. The copper from the site is on average more than ten times purer than a typical land-based copper mine. The site also contains significant quantities of high-grade gold (5 grams/tonne).

Other possible deep-sea mining sites are located in the Solomon Sea. In January 2011, Nautilus was granted a 20-year mining lease by the PNG Government.

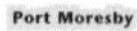

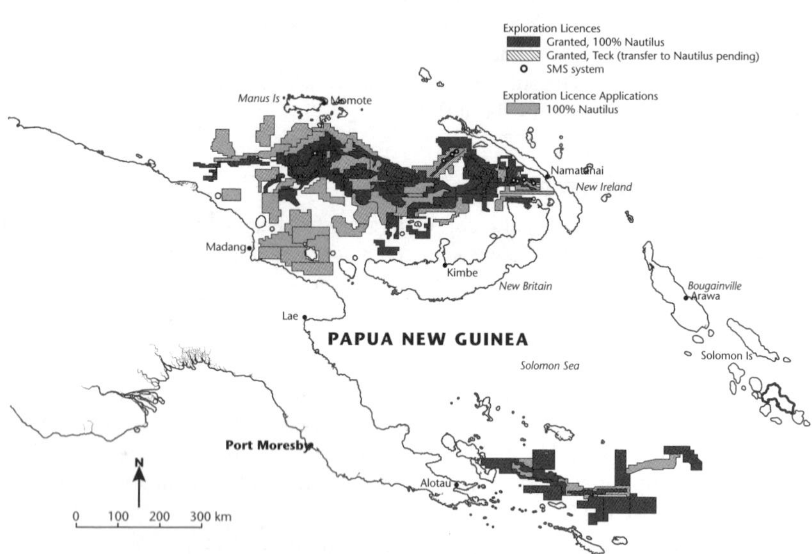

Proposed deep-sea mining sites around Papua New Guinea

Unit 12.5 Activity 3M: Ocean mining in PNG

1. Locate Solwara 1 on the map. Suggest one advantage in using this site as a trial for deep-sea mining.
2. Suggest a possible reason why mining in the Bismarck Sea is preferred over the Solomon Sea.
3. Suggest a reason for Nautilus's two-stage approach to deep-sea mining in the Bismarck Sea.

Seabed mineral resources and hydrothermal vents

Deep-sea **hydrothermal vents** are volcanic hot springs on the sea floor at depths of 1–6 kilometres. The hydrothermal vents are produced when very high pressures force cold, oxygen-rich seawater down through the porous sea bed.

The heat and high pressure causes the seawater to becomes acidic, low in oxygen and highly concentrated in metal sulfides dissolved from the surrounding rocks.

The metal-rich, superheated water (or vent fluid) becomes buoyant and rushes up through the seabed, forming the hydrothermal vents and mixing with cold water from the ocean bottom. This causes the minerals in the vent fluid to crystallise, forming massive sulfide deposits.

Hydrothermal vents are thought to support one of the rarest and most unique ecological communities known. Bacteria that can use sulfur compounds such as hydrogen sulfide (HS) are found there, as are long tube worms.

Over thousands of years, the metal sulfide particles settling around the vents have developed into huge mounds. These are known as **seafloor massive sulfides**, and can range from several thousand to 100 million tonnes. The seafloor massive sulfide deposits at Nautlius's Solwara 1 site in the Bismarck Sea could be over 5000 years old.

Location of seafloor massive sulfides

More than 250 deep-sea massive sulfide deposits have been identified worldwide.

Hydrothermal vents occur along underwater fault lines and volcanically active chains of mountain ranges. They are found in the Mid-Ocean Ridge – an 84 000 km chain of volcanic mountains that extends throughout the world's oceans. Most of the known sulfide deposits have been located along particular sections or rises of the Mid-Ocean Ridge.

Several deposits are also known at the Mid-Atlantic Ridge. Only one has so far been located in the Indian Ocean.

Extensive sulfide deposits were found to be associated with felsic volcanism – the most explosive type of volcanic activity – in the Manus and Woodlark Basins in Papua New Guinea and to the north of New Caledonia.

Mineral deposits in the Manus Basin

The Manus Basin seafloor massive sulfide deposits are located on the floor of the Bismarck Sea at approximately 1600 metres depth. The Solwara 1 project is located in the Eastern Manus Basin and was first discovered by the Australian Commonwealth Science and Industry Research Organisation (CSIRO) in 1996.

The Manus Basin includes about 40 000 hydrothermal vents that are over 0.25 metres high.

Unit 12.5 Activity 3N: Hydrothermal vents

1. Explain the origin of seafloor massive sulfide deposits.
2. Explain the formation of hydrothermal vents.
3. Describe the Mid-Ocean Ridge and the Mid-Atlantic Ridge.
4. Find out what unique forms of life exist around these hydrothermal vents.

Environmental consequences of mining in PNG

Although mining is important for Papua New Guinea's economy, it can also have environmental impacts. Soil erosion, contamination of soil and water systems by chemicals from the mine and loss of animal species due to habitat destruction are just some of the effects that need to be considered in any mining situation.

Legislation now exists to reduce the environmental effects of mining and ensure rehabilitation of the mining site after the mine is closed, ie restoring the land to close to its original state.

More information about the environmental consequences of mining in Papua New Guinea can be found in Unit 12.5, Topic 7.

Unit 12.5 Natural Resources and Chemical Industries in Papua New Guinea

Topic 4: Production of alcohol (ethanol) in PNG

Topic 4 covers the production of ethanol in Papua New Guinea:

- Fermentation (SP Brewery).
- Fermentation and distillation (Ramu Agri Industries).
- Chemical and physical properties of ethanol.
- Effects of ethanol and methanol.

Ethanol (C_2H_5OH) has been produced and consumed by humans for millennia, in the form of fermented and distilled alcoholic beverages, such as beer, wine, whisky, rum and brandy. It is a clear flammable liquid that boils at 78.4°C, and is also used as an industrial solvent, car fuel and raw material in the chemical industry. For more information on the chemistry and properties of ethanol, see Unit 12.4, Topic 7.

Beer

The only significant beer producer in Papua New Guinea is South Pacific Brewery Ltd, which is located in Port Moresby. It has been brewing beer, mostly lagers, for 60 years.

All beer is brewed from malted barley, hops, yeast and water, although other ingredients such as fruit, wheat and spices are sometimes used. The yeast turns sugars in the malt into alcohol. The process is called **fermentation**.

Example A

Fermentation reaction

Malt (sugar) + yeast → alcohol + carbon dioxide

The hops provide the bitter flavours in beer and the flowery aroma. The flavour of the beer depends on many things, including the types of malt and hops used, other ingredients and the yeast variety. Each yeast variety has its own distinctive effect on the beer. Carbon dioxide gives the sparkle (bubbles). Alcohol gives the intoxicating effect.

There are four main beer styles, which are determined by the variety of yeast used in their brewing:

- ales.
- lagers.
- spontaneously fermented beers.
- mixed origin beers.

Ales

Ale yeasts ferment at 15–20°C and occasionally as high as 24°C. Pure ale yeasts form a foam on the surface of the fermenting beer. Because of this they are often referred to as 'top-fermenting' yeasts, although some ale yeast strains settle at the bottom. Ales are generally ready to drink within three weeks of the beginning of fermentation, but some styles are aged for several months or years. Ales range in colour from very pale to black–opaque.

Lagers

Lager is fermented at much lower temperatures, around 10°C. Lager yeasts tend to collect at the bottom of the fermenter and so are often referred to as 'bottom-fermenting' yeasts. Lager is then stored for 30 days or longer at temperatures close to the freezing point. During the storing or 'lagering' process, the beer mellows and flavours become smoother. Sulfur components that developed during fermentation dissipate.

The popularity of lager was a major factor leading to the rapid introduction of refrigeration in the early 1900s. Today, lagers represent the vast majority of beers produced, and the most famous in PNG is the SP Lager beer.

Spontaneously fermented beers

Spontaneously fermented beers are nowadays primarily only brewed in Belgium. They are fermented by means of wild yeast strains that live in a part of the Zenne River, which flows through Brussels. These beers are also called Lambic beers.

Mixed origin beers

These beers are blends of spontaneously fermented beers and ales or lagers or they are ales or lagers that are also fermented by wild yeasts.

Unit 12.5 Activity 4A: Beer

1. Explain what alcohol is.
2. Write the formula for ethanol and methanol.
3. Explain what beer is.
4. Explain the different types of beers and how they are produced.

Beer production

The stages in beer brewing are malting, milling, mashing, brewing, cooling and fermentation, followed by maturation (racking), filtering (finishing) and packaging.

Beer is made from water and barley – a tall grass with seeds on the top of the stalk. Barley comes in many strains and varieties that ultimately influence the flavour of the beer. Barley and water are processed to create a sweetened liquid (called the wort), which is then flavoured with hops, and fermented with yeast.

Malting

Malting is the process of preparing the barley for brewing. Barley cannot be used to create the wort in its normal state, because the starch in the seed is insoluble. Each step of the malting process unlocks the starches in the barley.

The malt is first germinated in big tanks called steeps. Here the grain is steeped (soaked) in water for about 40 hours.

Next the barley grain is germinated. The grain is spread out on the floor of the germination room for 3–5 days, at which stage rootlets begin to form. During germination, a number of enzymes develop in the seed, which break down the starch into sugars that are easily fermentable.

Example B

Enzymes in barley

Enzymes in barley include alpha-amylase and beta-amylase, which are used to convert the starch in the grain into sugar.

The temperature determines the amount of enzymes produced. At high temperatures, low levels of enzyme are produced. At low temperatures, high levels of enzymes are produced.

When enzyme production has been maximised, the germinated barley is dried on metal racks in kilns at 50°C. A dark colour and malty flavour are produced. The enzymes stop working but are not destroyed.

The temperature is then raised to 85°C for a light malt, or higher for a dark malt. It is important that temperature increases are gradual so that the enzymes in the grain are not damaged. The root shoots are removed for cattle feed, and the dried malt is stored in silos.

Although malted barley is the primary ingredient, other grains can be used.

Example C

Unmalted corn, rice or wheat are sometimes added to beer to produce different beer flavours.

After kilning, the result is finished malt. The differences in the way the barley is malted will affect the flavor, colour and aroma of the beer. There are different types of malts.

Example D

Pale malts are dried at a low temperature and can give the beer a pale golden colour and a slightly bready flavour.

Milling

During the milling stage, the malted grain is cracked. Milling allows the grain to absorb the water it will eventually be mixed with in order for the water to extract sugars from the malt.

Malt needs to be milled correctly for maximum wort extraction to occur.

Example E

Over-milled malt results in overexposure, while under-milled malt results in loss of extracts due to some starch not being attacked by enzymes.

Mashing

Mashing converts the finely ground malt, the grist, to a sweetened liquid. Mashing converts the starches that were released during the malting stage to sugars that can be fermented.

The milled grain is dropped into warm water, and then gradually heated to around 75°C in a large cooking vessel called the mash tun. In this mash tun, the grain and heated water form a cereal mash and the starch dissolves in the water, transforming it into sugar, mainly maltose. Water is a key ingredient.

The sugar-rich water is then strained through the bottom of the mash in a process called **lautering** and is now called wort.

Lautering is achieved in either a lauter tun, a wide vessel with a false bottom, or a mash filter. Lautering has two stages:

- first wort run-off, when the extract is separated in an undiluted state from the mash.
- sparging, when the mash is rinsed with hot water.

Example F

Mashing

A mash filter is a plate-and-frame filter. The empty frames contain the mash, and have a capacity of about 100 litres. The plates contain a support structure for the filter cloth.

The plates, frames and filter cloths are arranged in a carrier frame so that there is a frame, cloth, plate, cloth and plates at each end of the structure. Newer mash filters have bladders that can press the liquid out of the grains between spargings. The wort travels through the cloth and feeds out through exit.

The next stage is the **brewing** stage. The spent grains are filtered out and the wort is ready for boiling. During this stage, important decisions will be made about the flavour, colour and aroma of the beer. Hops are added and the mixture is boiled for 1–2 hours to sterilise and concentrate it and extract the necessary essence from the hops.

Example G

Hops in beer

Certain types of hops are added at different times during the brewing stage to add bitterness and aroma and to help preserve it.

The wort is transferred quickly from the brew kettle through a device that filters out the hops, and then onto a heat exchanger to be cooled. The heat exchanger consists of tubing inside of a tub of cold water. It is important to cool the wort quickly so that yeast can be added in the next stage, otherwise the yeast will be killed. The hopped wort is saturated with air, essential for the growth of the yeast.

The cooled wort is transferred to the fermentation tank and yeast is added. Yeast is a microorganism that consumes the sugar in the wort and converts it into alcohol and carbon dioxide. This process of fermentation takes 10 days. The wort finally becomes beer.

Each brewery has its own strains of yeast, which will determine the character of the beer.

Example H

Yeasts in beer

Some yeast varieties rise to the top at the end of fermentation, and are then skimmed off. This is called top fermentation, and ales are brewed in this way.

In other varieties, at the end of fermentation the yeast cells sink to the bottom, the process is known as bottom fermentation, used for lager or pils.

The beer has now been brewed, but it can still be improved through maturation (also called racking). During this phase, the beer is moved, or racked, into a new tank called the conditioning tank. Over time, the taste ripens. The liquid becomes clearer as yeast and other particles settle.

Maturation is associated with lager beer. The beer is gradually cooled from 6–8°C to –0.5°C.

Another process in maturation is sedimentation, which is when solids precipitate in the tank. This can also be achieved by centrifuging.

The final stage in the brewing process is finishing – the beer is filtered and carbonated, which adds carbon dioxide to give the beer its sparkle. Carbonation is done using a carbon dioxide injector.

The beer is moved to a holding tank where it stays until it is bottled, canned or put into kegs. The beer is kept away from air. Beer is transferred into glass bottles or aluminium cans, which are packed into cardboard boxes and distributed to wholesalers and customers.

Unit 12.5 Activity 4B: Brewing

1. Which of the following is *not* a raw material for production of beer?
   - **A.** Water
   - **B.** Molasses
   - **C.** Barley
   - **D.** Hops
2. Explain what happens in the following stages.
   - **a.** Brewing
   - **b.** Malting
   - **c.** Mashing
   - **d.** Fermentation
   - **e.** Carbonation
3. What ingredient(s) in the processes give beer the following characteristics?
   - **a.** Brownish colour
   - **b.** Sparkle
   - **c.** Bubbles
   - **d.** Bitter taste
4. What substance in the beer causes intoxication? Write its formula and draw its chemical structure.

Fermentation and distillation

Ramu Agri Industries Ltd (formerly Ramu Sugar Ltd) is the main sugar refining company in Papua New Guinea. It supplies the refined sugars on shelves in trade stores and supermarkets. Ramu Agri Industries Ltd is the second-biggest producer of ethanol and rum in Papua New Guinea after SP Brewery.

Rum production

Molasses is a **by-product** of the production of sucrose from sugarcane. Molasses is the raw material for the production of ethanol and spirits such as rum.

The raw materials for rum are molasses, water and yeast. These are combined in the processes of fermentation, distillation, ageing and blending. During fermentation, rum flavours are also manufactured.

First, the molasses is diluted with water to reduce the sugar concentration. Then pure yeast culture is added to the mixture, which is called 'live wash'. The yeast converts the sucrose to ethanol, carbon dioxide and energy:

$$\text{Molasses} + \text{yeast} \rightarrow \text{ethanol} + \text{carbon dioxide} + \text{energy}$$

The fermentation process takes about 30 hours, during which time the yeast uses up the sugar in the molasses. The liquid left at the end of the fermentation process is called the 'dead wash'. After fermentation, ethanol is separated from the dead wash by distillation.

Unit 12.5 Activity 4C: Fermentation and distillation

1. What are the major products of Ramu Agri Industries Ltd?
2. What is the raw material for the production of ethanol by Ramu Agri Industries?
3. Explain the following processes carried out at Ramu Agri Industries.
   a. Fermentation.
   b. Distillation.
4. What is meant by 'live wash' and 'dead wash'?

Properties of ethanol

Ethanol is often called 'grain alcohol' and is made from fermented corn and barley.

Both ethanol and methanol are important renewable fuel alternatives to petroleum. Traditional fossil fuels such as petroleum and diesel were formed over millions of years. They are not renewable.

Ethanol is also used in thermometers and as a solvent, being **miscible** with water and with many organic solvents, such as short-chain hydrocarbons.

Example I

Ethanol as a solvent

Ethanol is soluble in acetic acid, acetone, benzene, carbon tetrachloride, chloroform, diethyl ether, **ethylene glycol**, glycerol, nitromethane, pyridine, toluene, pentane, hexane, trichloroethane and tetrachloroethylene.

The physical properties of ethanol stem primarily from the presence of its hydroxyl group (OH) and the shortness of its carbon chain.

The hydroxyl group can participate in hydrogen bonding, rendering it more viscous and less volatile than less polar organic compounds of similar molecular weight, such as propane.

Ethanol burns with a smokeless blue flame that is not always visible in normal light. It is slightly more refractive than water, having a refractive index of 1.36242 (at λ = 589.3 nm and 18.35°C).

The **triple point** for ethanol is 150 K (–123°C) at a pressure of 4.3×10^{-4} Pa. This is the point at which solid, liquid and gaseous ethanol co-exist in equilibrium.

Ethanol is a psychoactive drug and one of the oldest recreational drugs known – it produces a state known as alcohol intoxication when consumed.

Ethanol's miscibility with water contrasts with the immiscibility of longer-chain alcohols (five or more carbon atoms), whose water miscibility decreases sharply as the number of carbons increases. When ethanol and water are mixed, the final volume is less than the individual component volumes combined. The reaction is exothermic.

Example J

Mixing equal volumes of ethanol and water results in only 1.92 volumes of mixture. Mixing ethanol and water is exothermic, with up to 777 J mol^{-1} being released at 298 K.

Unit 12.5 Activity 4D: Ethanol

1. List and explain all the physical properties of ethanol.
2. List and explain all the solvent properties of ethanol.
3. What is the resulting volume when you mix 50 mL of water with 50 mL of ethanol?

Effects of ethanol and methanol

Alcohol was not a traditional beverage in Papua New Guinea. It was introduced by Europeans. The abuse of ethanol has become the biggest contributing factor to ill health and societal breakdown in Papua New Guinea.

Different types of alcoholic drink contain varying amounts of alcohol (ethanol). Beer is 4.5% ethanol, wine is 13.5% ethanol, and rum and vodka are 40% ethanol.

Ethanol is a central nervous system depressant and has significant psychoactive effects in sublethal doses. Ethanol is considered a psychoactive drug.

Death from ethanol consumption can occur when blood alcohol levels reach 0.4%. A blood level of 0.5% or more is commonly fatal. Levels of even less than 0.1% can cause intoxication, with unconsciousness often occurring at 0.3–0.4%.

Example K

At low dosages, ethanol can result in euphoria and relaxation. People experiencing these symptoms tend to become talkative and less inhibited, and may exhibit poor judgement. At higher dosages, ethanol acts as a central nervous system depressant, producing at progressively higher dosages, impaired sensory and motor function, slowed cognition, stupefaction, unconsciousness and possible death.

Prolonged heavy consumption of alcohol can cause significant permanent damage to the brain and other organs. Discontinuing consumption of alcohol after several years of heavy drinking can also be fatal.

Example L

Alcohol withdrawal

Alcohol withdrawal can cause anxiety, autonomic dysfunction, seizures and hallucinations. Delirium tremens is a condition that requires people with a long history of heavy drinking to undertake an alcohol detoxification regimen.

There is a strong correlation between high levels of alcohol consumption and an increased risk of developing alcoholism, cardiovascular disease, chronic pancreatitis, alcoholic liver disease and cancer. Damage to the central nervous system and peripheral nervous system can occur from chronic alcohol abuse.

The developing adolescent brain is particularly vulnerable to the toxic effects of alcohol, as is the developing brain of the unborn, possibly resulting in the foetal alcohol syndrome.

Drinking alcohol in moderate amounts has been linked to health benefits, such as reduced heart disease. However, some experts argue that the benefits of moderate alcohol consumption may be outweighed by other increased risks, including injury, violence, foetal damage, certain forms of cancer, liver disease and hypertension.

Example M

Papua New Guniea is experiencing a huge social and economic cost of excessive, irresponsible alcohol abuse: this includes thefts, burglaries, carjackings, domestic violence, rapes, car accidents and alcohol-related diseases. Alcohol abuse is hindering Papua New Guniea's development and may be a factor in derailing Vision 2050 – PNG's long-term plan for improving the country's health and wellbeing.

Methanol is also highly toxic to humans. It is a central nervous system depressant, causing headaches, dizziness, nausea, lack of coordiation and confusion. Large enough doses can cause blindness, unconsciousness and death.

Unit 12.5 Activity 4E: Effects of ethanol

1. What are the advantages of drinking ethanol?
2. What are the disadvantages of drinking ethanol?
3. Carry out some research to find out why homebrewed ethanol can be dangerous.
4. Carry out some research to find out some uses of methanol.

Unit 12.5 Natural Resources and Chemical Industries in Papua New Guinea

Topic 5: Production and uses of vegetable oils

Topic 5 covers the production and use of vegetable oils in Papua New Guinea:

- Extraction of oil from seeds and nuts.
- Soap production.
- Biodiesel production.

Fats and oils are **triglycerides**. A triglyceride is a **triester** formed from the reaction of three fatty acids molecules and glycerol (propane-1,2,3-triol) molecule:

$$\text{H-C(H)-OH},\ \text{H-C-OH},\ \text{H-C(H)-OH} + 3\ \text{HO-C(=O)-R} = \text{H-C(H)-O-C(=O)-R},\ \text{H-C-O-C(=O)-R},\ \text{H-C(H)-O-C(=O)-R} + 3H_2O$$

Glycerol **3 Fatty acids** **Triglyceride**

There are many different triglycerides – some are saturated and some are unsaturated (contain C=C double bonds). Triglycerides from plants are typically unsaturated and liquid at room temperature. Tricylcerides from animals tend to be unsaturated and so be solid at room temperature. Vegetable oils that are solid at room temperature are sometimes called vegetable fats.

Extraction of oil from seeds and nuts

Sources of vegetable oil in PNG

Although many plant parts may yield oil, commercially, oil is extracted mostly from seeds. There are more than 40 indigenous plant species with edible kernels in Papua New Guinea. Eleven species are used in village agriculture. Six of these have the potential for commercial development. All six species are marketed locally and sometimes sent to more distant markets within Papua New Guinea. In the global market, coconut oil, palm oil and peanut oil are most important.

Example A

PNG plants with commercial potential

Indigenous species with commercial potential are cut nut (*Barringtonia procera*), galip nut (*Canarium indicum*), karuka (*Pandanus julianettii*), okari (*Terminalia kaernbachii*), aila (*Inocarpus fagifer*) and talis (*Terminalia catappa*).

Unit 12.5 Activity 5A: Oil from plants

1. List and explain the plant species in PNG that have edible kernels.
2. What plant species in PNG have commercial potential?
3. If you had land, would you grow any of the plant species? Why or why not?

PNG coconut oil

Coconut oil is an edible tropical oil harvested from the coconut palm (*Cocos nucifera*). It is extracted from copra, which is the dried flesh or kernel of the coconut. Coconut oil is used in many products.

Coconut palms have been grown in Papua New Guinea for over a hundred years, having been introduced by German settlers in the late 19th century. Copra production declined in the late 20th century, but has since started to recover. Two large copra mills currently operate in Papua New Guinea, in Madang and near Rabaul. Copra is mostly processed into coconut oil and exported to Europe.

Example B

Uses of coconut oil

Coconut oil is used in margarine, soap, cosmetics, non-dairy cream, snack foods and industrial lubricants.

Coconut oil is high in saturated fatty acids – typically it contains 91% saturated fatty acids, 6% mono-unsaturated fatty acids (fatty acids with only one double bond) and 3% **polyunsaturated** (more than one double bond) fatty acids.

The only mono-unsaturated fatty acid in coconut oil is **oleic acid** ($CH_3(CH_2)_7CH{=}CH(CH_2)_7COOH$). The only polyunsaturated fatty acid is **linoleic acid** ($CH_3(CH_2)_4CH_2CH{=}CHCH_2CH{=}CH(CH_2)_7COOH$). The saturated fatty acids are made of varying amounts of different acids.

Example C

Coconut oil fatty acids

Typical saturated fatty acids in coconut oil are lauric acid ($C_{11}H_{23}COOH$) (45%), myristic acid ($C_{13}H_{27}COOH$) (17%), palmitic acid ($C_{15}H_{31}COOH$) (8%) and caprylic acid ($C_7H_{15}COOH$) (8%).

Copra production

At the coconut plantation, coconut trees are typically spaced 9 metres apart, giving a density of 100–160 coconut trees per hectare. Each tree produces 50–80 nuts per year.

Copra production is labour intensive. The fallen coconuts are collected and de-husked, and then the shells are removed, cracked and transported for drying in a dryer. The dried copra is packed in bags and transferred to a copra mill where the copra is crushed to produce oil (70%) and copra cake or copra meal (30%).

Extraction of oil from copra

The coconut oil is extracted from the copra either with a mechanical expeller or by solvent extraction with hexane.

Premium quality copra meal contains 20–22% crude protein and is used as a feed for horses and cattle. However, it can also contain trace amonts of aflatoxins, which are highly toxic and can pass through the food chain to humans.

Unit 12.5 Activity 5B: Coconut oil

1. What are the uses of coconut oil?

2. How much saturated fat is contained in coconut oil?

3. If there is a copra factory in your area, arrange to visit it.

PNG palm oil

Palm oil is an edible plant oil from the oil palms *Elaeis guineensis* and *Elaeis oleifera*. Neither species is native to Papua New Guinea. A mature tree produces reddish fruit that grows in large bunches weighing up to 40–50 kilograms. The palm fruit can take 5–6 months to mature and yields two different oils – palm oil and palm kernel oil. Palm oil is high in saturated fats and is semisolid at room temperature.

Example D

Uses of palm oil

Palm oil is used as a cooking oil and is a major ingredient in many foods, such as margarine and non-dairy spreads. It is also used in soaps, cosmetics, pharmaceuticals, lubricants and plastics. Palm biodiesel is also being considered as an alternative to diesel fuel.

Example E

Palm oil fatty acids

Typical saturated fatty acids in palm oil are palmitic acid ($C_{15}H_{31}COOH$) (44%), oleic acid ($C_{17}H_{33}COOH$) (39%), linoleic acid ($C_{17}H_{31}COOH$) (10%) and stearic acid ($C_{17}H_{35}COOH$) (5%).

In Papua New Guinea, oil palms are mostly grown in West New Britain, Oro and New Ireland Provinces. Also, a large area of oil palm trees is being planted in Bewani (Vanimo) in Sandaun Province. The major manufacturers of palm oil are NBPOL Ltd, Higaturu Oil Ltd and Hargy Oil Palm Ltd.

Palm oil production

At the oil palm plantation, palm trees are spaced about 8 metres apart, giving a density of 120–180 trees per hectare. Each tree produces about 40–50 kilograms of fruit per harvest and a very large amount each year.

Harvesting of oil palm is labour intensive. Bunches of fruit are weighed in boxes or in trucks on weighing bridges.

Fruit are manually removed from the bunches by axes and knives or in a rotating drum with a specially designed rotary beater. They are then sterilised or boiled with high-temperature and high-pressure steam or by cooking in hot water.

After sterilising, the fruit are beaten and pressed to extract the oil. Pressing can be by dry methods, in which the oil is mechanically squeezed out, or by wet methods, in which the oil is leached out by hot water.

Impurities such as phospholipids, pigments and free fatty acids are removed and the oil is dried.

Unit 12.5 Activity 5C: Extracting and processing palm oil

1. Outline the process of extracting oil from the oil palm tree.
2. List and explain the impurities that are separated by clarifying and drying.
3. What are the major uses of palm oil in PNG?
4. If there is a copra factory in your area, arrange to visit it.

Soap production (saponification)

When fats and oils are hydrolysed by an alkali, the products are salts of fatty acids and glycerol.

Example F

Soap fatty acids

The following fatty acids can be hydrolysed by NaOH to form salt (soap) and glycerol.

- Palmitic acid, $C_{15}H_{31}COOH$ forms sodium palmitate, $C_{15}H_{31}COONa$, and glycerol.
- Oleic acid, $C_{17}H_{33}COOH$ forms sodium oleate, $C_{17}H_{33}COONa$, and glycerol.
- Stearic acid, $C_{17}H_{35}COOH$ forms sodium stearate, $C_{17}H_{35}COONa$, and glycerol.

Soap making, or **saponification**, involves the alkaline hydrolysis of triglycerides. NaOH hydrolyses the three ester groups to produce the triol glycerol and three carboxylate ions.

Example G

Saponification to form sodium stearate soap

$$\begin{array}{ccccccc} CH_2OOC(CH_2)_{16}CH_3 & & & & CH_2OH & & \\ | & & & & | & & \\ CHOOC(CH_2)_{16}CH_3 & + & 3NaOH & \rightarrow & CHOH & + & 3CH_3(CH_2)_{16}COONa \\ | & & & & | & & \\ CH_2OOC(CH_2)_{16}CH_3 & & & & CH_2OH & & \\ \textbf{Triester of glycerol} & & \textbf{Base} & & \textbf{Glycerol} & & \textbf{Sodium stearate (soap)} \end{array}$$

Unit 12.5 Activity 5D: Saponification

1. What is a soap?
2. What are the raw materials for making soap?
3. Write a chemical reaction for the formation of sodium palmitate soap.
4. Write a chemical reaction for the formation of sodium oleate soap.

Soap production in PNG

The leading producers of commercial soap in Papua New Guinea are Colgate Palmolive Ltd and K.K. Kingston Ltd, which are located in Lae in the northern part of the country.

The raw material for soft soap in Papua New Guinea is vegetable oil – coconut oil or oil palm. Animal fat (from cows and pigs) is used to produce hard soap, which is slow to dissolve in water. Sodium hydroxide, salts, colours and perfume are added to make soap attractive.

The main step in soap making is saponification. After the oils and fats are blended, the feedstock is mixed with concentrated sodium hydroxide in large vessels called kettles. The mixture is heated by steam injectors, and soap and glycerol are formed during saponification the reaction.

The crude soap is extracted by washing the mixture with hot sodium chloride solution. This dissolves the glycerol and leaves the insoluble soap behind.

Excess salt and sodium hydroxide are removed from the soap by further washing with water. After removal of excess salt and alkali, the soap is vacuum dried and colours and perfumes are then added. Next, the solid soap is cut into bars or ground to powder. Finally the bars and powder are packaged for distribution.

Cleaning action of soap

Soaps remove dirt, oil and grease by dissolving them. They can do this because a soap molecule has two parts – a polar end and a non-polar hydrocarbon tail.

Example H

The sodium stearate molecule has a polar and non-polar part:

$$CH_3-CH_2-CH_2-CH_2-CH_2-CH_2-CH_2-CH_2-CH_2-CH_2-CH_2-CH_2-CH_2-CH_2-CH_2-CH_2-CH_2-C(=O)-O^-Na^+$$

Non-polar hydrocarbon tail **Polar end**

The polar end is soluble in water (a polar solvent) but insoluble in organic (non-polar) solvents. The non-polar end is soluble in organic solvents and fats but insoluble in water. Thus, the organic non-polar end sticks to oil and dirt while the polar end sticks to water molecules. This dual nature allows the almost water-insoluble dirt to be removed from dirty clothes and bodies.

Example I

Soap in action

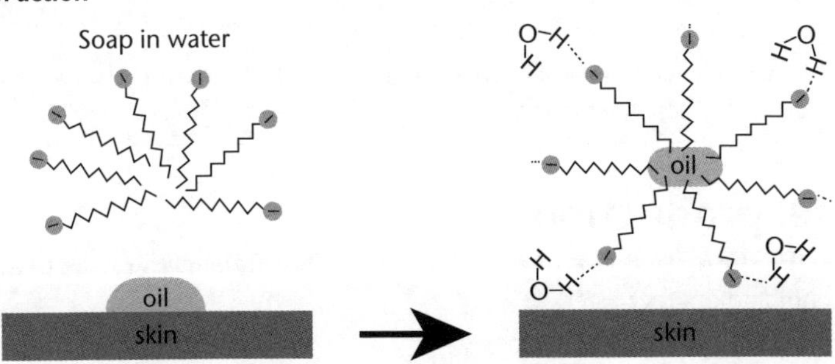

The fat soluble part of the soap molecule sticks to the oil on the skin's surface. The polar end of the soap molecule stocks to the water molecules. Agitation causes the oil and dirt to be cleaned from the skin and carried away in the water.

Synthetic detergents

Soap will not form a lather in every type of water. It depends on the type and quantity of ions dissolved in the water. If water contains calcium and magnesium ions, then soap forms an insoluble solid, or scum (eg calcium stearate). This water is called 'hard water'. Water that soap will lather in is called 'soft water'.

Detergents are popular in areas with hard water. Detergents are made from petroleum **by-products** rather than animal fats or vegetable oils. Instead of sodium and potassium salts of fatty acids, detergents contain alkylaryl sulfonates such as sodium dodecylbenzene sulfonate and sodium dodecylbenzene sulfate. Several types of detergents have been manufactured.

Example J

Detergent molecules

Detergents are alkylaryl sulfonates such as dodecylbenzene sulfonate, which is produced from benzene and propene.

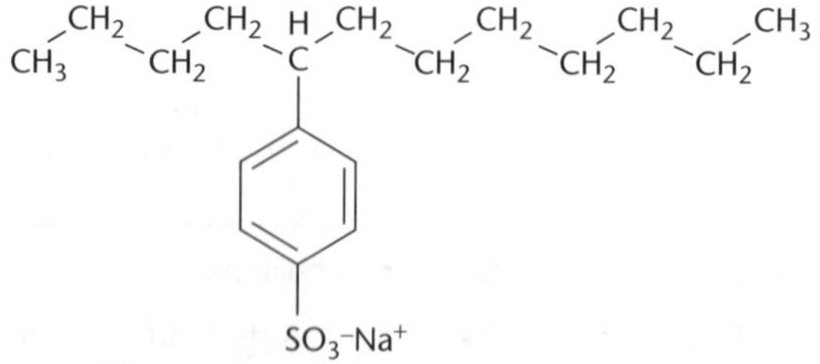

$CH_3-CH_2-CH_2-CH_2-C(H)(C_6H_4SO_3^-Na^+)-CH_2-CH_2-CH_2-CH_2-CH_2-CH_2-CH_3$

Dodecylbenzene sulfonate

Synthetic detergent works in the same way as soap, but detergents tend not to be used for personal hygiene products, such as shampoo, because they strip away too much oil from the hair and skin. Detergents are used in laundry detergents and dishwashing liquids.

Unit 12.5 Activity 5E: Soap and detergent

1. Who are the soap manufacturers in PNG?
2. Describe the steps in the production of soap.
3. Explain the cleansing action of soap.
4. Draw a comparison between soap and detergents.
5. If you know someone who makes their own soap, ask if you can observe or help them when they make their next batch.

Biodiesel production

A **transesterification** reaction is one in which an alcohol and an ester react to form another alcohol and another ester.

$$\underset{\text{Alcohol}}{R^1OH} + \underset{\text{Ester}}{R^2O-\overset{O}{\overset{\|}{C}}-R^3} \longrightarrow \underset{\text{Alcohol}}{R^2OH} + \underset{\text{Ester}}{R^1O-\overset{O}{\overset{\|}{C}}-R^3}$$

Biodiesel, or biofuel, is made by reacting vegetable or animal fats and oils (triglycerides) with alcohols such as methanol or ethanol in a base-catalysed reaction to produce ethyl or methyl esters of fatty acids and glycerol.

Under normal conditions, this reaction will proceed either exceedingly slowly or not at all, so heat and catalysts (acid and/or base) are used to speed up the reaction.

Example K

Biodiesel formation

Triglycerides react with ethanol in base to give biodiesel (ethyl esters) and glycerol.

$$\begin{array}{l} CH_2{-}O{-}C(=O){-}R^1 \\ CH{-}O{-}C(=O){-}R^2 \\ CH_2{-}O{-}C(=O){-}R^3 \end{array} + 3CH_3CH_2OH \underset{\text{catalyst}}{\overset{\text{NaOH}}{\rightleftharpoons}} \begin{array}{l} CH_3CH_2O{-}C(=O){-}R^1 \\ CH_3CH_2O{-}C(=O){-}R^2 \\ CH_3CH_2O{-}C(=O){-}R^3 \end{array} + \begin{array}{l} CH_2{-}OH \\ CH{-}OH \\ CH_2{-}OH \end{array}$$

Triglyceride **Ethanol** **Biodiesel (ethyl esters of fatty acids)** **Glycerol**

Triglycerides react with methanol to produce biodiesel (methyl esters) and glycerol.

$$\begin{array}{l} CH_2{-}O{-}C(=O){-}R^1 \\ CH{-}O{-}C(=O){-}R^2 \\ CH_2{-}O{-}C(=O){-}R^3 \end{array} + 3CH_3OH \underset{\text{catalyst}}{\overset{\text{NaOH}}{\rightleftharpoons}} \begin{array}{l} CH_3O{-}C(=O){-}R^1 \\ CH_3O{-}C(=O){-}R^2 \\ CH_3O{-}C(=O){-}R^3 \end{array} + \begin{array}{l} CH_2{-}OH \\ CH{-}OH \\ CH_2{-}OH \end{array}$$

Triglyceride **Methanol** **Biodiesel (methyl esters of fatty acids)** **Glycerol**

Biodiesel can be used in standard diesel engines, in which it performs better than standard diesel. It can be mixed with standard petroleum diesel fuel or used on its own. Biodiesel is a much cleaner fuel than standard diesel – its exhaust fumes contain much less unburnt hydrocarbons, carbon monoxide, particulate (solid) matter and aromatic hydrocarbon pollutants.

Biodiesel can also be used as a heating oil.

Common feedstocks for biodiesel production include recycled vegetable oil (eg that has been used to cook with) and 'virgin' vegetable oils (ie that have been extracted from the plant).

Recycled oil is processed to remove impurities from cooking, storage and handling, such as dirt, charred food and water. Virgin oils are refined, but not to a food-grade level.

Water is removed from the starting oils because it would hydrolyse the triglycerides, giving salts of the fatty acids (soaps) instead of producing biodiesel.

Almost all biodiesel is produced from virgin vegetable oils by the base-catalysed technique because it is the most economical process for treating virgin vegetable oils. It requires only low temperatures and pressures and results in over 98% yield (provided the starting oil is low in moisture and free fatty acids).

By-products of the reaction that need to be removed include soap, glycerol, excess alcohol and trace amounts of water. Glycerol is denser than the biodiesel esters so can be collected as the bottom layer of the product mixture or separated by centrifuging.

Excess methanol or ethanol can be recovered by distillation and used again. Soaps can be removed or converted into acids. Residual water is also removed from the fuel.

Biodiesel production in PNG

Biodiesl has been used in Papua New Guinea to power cars, boats and generators.

Example L

Biodiesel on Bougainville

During the ten-year crisis on Bougainville in the 1990s, violent disputes between residents and mining companies led to a blockade of the island by PNG. Forced to supply themselves with power and fuel, local people ran their trucks and cars on biodiesel that they made from coconut oil.

Research is continuing to determine how Papua New Guinea can best use biodiesel for its development agenda.

Unit 12.5 Activity 5F: Biodiesel

1. What is biodiesel?
2. Write the equation for the reaction of triglycerides with ethanol, noting all reaction products.
3. Why is the acid or base not consumed by the transesterification reaction?
4. Water has to be removed in the process of producing biodiesel. Why?
5. Do some research to find out about cetane numbers and how they compare for petroleum diesel and biodiesel.
6. Decide whether the following statements about biodiesel are true or false. You might need to carry out further research.
   a. Sulfur dioxide emissions are eliminated because biodiesel contains no sulfur.
   b. Biodiesel is plant-based so using it adds no extra CO_2 greenhouse gas to the atmosphere.
   c. The ozone-forming (smog) potential of biodiesel emissions is nearly 50% less than that of petroleum diesel emissions.
   d. Biodiesel is environmentally friendly: it is renewable, more biodegradable than sugar and less toxic than table salt.
   e. Biodiesel is a much better lubricant than petroleum diesel and extends engine life.

Unit 12.5 Natural Resources and Chemical Industries in Papua New Guinea

Topic 6: Traditional chemical practices in PNG

Topic 6 summarises some examples of applications of chemistry in making some useful products by ancient and current Papua New Guineans:

- Lime production.
- Salt production.
- Dyes and paint production.
- Traditional medicine.

Papua New Guinea has a large ethnic and biological diversity. It is home to 6% of the world's biodiversity and represents 1% of world's land mass. Traditional Papua New Guineans could live self-sufficiently by using the flora and other resources to produce all that they needed. Although not widely appreciated or documented, chemistry has been used by Papua New Guineans for thousands of years. The traditional knowledge has been passed on to younger generations by word of mouth.

Lime production

Lime powder is an essential ingredient for the custom of betelnut chewing – a social tradition that dates back thousands of years.

Betelnuts (fruit of the palm *Areca catechu*) and mustard sticks (daka) dipped in lime powder act as a mild stimulant, increasing feelings of well-being and reducing hunger. The nut stains teeth and lips bright red. There are health risks associated with chewing betelnuts – for example, it has been linked to throat and mouth diseases. Some users spit out the remnants, which leaves a health hazard wherever it lands.

Betelnuts contain chemicals called alkaloids, which contain basic nitrogen atoms. The lime (in the form of $Ca(OH)_2$) hydrolyses ester groups on the alkaloids to produce the chemicals that have a biological effect.

Traditional methods of preparation vary from one province to another. The common sources of lime are limestone, sea shells and sea corals. Harvesting of corals can become unsustainable if unregulated.

Example A

Lime from sea corals

The New Hanover people of New Ireland Province obtain lime from sea corals. Fresh coral is dried in the sun and then burnt in a fire for several hours. The hot lime is collected from the ash and wrapped in leaves to cool before use.

Example B

Lime from limestone

The Warea people of Morobe produce lime from limestone. The limestone is collected from the sides of rivers and ground into very fine particles. The powder is spread out on a lime bed. Leaves and ropes are placed underneath. At midnight, the Warea people light a fire under the lime bed and the burning continues until morning. At sunrise, the lime is collected from the ash while still hot and placed in bamboo tubes to cool before use.

Ancient Papua New Guineans knew that the decomposition of limestone to lime requires heat – it is an endothermic reaction:

$CaCO_3(s) + \text{heat} \rightarrow CaO(s) + CO_2(g)$

It was soon discovered that calcium oxide (CaO), the main product in the lime, is too corrosive to eat directly with the betelnut and mustard. It can cause severe irritation and burning if ingested or inhaled. The reaction of CaO with saliva is exothermic:

$CaO(s) + H_2O(\ell) \rightarrow Ca(OH)_2(s) + \text{heat}$

Therefore, when preparing the lime, they cool it in a vat of water, which hydrolyses the CaO to $Ca(OH)_2$ first. This product can now be used for the various purposes including betelnut chewing.

Unit 12.5 Activity 6A: Lime production

1. Report on traditional methods of producing lime by people in your province.
2. What is the effect of blowing air into limewater? Explain.
3. Find out some other traditional uses of lime.
4. What kind of reaction(s) takes place during the action of betelnut chewing with mustard and lime?
5. What are the advantages and disadvantages of chewing betelnut?

Salt production

In some parts of Papua New Guinea, salt is dug up from salt pits and consumed. But the traditional method of producing salt was to extract it from seawater by evaporating off the water to leave the salt.

People who live around the coastal areas use seawater for salt production by simply leaving the seawater in the sun and evaporating off the water.

Example C

Salt from seawater

The Warasua people of Gumini are well known for their salt-making skills. They fill hollow bamboo tubes with seawater and place them in the sun for a day. At sunset the bamboo is split in half and the salt crystals are collected and dried for three days.

Unit 12.5 Activity 6B: Salt production

1. Report on traditional methods of producing salt in your province.
2. Some modern Papua New Guineans prefer to consume salt bought at a supermarket rather than making it in the traditional way. Why do you think this is the case?

Dyes and paint production

Traditional dyes and paints are common throughout Papua New Guinea. Paints were made from soils, plants or animals, with **pigments** being extracted from local clays, natural plant dyes, insects, charcoal and ground shells.

- Yellow, red and brown pigments can be derived from local clays coloured by oxides.
- White pigments are obtained from ground, powdered lime.
- Black and grey pigments derive from charcoal or soot.

Sometimes the pigments were baked in leaf packets, which intensified the colours.

Most traditional dyes in Papua New Guinea come from plant sources. Dyes are used for painting either on traditional costumes or onto bodies. Methods of preparation and plant species used vary in different parts of Papua New Guinea.

Example D

The Mian people of Gumini and Warea people of Morobe use the same tubers of a native plant but have different methods of dye making.

Unit 12.5 Activity 6C: Dyes and paint production

1. Research and report on how people from your area make dyes.
2. Why do you think the paint industry in PNG doesn't use local pigments?

PNG traditional medicine

Documentation of PNG medicinal practices

Traditional medicines have been used in Papua New Guinea for centuries, with plants being the main source of medicinal ingredients. Each cultural group in Papua New Guinea knows about using plant materials for treating illnesses. Currently, 90% of Papua New Guineans rely on traditional medicine rather than modern medicine for their primary health care. Traditional medicine is used to treat sexually transmitted diseases, asthma, diarrhoea and dysentery, aches and pains including headaches, boils and sores, tuberculosis, colds and coughs, fever, malaria and insect bites.

Medicinal plants in Papua New Guinea were first recorded by the Russian botanist and explorer Nicholas Miklouho-Maclay in the late 19th century. He lived on the Rai coast of Madang.

Many other medicinal plants have since been recorded by missionaries, botanists and anthropologists.

Example E

Hortense Powdermaker, a US anthropologist, lived with the Lesu people of the east coast of New Ireland in the early 1930s. She recorded six medicinal plants.

Recently, scientists at the University of Papua New Guinea have recorded many more medicinal plants.

Example F

Alkaloids

Alkaloids are nitrogen-containing bases that are biologically active – they can be poisonous, but they can also be therapeutic in low dosages. Common medicines that are derived from plant alkaloids include quinine (antimalarial), morphine (pain reliever), atropine (muscle relaxant) and nicotinic acid (vitamin B3).

Chemist Dr David Holdsworth collected a large sample of medicinal plants of PNG. Of 600 samples collected in one of his surveys, 12% were found to contain alkaloids.

Unit 12.5 Activity 6D: Traditional medicine

1. Research and report on traditional medicines used by your people.
2. What is an alkaloid?
3. Research and discuss how PNG can learn from ancient civilisations such as the Chinese on the use of traditional medicines.
4. Explain whether traditional practices and modern medicine could complement each other in the PNG health system.

Current research on traditional medicine

Researchers at the University of Papua New Guinea are investigating ways of combining modern and traditional medicine to combat disease. This direction is required for the nation to achieve its vision of being smart and healthy by 2050.

Hundreds of examples of traditional medicinal plants have been documented in several references. The book *Medicinal Plants in Papua New Guinea* by Professor Prem Rai (World Health Organization, 2009) is a very useful reference. It arranges traditional medicinal plants in families, genus, species, vernacular name, uses and methods of preparation.

Example G

Species: *Rungia* sp.

Vernacular name: red flawa

Uses: treating gonorrhoea. The Engans use the leaves as poison antidote. The Highland provinces use the leaves and stem to combat diarrhoea.

Method of preparation: 5–6 leaves of the plant are boiled in water and strained into clean containers for drinking.

Example H

Species: *Alstonia scholaris*

Vernacular name: gwabim

Uses: treating malaria, headache, pain, toothache and malnutrition. On the Huon Peninsula, Morobe, the bark is used for malaria treatment. In Indonesia, Malaysia, Philippines and Central Province, PNG, the bark is used to cure fever. In New Ireland Province it is used to treat abdominal pain. In India the bark is used for malaria treatment. The Goilalas use the sap as a poison antidote.

The bark contains several alkaloids such as echitamine, echitamidine, pierinine and scholarine.

Method of preparation: the sap is collected in a container and mixed with an equal amount of water. The mixture is given to the patient to drink. For toothache, the leaves and juice are applied directly onto the painful tooth.

Unit 12.5 Activity 6E: Researching traditional medicine

1. Obtain a copy of the book *Medicinal Plants in Papua New Guinea* and verify the medicinal plants in your area.
2. What is your vernacular name for the two species listed in the previous two examples?

Unit 12.5 Natural Resources and Chemical Industries in Papua New Guinea

Topic 7: Environmental effects of mining and chemical industries in PNG

Topic 7 covers environmental pollution as a direct outcome of mining and use of chemicals in Papua New Guinea:

- Environmental consequences from mining activities.
- Environmental consequences from the use of chemicals.

Papua New Guinea has vast natural resources and a small but growing industrial sector. The population of seven million is increasing rapidly, so demand for more industries and consumer goods is rising. Increased mining activities and chemical production bring economic benefits, which means improved health care, schools and roads, but it will inevitably mean more environmental pollution. Soil erosion, contamination of soil and water systems by chemicals from the mine and loss of animal species due to habitat destruction are just some of the effects that need to be considered in any mining situation.

Mining companies in Papua New Guinea are required to follow environmental and rehabilitation codes, to reduce the environmental effects of mining and ensure the area mined is returned to close to its original state. Some mining methods may have significant adverse environmental and public health impacts in the locality of the mines. In Papua New Guinea, environmental codes have not always been strictly implemented.

Environmental consequences of mining

Open-cut mining is used to extract minerals that are located over a large area and relatively close to the surface. The land is cleared of vegetation, and the top soil is removed and stored for later use in the rehabilitation process. After the soil is removed, the next layer of rock is removed and placed in a pile. Then the mineral ore is removed by blasting with explosives, loaded onto large trucks and taken to the processing plant. Waste from the processing plant (tailings) should be collected in tailings dams and not released to the local river systems.

Underground mining occurs when the mineral ore is too deep for open-cut mining. People and vehicles enter the mine through a tunnel or a lift. This type of mining can disturb the surrounding area and some vegetation is cleared to allow for the mine infrastructure. Other effects include land subsidence, piles of waste rock and the need for tailings dams.

The methods used to monitor and control water flow at mine sites are diversion systems, containment ponds, groundwater pumping systems, subsurface drainage systems and subsurface barriers. Acidic mine drainage water is generally pumped to a treatment facility that neutralises the contaminants.

General effects of mining

Mining activities destroy wetlands, which are the natural habitats for birds, fish and other organisms. Mining activities also destroy forests because trees are removed to provide access to the area containing the gold, copper or iron ore.

Example A

Panguna, Ok Tedi, Porgera and Lihir mines are all open-cut mines. Large amounts of forest had to be removed in order to excavate the gold and copper.

Many mines require tailings dams to prevent waste being washed into the rivers. If dams are not built, perhaps to reduce costs, massive pollution can result down stream. In other cases, the tailings dam can overflow, and even breach, during periods of heavy rain.

Erosion of exposed hillsides, mine dumps and tailings dams and the resulting silting up of drainages, creeks and rivers can significantly affect the surrounding areas.

Underground gold or coal mining can require the removal of almost an entire layer of material deep under the surface.

Some mining involves the inadvertent dispersal of heavy metals, such as lead, into the atmosphere. Cyanide spills and mercury contamination also occur. This can have serious health effects, including mental retardation in children.

Example B

Both Ok Tedi and Tolukuma gold mines have had cyanide spills, which have killed fish and other organisms. A team of chemists from the University of Papua New Guinea has been assessing toxicity levels in the Tolukuma river system.

Unit 12.5 Activity 7A: Mining and the environment

1. List some general effects of mining on the environment.
2. Compare the effects of underground mining and open-cut mining.

Environmental effects of mining in PNG

The history of mining in Papua New Guinea dates back to the 1800s. Mining was largely land based until 2011 when the PNG Government allowed exploration of the seabed for possible deep-sea mining. If this venture proceeds, it will be the first deep-sea mining venture in the world. New technology has been developed and tested and new challenges are being faced.

Pollution from mining on land

Small-scale mining activities began in the 1800s with explorers and traders and culminated in the Bulolo gold-rush of the 1930s. Large gold-dredging equipment was air-lifted to the gold site. Environmental pollution was minimal and unrecorded.

Panguna Mine

Bougainville Copper Limited (BCL) developed the Panguna Mine on Bougainville Island in 1968. It has been closed since 1976 due to civil unrest in the province. This development marked one of the first major investments in Papua New Guinea, and was one of the world's largest copper and gold mines at the time.

Panguna Mine is Papua New Guinea's biggest mine and possibly the nation's first major mining disaster.

Mine wastes were dumped into the river system without proper tailing treatment. The river quality drastically deteriorated. Also, local communities were not happy with the quality of services and economic benefits from the mine.

The civil unrest that developed as a result cost more than 20 000 lives and forced the closure of the mine after only eight years of operation. The Autonomous Government of Bougainville is cautiously considering re-opening the mine but with strict controls and guidelines.

Ok Tedi Mine

Ok Tedi is Papua New Guinea's second biggest mine. It is a gold and copper mine situated in Tabubil, Western Province. It is also the scene of major environmental damage. The mine is located in a region of high rainfall. In 1984, when operations started, no existing mining technology could contain the waste tailings under such conditions. New technology was developed and tried, but the company convinced the PNG Government to relax tailings requirements.

All mine wastes were channelled into the Fly River system. Two leaching techniques were used:

- Cyanide leaching was used from 1984 to 1987 to extract the gold from the ore. Cyanide reacts with gold to produce a water-soluble gold cyanide complex:

$$4Au + 8NaCN + O_2 + 2H_2O \rightarrow 4Na[Au(CN)_2] + 4NaOH$$

- Froth flotation was used from 1988 to 2013 to extract the copper.

The annual load of waste dumped into the river system is calculated at 22 million tonnes. This had huge environmental impacts of silting, sedimentation, flooding and destruction of food gardens. Fish were killed or became contaminated, but were still consumed by the local people. The livelihoods of thousands of people who lived along the river system, and depended on it for sustenance, has been affected.

The Ok Tedi disaster is now a world-class example of mining companies failing to protect peoples' livelihood. In the 1990s, local communities sued the owner BHP for negligence.

Example C

The giant Ok Tedi Mine in Western Province is an example of a large-scale environmental disaster. When its tailings dams failed, all the mining waste was put into the surrounding creeks.

Porgera Mine

Papua New Guniea's third biggest mine is in Enga Province. Porgera Mine began as an underground gold mine in 1990 but changed to open cut in 1992. It has a tailings system and treats the waste before discharging it into the Pogera River, which drains into the Strickland River. There are still threats to the fish and animal life in the rivers. When the treatment plants are not functioning properly, contaminant levels increase. They should be detected and rectified. Strict monitoring systems are in place.

Example D

The Porgera gold and silver mine is both open cut and underground. The mine puts its toxic tailings (processed ore) directly into the local river. These tailings contain cyanide, mercury and other metals. Porgera Mine also has waste rock piles that are gradually washing into local rivers.

Lihir Mine

Papua New Guniea's fourth biggest open-cut mine is situated on Lihir Island in New Ireland Province. Forest clearing, the use of chemicals and the waste produced all contribute to environmental pollution. The gold mine has three methods of waste disposal:

- dumping waste rock at sea, mostly in shallow areas. This practice displaces marine organisms, causes sedimentation and makes water cloudy, which affects visibility.
- depositing tailings at sea after treatment, at depths of 125 metres. This is meant to immobilise the waste in areas of low current.
- stockpiling low-grade ores on land for later processing. This exposes the soil to wind and rain, and leads to erosion and redeposition.

Excessive pollution occurs when tailings are not properly treated before discharge.

Tolukuma Mine

The Tolukuma Mine is a small mine located in Central Province and runs both open-cut and underground mining for gold. Underground mining minimises damage to forests and vegetation, but the hilly surrounding countryside makes it difficult to dispose of the waste.

So far, 160 000 tonnes of waste has been disposed into the Auga-Angabanga River system. Communities along the river system have seen dead fish, and have experienced flooding and spillage of cyanide that was being transported by helicopter.

Kainantu Mine

This gold mine is located in Eastern Highlands Province. It is an underground mine, which means damage to forests and vegetation is minimised, but toxic chemicals used in the extraction processes still end up in the environment.

Ramu Mine

This nickel and cobalt mine in Madang Province is scheduled to start operation in 2013. It is operated by Metal Corporation of China. There are already environmental concerns about toxic chemicals used in the extraction processes ending up in the environment.

Environmental effects of mining at sea

Scientists first discovered deep-sea hydrothermal vents in 1976. Hydrothermal vents are commonly found near volcanically active sites. The water from the vent is rich in dissolved minerals, such as copper, gold, silver and zinc. The ecosystems at hydrothermal vents are produced by the rare combination of superheated, highly mineralised vent fluids, cold seawater and microorganisms that can use sulfur compounds as an energy source and a basis for making organic nutrients.

Example E

Hydrothermal vent ecosystems are rich in carbon dioxide, hydrogen sulfide, organic carbon compounds, methane, hydrogen and ammonium.

Hydrothermal vents provide habitat for thriving, abundant, ever-changing communities of organisms. Such ecosystems have been found to host over 500 previously unknown species that are thought to occur only at these vents.

Only 5% of the Earth's oceanic ridges have been surveyed in any detail. Seafloor volcanic systems and deep-sea ecosystems are still under-researched and not well understood. So we cannot predict the effects of individual deep-sea mining projects, or the combined effects of the many deep-sea mining projects proposed for the Bismarck Sea and throughout the Pacific.

Example F

Within three years, Nautilus Minerals' deep-sea pilot project Solwara 1 aims to produce more than 2 million tonnes of copper and gold ore each year. Nautilus will mine an area 1.6 kilometres beneath the Bismarck Sea, 50 kilometres off the coast of New Britain. Environmentalists and local fishing communities are concerned about where the ore will be processed and where the toxic waste will be treated and later dumped. The effect on marine life is also uncertain.

Unit 12.5 Activity 7B: Environmental effects of mining

1. Carry out some research and prepare a report on how PNG could have managed the BCL Panguna mining arrangement to produce a better environmental outcome.
2. Research the environmental impact of Ok Tedi mining activities on the Fly River system.
3. Discuss the statement: 'The Ok Tedi operation with regards to environmental pollution was another disaster for PNG.'
4. Research and discuss any improvements that have ben made to pollution management at the Porgera and Lihir mines.
5. Research and report on the current pollution status of the Ramu nickel and cobalt mine.
6. Explain how hydrothermal vents are formed.
7. Why is it not possible to predict the impact of the Solwara 1 project on the hydrothermal vent ecosystem at the site of the deep-sea mineral deposit?

Petroleum and gas production

Exploitation of oil and gas resources also increases environmental pollution. The process causes toxic gases and chemicals such as hexane, benzene, toluene, ethylbenzene and xylene to be emitted. If they are not treated or contained, then they represent a threat to humans – irritating the skin, eyes and respiratory tract, causing headaches, nausea and dizziness and other serious disorders. Massive oil spillages have occurred in other countries.

Chemicals used in the processing of petroleum and gas can cause major pollution problems to groundwater, seawater, land and air if left untreated. Pollution can be reduced by strict monitoring and treatment of wastes.

Example G

Pits are used to contain and immobilise wastes. Operators must ensure that groundwater does not become contaminated. Evaporation ponds are also useful.

Petroleum and gas production are also major sources of the greenhouse gases CO_2 and CH_4, which are contributing to global warming. Another, more drastic, option to eliminating pollution from petroleum and gas mining would be to develop alternative sources of power that are non-polluting.

Example H

Hydro power and solar energy are already being used in PNG. Geothermal energy is also being considered – licences have granted for energy companies to explore geothermal resources in West New Britain Province.

Mining in PNG: balancing the risks

Mining has been a valuable contributor to Papua New Guniea's economy. The industry has brought foreign investment, which in turn has created jobs and facilitated the building of infrastructure such as towns and roads. Mineral exports account for more than 70% of the country's earnings.

There is no reason for mining to be banned, but the environmental and social impact of each proposed mine should be analysed before licences are granted. Mining operators must be aware of the potential effects of their operations and ensure minimum risk to the environment.

Unit 12.5 Activity 7C: Petroleum and gas mining

1. List the pollutants found in the oil industry.
2. What is meant by the term 'greenhouse effect'?
3. What is a greenhouse gas? Give examples.
4. Discuss the statement: 'Fossil fuels cause more than 90% of greenhouse gas emissions.'

Environmental consequences of chemical use

The extensive use of chemicals has contributed to the increasing levels of pollution on our planet. To protect Papua New Guinea's environment, chemical use must be minimised.

Pollution from plastics

The manufacturing of plastics from petrochemicals produces greenhouse gases and causes the emission of other chemicals, such as styrene, and requires a lot of energy.

Plastics are made from polymers, which are very stable and long-lived compounds. Once they are disposed of in the environment, it is estimated that non-biodegradable plastics may take 1000 years to break down.

Plastic bags represent a problem on a global scale. Most end up in landfill (rubbish tips) and some are recycled. If they are disposed of as litter, they accumulate on streets and in parks, waterways and oceans where they can harm animal life. In an effort to reduce the amount of litter caused by plastic bags, Papua New Guinea has restricted the availability of non-biodegradable plastic bags.

Incineration at very high temperatures (more han 850°C) is the easiest way to dispose of plastics but this method produces toxic fumes, such as carcinogenic dioxins.

Example I

In the manufacture of polyvinyl chloride (PVC), factory workers have experienced health problems when exposed to the toxic vinyl chloride vapour.

Combustion of fossil fuels

Greenhouse gases

The complete combustion of the hydrocarbons in fossil fuels in the presence of sufficient oxygen yields carbon dioxide, water and energy. Carbon dioxide is a **greenhouse gas**, contributing to global warming.

Example J

Complete combustion of methane (natural gas)

$$CH_4(g) + 2O_2(g) \rightarrow CO_2(g) + 2H_2O(g) + \text{energy}$$

If there is not enough oxygen, incomplete combustion occurs, less energy is released and poisonous carbon monoxide is produced. Carbon monoxide acts as a poison by combining with haemoglobin in the blood. Haemoglobin transports oxygen around the body, so a person suffering from carbon monoxide poisoning dies from symptoms associated with lack of oxygen. The incomplete combustion of hydrocarbons can also produce soot or black carbon, a source of air pollution. Increased amounts of fine particulate matter in the air contribute to heart and lung problems, including asthma and bronchitis.

Example K

Incomplete combustion of octane and methane

$$2C_8H_{18}(g) + 17O_2(g) \rightarrow 16CO(g) + 18H_2O(g) + \text{energy}$$

$$CH_4(g) + O_2(g) \rightarrow C(s) + 2H_2O(g) + \text{energy}$$

Example L

Greenhouse gases

The following gases are considered greenhouse gases: water vapour, carbon dioxide, methane, nitrous oxide, **ozone** and chlorofluorocarbons (CFC).

The combustion of fossil fuels, particularly coal, produces nitrogen and sulfur oxides. These have negative effects on human health, especially on the respiratory system.

Acid rain

Rain water is naturally slightly acidic (pH 5.7–7) due to the presence of carbonic acid (H_2CO_3).

Example M

Formation of carbonic acid

Carbonic acid is formed by the reaction:

$$H_2O(\ell) + CO_2(g) \rightarrow H_2CO_3(aq)$$

Carbonic acid can then ionise in water to form low concentrations of carbonate and hydronium ions:

$$H_2O(\ell) + H_2CO_3(aq) \rightarrow HCO_3^-(aq) + H_3O^+(aq)$$

Other chemicals can also affect the pH of rain water, such as nitric acid produced in lightning strikes. This is a natural phenomenon. The very high temperature in the vicinity of a lightning bolt causes the oxygen and nitrogen in the air to react to form nitric dioxide, which hydrolyses to nitric acid.

In the atmosphere, nitrogen and sulfur oxides formed from the combustion of fossil fuels at power plants and in cars are converted into sulfuric acid and nitric acid.

Example N

Formation of sulfur and nitrogen oxides

Sulfur dioxide is oxidised by oxygen in the presence of sunlight to form sulfur trioxide:

$$2SO_2(g) + O_2(g) \rightarrow 2SO_3(g)$$

These two gases react readily with rain water to form sulfurous acid (H_2SO_3) and sulfuric acid (H_2SO_4):

$$SO_2(g) + H_2O(\ell) \rightarrow H_2SO_3(aq)$$
$$SO_3(g) + H_2O(\ell) \rightarrow H_2SO_4(aq)$$

Nitrogen oxide (NO) reacts with oxygen to form toxic nitrogen dioxide (NO_2):

$$2NO + O_2(g) \rightarrow 2NO_2(g)$$

NO_2 dissolves readily in rain water to form a mixture of nitrous acid (HNO_2) and nitric acid (HNO_3):

$$2NO_2(g) + H_2O(\ell) \rightarrow HNO_2(aq) + HNO_3(aq)$$

Nitrogen oxide and nitrogen dioxide are are known collectively as nitrogen oxides or NO_x.

The hydrolysis of sulfur and nitrogen oxides in the atmosphere is causing rain to become more acidic (pH of 5.5 or less). **Acid rain** has a detrimental effect on the environment – it increases soil and water acidity and hence impacts on insects and fish living there. Some microorganisms cannot tolerate changes to pH because their enzymes become denatured.

Acid rain also mobilises toxins such as aluminium, and leaches away essential nutrients and minerals such as magnesium. Soil chemistry can change dramatically when cations such as calcium and magnesium are leached by acid rain, thereby affecting sensitive species.

Example O

Acid rain and soil

$$2H^+(aq) + Mg^{2+}(clay) \rightarrow 2H^+(clay) + Mg^{2+}(aq)$$

Acid rain also affects vegetation and human health and damages buildings, especially those made of limestone and marble, which contain large amounts of calcium carbonate. Acids in the rain react with the calcium compounds in the stones to create calcium sulfate (gypsum), which then flakes off.

Example P

Acid rain and limestone buildings

$$CaCO_3(s) + H_2SO_4(aq) \rightarrow CaSO_4(aq) + CO_2(g) + H_2O(\ell)$$

Unit 12.5 Activity 7D: Combustion of fossil fuels

1. Write the complete combustion reaction of propane.
2. Find out the sources of all the greenhouse gases listed in Example L and whether any are being phased out.
3. Explain the consequences of increased concentrations of greenhouse gases in the atmosphere.
4. What does the term 'acid rain' refer to?
5. Write chemical equations for the formation of acid rain by nitrogen dioxide and sulfur dioxide.
6. Explain the origin of the nitrogen oxides that contribute to acid rain.

Persistent organic pollutants

Persistent organic pollutants (POPs) are toxic chemicals that have adverse effects on the environment and human health. POPs were used in industry or in agriculture to control weeds, pests and diseases. They can move around the world in the wind and water. This means that POPs produced in one country can affect another country. POPs are resistant to environmental degradation, so they persist for long periods in the environment. Even though they are now banned in many countries, traces of POPs are still found in many foods. POP spass though food chains from species to species.

Example Q

12 key POPs

In 2001, nearly 100 countries signed a treaty known as the Stockholm Convention, agreeing to reduce or eliminate the use and production of 12 key POPs: aldrin, chlordane, DDT (dichlorodiphenyltrichlorethane), dieldrin, endrin, heptachlor, hexachlorobenzene, mirex, toxaphene, PCBs (polychlorinated biphenyls), dioxins and furans.

POPs are highly toxic, and exposure can have a range of adverse effects, including cancer and death, at such low concentrations that they have been identified as the most toxic substances produced by humans.

DDT

DDT (dichlorodiphenyltrichloroethane) was first used during World War II to control insect-borne diseases such as malaria and typhus. Malaria is a disease caused by microorganisms called protists that are carried by mosquitoes. When an infected mosquito bites a human, the protist is transferred into the blood. After the World War II, DDT was also used as an agricultural insecticide. Many countries have banned the use of DDT. In Papua New Guinea, where malaria is endemic, and one of the major causes of illness and death, DDT was used by the Malaria Control Unit of the Department of Health, but it is no longer used. However, it is still being detected in human breast milk.

DDT affects the nervous system. Symptoms include dizziness, tremors, convulsion and neurological problems.

Cl Cl Cl Cl Cl

DDT (dichlorodiphenyl trichloroethane)

Example R

DDT

The systematic (IUPAC) name of DDT is 1,1,1-trichloro-2,2-di(4-chlorophenyl)ethane. DDT is an acronym for its common name **d**ichloro**d**iphenyl**t**richloroethane.

CFCs

Chlorofluorocarbons (**CFCs**) contain only carbon, chlorine, hydrogen and fluorine. They have been used as refrigerants in refrigerators and air conditioners, as propellants (in **aerosols**) and as industrial solvents, in the manufacture of foams and as cleaning agents for electronic equipment.

CFC	Formula	Uses
Trichlorofluoromethane (freon-11)	CCl_3F	Refrigeration, aerosols, foams
Dichlorofluoromethane (freon-12)	CCl_2F_2	Refrigeration, aerosols, foams, air conditioning
1,1,2-Trichloro-1,2,2-trifluoroethane (freon-12)	CCl_2FCClF_2	Electronics, dry cleaning, fire extinguishers
1,2-Dichloro-1,1,2,2-tetrafluoroethane (freon-14)	$CClF_2CClF_2$	Aerosols
1,2,2-Trichloro-1,1,2-trifluoroethane (freon-113)	$CClF_2CCl_2F$	Degreasing and electronics

The use of CFCs in Papua New Guinea and elsewhere has been banned because they destroy the **ozone** layer. The ozone layer acts as a filter, controlling the amount ot ultraviolet radiation that reaches the Earth. Although CFCs are not widely produced any more, it will be some time before their effect on the ozone layer reduces.

Example S

CFCs

CFCs destroy the ozone (O_3) layer, which protects the Earth against harmful ultraviolet radiation:

$$CCl_2F_2(g) \xrightarrow{\text{UV radiation}} CClF_2(g) + Cl^{\bullet}(g)$$

$$Cl^{\bullet}(g) + O_3(g) \rightarrow ClO^{\bullet}(g) + O_2(g)$$

$$ClO(g) + O^{\bullet}(g) \rightarrow O_2(g) + Cl^{\bullet}(g)$$

The chlorine atoms are not permanently used up in the reaction and can react again.

Unit 12.5 Activity 7E: Persistent organic pollutants

1. What is a persistent organic pollutant?
2. What does the acronym DDT stand for? Draw its structure.
3. What does the acronym CFC stand for? Draw the structure of a typical CFC.
4. Discuss pollution (air, water, land) in your province that results from domestic and chemical activities.
5. How does switching off unused electrical appliances minimise pollution and help the environment?
6. PNG has the third biggest rainforest in the world. How can PNG use this rainforest to help the planet?

Appendix

PERIODIC TABLE OF THE ELEMENTS

Atomic number: 1 — **H** — 1.0 :Atomic mass

Group 1	2	3	4	5	6	7	8	9	10	11	12	13	14	15	16	17	18
																	2 **He** 4.0
3 **Li** 6.9	4 **Be** 9.0											5 **B** 10.8	6 **C** 12.0	7 **N** 14.0	8 **O** 16.0	9 **F** 19.0	10 **Ne** 20.2
11 **Na** 23.0	12 **Mg** 24.3											13 **Al** 27.0	14 **Si** 28.1	15 **P** 31.0	16 **S** 32.0	17 **Cl** 35.5	18 **Ar** 40.0
19 **K** 39.1	20 **Ca** 40.1	21 **Sc** 45.0	22 **Ti** 47.9	23 **V** 50.9	24 **Cr** 52.0	25 **Mn** 54.9	26 **Fe** 55.9	27 **Co** 58.9	28 **Ni** 58.7	29 **Cu** 63.6	30 **Zn** 65.4	31 **Ga** 69.7	32 **Ge** 72.6	33 **As** 74.9	34 **Se** 78.9	35 **Br** 79.9	36 **Kr** 83.8
37 **Rb** 85.5	38 **Sr** 87.6	39 **Y** 88.9	40 **Zr** 91.2	41 **Nb** 92.9	42 **Mo** 95.9	43 **Tc** (98)	44 **Ru** 101.1	45 **Rh** 102.9	46 **Pd** 106.4	47 **Ag** 107.9	48 **Cd** 112.4	49 **In** 114.8	50 **Sn** 118.7	51 **Sb** 121.8	52 **Te** 127.6	53 **I** 126.9	54 **Xe** 131.3
55 **Cs** 132.9	56 **Ba** 137.3	71 **Lu** 175.0	72 **Hf** 178.5	73 **Ta** 180.9	74 **W** 183.9	75 **Re** 186.2	76 **Os** 190.2	77 **Ir** 192.2	78 **Pt** 195.1	79 **Au** 197.0	80 **Hg** 200.6	81 **Tl** 204.4	82 **Pb** 207.2	83 **Bi** 209.0	84 **Po** (209)	85 **At** (210)	86 **Rn** (222)
87 **Fr** (223)	88 **Ra** 226.0	103 **Lr** 262.1	104 **Rf**	105 **Db**	106 **Sg**	107 **Bh**	108 **Hs**	109 **Mt**	110 **DS**	111 **Rg**	112 **Cn**		114 **Fl**		116 **Lv**		

57 **La*** 138.9	58 **Ce** 149.1	59 **Pr** 140.9	60 **Nd** 144.2	61 **Pm** 146.9	62 **Sm** 150.4	63 **Eu** 152.0	64 **Gd** 157.3	65 **Tb** 159.0	66 **Dy** 162.5	67 **Ho** 164.9	68 **Er** 167.3	69 **Tm** 168.9	70 **Yb** 173.0
89 **Ac** 227.0	90 **Th** 232.0	91 **Pa** 231.0	92 **U** 238.0	93 **Np** 237.1	94 **Pu** 239.1	95 **Am** 241.1	96 **Cm** 247.1	97 **Bk** 249.1	98 **Cf** 251.1	99 **Es** 254.1	100 **Fm** 257.1	101 **Md** 258.1	102 **No** 255

Answers

Answers for many questions include a Marking Guide:

- ***A*** ('Achievement', meaning 'satisfactory achievement').
- ***M*** ('Merit', meaning 'high achievement').
- ***E*** ('Excellence', meaning 'very high achievement').

The Marking Guide has been made by the authors and the publishers and is not an official guide, but we hope it will help students who are striving for the best possible results.

Unit 12.1 Activity 1A: Testing for cations and anions (page 13)

1. Carbonate. (***A***)

2. SO_4^{2-} (***A***)

$Ba^{2+}(aq) + SO_4^{2-}(aq) \rightarrow BaSO_4(s)$ (***M***)

3. Cu^{2+}, NO_3^- (***A***)

$Cu^{2+}(aq) + 2OH^-(aq) \rightarrow Cu(OH)_2(s)$ (***M***)

$Cu^{2+}(aq) + 4NH_3(aq) \rightarrow [Cu(NH_3)_4]^{2+}(aq)$ (***E***)

4. **a.** A white precipitate would form when $AgNO_3$ was added, which would redissolve when dilute ammonia was added.

$Ag^+(aq) + Cl^-(aq) \rightarrow AgCl(s)$

$Ag^+(aq) + 2NH_3(aq) \rightarrow [Ag(NH_3)_2]^+$

(***A*** – observation; ***M*** – precipitation equation; ***E*** – complex ion equation)

b. A yellow precipitate would form which would not dissolve in ammonia.

$Ag^+(aq) + I^-(aq) \rightarrow AgI(s)$

(***A*** – observation; ***M*** – ionic equation)

5. **a.** Test X: Cl^- ions give a white precipitate which then redissolves; I^- ions give a yellow precipitate, which will not redissolve.

$Ag^+(aq) + Cl^-(aq) \rightarrow AgCl(s)$ (***M***) $Ag^+(aq) + I^-(aq) \rightarrow AgI(s)$ (***M***)

$AgCl(s) + 2NH_3(aq) \rightarrow [Ag(NH_3)_2]^+(aq) + Cl^-(aq)$ (***E***)

b. Test W: SO_4^{2-} ions give a white precipitate; NO_3^- ions do not give a precipitate.

$Ba^{2+}(aq) + SO_4^{2-}(aq) \rightarrow BaSO_4(s)$ (***M***)

c. Test Z: Al^{3+} ions give a white precipitate which does not redissolve; Zn^{2+} ions give a white precipitate, which redissolves.

$Al^{3+}(aq) + 3OH^-(aq) \rightarrow Al(OH)_3(s)$ (***M***)

$Zn^{2+}(aq) + 2OH^-(aq) \rightarrow Zn(OH)_2(s)$ (***M***)

$Zn(OH)_2(s) + 4NH_3(aq) \rightarrow [Zn(NH_3)_4]^{2+}(aq) + 2OH^-(aq)$ (***E***)

d. Test Y: Cu^{2+} ions give a pale blue gelatinous precipitate, which does not dissolve, Fe^{2+} ions give a pale green gelatinous precipitate (which does not dissolve and which turns brown if left in air). (For each of **a.–d.**, ***A*** – test correct)

$Cu^{2+}(aq) + 2OH^-(aq) \rightarrow Cu(OH)_2(s)$ (***M***)

$Fe^{2+}(aq) + 2OH^-(aq) \rightarrow Fe(OH)_2(s)$ (***M***)

6. $Zn^{2+}(aq) + 2OH^-(aq) \rightarrow Zn(OH)_2(s)$ (***M***)

$Zn(OH)_2(s) + 2OH^-(aq) \rightarrow [Zn(OH)_4]^{2-}(aq)$

or $Zn^{2+}(aq) + 4OH^-(aq) \rightarrow [Zn(OH)_4]^{2-}$ (***E***)

Then: $Zn^{2+}(aq) + 2OH^-(aq) \rightarrow Zn(OH)_2(s)$ (***M***)

$Zn(OH)_2(s) + 4NH_3(aq) \rightarrow [Zn(NH_3)_4]^{2+}(aq)$

or $Zn^{2+}(aq) + 4NH_3(aq) \rightarrow [Zn(NH_3)_4]^{2+}(aq)$ (***E***)

7. **a.** Mg^{2+}, Ba^{2+}, Al^{3+}, Zn^{2+}, Pb^{2+}. (***M***)

$Mg^{2+}(aq) + 2OH^-(aq) \rightarrow Mg(OH)_2(s)$ (***M***)

$Ba^{2+}(aq) + 2OH^-(aq) \rightarrow Ba(OH)_2(s)$ (***M***)

$Al^{3+}(aq) + 3OH^-(aq) \rightarrow Al(OH)_3(s)$ (***M***)

$Zn^{2+}(aq) + 2OH^-(aq) \rightarrow Zn(OH)_2(s)$ (***M***)

$Pb^{2+}(aq) + 2OH^-(aq) \rightarrow Pb(OH)_2(s)$ (***M***)

b. Add excess NaOH – if precipitate remains, ion is Mg^{2+} or Ba^{2+}.

$Al^{3+}(aq) + 4OH^-(aq) \rightarrow [Al(OH)_4]^-(aq)$ *or*

$Al(OH)_3(s) + OH^-(aq) \rightarrow [Al(OH)_4]^-$ (***E***)

$Zn^{2+}(aq) + 4OH^-(aq) \rightarrow [Zn(OH)_4]^{2-}(aq)$ *or*

$Zn(OH)_2(s) + 2OH^-(aq) \rightarrow [Zn(OH)_4]^{2-}$ (***E***)

$Pb^{2+}(aq) + 4OH^-(aq) \rightarrow [Pb(OH)_4]^{2-}(aq)$ *or*

$Pb(OH)_2(s) + 2OH^-(aq) \rightarrow [Pb(OH)_4]^{2-}$ (***E***)

To distinguish between Mg^{2+} or Ba^{2+}, add dil H_2SO_4 to the original solution – if a white precipitate forms, ion is Ba^{2+}; if precipitate dissolves, ion is Al^{3+}, Zn^{2+} or Pb^{2+}.

$Ba^{2+}(aq) + SO_4^{2-}(aq) \rightarrow BaSO_4(s)$ (***M***)

To the original solution, add NH_3 solution dropwise and then excess – if a white precipitate forms and dissolves in excess, the ion is Zn^{2+}; if the precipitate remains, the ion is Al^{3+} or Pb^{2+}.

$Zn^{2+}(aq) + 2OH^-(aq) \rightarrow Zn(OH)_2(s)$

$Zn^{2+}(aq) + 4NH_3(aq) \rightarrow [Zn(NH_3)_4]^{2+}(aq)$ *or*

$Zn(OH)_2(s) + 2OH^-(aq) \rightarrow [Zn(OH)_4]^{2-}(aq)$

To the original sample add dil H_2SO_4 – if a white precipitate forms, the ion is Pb^{2+}.

$Pb^{2+}(aq) + SO_4^{2-}(aq) \rightarrow PbSO_4(s)$ (***M***, ***E***)

8. **a.** NH_4^+, Cl^- (***A***)

b. $Ag^+(aq) + Cl^-(aq) \rightarrow AgCl(s)$ (***M***)

$AgCl(s) + 2NH_3(aq) \rightarrow [Ag(NH_3)_2]^+(aq) + Cl^-(aq)$

or $Ag^+(aq) + 2NH_3(aq) \rightarrow [Ag(NH_3)_2]^+(aq)$ (***E***)

$NH_4^+(aq) + OH^-(aq) \rightarrow NH_3(g) + H_2O(\ell)$ (with heat) (***M***)

9. **a.** Cu^{2+}, SO_4.

b. $Ba^{2+}(aq) + SO_4^{2-}(aq) \rightarrow BaSO_4(s)$ (***M***)

$Cu^{2+}(aq) + 2OH^-(aq) \rightarrow Cu(OH)_2(s)$ (***M***)

$Cu(OH)_2(s) + 4NH_3(aq) \rightarrow [Cu(NH_3)_4]^{2+}(aq) + 2OH^-(aq)$

or $Cu^{2+}(aq) + 4NH_3(aq) \rightarrow [Cu(NH_3)_4]^{2+}(aq)$ (***E***)

10. Add red litmus paper to the solutions – the ones which turn the litmus blue will be NaOH and Na_2CO_3. Add a few drops of dilute HCl to a sample of each solution. The sample that 'fizzes' will be Na_2CO_3:

($CO_3^{2-} + 2H^+ \rightarrow H_2O + CO_2$)

Add NaOH to the other two samples – a white precipitate forms with $MgSO_4$:

$Mg^{2+}(aq) + 2OH^-(aq) \rightarrow Mg(OH)_2(s)$ (***A*** – observation; ***M*** – equation)

11. Anion NO_3^- no reaction with any of the testing solutions.

Cation Al^{3+} $Al^{3+}(aq) + 3OH^-(aq) \rightarrow Al(OH)_3(s)$

$Al(OH)_3(s) + OH^-(aq) \rightarrow Al(OH)_4^-(aq)$

12.	Anion	SO_4^{2-}	$Ba^{2+}(aq) + SO_4^{2-}(aq) \rightarrow BaSO_4(s)$
	Cation	Fe^{3+}	$Fe^{3+}(aq) + 3OH^-(aq) \rightarrow Fe(OH)_3(s)$
			$Fe^{3+}(aq) + SCN^-(aq) \rightarrow [Fe(SCN)]^{2+}(aq)$
13.	Anion	I^-	$Ag^+(aq) + I^-(aq) \rightarrow AgI(s)$
	Cation	Zn^{2+}	$Zn^{2+}(aq) + 2OH^-(aq) \rightarrow Zn(OH)_2(s)$
			$Zn(OH)_2(s) + 2OH^-(aq) \rightarrow [Zn(OH)_4]^{2-}(aq)$
			$Zn^{2+}(aq) + 2OH^-(aq) \rightarrow Zn(OH)_2(s)$
			$Zn(OH)_2(s) + 4NH_3(aq) \rightarrow [Zn(NH_3)_4]^{2+}(aq) + 2OH^-(aq)$

Unit 12.1 Activity 1B: Qualitative analysis – multiple choice (page 15)

1. A

2. C

3. C

4. D

Unit 12.1 Activity 2A: Molar mass (page 21)

1.	36.5 g mol^{-1} (**A**)	**2.**	32.0 g mol^{-1} (**A**)	**3.**	46.0 g mol^{-1} (**A**)
4.	63.0 g mol^{-1} (**A**)	**5.**	98.1 g mol^{-1} (**A**)	**6.**	342.0 g mol^{-1} (**A**)
7.	44.0 g mol^{-1} (**A**)	**8.**	32.0 g mol^{-1} (**A**)	**9.**	40.3 g mol^{-1} (**A**)
10.	111.0 g mol^{-1} (**A**)	**11.**	136.1 g mol^{-1} (**A**)	**12.**	249.7 g mol^{-1} (**A**)
13.	286.0 g mol^{-1} (**A**)	**14.**	59.0 g mol^{-1} (**A**)	**15.**	95.0 g mol^{-1} (**A**)

Unit 12.1 Activity 2B: Mole calculations (page 22)

1. **a.** $M(CH_4) = M(C) + 4M(H)$
$= 12.0 + 4 \times 1.0 = 16.0 \text{ g mol}^{-1}$

b. $m = nM = 2 \times 16.0 = 32.0 \text{ g}$

c. $n = \frac{m}{M} = \frac{64.0 \text{ g}}{16.0 \text{ g mol}^{-1}} = 4.00 \text{ mol}$

d. $m = nM = 3 \text{ mol} \times 16.0 \text{ g mol}^{-1}$
$= 48.0 \text{ g}$

2. **a.** $M(CaCO_3) = M(Ca) + M(C) + 3M(O)$
$= 40.1 + 12.0 + 3 \times 16.0$
$= 100.1 \text{ g mol}^{-1}$

b. $m = nM = 0.0500 \text{ mol} \times 100.1 \text{ g mol}^{-1}$
$= 5.00 \text{ g}$

c. $n = \frac{m}{M} = \frac{10.0 \text{ g}}{100.1 \text{ g mol}^{-1}} = 0.10 \text{ mol}$

d. $n = \frac{m}{M} = \frac{50.0 \text{ g}}{100.1 \text{ g mol}^{-1}} = 0.500 \text{ mol}$

1 mol = 6.02×10^{23} Ca^{2+} ions

0.5 mol = 3.01×10^{23} Ca^{2+} ions

3. $M(CO_2) = M(C) + 2M(O)$
$= 12.0 + 2 \times 16.0 = 44.0 \text{ g mol}^{-1}$

4. $n = \frac{m}{M} = \frac{6\,400 \text{ g}}{64.1 \text{ g mol}^{-1}} = 99.8 \text{ mol}$

5. $n = \frac{m}{M} = \frac{8\,000 \text{ g}}{80.1 \text{ g mol}^{-1}} = 99.9 \text{ mol}$

6. $M(C_3H_8) = 44.0 \text{ g mol}^{-1}$

$n(C_3H_8) = \frac{88.0 \text{ g}}{44.0 \text{ g mol}^{-1}} = 2.00 \text{ mol}$

$n(C) = 3 \times 2.00 = 6.00 \text{ mol}$

7. $m = nM$

a. $m = 5.50 \text{ mol} \times 46.0 \text{ g mol}^{-1} = 253 \text{ g}$ (***A***)

b. $m = 0.150 \text{ mol} \times 84.0 \text{ g mol}^{-1} = 12.6 \text{ g}$ (***A***)

c. $m = 10.7 \text{ mol} \times 58.0 \text{ g mol}^{-1} = 620.6 \text{ g}$ (***A***)

d. $m = 1.25 \text{ mol} \times 143.5 \text{ g mol}^{-1} = 179.4 \text{ g}$ (***A***)

e. $m = 0.600 \text{ mol} \times 180 \text{ g mol}^{-1} = 108 \text{ g}$ (***A***)

f. $m = 25.0 \text{ mol} \times 254 \text{ g mol}^{-1} = 6350 \text{ g}$ (***A***)

g. $M(CCl_4) = 154 \text{ g mol}^{-1}$ (***A***)

$m = 0.50 \text{ mol} \times 154 \text{ g mol}^{-1} = 77 \text{ g}$ (***A***)

h. $M(CuO) = 79.6 \text{ g mol}^{-1}$ (***A***)

$m = 0.250 \text{ mol} \times 79.6 \text{ g mol}^{-1} = 19.9 \text{ g}$ (***A***)

8. **a.** $M(Al_2O_3) = 102.0 \text{ g mol}^{-1}$

b. $n = \frac{m}{M} = \frac{1.02 \text{ g}}{102 \text{ g mol}^{-1}} = 0.0100 \text{ mol}$

c. $n(Al^{3+}) = 2 \times n(Al_2O_3) = 0.0200 \text{ mol}$

d. $n(O^{2-}) = 3 \times n(Al_2O_3) = 0.0300 \text{ mol}$

9. **a.** $M((NH_4)_2SO_4) = 132.1 \text{ g mol}^{-1}$

b. $n((NH_4)_2SO_4) = \frac{1320 \text{ g}}{132.1 \text{ g mol}^{-1}} = 10.0 \text{ mol}$

c. $n(NH_4^+) = 2 \times n((NH_4)_2SO_4) = 2 \times 10.0 = 20.0 \text{ mol}$

d. $n(SO_4^{2-}) = n(NH_4)_2SO_4 = 10.0 \text{ mol}$

e. $n(S) = n(SO_4^{2-}) = 10.0 \text{ mol}$

f. $n(N) = n(NH_4^+) = 20.0 \text{ mol}$

10. $n = \frac{m}{M}$

a. $n = \frac{15.2 \text{ g}}{58.5 \text{ g mol}^{-1}} = 0.260 \text{ mol}$

b. $n = \frac{10.9 \text{ g}}{121 \text{ g mol}^{-1}} = 0.0901 \text{ mol}$

c. $n = \frac{25.0 \text{ g}}{99.4 \text{ g mol}^{-1}} = 0.252 \text{ mol}$

d. $n = \frac{21.5 \text{ g}}{106 \text{ g mol}^{-1}} = 0.203 \text{ mol}$

e. $n = \dfrac{1.95\text{ g}}{17.0\text{ g mol}^{-1}} = 0.115\text{ mol}$

f. $n = \dfrac{10.0\text{ g}}{159.7\text{ g mol}^{-1}} = 0.626\text{ mol}$

g. $M(ZnCO_3) = 125.4\text{ g mol}^{-1}$

$n = \dfrac{282\text{ g}}{125.4\text{ g mol}^{-1}} = 2.25\text{ mol}$

h. $M((NH_4)_2CO_3) = 96.0\text{ g mol}^{-1}$

$n = \dfrac{34\text{ g}}{96.0\text{ g mol}^{-1}} = 0.35\text{ mol}$

Unit 12.1 Activity 2C: Balancing equations (page 24)

1. $H_2(g) + Cl_2(g) \rightarrow 2HCl(g)$

2. $2H_2(g) + O_2(g) \rightarrow 2H_2O(\ell)$

3. $4Al(s) + 3O_2(g) \rightarrow 2Al_2O_3(s)$

4. $Fe_2O_3(s) + 3C(s) \rightarrow 2Fe(s) + 3CO(g)$

5. $4Na(s) + O_2(g) \rightarrow 2Na_2O(s)$

6. $N_2(g) + 3H_2(g) \rightarrow 2NH_3(g)$

7. $2SO_2(g) + O_2(g) \rightarrow 2SO_3(g)$

8. $CH_4(g) + 2O_2(g) \rightarrow CO_2(g) + 2H_2O(g)$

9. $2C(s) + O_2(g) \rightarrow 2CO(g)$

10. $C_3H_8(g) + 5O_2(g) \rightarrow 3CO_2(g) + 4H_2O(g)$

11. $C_3H_8(g) + 2O_2(g) \rightarrow 3C(s) + 4H_2O(g)$

12. $2Fe(s) + 3Cl_2(g) \rightarrow 2FeCl_3(s)$

13. $2PbO(g) + C(s) \rightarrow 2Pb(s) + CO_2(g)$

Unit 12.1 Activity 2D: The mole and chemical equations (page 27)

1. $n(Zn) = \dfrac{m}{M} = \dfrac{1.65\text{ g}}{65.4\text{ g mol}^{-1}} = 2.53 \times 10^{-2}\text{ mol}$

$n(ZnI_2) = n(Zn) = 2.53 \times 10^{-2}\text{ mol}$

$m(ZnI_2) = nM = 2.53 \times 10^{-2}\text{ mol} \times 319.4\text{ g mol}^{-1}$

$= 8.06\text{ g}$ (**E**)

2. **a.** $m(CO_2) = 44.0\text{ g mol}^{-1}$

b. $n(CO_2) = \dfrac{m}{M} = \dfrac{8.80\text{ g}}{44.0\text{ g mol}^{-1}} = 0.200\text{ mol}$

$n(MgCO_3) = n(CO_2) = 0.200\text{ mol}$

$m(MgCO_3) = nM = 0.200 \times 84.3\text{ g}$

$= 16.9\text{ g}$ (**E**)

3. $n(Mg) = \dfrac{m}{M} = \dfrac{2.40\text{ g}}{24.3\text{ g mol}^{-1}} = 9.88 \times 10^{-2}\text{ mol}$

$n(Mg) = n(MgO) = 9.88 \times 10^{-2}\text{ mol}$

$m(MgO) = nM = 9.88 \times 10^{-2}\text{ mol} \times 40.3\text{ g mol}^{-1}$

$= 3.98\text{ g}$ (**E**)

4. **a.** $n(CO_2) = \dfrac{m}{M} = \dfrac{11.0\text{ g}}{44.0\text{ g mol}^{-1}} = 0.25\text{ mol}$

$n(CO_2) = n(CuCO_3) = 0.25\text{ mol}$

$m(CuCO_3) = 0.25\text{ mol} \times 123.6\text{ g mol}^{-1}$

$= 30.9\text{ g}$ (**E**)

b. $n(CuCO_3) = \frac{m}{M} = \frac{247\ g}{123.6\ g\ mol^{-1}} = 2.00\ mol$

$n(CuO) = n(CuCO_3) = 2.00\ mol$
$m(CuO) = nM = 2.00\ mol \times 79.6\ g\ mol^{-1}$
$= 159\ g$ (**E**)

c. $n(CuO) = \frac{m}{M} = \frac{318\ g}{79.6\ g\ mol^{-1}} = 3.99\ mol$

$n(CO_2) = n(CuO) = 3.99\ mol$
$m(CO_2) = 3.99\ mol \times 44.0\ g\ mol^{-1}$
$= 176\ g$ (**E**)

5. $n(CaCO_3) = \frac{m}{M} = \frac{100\ 000}{100.1} = 999\ mol$

$n(CaO) = n(CaCO_3) = 999\ mol$
$m(CaO) = nM = 999\ mol \times 56.1\ g\ mol^{-1}$
$= 56\ kg$ (**E**)

6. $n(O_2) = \frac{m}{M} = \frac{8000\ g}{32\ g\ mol^{-1}} = 250\ mol$

$n(Ca) = 2n(O_2) = 500\ mol$
$m(Ca) = nM = 500\ mol \times 40.1\ g\ mol^{-1}$
$= 20.1\ kg$ (**E**)

7. $n(SO_3) = \frac{m}{M} = \frac{8.0 \times 10^6\ g}{80.1\ g\ mol^{-1}} = 1 \times 10^5\ mol$

$n(H_2SO_4) = n(SO_3) = 1 \times 10^5\ mol$
$m(H_2SO_4) = nM = 1 \times 10^5\ mol \times 98.1\ g\ mol^{-1}$
$= 9.8$ tonnes (**E**)

8. $n(Fe_2O_3) = \frac{m}{M} = \frac{1.6 \times 10^6\ g}{159.8\ g\ mol^{-1}} = 10\ 000\ mol$

$n(Fe) = 2n(Fe_2O_3) = 20\ 000\ mol$
$m(Fe) = nM = 20\ 000\ mol \times 55.9\ g\ mol^{-1}$
$= 1.12 \times 10^6\ g = 1.12$ tonnes (**E**)

9. $n(CH_4) = \frac{m}{M} = \frac{32\ g}{16.0\ g\ mol^{-1}} = 2\ mol$

$n(H_2O) = 2n(CH_4) = 2 \times 2 = 4\ mol$ (**M**)

10. $n(Ca) = \frac{m}{M} = \frac{4.00\ g}{40.1\ g\ mol^{-1}} = 0.0998\ mol$

$n(H_2O) = 2n(Ca) = 0.200\ mol$ (**M**)
$m(H_2O) = 0.200\ mol \times 18.0\ g\ mol^{-1}$
$= 3.60\ g$

11. **a.** $\frac{n(O_2)}{1} = \frac{n(X)}{4} = 0.25\ n(X)$

b. $m(O_2) = 6.2\text{ g} - 4.6\text{ g} = 1.6\text{ g}$

$$n(O_2) = \frac{m}{M} = \frac{1.6\text{ g}}{32\text{ g mol}^{-1}} = 0.050\text{ mol } (\textbf{\textit{E}})$$

12. $n(CO(NH_2))_2 = \frac{m}{M} = \frac{182\,000\text{ g}}{60.0\text{ g mol}^{-1}} = 3033\text{ mol}$

$n(NH_3) = 2n(CO(NH_2)_2) = 2 \times 3033\text{ mol}$
$= 6066\text{ mol}$

$n(NH_3) = nM = 6066\text{ mol} \times 17.0\text{ g mol}^{-1}$
$= 103\text{ kg } (\textbf{\textit{E}})$

13. $n(CaF_2) = \frac{m}{M} = \frac{15.6\text{ g}}{78.1\text{ g mol}^{-1}} = 0.200\text{ mol}$

$n(HF) = 2n(CaF_2) = 0.400\text{ mol}$
$m(HF) = nM = 0.400\text{ mol} \times 20.0\text{ g mol}^{-1}$
$= 8.00\text{ g } (\textbf{\textit{E}})$

14. $n(C_4H_8) = \frac{m}{M} = \frac{21.2\text{ g}}{56.0\text{ g mol}^{-1}} = 0.379\text{ mol}$

$n(CO_2) = 4n(C_4H_8) = 4 \times 0.379\text{ mol} = 1.52\text{ mol}$
$m(CO_2) = nM = 1.52\text{ mol} \times 44.0\text{ g mol}^{-1}$
$= 66.9\text{ g } (\textbf{\textit{E}})$

15. $n(C_6H_{10}) = \frac{m}{M} = \frac{9.90\text{ g}}{82\text{ g mol}^{-1}} = 0.121\text{ mol}$

$n(H_2O) = 5n(C_6H_{10}) = 5 \times 0.121\text{ mol} = 0.605\text{ mol}$
$m(H_2O) = 0.605\text{ mol} \times 18.0\text{ g mol}^{-1} = 10.9\text{ g } (\textbf{\textit{E}})$

16. $n(SO_2) = \frac{m}{M} = \frac{100\,000\text{ g}}{64.1\text{ g mol}^{-1}} = 1560\text{ mol}$

$n(SO_2) = n(SO_3) = 1560\text{ mol}$
$m(SO_3) = nM = 1560\text{ mol} \times 80.1\text{ g mol}^{-1} = 125\text{ kg } (\textbf{\textit{E}})$

Unit 12.1 Activity 2E: Percentage composition, empirical and molecular formulae (page 33)

1. $M(KNO_3) = 101.1\text{ g mol}^{-1}$

$$\%\text{ K} = \frac{39.1}{101.1} \times 100 = 38.7\%$$

$$\%\text{ N} = \frac{14.0}{101.1} \times 100 = 13.9\%$$

$$\%\text{ O} = \frac{3 \times 16.0}{101.1} \times 100 = 47.5\%$$

2. $M(MgO) = 40.3\text{ g mol}^{-1}$

$$\%\text{ Mg} = \frac{24.3}{40.3} \times 100 = 60.3\%$$

3. $M(C_2H_2) = 26\text{ g mol}^{-1}$

$$\%\text{ C} = \frac{2 \times 12}{26} \times 100 = 92.3\%$$

4. **a.** $n(\text{Cu}) = \dfrac{80}{63.6} = 1.26$

$n(\text{O}) = \dfrac{20}{16.0} = 1.25$

Ratio Cu : O = 1.26 : 1.25 = $\dfrac{1.26}{1.25} : \dfrac{1.25}{1.25}$ = 1 : 1, ie CuO

b. $n(\text{Al}) = \dfrac{53}{27.0} = 1.96$

$n(\text{O}) = \dfrac{47}{16.0} = 2.94$

Al : O = 1.96 : 2.94 = $\dfrac{1.96}{1.96} : \dfrac{2.94}{1.96}$ = 1 : 1.5 = 2 : 3, ie Al_2O_3

c. $n(\text{H}) = \dfrac{1.6}{1.0} = 1.6$

$n(\text{N}) = \dfrac{22.2}{14.0} = 1.59$

$n(\text{O}) = \dfrac{76.2}{16.0} = 4.76$

H : N : O = 1.6 : 1.59 : 4.76 = $\dfrac{1.6}{1.59} : \dfrac{1.59}{1.59} : \dfrac{4.76}{1.59}$ = 1 : 1 : 3, ie HNO_3

5. **a.** $Mg^{2+} : Cl^{-}$ = 1 : 2

b. $M(MgCl_2) = 95.3\ \text{g mol}^{-1}$

$\%\ \text{Mg} = \dfrac{24.3}{95.3} \times 100 = 25.5\%$

6. $n(\text{S}) = \dfrac{40}{32.1} = 1.25\ \text{mol}$

$n(\text{O}) = \dfrac{60}{16.0} = 3.75\ \text{mol}$

$n(\text{S}) : n(\text{O})$ = 1 : 3, ie SO_3

7. $n(\text{C}) = \dfrac{90}{12.0} = 7.5$

$n(\text{H}) = \dfrac{10}{1} = 10.0$

$n(\text{C}) : n(\text{H})$ = 7.5 : 10 = $\dfrac{7.5}{7.5} : \dfrac{10.00}{7.5}$ = 1 : 1.33 = 1 : $\dfrac{4}{3}$ = 3 : 4, ie C_3H_4

8. $n(\text{Si}) = \dfrac{0.28\ \text{g}}{28.1\ \text{g mol}^{-1}} = 0.009\,96\ \text{mol}$

$m(\text{O}) = 0.60\ \text{g} - 0.28\ \text{g} = 0.32\ \text{g}$

$n(\text{O}) = \dfrac{0.32\ \text{g}}{16.0\ \text{g mol}^{-1}} = 0.02\ \text{mol}$

$n(\text{Si}) : n(\text{O})$ = 1 : 2, ie SiO_2

9. $n(\text{V}) = \dfrac{10.2\ \text{g}}{50.9\ \text{g mol}^{-1}} = 0.200\ \text{mol}$

$n(\text{Cl}) = \dfrac{21.3\ \text{g}}{35.5\ \text{g mol}^{-1}} = 0.600\ \text{mol}$

$n(\text{V}) : n(\text{Cl})$ = 1 : 3, ie VCl_3

10. $M(CH_2) = 14 \text{ g mol}^{-1}$ $M(\text{compound}) = 84 \text{ g mol}^{-1}$

$$\frac{M(\text{compound})}{M(CH_2)} = \frac{84}{14}$$

Molecular formula = $6(CH_2) = C_6H_{12}$

11. **a.** $n(C) = \frac{85.7}{12.0} = 7.14$ $n(H) = \frac{14.3}{1.0} = 14.3$

$n(C) : n(H) = 1 : 2$ EF = CH_2

$M = 28 \text{ g mol}^{-1}$, $M(CH_2) = 14$, ∴ Molecular formula = C_2H_4

b. $n(N) = \frac{30.4}{14.0} = 2.17$ $n(O) = \frac{69.6}{16.0} = 4.35$

$n(N) : n(O) = 1 : 2$ EF = NO_2

$M(\text{compound}) = 92 \text{ g mol}^{-1}$, $M(NO_2) = 46 \text{ g mol}^{-1}$

Molecular formula = $2(NO_2) = N_2O_4$

c. $n(H) = \frac{2}{1.0} = 2$ $n(S) = \frac{33}{32.1} = 1.03$ $n(O) = \frac{65}{16.0} = 4.06$

$n(H) : n(S) : n(O) = 2 : 1 : 4$ EF = H_2SO_4

$M(\text{compound}) = 98 \text{ g mol}^{-1}$, $M(H_2SO_4) = 98 \text{ g mol}^{-1}$

Molecular formula = H_2SO_4

12. **a.** $n(C) = \frac{82.7}{12.0} = 6.89$ $n(H) = \frac{17.3}{1.0} = 17.3$

$n(C) : n(H) = 1 : 2.5 = 2 : 5$

EF = C_2H_5 $M(C_2H_5) = 29 \text{ g mol}^{-1}$

b. $M(\text{compound}) = 58 \text{ g mol}^{-1}$

Molecular formula = $2(C_2H_5) = C_4H_{10}$

13. **a.** $n(C) = \frac{80}{12.0} = 6.67$ $n(H) = \frac{20}{1.0} = 20$

$n(C) : n(H) = 1 : 3$ EF = CH_3

b. $M(CH_3) = 15 \text{ g mol}^{-1}$ $M(\text{compound}) = 30 \text{ g mol}^{-1}$

Molecular formula = $2(CH_3) = C_2H_6$

14. **a.** $n(N) = \frac{30.3}{14.0} = 2.16$ $n(O) = \frac{69.7}{16.0} = 4.3$

$n(N) : n(O) = 1 : 2$ EF = NO_2

b. $M(NO_2) = 46 \text{ g mol}^{-1}$ $M(\text{compound}) = 92 \text{ g mol}^{-1}$

Molecular formula = $2(NO_2) = N_2O_4$

15. **a.** $n(C) = \frac{40.92}{12.0} = 3.41$ $n(H) = \frac{4.58}{1.0}$ $n(O) = \frac{54.5}{16.0} = 3.40$

$n(C) : n(H) : n(O) = 1 : 1.34 : 1$

$= 3 : 4 : 3$

EF = $C_3H_4O_3$

b. $M(C_3H_4O_3) = 88 \text{ g mol}^{-1}$, $M(\text{compound}) = 176 \text{ g mol}^{-1}$

$$\frac{M(\text{compound})}{M(C_3H_4O_3)} = \frac{176}{88} = 2$$

Molecular formula = $2(C_3H_4O_3)$

$= C_6H_8O_6$

16. **a.** $n(\text{C}) = \frac{26.67}{12.0} = 2.22$ $n(\text{H}) = \frac{2.22}{1.0} = 2.22$ $n(\text{O}) = \frac{71.11}{16.0} = 4.44$

$n(\text{C}) : n(\text{H}) : n(\text{O}) = 2.22 : 2.22 : 4.44$
$= 1 : 1 : 2$

EF = CHO_2

b. $M(CHO_2) = 45\ \text{g mol}^{-1}$ M (compound) = 90 g mol^{-1} Molecular formula $C_2H_2O_4$

17. **a.** $n(\text{Pb}) = \frac{2.95\ \text{g}}{207\ \text{g mol}^{-1}} = 0.0143\ \text{mol}$

b. $m(\text{S}) = 3.42 - 2.95\ \text{g} = 0.47\ \text{g}$

$n(\text{S}) = \frac{0.47\ \text{g}}{32.1\ \text{g mol}^{-1}} = 0.0146\ \text{mol}$

c. PbS

Unit 12.1 Activity 2F: Water of crystallisation (page 35)

1. **a.** $m(NiCl_2) = 23.05 - 21.53\ \text{g} = 1.52\ \text{g}$

b. $n(NiCl_2) = \frac{m}{M} = \frac{1.52\ \text{g}}{129.7\ \text{g mol}^{-1}} = 0.0117\ \text{mol}$

c. $m(H_2O) = 24.09 - 22.84\ \text{g} = 1.25\ \text{g}$

$n(H_2O) = \frac{m}{M} = \frac{1.25\ \text{g}}{18.0\ \text{g mol}^{-1}} = 0.0694\ \text{mol}$

d. $n(NiCl_2) : n(H_2O) = 0.0117 : 0.0694$
$= 1 : 5.9$

Formula of salt $NiCl_2.6H_2O$ (**E**)

2. **a.** $m(Na_2CO_3) = 0.77\ \text{g}$ $n(Na_2CO_3) = \frac{0.77\ \text{g}}{106\ \text{g mol}^{-1}} = 0.007\,26\ \text{mol}$

b. $m(H_2O) = 2.07\ \text{g} - 0.77\ \text{g} = 1.30\ \text{g}$

$n(H_2O) = \frac{1.30\ \text{g}}{18.0\ \text{g mol}^{-1}} = 0.0722\ \text{mol}$

$n(Na_2CO_3) : n(H_2O) = 0.007\,26 : 0.0722$
$= 1 : 10$

Formula $Na_2CO_3.10H_2O$ (**E**)

3. $m(CaCl_2) = 5.00\ \text{g}$ $n(CaCl_2) = \frac{5.00\ \text{g}}{111.1\ \text{g mol}^{-1}} = 0.0450\ \text{mol}$

$m(H_2O) = 9.86\ \text{g} - 5.00\ \text{g} = 4.86\ \text{g}$

$n(H_2O) = \frac{4.86\ \text{g}}{18.0\ \text{g mol}^{-1}} = 0.270\ \text{mol}$

$n(CaCl_2) : n(H_2O) = 0.0450 : 0.270$
$= 1 : 6$

Formula $CaCl_2.6H_2O$ (**E**)

4. $m(Na_2S_2O_3) = 23.66\ \text{g} - 20.26\ \text{g} = 3.40\ \text{g}$

$n(Na_2S_2O_3) = \frac{3.40\ \text{g}}{158\ \text{g mol}^{-1}} = 0.0215\ \text{mol}$

$m(H_2O) = 25.58 - 23.66\ g = 1.92\ g$

$n(H_2O) = \dfrac{1.92\ g}{18.0\ g\ mol^{-1}} = 0.107\ mol$

$n(Na_2S_2O_3) : n(H_2O) = 0.0215 : 0.107 = 1 : 4.98$

Formula $Na_2S_2O_3.5H_2O$

5. $m(BaCl_2) = 4.26\ g$ $n(BaCl_2) = \dfrac{4.26\ g}{208\ g\ mol^{-1}} = 0.0205\ mol$

$m(H_2O) = 5.00 - 4.26\ g = 0.74\ g$

$n(H_2O) = \dfrac{0.74\ g}{18.0\ g\ mol^{-1}} = 0.0411\ mol$

$n(BaCl_2) : n(H_2O) = 0.0205 : 0.0411$

$= 1 : 2$

Formula $BaCl_2.2H_2O$

$x = 2$

Unit 12.1 Activity 2G: Quantitative analysis – multiple choice (page 35)

1. C **2.** B **3.** A

4. B **5.** C **6.** D

7. B **8.** A **9.** C

Unit 12.1 Activity 3A: Concentrations (page 40)

1. **a.** D **b.** D

c.

	Solutions				
	A	B	C	D	E
Concentration (mol L^{-1})	8	4	2	2	1

(A)

d. A **e.** E **f.** E

2. **a.** $c = \dfrac{n}{V} = \dfrac{0.004\ mol}{0.020\ L} = 0.2\ mol\ L^{-1}$ (**A**)

b. $c = \dfrac{n}{V} = \dfrac{0.5\ mol}{0.040\ L} = 12.5\ mol\ L^{-1}$ (**A**)

c. $c = \dfrac{n}{V} = \dfrac{0.15\ mol}{0.25\ L} = 0.60\ mol\ L^{-1}$ (**A**)

3. **a.** **i.** $c = \dfrac{m}{V} = \dfrac{18\ g}{0.20\ L} = 90\ g\ L^{-1}$ (**A**)

ii. $c = \dfrac{m}{V} = \dfrac{180\ g}{0.400\ L} = 450\ g\ L^{-1}$ (**A**)

b. $c = \dfrac{m}{V} = \dfrac{23.0\ g}{0.200\ L} = 115\ g\ L^{-1}$ (**A**)

c. $c = \dfrac{m}{V} = \dfrac{39.9\ g}{0.500\ L} = 79.8\ g\ L^{-1}$ (**A**)

4. **a.** $c = \frac{m}{V} = \frac{58.5\text{ g}}{2.00\text{ L}} = 29.3\text{ g L}^{-1}$ (***A***)

$n(\text{NaCl}) = \frac{m}{M} = \frac{58.5\text{ g}}{58.5\text{ g mol}^{-1}} = 1.00\text{ mol}$

$c = \frac{n}{V} = \frac{1.00\text{ mol}}{2.00\text{ L}} = 0.500\text{ mol L}^{-1}$ (**M**)

b. $c = \frac{m}{V} = \frac{10.6\text{ g}}{0.100\text{ L}} = 106\text{ g L}^{-1}$ (***A***)

$n(\text{Na}_2\text{CO}_3) = \frac{m}{M} = \frac{10.6\text{ g}}{106\text{ g mol}^{-1}} = 0.100\text{ mol}$

$n = \frac{\text{c}}{\text{V}} = \frac{0.100\text{ mol}}{0.100\text{ L}} = 1.00\text{ mol L}^{-1}$ (**M**)

c. $c = \frac{m}{V} = \frac{0.2\text{ g}}{0.01\text{ L}} = 20\text{ g L}^{-1}$ (***A***)

$n(\text{NaOH}) = \frac{m}{M} = \frac{0.2\text{ g}}{40\text{ g mol}^{-1}} = 0.005\text{ mol}$

$c = \frac{n}{V} = \frac{0.005\text{ mol}}{0.01\text{ L}} = 0.5\text{ mol L}^{-1}$ (**M**)

d. $c = \frac{m}{V} = \frac{1.7\text{g}}{0.500\text{ L}} = 3.4\text{ g L}^{-1}$ (***A***)

$n(\text{AgNO}_3) = \frac{m}{M} = \frac{1.7\text{ g}}{170\text{ g mol}^{-1}} = 0.01\text{ mol}$

$c = \frac{n}{V} = \frac{0.01\text{ mol}}{0.500\text{ L}} = 0.02\text{ mol L}^{-1}$ (**M**)

e. $c = \frac{m}{V} = \frac{5.3\text{ g}}{1\text{ L}} = 5.3\text{ g L}^{-1}$ (***A***)

$n(\text{NH}_4\text{Cl}) = \frac{m}{M} = \frac{5.3\text{ g}}{53.5\text{ g mol}^{-1}} = 0.099\text{ mol}$

$c = \frac{n}{V} = \frac{0.099\text{ mol}}{1.00\text{ L}} = 0.099\text{ mol L}^{-1}$ (**M**)

5. **a.** Before dilution: $c = \frac{n}{V} = \frac{0.5}{0.5} = 1\text{ mol L}^{-1}$

After dilution: $c = \frac{0.5}{1} = 0.5\text{ mol L}^{-1}$

b. Before dilution: $c = \frac{0.1}{0.25} = 0.4\text{ mol L}^{-1}$

After dilution: $c = \frac{0.1}{1} = 0.1\text{ mol L}^{-1}$

c. Before dilution: $c = \frac{0.02}{0.005} = 4\text{ mol L}^{-1}$

After dilution: $c = \frac{0.02}{0.5} = 0.04\text{ mol L}^{-1}$

d. Before dilution: $c = \frac{0.04}{0.01} = 4 \text{ mol L}^{-1}$

After dilution: $c = \frac{0.04}{0.1} = 0.4 \text{ mol L}^{-1}$

6. **a.** $n = cV = 0.0100 \text{ mol L}^{-1} \times 0.250 \text{ L} = 0.002\,50 \text{ mol}$ (***A***)

$m = nM = 0.00250 \text{ mol} \times 170 \text{ g mol}^{-1} = 0.425 \text{ g}$ (***M***)

b. $n = cV = 0.500 \text{ mol L}^{-1} \times 0.500 \text{ L} = 0.250 \text{ mol}$ (***A***)

$m = nM = 0.250 \text{ mol} \times 294.2 \text{ g mol}^{-1} = 73.6 \text{ g}$ (***M***)

c. $n = cV = 0.100 \text{ mol L}^{-1} \times 0.100 \text{ L} = 0.0100 \text{ mol}$ (***A***)

$m = nM = 0.0100 \text{ mol} \times 208 \text{ g mol}^{-1} = 2.08 \text{ g}$ (***M***)

d. $n = cV = 1.00 \text{ mol L}^{-1} \times 0.200 \text{ L} = 0.200 \text{ mol}$ (***A***)

$m = nM = 0.200 \text{ mol} \times 277.9 \text{ g mol}^{-1} = 55.6 \text{ g}$ (***M***)

7. $n = cV = 0.00100 \text{ mol L}^{-1} \times 0.250 \text{ L}$

$= 2.50 \times 10^{-4} \text{ mol}$

$m(COCl_2.6H_2O) = nM = 2.50 \times 10^{-4} \text{ mol} \times 237.8 \text{ g mol}^{-1}$

$= 0.0595 \text{ g}$

8. $n = cV = 0.0125 \text{ mol L}^{-1} \times 0.500 \text{ L}$

$= 6.25 \times 10^{-3} \text{ mol}$

$m(Fe(NO_3)_3.9H_2O) = nM = 6.25 \times 10^{-3} \text{ mol} \times 403.9 \text{ g mol}^{-1}$

$= 2.52 \text{ g}$

Unit 12.1 Activity 3B: Standard solutions (page 45)

1. **a.** Sodium ions, 0.0005 mol; chloride ions, 0.0005 mol (***A***)
b. Hydrogen ions, 0.0125 mol; chloride ions, 0.0125 mol (***A***)
c. Sodium ions, 0.004 mol; carbonate ions, 0.002 mol (***A***)
d. Sodium ions, 0.1 mol; hydroxide ions, 0.1 mol (***A***)
e. Hydrogen ions, 0.1 mol; nitrate ions, 0.1 mol (***A***)
f. Iron(III) ions, 0.0088 mol; chloride ions 0.026 mol (***A***)
g. Sodium ions, 0.0003 mol; sulfate ions, 0.00015 mol (***A***)
h. Calcium ions, 0.0003 mol; hydroxide ions, 0.0006 mol (***A***)
i. Hydrogen ions, 0.01 mol; sulfate ions, 0.005 mol (***A***)

2. **a.** Rinse the volumetric flask and funnel with tap water, then with distilled water from the wash-bottle. Transfer sodium carbonate to the volumetric flask using the spatula and funnel. Rinse with distilled water, making sure no solid is left in the beaker, on the spatula, or in the funnel. Remove funnel, dissolve solid in a little distilled water, then carefully make up to the 'mark' on the volumetric flask. Mix well.

b. $n(Na_2CO_3) = \frac{14.32 \text{ g}}{286.1 \text{ g mol}^{-1}} = 0.0501 \text{ mol}$

$c(Na_2CO_3) = \frac{0.0501 \text{ mol}}{0.250 \text{ L}} = 0.200 \text{ mol L}^{-1}$ (***M***)

c. Primary.

3. **a.** **i.** $122.05 - 116.48 \text{ g} = 5.57 \text{ g}$ **ii.** $n = \frac{m}{M} = \frac{5.57 \text{ g}}{106 \text{ g mol}^{-1}} = 0.0525 \text{ mol}$

b. $c = \frac{n}{V} = \frac{0.0525 \text{ mol}}{0.250 \text{ L}} = 0.210 \text{ mol L}^{-1}$

c. $n = cV = 0.210 \text{ mol L}^{-1} \times 0.020\,00 \text{ L} = 0.004\,20 \text{ mol}$

4. **a.** $n = \frac{m}{M} = \frac{5.30\ \text{g}}{160.0\ \text{g mol}^{-1}} = 0.0500\ \text{mol}$

b. $c = \frac{n}{V} = \frac{0.0500\ \text{mol}}{0.250\ \text{L}} = 0.200\ \text{mol L}^{-1}$

c. $n = cV = 0.200\ \text{mol L}^{-1} \times 0.02000\ \text{L}$
$= 0.004\ \text{mol}$

5. **a.** $40.0\ \text{g mol}^{-1}$

b. $n = \frac{m}{M} = \frac{10.00\ \text{g}}{40.0\ \text{g mol}^{-1}} = 0.250\ \text{mol}$

c. $c = \frac{n}{V} = \frac{0.250\ \text{mol}}{0.500\ \text{L}} = 0.500\ \text{mol L}^{-1}$

d. $n = cV = 0.500\ \text{mol L}^{-1} \times 0.07500\ \text{L}$
$= 0.0375\ \text{mol}$

6. **a.** $n = \frac{m}{M} = \frac{0.160\ \text{g}}{40.0\ \text{g mol}^{-1}} = 0.00400\ \text{mol}$

$c = \frac{n}{V} = \frac{0.00400\ \text{mol}}{0.0500\ \text{L}} = 0.0800\ \text{mol L}^{-1}$

b. The NaOH does not have a constant composition. It can absorb water from the air, so its mass can change between weighing and making up the solution.

Unit 12.1 Activity 4A: Atoms, elements and isotopes (page 50)

1. **a.** 6 **b.** 6 **c.** 14

2. 34

3. **a.** 39 **b.** 20

4.

Symbol	Mass number	Atomic number	Number of neutrons	Number of protons
${}^{3}_{1}H$	3	1	2	1
${}^{13}_{6}C$	13	6	7	6
${}^{56}_{26}Fe$	56	26	30	26
${}^{37}_{17}Cl$	37	17	20	17
${}^{16}_{8}O$	16	8	8	8

5. **a.** 42 **b.** 22

6. ${}^{3}_{1}H$ has two neutrons, ${}^{1}_{1}H$ has no neutrons.

7. **a.** Beryllium. **b.** Sodium. **c.** Lead. **d.** Iron. **e.** Copper.
f. Lithium. **g.** Silver. **h.** Gold. **i.** Potassium. **j.** Silicon.

8. Isotopes.

9. **a.** Isotopes are atoms of the same element (ie have the same number of protons or same atomic number) which have a different atomic mass due to different numbers of neutrons in the nucleus.

Examples – ${}^{1}_{1}H$, ${}^{2}_{1}H$, ${}^{3}_{1}H$ (hydrogen-1, deuterium, tritium)
– ${}^{16}_{6}O$, ${}^{18}_{6}O$ (oxygen-16, oxygen-18)

b. Since isotopes of an element have the same atomic number (Z), they have the same number of protons and electrons. It is electrons that are involved in chemical reactions, hence isotopes react in the same way.

Unit 12.1 Activity 4B: Electron arrangements (page 53)

1. **a.** Li (2, 1) **b.** N (2, 5) **c.** Na (2, 8, 1) **d.** Ar (2, 8, 8)
e. K (2, 8, 8, 1) **f.** O (2, 6) **g.** O (2, 6) **h.** Ca (2, 8 ,8 ,2)

2. Chlorine.

3. **a.** H (1 dot) **b.** •N• (5 dots) **c.** •F: (7 dots) **d.** Ca (2 dots) **e.** •S: (6 dots)
f. K (1 dot) **g.**  :Ne: (8 dots)

4. **a.** C **b.** A, D **c.** C **d.** E **e.** B

Unit 12.1 Activity 4C: The periodic table (page 55)

1. The group number indicates the number of valence electrons.

For the first 20 elements:

- Groups 1 and 2 equate to the number of valence electrons (ie 1 and 2).
- Groups 13–18 equate to 3–8 valence electrons.

Elements in the same group have the same number of valence electrons in their atoms (eg Li, Na, K etc all have one valence electron).
The period defines the number of energy levels present in the atoms of that group. For the third period (Na, Mg, Al, Si etc), all atoms have three energy shells.

The first two shells are full, the third shell contains the valence electrons.
Going down a group, additional energy levels are added successively to the atoms of elements in the first period down to the seventh period.
Going across a group, electrons are progressively added to the same (valence) energy shell.

2. Metal atoms have only 1–3 valence electrons; non-metal atoms have 4–8 valence electrons.

Unit 12.1 Activity 4D: Ions (page 58)

1. **a.** Li^+ **b.** Br^- **c.** Ba^{2+} **d.** S^{2-} **e.** Ag^+ **f.** F^-
g. Fe^{2+} **h.** Cu^+ **i.** NO_3^- **j.** OH^- **k.** HCO_3^- **l.** SO_4^{2-}

2. **a.** Magnesium ion, 2 **b.** Zinc ion, 2 **c.** Lead ion, 2
d. Oxide ion, 2 **e.** Chloride ion, 1 **f.** Iron(III) ion, 3
g. Aluminium ion, 3 **h.** Iodide ion, 1 **i.** Copper(II) ion, 2
j. Fluoride ion, 1

3. **a.** **i.** Potassium ion; K^+ (2, 8, 8) **ii.** Loss of electrons.
b. **i.** Oxide ion; O^{2-} (2, 8) **ii.** Gain of electrons.
c. **i.** Fluoride ion; F^- (2, 8) **ii.** Gain of electrons.
d. **i.** Aluminium ion; Al^{3+} (2, 8) **ii.** Loss of electrons.

4. **a.** Ca (2, 8, 8, 2) **b.** Ca^{2+} (2, 8, 8) **c.** S^{2-} (2, 8, 8)

Unit 12.1 Activity 5A: Lewis structures (page 63)

1. **a.** **i.** 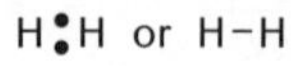

b. **i.** 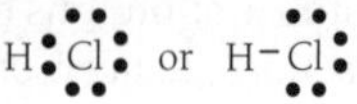

a. **ii.** 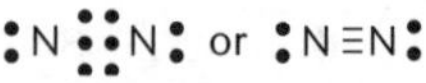

b. **ii.** 

a. **iii.** 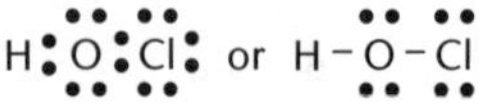

b. **iii.** 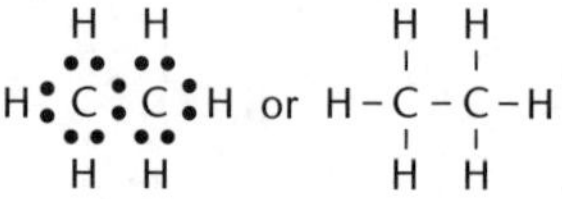

c. **i.** 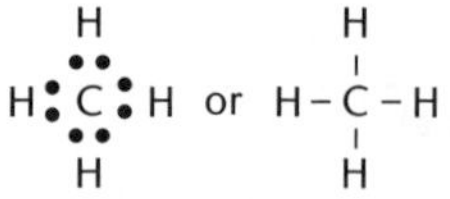

d. **i.** 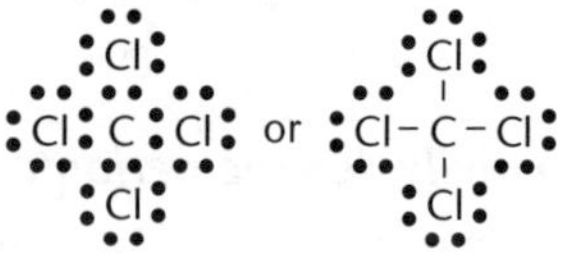

c. **ii.** 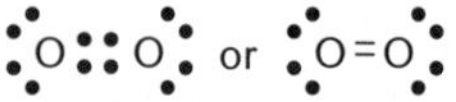

d. **ii.** 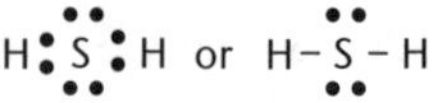

c. **iii.** 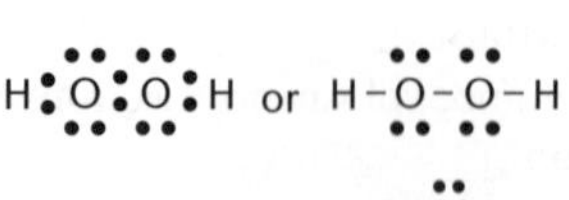

d. **iii.** 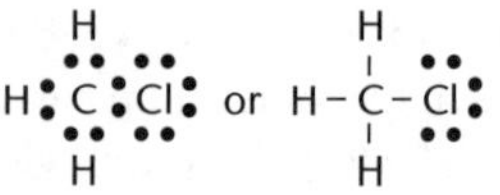

2. **a.** 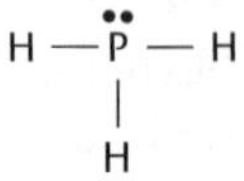

b. 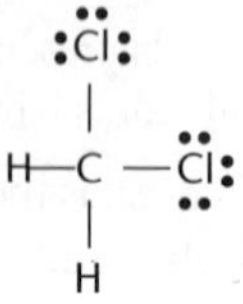

c. 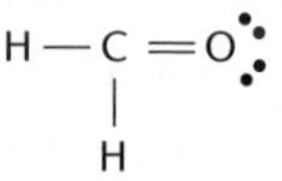

d. 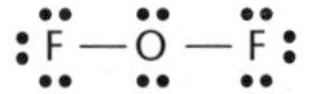

e. 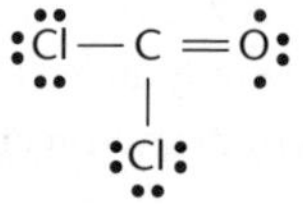

Unit 12.1 Activity 5B: Polar bonds and bonding as a continuum (page 65)

1. **a.** N **b.** O **c.** C **d.** O **e.** F (***A*** – all correct)

2. **a.** and **b.**

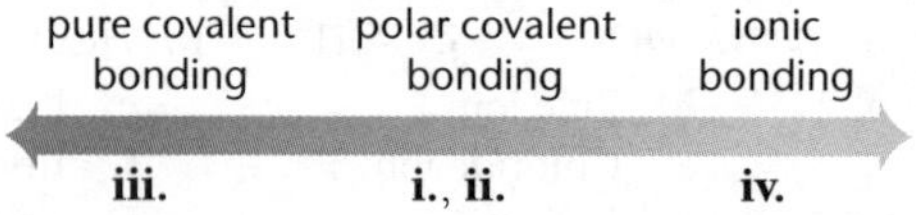

3. **a.** Ionic. **b.** Polar covalent. **c.** Polar covalent. **d.** Ionic.
e. Ionic. **f.** Polar covalent. **g.** Covalent. **h.** Ionic.
i. Polar covalent. **j.** Covalent.

Unit 12.1 Activity 5C: Shapes of molecules (page 69)

1. **a.** CH_4 **b.** NH_3 **c.** H_2O **d.** 4 **e.** 4
f. 4 **g.** 0 **h.** 1 **i.** 2 **j.** Tetrahedral.
k. Trigonal pyramid. **l.** V-shaped or bent. (***M***)

2. **a.** H_2S – 4 sets of electrons on S atom arranged tetrahedrally
2 non-bonding ⇒ V-shaped or bent (***M***)

b. CF_4 – 4 sets of electrons on C, all are bonding
$\Rightarrow$ tetrahedral (**M**)

c. NF_3 – 4 sets of electrons on N arranged tetrahedrally
1 non-bonding set of electrons $\Rightarrow$ trigonal pyramid (**M**)

d. CO_2 – 2 sets of electrons on C arranged linearly
$\Rightarrow$ linear shape (**M**)

3. **a.** 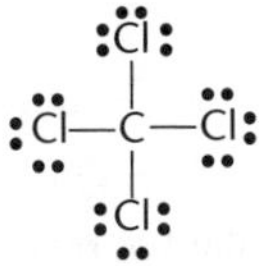 Cl—C(—Cl)(—Cl)—Cl **b.** Tetrahedral.

c. Four groups of electrons around central atom, all are bonding.

d. 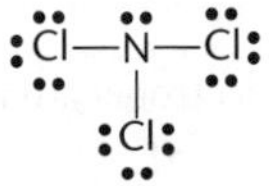 Cl—N(—Cl)—Cl **e.** Trigonal pyramid.

f. 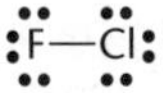 F—Cl **g.** Linear. **h.** Diatomic molecule. (**M**)

4. CO_2. (**A**)
Only two areas of negative charge around the central atom. O=C=O (**M**)

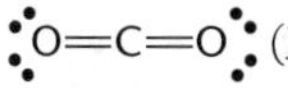

5. **a.** Linear. H—Cl

b. Trigonal pyramid. Cl—P(—Cl)—Cl 4 sets of electrons, 3 bonds, so trigonal pyramid.

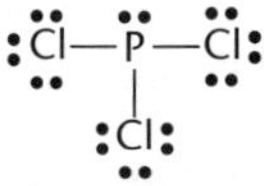

c. Tetrahedral. Cl—C(—Cl)(—Cl)—Cl 4 sets of electrons, all bonding, so tetrahedral.

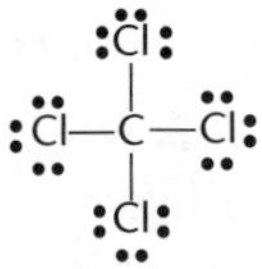

d. V-shaped Cl—O—Cl 4 sets of electrons, 2 bonds, so V-shaped

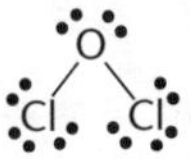

e. V-shaped. H—O—Cl 4 sets of electrons (on O), 2 bonds, so V-shaped. (**M**)

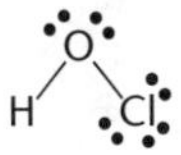

6. **a.** 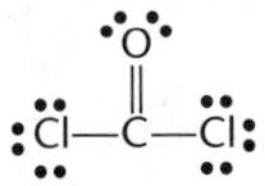 Cl—C(=O)—Cl

b. Trigonal planar. 3 electron groups (regions of negative charge) around the central C atom. (**M**)

7. Lewis structures O—S=O and H—O—H.
For H_2O, there are 4 sets of electrons on the central atom. These will be arranged tetrahedrally with an angle of 109° between them. For SO_2, there are 3 sets of electrons on the central atom, so the angle between them will be 120°.

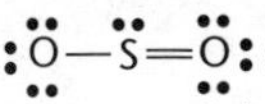

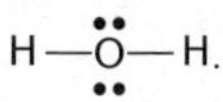

Unit 12.1 Activity 5D: Polarity of molecules (page 73)

1. D.

The four C–Cl bonds in CCl_4 are polar, as Cl is more electronegative than C, and the bonding electrons are closer to the Cl nucleus. However, the symmetry of the molecule around the central C atom means the polarities cancel each other, so the molecule is, overall, non-polar. (***A*** – identifying CCl_4; ***M*** – explanation)

2.

Non-polar	Polar
a, d	b, c, e, f, g

- F_2 – non-polar bonds.
- CF_4 – polarity of bonds cancel, because of symmetry around C.

All others have polar bonds, and shape of the molecule means that the polarity of the bonds do not cancel.

3. **a.** B and D.

b. B. Polar because the bond is formed by atoms of different electronegativity.
D. Non-polar because the bond is formed by identical atoms. (***M***)

4. NCl_3 4 sets of electrons on N – 3 bonds – trigonal pyramid shape. Bonds are polar. Molecule is not symmetrical about the central atom, so polarity of bonds doesn't cancel – polar molecule.

CCl_4 4 sets of electrons – tetrahedral shape. C–Cl bonds are polar, but symmetry around C means polarity cancels – molecule is non-polar. (***E***)

5. **Solid potassium iodide** **Solution of potassium iodide**

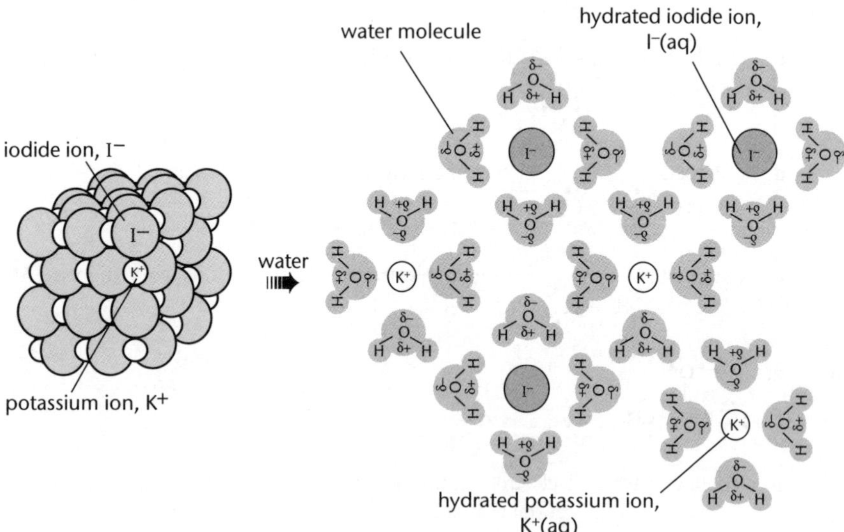

b. $K^+(aq)$ $I^-(aq)$

6. CO_2 is non-polar – C–O bonds polar because O more electronegative than C; symmetrical shape means polarity of bonds cancel, SO_2 is polar. O more electronegative than S, so S–O bonds are polar. Polar S–O bonds are arranged in V-shape around S, so bond polarity does not cancel.

7. H_2O (Lewis structure: O with two lone pairs, bonded to two H) Is a polar molecule, so has 'charge separation' – ie one end of molecule is δ+ and the other δ–. SO_2 is also a polar molecule. SO_2 dissolves in H_2O because the opposite 'poles' of the molecules will be attracted to each other.

8. (Lewis structure of CCl_4: C bonded to four Cl, each with three lone pairs) Non-polar, because polar C–Cl bonds are arranged symmetrically, so polarity of bonds cancel.

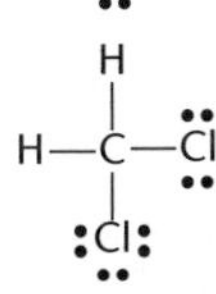

C–Cl bonds are more polar than C–H because Cl is more eletronegative than C and H. This means that there is more electron density around the Cl atoms, so the molecule is polar.

Unit 12.1 Activity 5E: Molecules – multiple choice (page 74)

1.	C	**2.**	A	**3.**	B	**4.**	D
5.	C	**6.**	C	**7.**	B	**8.**	D

Unit 12.2 Activity 1A: Acids and bases (page 83)

1. **a.** Proton donor. **b.** Proton acceptor. **c.** Proton donor and acceptor.

2. **a** and **i**; **j** and **b**; **f** and **c**; **d** and **g**; **e** and **h**.

3. **a.** HSO_4^- **b.** CH_3NH_2

4. **a.** $C_2H_5NH_3^+$ **b.** CH_3COOH **c.** HNO_3

5. **a.** $HNO_3(\ell) + H_2O(\ell) \rightarrow H_3O^+(aq) + NO_3^-(aq)$ (**A**)

proton transfer

b. Hydronium ion, nitrate ion. **c.** Nitric acid solution.

6. **a.** Sulfuric acid donates hydrogen ions to water. **b.** 0.4 mol

7. **a.** $KOH(s) \xrightarrow{H_2O} K^+(aq) + OH^-(aq)$ (**M**)

b. Is completely ionised to produce hydroxide, OH^-, ions in solution. (**M**)

8. **a.** $HNO_3 + H_2O \rightarrow H_3O^+ + NO_3^-$

b. $CH_3CH_2COOH + H_2O \rightleftharpoons CH_3CH_2COO^- + H_3O^+$

c. $NH_4Cl \xrightarrow{H_2O} NH_4^+ + Cl^-$

$NH_4^+ + H_2O \rightleftharpoons NH_3 + H_3O^+$

d. $KOH \xrightarrow{H_2O} K^+ + OH^-$

e. $CH_3NH_2 + H_2O \rightleftharpoons CH_3NH_3^+ + OH^-$ (**M**)

9. **a.** HCl **b.** NH_4^+ **c.** H_3O^+ **d.** H_2CO_3 **e.** HCO_3^-
f. H_2O **g.** CH_3COOH (**A**)

10. **a.** HCO_3^- **b.** Cl^- **c.** NH_3 **d.** OH^- **e.** NO_3^-
f. SO_4^{2-} **g.** H_2O (**A**)

11. **a.** $Zn(OH)_2$ is amphoteric – reacts with acid and base:

$Zn(OH)_2 + HCl \rightarrow ZnCl_2 + H_2O$; $Zn(OH)_2 + 2OH^- \rightarrow Zn(OH)_4^{2-}$ (***M***)

b. NaOH is basic – dissociates in water to produce OH^-:

$NaOH \xrightarrow{H_2O(\ell)} Na^+ + OH^-$ (***M***)

c. $NaHCO_3$ – contains an amphiprotic ion, HCO_3^-, which can both donate and accept a proton:

$HCO_3^- + H_2O \rightarrow CO_3^{2-} + H_3O^+$

$HCO_3^- + H_2O \rightarrow H_2CO_3 + OH^-$ (***M***)

12. **a.** **i.** $H_2O + CH_3COOH \rightarrow H_3O^+ + CH_3COO^-$ acidic

ii. $H_2O + NH_4^+ \rightarrow H_3O^+ + NH_3$ acidic (***A***)

b. **i.** $H_2O + NH_3 \rightarrow NH_3^+ + OH^-$ basic

ii. $H_2O + CH_3COO^- \rightarrow CH_3COOH + OH^-$ basic (***A***)

13. **a.** Acid–base reaction; proton transfer from NH_4^+ to O^{2-}. (***M***)

b. No proton transfer

c. Acid–base reaction; proton transfer from HCl to CO_3^{2-}. (***M***)

14. **a.** **i.** HCOOH – donates a proton.

ii. NH_4^+ – donates a proton.

iii. H_2O – donates a proton.

b. **i.** $HCOOH / HCOO^-$

H_3O^+ / H_2O

iii. H_2CO_3 / HCO_3^-

H_2O / OH^-

15. **a.** Ammonium chloride completely dissociates when it is dissolved in water:

$NH_4Cl \xrightarrow{H_2O} NH_4^+ + Cl^-$

The ammonium ions are weakly acidic and partially dissociate with water:

$NH_4^+ + H_2O \rightleftharpoons NH_3 + H_3O^+$

Since H_3O^+ is formed, the solution is acidic. (***E***)

b. Sodium carbonate completely dissociates when it dissolves in water:

$Na_2CO_3 \xrightarrow{H_2O} 2Na^+ + CO_3^{2-}$

The CO_3^{2-} ions can accept a proton from water:

$CO_3^{2-} + H_2O \rightleftharpoons HCO_3^- + OH^-$

Since OH^- ions are formed, the resulting solution is basic. (***E***)

16. A base.

HCO_3^- accepts a proton from the water to become H_2CO_3. (***M***)

17. Both $NaHCO_3$ and NaCl dissociate into ions when dissolved in water:

$NaHCO_3 \xrightarrow{H_2O} Na^+ + HCO_3^-$

$NaCl \xrightarrow{H_2O} Na^+ + Cl^-$

For NaCl, neither ion has any reaction with water, so the solution formed is neutral.
For $NaHCO_3$, the HCO_3^- ion can accept a proton from water:

$HCO_3^- + H_2O \rightleftharpoons H_2CO_3 + OH^-$

Since OH^- ions are formed, the resulting solution is basic. (***E***)

18. **a.** $H_3PO_4 / H_2PO_4^-$
$H_2PO_4^- / HPO_4^{2-}$
HPO_4^{2-} / PO_4^{3-}
H_3O^+ / H_2O

b. Equation 3.
HPO_4^{2-} donates a proton to water. (***M***)

Unit 12.2 Activity 2A: Acid–base titrations (page 94)

1. **a.** $H^+(aq) + OH^-(aq) \rightarrow H_2O(\ell)$ *or* $HCl(aq) + NaOH(aq) \rightarrow NaCl(aq) + H_2O(\ell)$

b. 14.99 mL (***A***)

c. $n(OH^-) = cV$
$= 0.400 \text{ mol L}^{-1} \times 0.02500 \text{ L}$
$= 0.0100 \text{ mol}$

d. 0.0100 mol (***A***)

e. $c(HCl) = \dfrac{n}{V}$
$= \dfrac{0.0100 \text{ mol}}{0.01499 \text{ L}}$
$= 0.667 \text{ mol L}^{-1}$

2. **a.** To show when the equivalence point is reached.

b. $n = cV$
$= 0.0500 \text{ mol L}^{-1} \times 0.0100 \text{ L}$
$= 5.00 \times 10^{-4} \text{ mol}$

c. $2HCl(aq) + K_2CO_3(aq) \rightarrow 2KCl(aq) + CO_2(g) + H_2O(\ell)$

d. $n(HCl) = 2n(K_2CO_3)$
$= 2 \times 5.00 \times 10^{-4}$ mol (or 0.0005 mol)
$= 1.00 \times 10^{-3}$ mol (***M***)

e. $c = \dfrac{n}{V}$
$= \dfrac{1.00 \times 10^{-3} \text{ mol}}{0.02500 \text{ L}}$
$= 0.0400 \text{ mol L}^{-1}$ (***E***)

3. **a.** 10 mL pipette.

b. Rough titration not done accurately.

c. $\dfrac{23.92 + 24.10 + 24.06}{3} = 24.03 \text{ mL}$
$n(HCl) = cV = 0.125 \text{ mol L}^{-1} \times 0.02403 \text{ L}$
$= 0.00300 \text{ mol}$

d. $Ca(OH)_2 + 2HCl \rightarrow CaCl_2 + 2H_2O$

e. $n(Ca(OH)_2) = \dfrac{n(HCl)}{2} = \dfrac{0.00300}{2} = 0.00150 \text{ mol}$

f. $c(Ca(OH)_2) = \dfrac{n}{V} = \dfrac{0.00150 \text{ mol}}{0.0100 \text{ L}} = 0.150 \text{ mol L}^{-1}$ (***E***)

4. **a.** $n(CH_3COOH) = cV = 0.100 \text{ mol L}^{-1} \times 0.0200 \text{ L}$
$= 0.00200 \text{ mol}$

b. $n(OH^-) = n(CH_3COOH) = 0.002\,00$ mol

c. $c(NaOH) = \frac{n}{V} = \frac{0.002\,00 \text{ mol}}{0.016\,00 \text{ L}} = 0.125 \text{ mol L}^{-1}$

5. $CH_3COOH + NaOH \rightarrow CH_3COONa + H_2O$

$n(CH_3COOH) = cV = 0.252 \text{ mol L}^{-1} \times 0.015\,00 \text{ L}$
$= 0.003\,78$ mol

$n(CH_3COOH) = n(NaOH) = 0.003\,78$ mol

$c(NaOH) = \frac{n}{V} = \frac{0.003\,78 \text{ mol}}{0.024\,96 \text{ L}} = 0.151 \text{ mol L}^{-1}$ (***M***)

6. $H_2SO_4 + 2NaOH \rightarrow Na_2SO_4 + 2H_2O$

$n(NaOH) = cV = 0.0820 \text{ mol L}^{-1} \times 0.022\,50 \text{ L}$
$= 1.845 \times 10^{-3}$ mol

$n(H_2SO_4) = \frac{n(NaOH)}{2} = \frac{1.845 \times 10^{-3} \text{ mol}}{2} = 9.225 \times 10^{-4}$ mol

$c(H_2SO_4) = \frac{n}{V} = \frac{9.225 \times 10^{-4} \text{ mol}}{0.025\,00 \text{ L}} = 0.0369 \text{ mol L}^{-1}$ (***E***)

7. **a.** $m(Na_2CO_3) = 131.10 - 128.45$ g
$= 2.65$ g

b. **i.** $c = \frac{m}{V} = \frac{2.65 \text{ g}}{0.100 \text{ L}} = 26.5 \text{ g L}^{-1}$

ii. $n(Na_2CO_3) = \frac{m}{M} = \frac{2.65 \text{ g}}{106.0 \text{ g mol}^{-1}} = 0.0250$ mol

$c(Na_2CO_3) = \frac{n}{V} = \frac{0.025 \text{ mol}}{0.100 \text{ L}} = 0.250 \text{ mol L}^{-1}$

c. 20 mL pipette.

d. Burette.

e. Indicator changes colour from yellow to red.

f. $2H^+ + CO_3^{2-} \rightarrow CO_2 + H_2O$ *or*
$2HCl(aq) + Na_2CO_3(aq) \rightarrow CO_2(g) + H_2O(\ell) + 2NaCl(aq)$

g. $n(Na_2CO_3) = cV = 0.250 \text{ mol L}^{-1} \times 0.0200 \text{ L}$
$= 0.005\,00$ mol

$n(HCl) = 2n(Na_2CO_3) = 0.005\,00 \times 2 = 0.0100$ mol

$c(HCl) = \frac{n}{V} = \frac{0.0100 \text{ mol}}{0.005\,16 \text{ L}} = 1.94 \text{ mol L}^{-1}$

8. $Na_2CO_3(aq) + 2HCl(aq) \rightarrow 2NaCl(aq) + CO_2(g) + H_2O(\ell)$

$n(Na_2CO_3) = cV = 0.208 \text{ mol L}^{-1} \times 0.0200 \text{ L}$
$= 0.004\,16$ mol

$n(HCl) = 2n(Na_2CO_3) = 0.008\,32$ mol

$c(HCl) = \frac{n}{V} = \frac{0.008\,32 \text{ mol}}{0.031\,22 \text{ L}} = 0.266 \text{ mol L}^{-1}$ (***E***)

9. **a.** $n(Na_2CO_3) = \frac{m}{M} = \frac{3.02\ g}{106.0\ g\ mol^{-1}} = 0.0285\ mol$

b. $c(Na_2CO_3) = \frac{n}{V} = \frac{0.0285\ mol}{0.500\ L} = 0.0570\ mol\ L^{-1}$

c. $Na_2CO_3 + 2HCl \rightarrow 2NaCl + H_2O + CO_2$

d. $V(HCl) = \frac{15.02 + 15.12 + 15.00}{3} = 15.05\ mL$

$n(HCl) = 2n(Na_2CO_3) = 0.0570\ mol\ L^{-1} \times 0.0100\ L \times 2$
$= 1.140 \times 10^{-3}\ mol$

$c(HCl) = \frac{n}{V} = \frac{1.140 \times 10^{-3}\ mol}{0.01505\ L} = 0.0757\ mol\ L^{-1}$

e. $Ca(OH)_2(aq) + 2HCl(aq) \rightarrow CaCl_2(aq) + H_2O(\ell)$

f. Average titre HCI = 13.78 mL = 0.013 78 L
$n(HCl) = 0.0757\ mol\ L^{-1} \times 0.01378\ L = 1.043 \times 10^{-3}\ mol$

$\frac{n(HCl)}{2} = \frac{n(Ca(OH)_2)}{1}$

$n(Ca(OH)_2) = \frac{1}{2}n(HCl) = \frac{1.043 \times 10^{-3}}{2} = 5.216 \times 10^{-4}\ mol$

$c(Ca(OH)_2) = \frac{5.216 \times 10^{-4}\ mol}{0.0100\ L} = 0.0522\ mol\ L^{-1}$

Unit 12.2 Activity 3A: Strong and weak acids (page 101)

1. **a.** pH = –log(0.015) = 1.82
b. pH = –log(0.500) = 0.301
c. pH = –log(0.00213) = 2.67 (***A***)

2. **a.** A strong acid fully dissociates in water, a weak acid only partially dissociates.
b. A concentrated acid has a high ratio of acid to water particles, a dilute acid has a low ratio of acid to water particles. (***A***)

3. **a.** $K_a = \frac{[H_3O^+][CH_3COO^-]}{[CH_3COOH]}$ **b.** $K_a = \frac{[H_3O^+][NH_3]}{[NH_4^+]}$ **c.** $K_a = \frac{[H_3O^+][F^-]}{[HF]}$ (***A***)

4. Low percentage dissociation of weak acid species, so change in concentration of acid species is negligible. (***M***)

5. **a.** $K_a = \frac{[H_3O^+][CH_3COO^-]}{[CH_3COOH]} = 1.74 \times 10^{-5}$

$[CH_3COOH] = 0.125\ mol\ L^{-1}$

$[H_3O^+]^2 = 1.74 \times 10^{-5} \times 0.125$

$[H_3O^+] = 1.47 \times 10^{-3}$

pH = 2.83

b. $[H_3O+]^2 = 1.74 \times 10^{-5} \times 0.500$
$[H_3O^+] = 2.94 \times 10^{-3}$
pH = 2.53

c. $K_a = \frac{[H_3O^+][C_2H_5COO^-]}{[C_2H_5COOH]} = 1.35 \times 10^{-5}$

$[C_2H_5COOH] = 0.002\,50 \text{ mol L}^{-1}$

$[H_3O^+]^2 = 1.35 \times 10^{-5} \times 0.002\,50$

$[H_3O^+] = 1.84 \times 10^{-4}$

pH = 3.74

d. $K_a = \frac{[H_3O^+][NH_3]}{[NH_4^+]} = 5.75 \times 10^{-10}$

$[NH_4^+] = 0.0100 \text{ mol L}^{-1}$

$[H_3O^+]^2 = 5.75 \times 10^{-10} \times 0.0100$

$[H_3O^+] = 2.40 \times 10^{-6}$

pH = 5.62 (***M***)

6. pH = 3.20

$[H_3O^+] = 10^{-3.2} = 6.31 \times 10^{-4}$

$K_a = \frac{[H_3O^+][Vit^-]}{[HVit]} = 7.94 \times 10^{-5}$

[HVit] = concentration of vitamin C

$[HVit] = \frac{[H_3O^+]^2}{K_a} = \frac{(6.31 \times 10^{-4})^2}{7.94 \times 10^{-5}} = 5.01 \times 10^{-3} \text{ mol L}^{-1}$ (***M***)

Unit 12.2 Activity 3B: K_a and pK_a (page 103)

1. In order of decreasing K_a values: **c. a. f. e. d. b.** (***A***)

2. **a.** **i.** $pK_a = -\log(1.82 \times 10^{-4}) = 3.74$

ii. $pK_a = -\log(6.76 \times 10^{-4}) = 3.17$

iii. $pK_a = -\log(1.10 \times 10^{-5}) = 4.96$

iv. $pK_a = -\log(4.57 \times 10^{-8}) = 7.34$

v. $pK_a = -\log(2.24 \times 10^{-12}) = 11.65$

vi. $K_a = 10^{-9.24} = 5.75 \times 10^{-10}$

vii. $K_a = 10^{-(-1.30)} = 20.0$

viii. $K_a = 10^{-3.15} = 7.07 \times 10^{-4}$

ix. $K_a = 10^{-2.17} = 6.76 \times 10^{-3}$

x. $K_a = 10^{-9.21} = 6.17 \times 10^{-10}$ (***A***)

b. HNO_3, $[Fe(H_2O)_6]^{3+}$, HNO_2, HF, HCOOH, $[Al(H_2O)_6]^{3+}$, $[Cu(H_2O)_4{}^{2+}]$, HCN, NH_4^+, H_2O_2 (***A***)

3. **a.** $K_a = 10^{-3.86} = 1.38 \times 10^{-4}$

$HLac + H_2O \rightleftharpoons Lac^- + H_3O^+$

$K_a = \frac{[Lac^-][H_3O^+]}{[HLac]}$

$[HLac] = 0.100 \text{ mol L}^{-1}$

$[H_3O^+]^2 = 1.38 \times 10^{-4} \times 0.100$

$[H_3O^+] = 3.71 \times 10^{-3}$

pH = 2.43 (**M**)

b. $K_a = 10^{-9.21} = 6.17 \times 10^{-10}$

$[H_3O^+] = \sqrt{K_a \times [HCN]}$

$= \sqrt{6.17 \times 10^{-10} \times 0.0200}$

$= 3.5 \times 10^{-6}$

pH = 5.45 (**M**)

Unit 12.2 Activity 3C: pH of bases (page 107)

1. **a.** $[H_3O^+] = \dfrac{1 \times 10^{-14}}{0.1} = 1 \times 10^{-13}$ and pH = 13

b. $[H_3O^+] = \dfrac{1 \times 10^{-14}}{0.5} = 2 \times 10^{-14}$ and pH = 13.7 (**A**)

2. **a.** $[OH^-] = \dfrac{1 \times 10^{-14}}{0.1} = 1 \times 10^{-13}$

b. $[OH^-] = \dfrac{1 \times 10^{-14}}{0.2} = 5 \times 10^{-14}$ (**A**)

3. **a.** $NH_3 + H_2O \rightleftharpoons NH_4^+ + OH^-$

$[H_3O^+] = \sqrt{\dfrac{1 \times 10^{-14} \times 5.75 \times 10^{-10}}{0.0100}} = 2.40 \times 10^{-11}$

pH = 10.6 (**E**)

b. $CH_3NH_2 + H_2O \rightleftharpoons CH_3NH_3^+ + OH^-$

$[H_3O^+] = \sqrt{\dfrac{1 \times 10^{-14} \times 2.78 \times 10^{-11}}{0.200}} = 1.18 \times 10^{-12}$

pH = 11.9 (**E**)

c. $pK_a(CH_3COOH) = 10^{-4.76} = 1.74 \times 10^{-5}$

$CH_3COO^- + H_2O \rightleftharpoons CH_3COOH + OH^-$

$[H_3O^+] = \sqrt{\dfrac{1 \times 10^{-14} \times 1.74 \times 10^{-5}}{0.100}} = 1.32 \times 10^{-9}$

pH = 8.88 (**E**)

d. $pK_a(HF) = 10^{-3.17} = 6.76 \times 10^{-4}$

$F^- + H_2O \rightleftharpoons HF + OH^-$

$[H_3O^+] = \sqrt{\dfrac{1 \times 10^{-14} \times 6.76 \times 10^{-4}}{0.05}} = 1.16 \times 10^{-8}$

pH = 7.93 (**E**)

e. $K_a(HCOOH) = 10^{-3.74} = 1.82 \times 10^{-4}$

$HCOO^- + H_2O \rightleftharpoons HCOOH + OH^-$

$$[H_3O^+] = \sqrt{\frac{1 \times 10^{-14} \times 1.82 \times 10^{-4}}{0.00250}} = 2.70 \times 10^{-8}$$

pH = 7.57 (**E**)

Unit 12.2 Activity 4A: Reactions of acids and bases (page 111)

1. **a.** No change.

b. Blue litmus paper goes red.

(***A*** – both correct)

2. **a.** Medium.

b. High.

c. Low.

(***A*** – two correct; ***M*** – three correct)

3. Test tubes; magnesium ribbon; burning splint; dilute acid.

(***A*** – three correct; ***M*** – four correct)

4. Reactions **a.**, **c.**, **e.**

All alkalis (eg sodium hydroxide) are bases, but for a base to be an alkali, it must be soluble in water. (***A*** – two correct)

5. **a.** Copper oxide *or* copper hydroxide *or* copper carbonate plus nitric acid.

b. Zinc oxide *or* zinc hydroxide *or* zinc carbonate plus sulfuric acid.

c. Magnesium oxide *or* magnesium hydroxide *or* magnesium carbonate *or* magnesium hydrogen carbonate plus hydrochloric acid.

Magnesium hydrogen carbonate is accepted but not copper hydrogen carbonate or zinc hydrogen carbonate (they do not exist).

(***A*** – two correct; ***M*** – three correct)

6. Test tubes; sodium bicarbonate; limewater; dilute acid.

(***A*** – three correct; ***M*** – four correct)

Unit 12.2 Activity 4B: Reactions of alkalis (page 113)

1. **a.** The solution would change colour from red to blue.

b. No change.

(***A*** – both correct)

2. **a.** 11–12 (***A***)

b. 9–10 (***A***)

3. Ammonium chloride; sodium hydroxide solution; test tube; Bunsen burner; litmus paper (for detection). (**A** – three correct; **M** – four correct)

4. Take a known volume of the acid. Add a coloured indicator. Carefully add sodium hydroxide solution until the indicator just changes colour. Repeat the process without the indicator but using the same volume of solutions. Evaporate the solution produced until only solid sodium chloride remains.

$NaOH(aq) + HCl(aq) \rightarrow NaCl(aq) + H_2O(\ell)$

(**A** – mixing acid and sodium hydroxide solutions and adding an indicator; **M** – explaining that care and accuracy are needed in the mixing; **E** – repeating process without indicator, and the evaporation of water)

Unit 12.2 Activity 4C: Word and formula equations (page 116)

1. **a.** Magnesium + hydrochloric acid → magnesium chloride + *hydrogen*.
b. Iron(III) oxide + sulfuric acid → *iron(III) sulfate* + water.
c. Calcium hydroxide + *nitric acid* → calcium nitrate + water.
d. Zinc carbonate + nitric acid → *zinc nitrate* + water + carbon dioxide.
e. Calcium bicarbonate + hydrochloric acid → *calcium chloride* + water + *carbon dioxide*.
f. Sodium hydroxide + ethanoic acid → *sodium ethanoate* + water.
(**A** – two equations correct; **M** – four equations correct; **E** – six equations correct)

2. **a.** **i.** Magnesium + sulfuric acid → *magnesium sulfate* + *hydrogen*.
ii. Copper hydroxide + hydrochloric acid → *copper chloride* + *water*.
iii. Magnesium hydrogen carbonate + sulfuric acid → *magnesium sulfate* + *water* + *carbon dioxide*.
b. **i.** *Zinc oxide* (or *zinc hydroxide*) + *nitric acid* → zinc nitrate + water.
ii. *Sodium carbonate* (or *sodium hydrogen carbonate*) + *ethanoic acid* → sodium ethanoate + water + carbon dioxide.
iii. *Calcium carbonate* (or *calcium hydrogen carbonate*) + *nitric acid* → calcium nitrate + water + carbon dioxide.
(**A** – two equations correct; **M** – four equations correct; **E** – six equations correct)

3. **a.** $Mg(s) + H_2SO_4(aq) \rightarrow MgSO_4(aq) + H_2(g)$.
b. $Cu(OH)_2(s) + 2HCl(aq) \rightarrow CuCl_2(aq) + 2H_2O(\ell)$.
c. $Mg(HCO_3)_2(aq) + H_2SO_4(aq) \rightarrow MgSO_4(aq) + 2H_2O(\ell) + 2CO_2(g)$.
d. $Zn(OH)_2(s) + 2HNO_3(aq) \rightarrow Zn(NO_3)_2(aq) + 2H_2O(\ell)$.
e. $NaHCO_3(s) + CH_3COOH(aq) \rightarrow CH_3COONa(aq) + H_2O(\ell) + CO_2(g)$.
f. $Ca(HCO_3)_2(aq) + 2HNO_3(aq) \rightarrow Ca(NO_3)_2(aq) + 2H_2O(\ell) + 2CO_2(g)$.
(**A** – two equations from parts **a.**, **b.**, and **c.** correct; **M** – all equations from parts **a.**, **b.**, and **c.** correct plus one equation from part **d.**, **e.** or **f.** correct; **E** – all six equations correct)

4. **a.** $Mg(s) + 2HNO_3(aq) \rightarrow Mg(NO_3)_2(aq) + H_2(g)$.
b. $Na_2CO_3(s) + H_2SO_4(aq) \rightarrow Na_2SO_4(aq) + H_2O(\ell) + CO_2(g)$.
c. $2Fe(OH)_3(s) + 3H_2SO_4(aq) \rightarrow Fe_2(SO_4)_3(aq) + 6H_2O(\ell)$
d. $ZnO(s) + 2HCl(aq) \rightarrow ZnCl_2(aq) + H_2O(\ell)$.
e. $NaOH(aq) + CH_3COOH(aq) \rightarrow CH_3COONa(aq) + H_2O(\ell)$.
f. $Ca(HCO_3)_2(aq) + 2HCl(aq) \rightarrow CaCl_2(aq) + 2H_2O(\ell) + 2CO_2(g)$.
(**A** – one equation correct; **M** – three equations correct; **E** – six equations correct)

Unit 12.2 Activity 4D: Preparation of salts, water of crystallisation (page 118)

1. Place a volume (eg 20 mL) of dilute sulfuric acid in a beaker and heat the beaker on a tripod and gauze until the acid is just boiling. Add samples of zinc oxide, with stirring and continued heating, until no more zinc oxide will dissolve. Filter off the excess zinc oxide and leave the hot solution to cool. If no crystals appear on heating, then evaporate the solution to half bulk. Allow the solution to cool. Filter off the crystals, wash with distilled water and dry with absorbent (filter) paper. The crystals will be a sample of zinc sulfate heptahydrate.

(***A*** – indication that chemicals must be mixed and the reaction heated; **M** – ***A*** plus indication that excess zinc oxide required and that excess must be filtered off; **E** – **M** plus clear indication for obtaining crystals and their isolation in a pure, dry state)

2. **a.** Neutralisation. (***A***)
b. Crystals. (***A***)
c. Heating. (***A***)
d. Anhydrous. (***A***)
e. Hydrated. (***A***)

3. If a *crystal* contains *water of crystallisation*, it is described as a *hydrate*. *Anhydrous* means that the *crystal* does not contain *water of crystallisation*.
(***A*** – one sentence correct; **M** – both sentences correct)

4. **a.** $CuSO_4.5H_2O$ *or* $CaSO_4.2H_2O$.
b. NaCl *or* $CuSO_4$.
c. NaCl *or* $CuSO_4$.
d. $CuSO_4$.
(***A*** – two correct; **M** – four correct)

Unit 12.3 Activity 1A: Oxidation numbers (page 123)

1. **a.** +4 (***A***) **b.** 0 (***A***) **c.** +2 (***A***) **d.** +4 (***A***) **e.** +4 (***A***)

2. **a.** –1 (***A***) **b.** 0 (***A***) **c.** +1 (***A***) **d.** +7 (***A***)

3. **a.** –3 **b.** –3 **c.** 0 **d.** +1 **e.** +2
f. +3 **g.** +4 **h.** +5

4. **a.** Mg: 0 **b.** Cl: 0 (***A***) **c.** H: +1; O: –2 (***A***)
d. Ca: +2; H: –1 (***A***) **e.** Ag: +1 (***A***) **f.** Fe: +3 (***A***)
g. S: +4; O: –2 (***A***) **h.** S: +6; O: –2 (***A***) **i.** C: –4; H: +1 (***A***)
j. Mg: +2; O: –2 (***A***) **k.** C: +4; O: –2 (***A***) **l.** S: 0 (***A***)
m. C: +2; O: –2 (***A***) **n.** Mn: +7; O: –2 (***A***) **o.** C: +4; O: –2 (***A***)
p. H: +1; O: –1 (***A***) **q.** S: +2; O: –2 (***A***) **r.** Cr: +6; O: –2 (***A***)
s. Na: +1; H: +1; C: +4; O = –2 (***A***) **t.** Cu: +2; C: +4; O: –2 (***A***)
u. Na: +1; P: +5; O: –2 (***A***) **v.** H: +1; Si: +4; O: –2 (***A***)
w. H: +1; P: +5; O: –2 (***A***) **x.** Cu: +2; S: +6; O: –2 (***A***)
y. H: +1; S: +6; O: –2 (***A***)

5. **a.** H oxidised $0 \rightarrow +1$; Cl reduced $0 \rightarrow -1$.
b. Zn oxidised $0 \rightarrow +2$; Cu reduced $+2 \rightarrow 0$.
c. C oxidised $+2 \rightarrow +4$; H reduced $+1 \rightarrow 0$.
d. No change.

e. S oxidised $-2 \rightarrow +4$; O reduced $0 \rightarrow -2$.
f. No change.
g. Fe oxidised $+2 \rightarrow +3$; Mn reduced $+7 \rightarrow +2$.
h. Fe oxidised $0 \rightarrow +2$; O reduced $0 \rightarrow -2$.
(***A*** – species identified; ***M*** – correct oxidation numbers *or* 'No change')

6. a. Not oxidised or reduced. No change in oxidation number.
b. Oxidised; Na to Na^+. Oxidation number increases from 0 to +1.
c. Oxidised; Co^{2+} to Co^{3+}. Oxidation number increases from +2 to +3.
d. Reduced; O_2 to H_2O. Oxidation number goes down from 0 to –2.
e. Not oxidised or reduced. No change in oxidation number.
f. Reduced; Pb^{2+} to Pb. Oxidation number is reduced from +2 to 0.
g. No change. (***M***)

Unit 12.3 Activity 1B: Oxidants and reductants (page 125)

1. a. Pb^{2+} is the oxidant (***A***); because the oxidation number has decreased from +2 to 0 *or* because electrons have been gained by the Pb^{2+} *or* PbO has lost oxygen. (***M***)
b. C is the reductant (***A***); because the oxidation number has increased from 0 to +4 *or* because electrons have been lost by the carbon *or* C has gained oxygen. (***M***)

2. a. Cu^{2+} is the oxidant (***A***); because the oxidation number has decreased from +2 to 0 *or* because electrons have been gained by the Cu^{2+}. (***M***)
b. Fe is the reductant (***A***); because the oxidation number has increased from 0 to +2 *or* because electrons have been lost by the iron. (***M***)

3. a. $Zn(s) + 2H^+(aq) \rightarrow Zn^{2+}(aq) + H_2(g)$ (***M***)
b. Zinc metal; is oxidised to zinc ions. (***A***)
c. Hydrogen ions. (***A***)
d. Hydrogen ions; reduced to hydrogen. (***A***)
e. Zinc metal. (***A***)
f. From zinc metal to hydrogen ions. (***M***)
g. Chloride ions. (***A***)

4. a. Magnesium metal. (***A***)
b. Chlorine gas. (***A***)
c. Chlorine gas is the oxidant (***A***): it gets reduced as it oxidises the magnesium. (***M***)
d. From magnesium to chlorine. (***M***)

5. a. Cu – reductant, oxidised to Cu^{2+} (ON of Cu goes up from 0 to +2).
NO_3^- – oxidant, reduced to NO_2 (ON of N goes down from +5 to +4). (***M***)
b. Fe^{3+} – oxidant, reduced to Fe^{2+} (ON of Fe goes down from +3 to +2).
SO_2 – reductant, oxidised to SO_4^{2-} (ON of S goes up from +4 to +6). (***M***)
c. I^- – reductant, oxidised to I_2 (ON of I goes up from –1 to 0).
H_2O_2 – oxidant, reduced to H_2O (ON of O goes down from –1 to –2). (***M***)
d. $Cr_2O_7^{2-}$ – oxidant, reduced to Cr^{3+} (ON of Cr goes down from +7 to +3).
SO_3^{2-} – reductant, oxidised to SO_4^{2-} (ON of S goes up from +4 to +6). (***M***)
e. Mg – reductant, oxidised to Mg^{2+} (ON of Mg goes up from 0 to +2).
H^+ – oxidant, reduced to H_2 (ON of H goes down from +1 to 0). (***M***)
f. C – reductant, oxidised to CO_2 (ON of C goes up from 0 to +4).
O – oxidant, reduced (ON goes from 0 to –2. (***M***)

g. O_2 – oxidant, reduced to O^{2-} (ON of O goes down from 0 to –2).
S – reductant, oxidised to SO_2 (ON of S goes up from 0 to +4). (***M***)

h. C – reductant, oxidised to CO (ON of C goes up from 0 to +2).
Cu^{2+} oxidant, reduced to Cu (ON of Cu goes up from +2 to 0). (***M***)

Reason could also describe gain or loss of electrons.

Unit 12.3 Activity 1C: Redox equations (page 130)

1.
a. $Na^+ + e^- \rightarrow Na$ (***A***) Reduction. (***A***)
b. $2H^+ + 2e^- \rightarrow H_2$ (***A***) Reduction. (***A***)
c. $Fe^{2+} \rightarrow Fe^{3+} + e^-$ (***A***) Oxidation. (***A***)
d. $Cu \rightarrow Cu^{2+} + 2e^-$ (***A***) Oxidation. (***A***)
e. $SO_2 + 2H_2O \rightarrow SO_4^{2-} + 4H^+ + 2e^-$ Oxidation. (***A***)
f. $Cr_2O_7^{2-} + 14H^+ + 6e^- \rightarrow 2Cr^{3+} + 7H_2O$ Reduction. (***A***)
g. $MnO_4^- + 8H^+ + 5e^- \rightarrow Mn^{2+} + 4H_2O$ Reduction.
h. $H_2O_2 + 2H^+ + 2e^- \rightarrow 2H_2O$ Reduction.
i. $2Br^- \rightarrow Br_2 + 2e^-$ Oxidation.
j. $I_2 + 2e^- \rightarrow 2I^-$ Reduction.

2.
a. $Fe^{2+} \rightarrow Fe^{3+} + e^-$ (×5); $MnO_4^- + 8H^+ + 5e^- \rightarrow Mn^{2+} + 4H_2O$ (***A***)
Overall: $MnO_4^- + 5Fe^{2+} + 8H^+ \rightarrow Mn^{2+} + 5Fe^{3+} + 4H_2O$ (***M***)
b. $Mg \rightarrow Mg^{2+} + 2e^-$; $Ag^+ + e^- \rightarrow Ag$ (×2) (***A***)
Overall: $Mg + 2Ag^+ \rightarrow Mg^{2+} + 2Ag$ (***M***)
c. $Fe^{2+} \rightarrow Fe^{3+} + e^-$ (×6); $Cr_2O_7^{2-} + 14H^+ + 6e^- \rightarrow 2Cr^{3+} + 7H_2O$ (***A***)
Overall: $Cr_2O_7^{2-} + 6Fe^{2+} + 14H^+ \rightarrow 2Cr^{3+} + 6Fe^{3+} + 7H_2O$ (***M***)
d. $SO_2 + 2H_2O \rightarrow SO_4^{2-} + 4H^+ + 2e^-$ (×5); $MnO_4^- + 8H^+ + 5e^- \rightarrow Mn^{2+} + 4H_2O$ (×2)
Overall: $5SO_2 + 2H_2O + 2MnO_4^- \rightarrow 5SO_4^{2-} + 4H^+ + 2Mn^{2+}$
e. $2Cl^- \rightarrow Cl_2 + 2e^-$; $H_2O_2 + 2H^+ + 2e^- \rightarrow 2H_2O$
Overall: $2Cl^- + H_2O_2 + 2H^+ \rightarrow Cl_2 + 2H_2O$
f. $MnO_4^- + 5e^- + 8H^+ \rightarrow Mn^{2+} + 4H_2O$ (×2); $2I^- \rightarrow I_2 + 2e^-$ (×5)
Overall: $2MnO_4^- + 16H^+ + 10I^- \rightarrow 2Mn^{2+} + 8H_2O + 5I_2$
g. $Cr_2O_7^{2-} + 14H^+ + 6e^- \rightarrow 2Cr^{3+} + 7H_2O$; $H_2S \rightarrow S + 2H^+ + 2e^-$ (×3)
Overall: $Cr_2O_7^{2-} + 8H^+ + 3H_2S \rightarrow 2Cr^{3+} + 7H_2O + 3S$
h. $Zn \rightarrow Zn^{2+} + 2e^-$; $2H^+ + VO_2^+ + e^- \rightarrow VO^{2+} + H_2O$ (×2)
Overall: $Zn + 4H^+ + 2VO_2^+ \rightarrow Zn^{2+} + 2VO^{2+} + 2H_2O$
i. $2I^- \rightarrow I_2 + 2e^-$ (×5); $10e^- + IO_3^- + 12H^+ \rightarrow I_2 + 6H_2O$
Overall: $10I^- + 2IO_3^- + 12H^+ \rightarrow 6I_2 + 6H_2O$
j. $Cu \rightarrow Cu^{2+} + 2e^-$; $NO_3^- + 2H^+ + e^- \rightarrow NO_2 + H_2O$ (×2)
Overall: $Cu + 2NO_3^- + 4H^+ \rightarrow Cu^{2+} + 2NO_2 + 2H_2O$
k. $H_2S \rightarrow S + 2H^+ + 2e^-$ (×2); $4e^- + SO_2 + 4H^+ \rightarrow S + 2H_2O$
Overall: $2H_2S + SO_2 \rightarrow 3S + 2H_2O$

Unit 12.3 Activity 1D: Common oxidants and reductants (page 131)

1.
a. Oxidant: permanganate ion (***A***); reductant: iodide ion. (***A***)
b. K^+ (and SO_4^{2-} ions from the sulfuric acid). (***A***)

c. Oxidation: $2I^- \rightarrow I_2 + 2e^-$
Reduction: $MnO_4^- + 8H^+ + 5e^- \rightarrow Mn^{2+} + 4H_2O$
(***A*** – one only correct; ***M*** – both correct)
d. $2MnO_4^- + 10I^- + 16H^+ \rightarrow 2Mn^{2+} + 5I_2 + 8H_2O$ (***E***)

2. **a.** $SO_2 + 2H_2O \rightarrow SO_4^{2-} + 4H^+ + 2e^-$ (***M***) **b.** $O_2 + 4e^- \rightarrow 2O^{2-}$ (***M***)
c. $Fe^{3+} + e^- \rightarrow Fe^{2+}$ (***M***)

3. **a.** $2Br^- \rightarrow Br_2 + 2e^-$
b. $H_2 \rightarrow 2H^+ + 2e^-$
c. $I_2 + 2e^- \rightarrow 2I^-$
d. $H_2O_2 + 2H^+ + 2e^- \rightarrow 2H_2O$

4. **a.** $Cu^{2+} + Mg \rightarrow Mg^{2+} + Cu$
b. $2Na + Cl_2 \rightarrow 2Na^+ + 2Cl^-$
c. $2I^- + 2Fe^{3+} \rightarrow 2Fe^{2+} + I_2$
d. $SO_2 + H_2O_2 \rightarrow SO_4^{2-} + 2H^+$

5. **a.** Orange colour disappears:

$$2e^- + Br_2 \rightarrow 2Br^-$$
$$SO_2 + 2H_2O \rightarrow SO_4^{2-} + 4H^+ + 2e^-$$

Overall: $Br_2 + SO_2 + 2H_2O \rightarrow 2Br^- + SO_4^{2-} + 4H^+$

b. Purple colour disappears, red-brown solution forms:
$MnO_4^- + 8H^+ + 5e^- \rightarrow Mn^{2+} + 4H_2O$; $Fe^{2+} \rightarrow Fe^{3+} + e^-$
Overall: $MnO_4^- + 8H^+ + 5Fe^{2+} \rightarrow Mn^{2+} + 4H_2O + 5Fe^{3+}$

6. **a.** **i.**

$$(SO_2 + 2H_2O \rightarrow SO_4^{2-} + 4H^+ + 2e^-) \times 3$$
$$Cr_2O_7^{2-} + 14H^+ + 6e^- \rightarrow 2Cr^{3+} + 7H_2O$$
$$3SO_2 + Cr_2O_7^{2-} + 2H^+ \rightarrow 3SO_4^{2-} + 2Cr^{3+} + H_2O$$

ii. Orange colour of dichromate will disappear and solution turns green because of the presence of Cr^{3+}.

b. **i.**

$$(Fe^{2+} \rightarrow Fe^{3+} + e^-) \times 5$$
$$(MnO_4^- + 8H^+ + 5e^- \rightarrow Mn^{2+} + 4H_2O) \times 2$$
$$5Fe^{2+} + 2MnO_4^- + 16H^+ \rightarrow Fe^{3+} + 2Mn^{2+} + 8H_2O$$

ii. Purple colour of MnO_4^- is replaced by yellow-brown colour of Fe^{3+}.

c. **i.** $Cu^{2+} + 2e^- \rightarrow Cu$

$$Mg \rightarrow Mg^{2+} + 2e^-$$
$$Cu^{2+} + Mg \rightarrow Cu + Mg^{2+}$$

ii. Magnesium metal disappears and red-brown solid is formed – copper metal.

d. **i**

$$2Br^- \rightarrow Br_2 + 2e^-$$
$$H_2O_2 + 2H^+ + 2e^- \rightarrow 2H_2O$$
$$2Br^- + H_2O_2 + 2H^+ \rightarrow Br_2 + 2H_2O$$

ii. Colourless solution is replaced by orange colour of bromine in solution.

7. **a.** MnO_4^- is reduced to colourless Mn^{2+}(aqueous). The black solid is iodine, I_2, formed by the oxidation of the iodide, I^-.

$$(2I^- \rightarrow I_2 + 2e^-) \times 5$$
$$(MnO_4^- + 8H^+ + 5e^- \rightarrow Mn^{2+} + 4H_2O) \times 2$$
$$10I^- + 2MnO_4^- + 16H^+ \rightarrow 5I_2 + 2Mn^{2+} + 8H_2O$$

b. Silver needles are silver metal forming as silver ions are reduced. The blue colour is due to the copper metal being oxidised to Cu^{2+} ions.

$$Cu \rightarrow Cu^{2+} + 2e^-$$
$$2Ag^+ + 2e^- \rightarrow 2Ag$$
$$Cu + 2Ag^+ \rightarrow Cu^{2+} + 2Ag$$

c. Red-brown colour of solution due to the formation of Fe^{3+} ions formed when Fe^{2+} is oxidised by peroxide.

$$(Fe^{2+} \rightarrow Fe^{3+} + e^-) \times 2$$
$$H_2O_2 + 2H^+ + 2e^- \rightarrow 2H_2O$$
$$2Fe^{2+} + H_2O_2 + 2H^+ \rightarrow 2Fe^{3+} + 2H_2O$$

d. Yellow-brown colour of Fe^{3+} disappears when Fe^{3+} ions are reduced to Fe^{2+} ions, which are colourless in aqueous solution.

$$(Fe^{3+} + e^- \rightarrow Fe^{2+}) \times 2$$
$$SO_2 + 2H_2O \rightarrow SO_4^{2-} + 4H^+ + 2e^-$$
$$2Fe^{3+} + SO_2 + 2H_2O \rightarrow 2Fe^{2+} + SO_4^{2-} + 4H^+$$

Unit 12.3 Activity 1E: Halogens as oxidants (page 134)

1. **a.** $2I^-(aq) + Cl_2 \rightarrow I_2 + 2Cl^-(aq)$ (**A**) **b.** No reaction. (**A**)
c. $2Br^-(aq) + Cl_2 \rightarrow Br_2 + 2Cl^-(aq)$ (**A**) **d.** No reaction. (**A**)

2. **a.** $2Fe + 3Cl_2 \rightarrow 2FeCl_3$; $2Fe + 3Br_2 \rightarrow 2FeBr_3$ (**A**)
b. $2Fe + 3Cl_2 \rightarrow 2FeCl_3$ (**A**)

3. It is too powerful an oxidant and is too reactive to be used in normal laboratory conditions.

4. **a.** **i.** No change – no reaction. (**A**) **ii.** Orange – bromine liberated. (**A**)
iii. Yellow-brown – iodine liberated. (**A**)
b. **ii.** $2Br^- + Cl_2 \rightarrow 2Cl^- + Br_2$ (**M**) **iii.** $2I^- + Br_2 \rightarrow 2Br^- + I_2$ (**M**)

Unit 12.3 Activity 2A: Cell diagrams (page 138)

1. **a.** $Zn(s) \mid Zn^{2+}(aq) \parallel IO_3^-(aq), I_2(aq) \mid C(s)$
b. $Fe(s) \mid Fe^{2+}(aq) \parallel Ag^+(aq) \mid Ag(s)$
c. $Zn(s) \mid Zn^{2+}(aq) \parallel Fe^{3+}(aq), Fe_{2+}(aq) \mid Pt(s)$
d. $Pt(s) \mid I^-(aq), I_2(aq) \parallel Cr_2O_7^{2-}(aq), Cr^{3+}(aq) \mid C(s)$ (**A**)

Unit 12.3 Activity 2B: Using reduction potentials (page 144)

1. **a.** Ca **b.** OCl^- **c.** I_2 **d.** Fe^{2+} (**A**)

2. H_2O_2 – it is the oxidant in the redox couple with the most positive $E°$ value. (**A**)

Unit 12.3 Activity 2C: Calculating $E°_{cell}$ values (page 145)

1. **a.** **i.** +0.62 V **ii.** +1.17 V **iii.** +1.24 V **iv.** +1.30 V (***A***)
   b. **i.** $Sn^{2+} + Zn \rightarrow Sn + Zn^{2+}$
   ii. $5Cu + 2MnO_4^- + 16H^+ \rightarrow 2Mn^{2+} + 8H_2O + 5Cu^{2+}$
   iii. $2Ag^+ + Fe \rightarrow Fe^{2+} + 2Ag$
   iv. $Zn + I_2 \rightarrow Zn^{2+} + 2I^-$ (***A***)
2. +0.81 V (***A***)
3. **a.** and **b.**

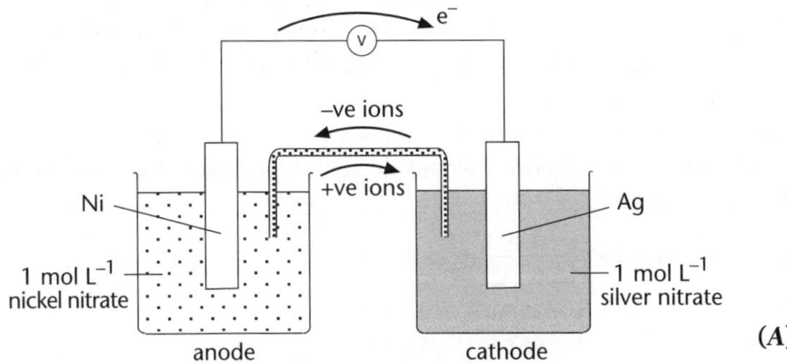

(***A***)

   c. $E°_{cell} = +1.04$ V (***A***)
4. **a.**

Cell	Strongest oxidant	Strongest reductant
i.	Ag^+	Zn
ii.	Fe^{3+}	Sn
iii.	MnO_4^-	Cl^-

(***A***)

   b. **i.** $Ag^+ + e^- \rightarrow Ag$ and $Zn \rightarrow Zn^{2+} + 2e^-$
   ii. $Fe^{3+} + e^- \rightarrow Fe^{2+}$ and $Sn \rightarrow Sn^{2+} + 2e^-$
   iii. $2Cl^- \rightarrow Cl_2 + 2e^-$ and $MnO_4^- + 8H^+ + 5e^- \rightarrow Mn^{2+} + 4H_2O$ (***A***)
   c. **i.** $2Ag^+ + Zn \rightarrow Zn^{2+} + 2Ag$
   ii. $2Fe^{3+} + Sn \rightarrow Sn^{2+} + 2Fe^{2+}$
   iii. $10Cl^- + 2MnO_4^- + 16H^+ \rightarrow 2Mn^{2+} + 8H_2O + 5Cl_2$ (***A***)
   d. **i.** $E°_{cell} = +0.80\text{ V} - (-0.76\text{ V}) = +1.56\text{ V}$
   ii. $E°_{cell} = +0.77\text{ V} - (-0.14\text{ V}) = +0.91\text{ V}$
   iii. $E°_{cell} = 1.51\text{ V} - 1.36\text{ V} = +0.15\text{ V}$ (***M***)
   e. **i.** $Zn(s)Zn^{2+}(aq) \| Ag^+(aq) | Ag(s)$
   ii. $Sn(s) | Sn^{2+}(aq) \| Fe^{3+}(aq),Fe^{2+}(aq) | Pt(s)$
   iii. $Pt(s) | Cl^-(aq),Cl_2(aq) \| MnO_4^-(aq),Mn^{2+}(aq) | C(s)$ (***A***)
5. **a.** In order of decreasing reactivity: B, A, C, D. The most reactive metal will have the most negative reduction potential (strongest reductant). (***M***)

b. **i.** and **ii.** (**E**)

SHE connected to:	$E°_{cell}$ (V)	SHE is:
B	0.47	Positive (B is negative)
C	0.09	Positive (C is negative)
D	0.34	Negative (D is positive)

6. **a.** +0.46 V (***A***) **b.** Cu(s) | Cu^{2+}(aq) || Ag^{+}(s) | Ag(s) (***A***)

c. Electrons flow from Cu to Ag through the wires. Ions move through the salt bridge, positive ions (cations) moving to the Ag^{+}/Ag half-cell from the Cu^{2+}/Cu half-cell, and negative ions (anions) moving in the opposite direction. (***M***)

d. **i.** $2Ag^{+} + Cu \rightarrow Cu^{2+} + 2Ag$

ii. Silver ions would be deposited on the silver electrode as silver metal (grey solid). The copper electrode would slowly disappear and the solution would become a darker blue. (***M***)

e. $n(Ag) = 2n(Cu) = 2 \times \dfrac{3.20\ g}{63.6\ g\ mol^{-1}} = 0.101\ mol$

$m(Ag) = 0.101\ mol \times 108\ g\ mol^{-1} = 10.9\ g$

Mass increases by 10.9 g. (***E***)

f. SHE has a standard reduction potential of 0.00 V so when it is connected to the Cu^{2+}/Cu half-cell it would be the negative electrode where oxidation occurs. Cu^{2+}/Cu would be positive so Cu (pink metal) would be deposited on the electrode and the blue colour would fade. $E°_{cell}$ = +0.34 V. H^{+} ions are produced in the other electrode, so the acidity increases. (***E***)

Unit 12.3 Activity 2D: Predicting redox reactions (page 148)

1. **a.** Yes ($E°_{cell}$ = +0.26 V).

b. Yes ($E°_{cell}$ = +0.59 V).

c. No ($E°_{cell}$ = –0.78 V). (***A***)

2. $E°_{cell}$ for the reaction of Fe^{3+} with Cl^{-} is negative so no reaction occurs. The $E°_{cell}$ for the reaction of Fe^{3+} with I^{-} is positive, so Fe^{2+} and I_2 are formed spontaneously. (***M***)

3. $E°_{cell}$ for reaction $2Ag^{+} + Fe \rightarrow Fe^{2+} + 2Ag$ is positive (+1.25 V), so the reaction occurs. (***M***)

4. $E°_{cell}$ for reaction $MnO_4^{-} + Cl^{-}$ is +0.31 V so the chloride can be oxidised to chlorine – a reaction which may not be wanted. (***M***)

5. **a.** $E°_{cell}$ for this reaction is +1.01 V. Since $E°_{cell}$ positive, the reaction occurs spontaneously.

b. $2H^{+} + H_2O_2 + 2Fe^{2+} \rightarrow 2H_2O + 2Fe^{3+}$

c. Colourless (or pale green) solution will turn orange. (***E***)

Unit 12.3 Activity 2E: Batteries (page 152)

1. **a.** $Zn \rightarrow Zn^{2+} + 2e^{-}$ **b.** $HgO + H_2O + 2e^{-} \rightarrow Hg + 2OH^{-}$

c. Zn electrode.

d. Mercury is a hazardous material that must be disposed of safely. (***A***)

2. **a.** $PbO_2 + 4H^{+} + 2SO_4^{2-} + Pb \rightarrow 2PbSO_4 + 2H_2O$ (***A***)

b. Anode = Pb (negative); cathode = PbO_2 (positive). (***A***)

c. +2.05 V (***A***)

d. Water is produced in the reaction and dilutes the solution. Also, $PbSO_4$ is precipitated. (***M***)

3. **a.** $2H_2 + O_2 \rightarrow 2H_2O$ (***A***)

b. **i.** and **ii.**

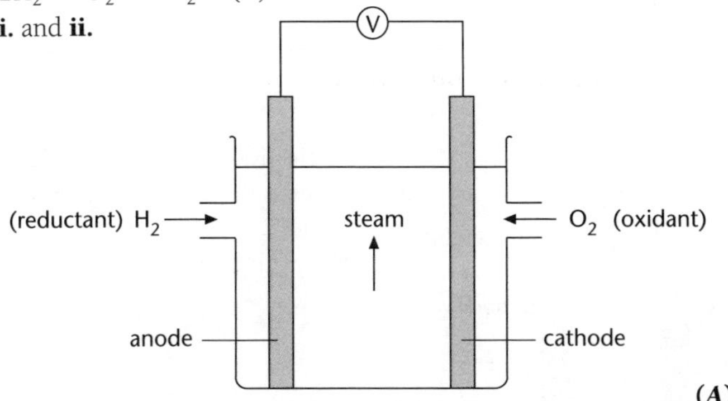

(***A***)

c. –0.83 V (***M***)

Unit 12.3 Activity 2F: Iron (page 155)

1. **a.** Iron(II) ions are oxidised by air to iron(III) ions. Iron(III) hydroxide is red-brown.
$4Fe^{2+} + O_2 + 2H_2O \rightarrow 4Fe^{3+} + 4OH^-$
and $Fe^{3+}(aq) + 3OH^-(aq) \rightarrow Fe(OH)_3$

b. Iron(III) aqueous is an acidic ion releasing H+(aq) in aqueous solution to form H_3O^+ ions.
$[Fe(H_2O)_6]^{3+} + H_2O \rightarrow [Fe(H_2O)_5OH]^{2+} + H_3O^+$
The H_3O^+ reacts with CO_3^{2-} to produce CO_2 gas and H_2O.

c. Fe^{3+} is reduced to Fe^{2+} and I^- is oxidised to I_2. The grey-brown precipitate is a mixture of FeI_2 and I_2. (***M***)

2. **a.** Formation of orange-brown precipitate: $Fe^{3+} + 3OH^- \rightarrow Fe(OH)_3$
Formation of dark-red solution: $Fe^{3+} + SCN^- \rightarrow [FeSCN]^{2+}$

b. Equation showing why the solution is acidic:
$[Fe(H_2O)_6]^{2+} + H_2O \rightarrow [Fe(H_2O)_5OH]^{2+} + H_3O^+$ (***M***)

Unit 12.3 Activity 2G: Copper (page 158)

1. **a.** $Cu(OH)_2(s) \rightarrow CuO(s) + H_2O(g)$; copper(II) oxide.

b. This is not a redox reaction as the oxidation number of Cu is +2 in both reactants and products. Oxidation numbers of oxygen and hydrogen do not change. (***A***)

2. Copper is oxidised by air, and reacts with water and carbon dioxide to form basic copper(II) carbonate (green). (***M***)

3. Copper(II) ions oxidise iodide ions, forming copper(I) iodide and iodine.
$2Cu^{2+}(aq) + 4I^-(aq) \rightarrow 2CuI(s) + I_2(aq)$ (***M***)

Unit 12.3 Activity 2H: Manganese (page 159)

1. **a.** $MnSO_4$, almost white (very pale pink). **b.** MnO_2, black (or dark brown ppt).
c. K_2MnO_4, green. **d.** $KMnO_4$, purple. (***A***)

2. **a.** +7 **b.** +3 **c.** +4 **d.** +3 **e.** +6 (***A***)
3. MnO_2, manganese dioxide/manganese(IV) oxide. (***A***)

Unit 12.3 Activity 2I: Chromium and vanadium (page 160)

1. **a.** CrO, black **b.** Cr_2O_3, green **c.** CrO_3, orange
2. **a.** +6 **b.** +6 **c.** +3

Unit 12.3 Activity 3A: Electrolysis (page 168)

1. **a.** Anode. **b.** Cathode. (***A***)
2. Negative electrode: $Mg^{2+} + 2e^- \rightarrow Mg$. Positive electrode: $2Cl^- \rightarrow Cl_2 + 2e^-$ (***M***)
3. **a.**

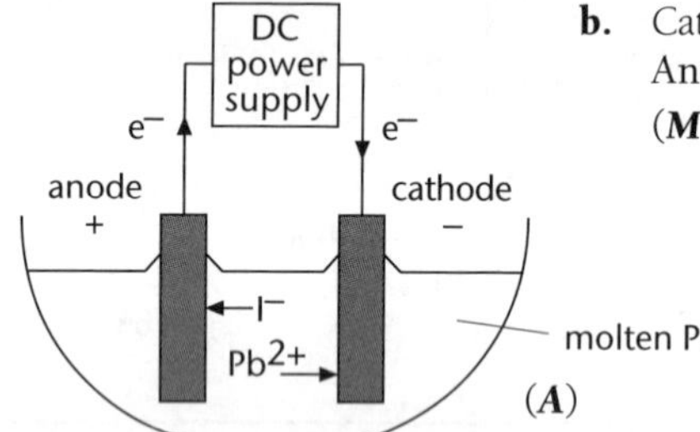

(***A***)

b. Cathode: $Pb^{2+}(\ell) + 2e^- \rightarrow Pb(\ell)$
Anode: $2I^-(\ell) \rightarrow I_2(g) + 2e^-$
(***M*** – both correct)

4. **a.** $Al^{3+}(\ell) + 3e^- \rightarrow Al(\ell)$ (***M***) **b.** $2O^{2-}(\ell) \rightarrow O_2(g) + 4e^-$
   c. $4Al^{3+}(\ell) + 6O^{2-}(\ell) \rightarrow 4Al(\ell) + 3O_2(g)$
5. **a.** **i.** Purple spot moves to the left; MnO_4^- ion attracted to positive electrode.
   ii. Blue spot moves to the right; Cu^{2+} ion attracted to negative electrode.
   iii. Orange spot moves to the left; $Cr_2O_7^{2-}$ ion attracted to positive electrode.
   iv. Green spot moves to the right (***M***); Cr^{3+} ion attracted to negative electrode. (***M***)
   b. Provides a means for ions and hence charge to move / improves the conductivity of the aqueous solutions. (***M***)
6. **a.** Sodium ions and calcium ions. (***A***) **b.** $Na^+(\ell) + e^- \rightarrow Na(\ell)$ (***M***)
   c. Chloride ions. (***A***) **d.** $2Cl^-(\ell) \rightarrow Cl_2(g) + 2e^-$ (***M***)
   e. $2Na^+(\ell) + 2Cl^-(\ell) \rightarrow 2Na(\ell) + Cl_2(g)$ (***M***)
7. **a.** **i.** $Cu^{2+} + 2e^- \rightarrow Cu$
   ii. $2Cl^- \rightarrow Cl_2 + 2e^-$ (***M***)
   b. Cu^{2+} ions (cations) in the electrolyte solution migrate to the negative electrode (cathode) where reduction will occur; Cu^{2+} ions gain electrons, and are reduced to Cu metal. (See equation **i.**). At the anode (the positive electrode) where oxidation will occur, the anions (Cl^- ions), will be oxidised to chlorine gas. (***E***)
8. **a.** Al^{3+} ions (cations) will migrate to the cathode. O^{2-} ions (anions) will migrate to the anode (the positive electrode). (***M***)
   b. **i.** $Al^{3+} + 3e^- \rightarrow Al$
   ii. $2O^{2-} \rightarrow O_2 + 4e^-$ (***A***)
9. When water is electrolysed, the gases oxygen and hydrogen are formed. At the anode, water is oxidised:
   $2H_2O \rightarrow O_2 + 4H^+ + 4e^-$
   At the cathode, water is reduced:
   $2e^- + 2H_2O \rightarrow H_2 + 2OH^-$ (***E***)

10. **a.** **i.–iii.**

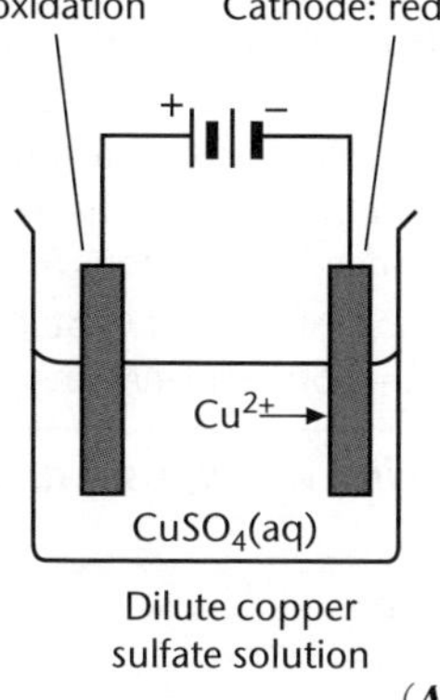

(***A***)

iv. Cu^{2+} produced at the anode and migrates to the cathode.

b. Anode: $Cu(s) \rightarrow Cu^{2+}(aq) + 2e^-$
Cathode: $Cu^{2+}(aq) + 2e^- \rightarrow Cu(s)$ (***M***)

Unit 12.4 Activity 1A: Physical properties of carbon (page 175)

1. **a.** 'Hard' indicates that it is not easy to penetrate the substance; not easily scratched.
b. 'Insoluble' refers to a solute that does not dissolve in a solvent.
c. 'Shiny' refers to a substance (usually solid) that has a polished surface that reflects light readily.
d. 'Melting point' refers to the temperature at which a solid changes to a liquid.
(***A*** – two correct; ***M*** – four correct)

2. **a.** A natural diamond is like a dull glass, but if faces have been cut into the diamond then it sparkles.
b. A grey-black solid that shines.
(***A*** – one description correct; ***M*** – both descriptions correct)

3. **a.** The diamond structure is composed of carbon atoms that are joined, in three dimensions, to four other carbon atoms. This arrangement produces a network lattice that is very strong and does not break easily.
b. The graphite structure is composed of carbon atoms that are joined, in two dimensions, to three other carbon atoms. This arrangement produces a layer lattice that is weakly bonded to the neighbouring layers. The layers are able to slide over each other, creating a soft and slippery material.
(***A*** – an indication of the bonding between the carbon atoms in both substances; ***M*** – an explanation of the arrangement of atoms in each substance, ie a three-dimensional arrangement for diamond and a two-dimensional arrangement for graphite; ***E*** – the arrangements explaining the property of the substance)

4. As diamond, carbon is used for the tips of drills due to its hardness. As diamond, carbon is used for jewellery due to the hardness of the substance allowing faces to be cut into the surface to reflect light to a maximum. As graphite, carbon is used for pencil leads due to the slipperiness of the substance.
(***A*** – three uses without explanation; ***M*** – three uses with one explanation; ***E*** – three uses with three explanations)

5. An allotrope is a form of an element in the same state as another form but with a different arrangement of atoms, eg diamond and graphite are both composed of carbon atoms but the atoms are arranged in different ways. An isotope is an atom of an element with a different number of neutrons in the nucleus to other atoms of the same element, eg carbon-12 is an atom of carbon that contains 6 neutrons and carbon-13 is an atom of carbon that contains 7 neutrons.

(**A** – allotrope related to form of the element, isotope related to the atom of the element; **M** – one good explanation; **E** – both explanations good)

Unit 12.4 Activity 1B: Carbon in the living world, as a fuel and as a reducing agent (page 178)

1. Coal; very abundant. Diamond; very rare.

(**A** – two examples with abundances correct)

2. Coal heated in the absence of air produces *coke*, gases and oils. *Coke* is a purer form of the element *carbon* than coal. (**A** – all three words correct)

3. **a.** Carbon + oxygen → *carbon dioxide*

b. Carbon dioxide + water → *glucose* + *oxygen*

c. $2ZnO(s) + C(s) \rightarrow 2Zn(s) + CO_2(g)$

(**A** – one equation correct; **M** – three equations correct)

4. Fuel – any substance that combines with oxygen from the air to release energy.

Exothermically – a process that occurs with the evolution of energy.

Coal – a product of dead vegetation being covered, compressed and heated over a period of millions of years.

Petroleum – a naturally occurring substance that contains hydrocarbons and is found under the ground trapped by a hard, impervious, cap rock. (**A** – one explanation correct; **M** – two explanations correct; **E** – four explanations correct)

Unit 12.4 Activity 2A: Properties of carbon dioxide (page 181)

1. **a.** Colourless; no smell; denser than air.

b. Weakly acidic; does not support combustion; turns limewater milky.

(**A** – two physical properties and one chemical property; **M** – three physical properties and two chemical properties)

2. **a.** **i.** The limewater added to the lower jar would turn milky but that added to the upper jar would be unchanged.

ii. The limewater added to both jars would turn milky.

b. Limewater detects the presence of carbon dioxide. After a few seconds of mixing, the carbon dioxide has not had time to spread into the upper jar. Although carbon dioxide is denser than air, it will mix with air by trying to occupy all available space. After an hour, carbon dioxide will be present throughout both gas jars.

(**A** – observations correct for both jars for both periods of time; **M** – recognition that after a few seconds, there was insufficient time for the gases to mix; **E** – explanation that the carbon dioxide would move to the upper jar despite its greater density than air)

3. **a.** Copper carbonate, sodium carbonate, sodium hydrogen carbonate.

b. Copper carbonate. Copper chloride.

c. Sodium hydrogen carbonate.

(Copper carbonate produces carbon dioxide gas when heated but its colour is blue-green.)

d. Tartaric acid.

(***A*** – four substances correctly identified; ***M*** – six substances correctly identified)

4. **a.**

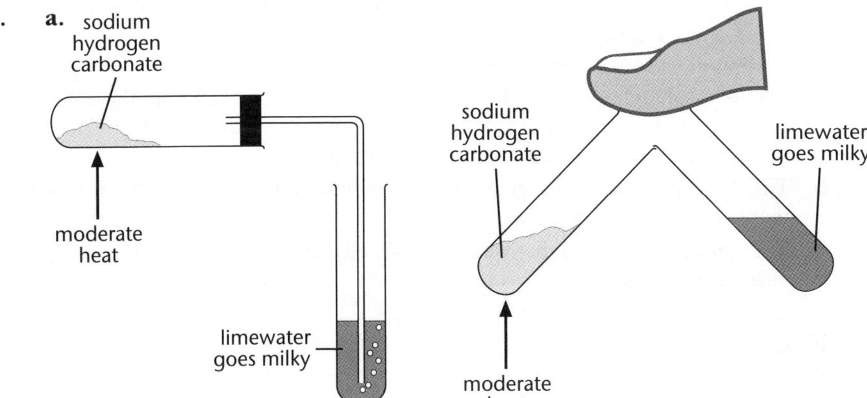

Method 1 – bubbling the gas through limewater

Method 2 – using the density of the gas to allow it to fall into a test tube of limewater

(***A*** – apparatus is workable, ie a solid can be heated and a gas led through a solution; ***M*** – apparatus has reasonable proportions and there is no risk to safety, ie limewater could not suck back into the heated test tube; ***E*** – complete, accurate and well-drawn diagram)

b. The limewater would go milky initially and then slowly become a clear solution. The milkiness is due to the formation of insoluble calcium carbonate:

$Ca(OH)_2(aq) + CO_2(g) \rightarrow CaCO_3(s) + H_2O(\ell)$.

When more carbon dioxide is passed through the milky liquid, the insoluble calcium carbonate is converted into soluble calcium hydrogen carbonate:

$CaCO_3(s) + H_2O(\ell) + CO_2(g) \rightarrow Ca(HCO_3)_2(aq)$.

(***M*** – recognition of the formation of insoluble calcium carbonate which is then converted into soluble calcium hydrogen carbonate; ***E*** – ***M*** plus equations)

5. **a.** Calcium carbonate → calcium oxide + *carbon dioxide*

b. *Sodium hydrogen carbonate* → sodium carbonate + carbon dioxide + water

c. $MgCO_3(s) \rightarrow \mathit{MgO(s)} + CO_2(g)$

d. $Na_2CO_3(s) + 2HCl(aq) \rightarrow \mathit{2NaCl(aq)} + H_2O(\ell) + CO_2(g)$

e. $\mathit{NaHCO_3(s)} + HCl(aq) \rightarrow NaCl(aq) + H_2O(\ell) + CO_2(g)$

f. $Ca(OH)_2(aq) + CO_2(g) \rightarrow \mathit{CaCO_3(s)} + \mathit{H_2O(\ell)}$.

(***A*** – **a.** and **b.** both correct; ***M*** – ***A*** plus one symbol equation correct; ***E*** – all equations correct)

6. The absence of oxygen would cause death. (***A***) Carbon dioxide is not an active poison to the human system. (***M***)

Unit 12.4 Activity 2B: Solubility in water and uses of carbon dioxide (page 183)

1. Aeration of drinks to make them fizzy. (***A***) The gas is much more soluble in water at high pressure and the gas escapes when pressure returns to normal. (***M***)

Fire extinguisher. (***A***) The gas is stable to heat energy and denser than air (oxygen). (***M***)

Stage effect, eg to create mist. (***A***) The density of the gas is greater than air and the compressed gas cools when it expands. (***M***)

2. The flame would be extinguished. (***A***) Carbon dioxide gas is denser than air. The gas would 'fall' onto the flame and exclude air. (***M***) The candle cannot burn without a supply of oxygen (air). (***E***)

3. **a.** Carbon dioxide gas is much more soluble in water when under pressure greater that atmospheric pressure. (***A***) When the stopper is removed from the bottle, the pressure inside the bottle is much greater than the pressure outside the bottle and the gas escapes to the lower pressure region. (***M***)

b. The shaking of the bottle agitates the gas dissolved in the wine and consequently more gas is released in a given time. (***E***)

c. The carbon dioxide gas that was in the sealed can (***A***), under pressure, would have escaped. (***M***)

4. **a.** $CO_2(g)$.

b. $CO_2(aq)$.

c. $CO_2(aq) \rightarrow CO_2(g) + H_2O(\ell)$

(***A*** – **a.** or **b.** correct; ***M*** – both **a.** and **b.** correct; ***E*** – **a.**, **b.** and **c.** correct)

5. Place a piece of damp blue litmus paper in the gas jar. By shaking the test tube, the litmus paper will become pink. Alternatively, invert the test tube of gas over a beaker that contains sodium hydroxide solution and observe the solution rise to the top of the test tube.

(***A*** – description of either experiment; ***M*** – result of either experiment)

Unit 12.4 Activity 2C: Carbon compounds in the environment (page 186)

1. **a.** Methane, (***A***); plus other hydrocarbons (eg from petrol, diesel oil). (***M***)

b. The composition of the atmosphere is changing due to human activities. The percentage of carbon dioxide and hydrocarbons in the atmosphere is increasing. These gases retain heat more than the usual components of air, especially oxygen and nitrogen. Thus, heat escaping from Earth's surface is retained in the atmosphere. The temperatures of the atmosphere are increasing as a consequence of this retained heat energy.

(***A*** – carbon dioxide, or any other gas, is recognised as contributing to global warming; (***M*** – ***A*** plus two sentences illustrating different points; ***E*** – ***A*** plus four or more sentences illustrating different points)

2. Carbon monoxide is absorbed by the red cells of blood more strongly than oxygen, and irreversibly. Thus, the supply of oxygen carried by the blood is reduced (or possibly eliminated). Energy is released in the body by the reaction between oxygen and glucose, both carried by the blood, at muscle sites. No oxygen, no life.

(***A*** – carbon monoxide takes the place of oxygen in red blood cells; ***M*** – the absorption of the carbon monoxide is irreversible; ***E*** – the lack of oxygen in the blood results in no energy being released at muscle sites, resulting in death)

3. 0.03%. Photosynthesis removes carbon dioxide from the atmosphere; water in the form of seas, lakes, rivers, etc removes carbon dioxide acting as a reservoir for dissolved carbon dioxide; respiration by plants and animals adds carbon dioxide to the atmosphere.

(***A*** – percentage correct (ie less than 0.1%); ***M*** – two explanations for carbon dioxide entering or leaving the atmosphere; ***E*** – all three factors explained)

4. *Refer* Figure on page 184. Stalagmites grow from the ground and stalactites grow from the ceiling. Calcium carbonate dissolves in water that contains carbon dioxide to produce soluble calcium hydrogen carbonate. The warmth in a cave causes the calcium hydrogen carbonate solution to evaporate and the unstable hydrogen carbonate reverts to calcium carbonate. The slow and steady deposition of calcium carbonate in the cave produces the two types of growth.

(***A*** – recognition that rainwater can convert insoluble calcium carbonate to soluble calcium hydrogen carbonate; ***M*** – ***A*** plus indication of the difference between stalactites and stalagmites; ***E*** – ***M*** plus complete, accurate and well-drawn diagram and equations for the dissolving of calcium carbonate and the re-formation of calcium carbonate)

Unit 12.4 Activity 3A: Alkanes (page 194)

1. C_nH_{2n+2}. (***A***)

2. **a.** The molecular formula of a compound indicates the number and type of atoms in one molecule of that compound. The molecular formula C_2H_6 indicates that one molecule of ethane contains two carbon atoms and six hydrogen atoms.

b. The structural formula of a compound indicates how the atoms in one molecule of the compound are bonded to each other, eg:

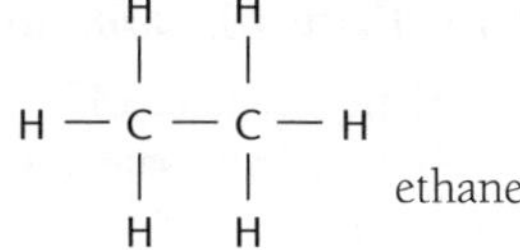

The two carbon atoms are joined to each other and to three hydrogen atoms; each hydrogen atom is joined to one carbon atom.

(***M*** – one formula explained correctly; ***E*** – both formulae explained correctly, plus correct structural formula)

3. **a.** *Diane*.

b. *Tetrane*. (***A*** – both names correct)

4. **a.** **i.** Gas.

ii. Tertiary.

b. Liquid.

(***A*** – any one answer correct; ***M*** – all three correct)

5. A secondary carbon atom is a carbon atom that is bonded to two other carbon atoms. (***M***)

6. Natural gas – methane, ethane, propane, butane. Petroleum – a range of alkanes from methane, through pentane and hexane (petrol), to alkanes with larger molecules.

(***A*** – two sources named; ***M*** – two components named for each source)

7. **a.** C (2, 4) H (1) **b.** $\cdot\dot{\underset{\cdot}{C}}\cdot$ H$\cdot$ **c.**

$$\begin{array}{c} H \\ | \\ H-C-H \\ | \\ H \end{array} \qquad \begin{array}{c} H\ \ H \\ |\ \ \ | \\ H-C-C-H \\ |\ \ \ | \\ H\ \ H \end{array}$$

Methane Ethane

8. **a.** **i.**

$$\begin{array}{c} H\ \ H \\ |\ \ \ | \\ H-C-C-H \\ |\ \ \ | \\ H\ \ H \end{array}$$

ii.

$$\begin{array}{c} H\ \ H\ \ H \\ |\ \ \ |\ \ \ | \\ H-C-C-C-H \\ |\ \ \ |\ \ \ | \\ H\ \ H\ \ H \end{array}$$

iii.

$$\begin{array}{c} H\ \ H\ \ H\ \ H\ \ H\ \ H \\ |\ \ \ |\ \ \ |\ \ \ |\ \ \ |\ \ \ | \\ H-C-C-C-C-C-C-H \\ |\ \ \ |\ \ \ |\ \ \ |\ \ \ |\ \ \ | \\ H\ \ H\ \ H\ \ H\ \ H\ \ H \end{array}$$

b. $CH_3—CH_2—CH_2—CH_2—CH_2—CH_3$ or $CH_3(CH_2)_4CH_3$

9. **a.** C_2H_6 **b.** C_4H_{10} **c.** $C_{10}H_{22}$ **d.** $C_{32}H_{66}$

10. **a.** Pentane. **b.** Butane. **c.** Heptane. **d.** Methane. **e.** Octane. (***A***)

11. **a.** **i.** Family of compounds with the same functional group and where each new member contains an additional $—CH_2—$ group. (***A***)

ii. A compound of hydrogen and carbon with only single bonds between atoms. (***A***)

b. The alkanes form a homologous series. The alkanes are saturated hydrocarbons. (***A***)

Unit 12.4 Activity 3B: Naming alkanes (page 201)

1. **a.** 3-Methylpentane. (***A***) **b.** 3-Methylhexane. (***A***)
c. 3,4-Dimethylhexane. (***A***) **d.** 2,3-Dimethylpentane. (***A***)

2. **a.**

$$\begin{array}{l} \qquad\ \ CH_3 \\ \qquad\quad | \\ CH_3—CH—CH_3 \end{array}$$ (***A***)

b.

$$\begin{array}{l} \qquad\ \ CH_3 \\ \qquad\quad | \\ CH_3—CH—CH_2—CH_3 \end{array}$$ (***A***)

c.

$$\begin{array}{l} \qquad\qquad\ CH_3 \\ \qquad\qquad\quad | \\ CH_3CH_2CHCH_2CH_2CH_3 \end{array}$$ (***A***)

d.

$$\begin{array}{l} \qquad\ \ CH_3 \\ \qquad\quad | \\ CH_3—C—CH_3 \\ \qquad\quad | \\ \qquad\ \ CH_3 \end{array}$$ (***A***)

e. $CH_3CH_2CH_2CH_2CH_2CH_2CH_2CH_3$ (***A***)

f.

$$\begin{array}{l} \qquad\ \ CH_3\ CH_3 \\ \qquad\quad |\qquad | \\ CH_3—C—CH—CH_3 \\ \qquad\quad | \\ \qquad\ \ CH_3 \end{array}$$ (***A***)

3.

$$\begin{array}{l} CH_3—CH—CH—CH_2\quad CH_3 \\ \qquad\quad |\qquad | \\ \qquad\ CH_3\ \ CH_2 \\ \qquad\qquad\quad | \\ \qquad\qquad\ CH_3 \end{array}$$ (***A***)

4. **a.** Pentane $CH_3CH_2CH_2CH_2CH_3$ (***A***) Methylbutane $CH_3CH(CH_3)CH_2CH_3$ (***A***)

$CH_3—C(CH_3)_2—CH_3$ Dimethylpropane (***M***)

b. Hexane $CH_3CH_2CH_2CH_2CH_2CH_3$ 2-Methylpentane $CH_3CH(CH_3)CH_2CH_2CH_3$

3-Methylpentane $CH_3CH_2CH(CH_3)CH_2CH_3$

2,2-Dimethylbutane $CH_3—C(CH_3)_2—CH_2CH_3$

2,3-Dimethylbutane $CH_3—CH(CH_3)—CH(CH_3)—CH_3$ (***M***)

5. **a.** Cyclopentane C_5H_{10} (ring of five C atoms, each bonded to two H) **b.** Cyclopropane C_3H_6 (ring of three C atoms, each bonded to two H) (***A***)

Unit 12.4 Activity 3C: Physical properties of alkanes (page 204)

1. **a.** Molecules of petrol are non-polar and molecules of water are polar.
   b. Petroleum oil and petrol molecules are both non-polar.
   c. Molecules of candle wax and diesel oil are both non-polar. (***A***)
2. **a.** Candle wax. **b.** Candle wax. (***A***)

3. **a.**

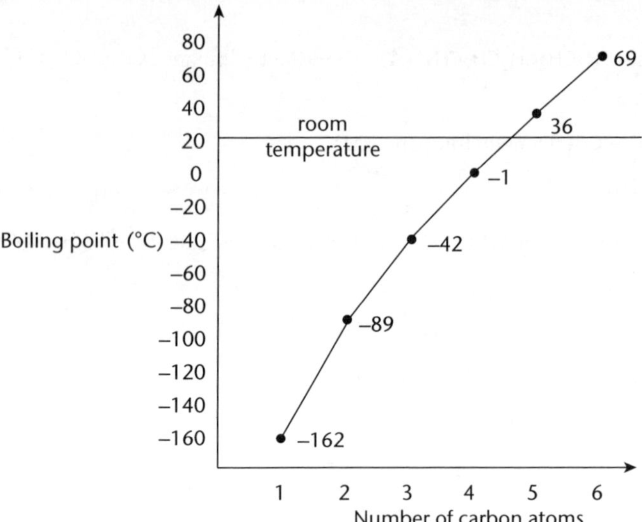

b. **i.** Methane, ethane, propane, butane. **ii.** Pentane, hexane.

4. Diesel oil is less dense than water. For the oil and water to mix, the molecules in diesel oil and the molecules in water must be separated. These separation processes require energy. The energy to separate the water molecules is much greater than that required to separate the hydrocarbon molecules. Enough energy must be released when the hydrocarbon molecules attract the water molecules to repay this energy debt. The hydrocarbon molecules do not attract the water molecules strongly enough to repay the energy needed to separate the water molecules.

(***A*** – diesel has lower density than water; ***M*** – dissolving of oil in water requires separation of molecules and attraction between water and hydrocarbon molecules; ***E*** – different strengths of attraction of hydrocarbon and water molecules)

Unit 12.4 Activity 3D: Complete and incomplete combustion (page 206)

1. **a.** **i.** Carbon dioxide, water.

ii. Carbon monoxide and/or carbon and water.

b. **i.** Blue.

ii. Yellow.

(***A*** – four of seven answers correct; ***M*** – all seven correct)

2. **a.** The wick allows molten wax to rise. (***A***)

b. Wax vapour. (***M***) When the molten wax reaches the top of the wick, it vaporises. The vapour mixes with the oxygen of the air and combustion occurs. (***E***)

c. There is sufficient air for the wax vapour to undergo complete combustion. (***A***)

d. The wax vapour is getting insufficient air for complete combustion. (***A***) Carbon is being formed which becomes red hot and is responsible for the yellow colour of the flame. (***M***)

3. **a.** The first equation shows that the products of combustion are carbon dioxide and steam. (***A***) These products are formed when a hydrocarbon undergoes complete combustion. (***M***)

b. The second and third equations show that the products of combustion are carbon monoxide or carbon, together with steam. (***A***) These products are formed when a hydrocarbon undergoes incomplete combustion. (***M***)

c. One product of incomplete combustion is unburnt carbon. This black material produced during incomplete combustion is commonly referred to as 'soot'. (***A***)

4. Incomplete combustion releases less heat for a given quantity of gas burnt than does complete combustion. Thus, a given quantity of water takes longer to boil when incomplete combustion is occurring in the flame (yellow).
(***A*** – incomplete combustion releases less heat; ***M*** – ***A*** plus for a given supply of gas)

Unit 12.4 Activity 3E: Haloalkanes (page 209)

1. **a.** Reactants: methylbutane, bromine.
Products: 2-bromo-3-methylbutane, hydrogen bromide. (***M***)

b. Substitution. (***A***)

c. Sunlight (ultraviolet light). (***A***)

d. The orange-brown colour of bromine disappears over time (slowly) (bromine is decolorised). (***A***)

2. **a.** $CH_4 + Br_2 \rightarrow CH_3B$ + HBr
bromomethane, hydrogen bromide (***A***)

b. The orange-brown colour of bromine will disappear. (***A***)

3. **a.** **i.** 1-Bromopropane (***A***) **ii.** 1,1,1-Tribromoethane (***A***)
iii. 2,5-Dichlorohexane (***A***)
iv. Tetrachloromethane (***A***)

b. **i.** Primary. **ii.** Primary.

4. **a.** **i.** 1,3-Dibromopropane **ii.** Primary. (***A***, ***M***)

b. **i.** 3-Chloro-3-ethylpentane **ii.** Tertiary. (***A***, ***M***)

5. **a.**
```
    H  H
    |  |
H—C—C—H
    |  |
   Br Br
```

b.
```
   H  H  H
   |  |  |
H—C—C—C—H
   |  |  |
   H  Cl H
```

c.
```
         H
         |
       H—C—H
    H    |    H   H
    |    |    |   |
H—C———C———C———C—H
    |    |    |   |
    Cl   Cl   Cl  H
```
or
```
           CH3
           |
CH2———C———CH—CH3
 |    |    |
 Cl   Cl   Cl
```
(***A***, ***M***)

6. **a.** $BrCH_2CH_2CH_2CH_3$ Primary.
```
CH3CHCH2CH3
   |
   Br
```
Secondary.

```
        Br
        |
CH3—C—CH3
        |
        CH3
```
Tertiary.

b. **i.** $CH_3-CH(CH_3)-CH_2Br$ 1-Bromo-2-methylpropane

ii. Primary.

Unit 12.4 Activity 4A: Alkenes (page 214)

1. C_nH_{2n}. (***A***)

2. C_2H_4 is the *molecular formula* of ethene – indicates the number and type of atoms in one molecule of the compound. The second formula is the *structural formula* of ethene – shows the bonding of the atoms in one molecule of the compound.

(***A*** – correct names of each formula; ***M*** – correct explanation of one term; ***E*** – correct explanation of both terms)

3. **a.** Butene. (***A***)

b. *Any one of*:

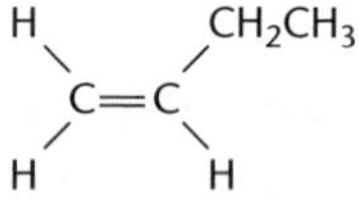

$(H)_2C=C(H)(CH_2CH_3)$ or $(CH_3)(H)C=C(CH_3)(H)$ or $(H)_2C=C(CH_3)_2$

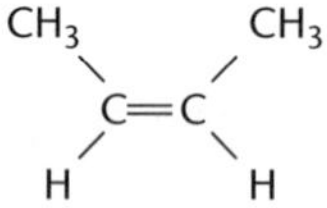

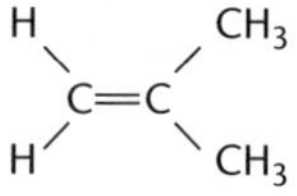

(***E*** – one structure correct)

4. **a.** Simplest whole number ratio of atoms: CH_2.

b. Actual number of each atom in molecule: C_3H_6.

c. Shows the specific arrangement and bonds between the atoms in a molecule: $H_2C=CH-CH_3$ (***M***)

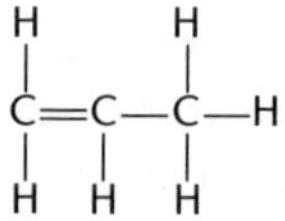

5. **a.** C_4H_8 **b.** $C_{12}H_{24}$ **c.** C_5H_{10} (***A***)

6. **a.** Ethene. **b.** Propene. (***A***) **c.** Propene. (***A***) **d.** 3-Methylpent-2-ene (***A***)

e. 2,3-Dimethylpent-2-ene (***A***)

7. **a.** $CH_3CH=CHCH_2CH_3$ **b.** $CH_3CH(CH_3)CH=CHCH_2CH_3$

c. $CH_2=C(CH_3)CH_3$ **d.** $CH_3C(CH_3)=C(CH_3)CH_3$

8. $CH_2=CHCH_2CH_2CH_3$ Pent-1-ene

$CH_3CH=CHCH_2CH_3$ Pent-2-ene

$CH_2=C(CH_3)CH_2CH_3$ 2-Methylbut-1-ene

$CH_2=CHCH(CH_3)CH_3$ 3-Methylbut-1-ene

$CH_3-C(CH_3)=CHCH_3$ 2-Methylbut-2-ene (***M***)

Unit 12.4 Activity 4B: Isomers of alkenes (page 215)

1. **a.** $H \quad CH_3$ / $C{=}C$ / $H \quad CH_2CH_3$

b. $H \quad H$ / $C{=}C$ / $H \quad CH_2CH_2CH_3$

c. $CH_3CH_2 \quad CH_2CH_3$ / $C{=}C$ / $H \quad H$

cis-Hex-3-ene

$CH_3CH_2 \quad H$ / $C{=}C$ / $H \quad CH_2CH_3$

trans-Hex-3-ene

d. $CH_3 \quad H$ / $C{=}C$ / $CH_3 \quad CH_2CH_3$ (**M**)

2. $CH_2{=}CHCH_2CH_3$ But-1-ene – structural isomer.

$CH_2{=}C(CH_3)CH_3$ Methylpropene – structural isomer.

$CH_3 \quad CH_3$ / $C{=}C$ / $H \quad H$

cis-But-2-ene

$CH_3 \quad H$ / $C{=}C$ / $H \quad CH_3$

trans-But-2-ene

structural and geometric isomers. (**M**)

3. **a.** **i.** 2-Methylpent-2-ene. **ii.** C_6H_{12} **iii.** No geometric isomers.

b. **i.** Pent-2-ene. **ii.** C_5H_{10}

iii. *cis*-Pent-2-ene. *trans*-Pent-2-ene (**M**)

$CH_3 \quad CH_2CH_3$ / $C{=}C$ / $H \quad H$

$CH_3 \quad H$ / $C{=}C$ / $H \quad CH_2CH_3$ (**M**)

4. **a.** $CH_3 \quad CH_3$ / $C{=}C$ / $Cl \quad Cl$

cis

$CH_3 \quad Cl$ / $C{=}C$ / $Cl \quad CH_3$

trans

b. $Br \quad H$ / $C{=}C$ / $Br \quad CH_3$ (**M**)

5. Geometrical isomerism can only occur in alkenes when there are two different groups attached to each of the carbon atoms in the double bond. Alkane molecules cannot exhibit geometrical isomerism, because free rotation about the C–C single bonds is possible and so the groups around the bonds are not fixed in space. (**M**)

6. **a.** $H \quad H$ / $C{=}C$ / $Cl \quad Cl$

cis-1,2-Dichloroethene

$Cl \quad H$ / $C{=}C$ / $H \quad Cl$

trans-1,2-Dichloroethene

b. $Cl \quad H$ / $C{=}C$ / $Cl \quad H$ 1,1-Dichloroethene (**M**)

c. The molecule in **a**. has two different groups on each C of the double bond. The molecule in **b**. has identical groups on the C of the double bond, so swapping them around does not change the geometry of the molecule. (**M**)

Unit 12.4 Activity 4C: Reactions of alkenes (page 219)

1. Addition (hydration). (*A*)
2. Bromine would decolorise (change from orange to colourless).

   $CH_2{=}CH_2 + Br_2 \rightarrow \underset{Br}{CH_2}{-}\underset{Br}{CH_2}$ (**M**)

3. **a.** $CH_2{=}CH_2 + H_2 \rightarrow CH_3CH_3$ (**M**)

   b. $CH_3CH{=}CHCH_3 + H_2 \xrightarrow{Pt} CH_3CH_2CH_2CH_3$ (**M**)

   c. $CH_3\underset{CH_3}{C}{=}CHCH_3 + H_2 \xrightarrow{Pt} CH_3\underset{CH_3}{C}HCH_2CH_3$ (**M**)

4. $CH_3\underset{Br}{C}HCH_3$ 2-Bromopropane (**M**)

5. $CH_3{-}\overset{CH_3}{\underset{OH}{C}}{-}CH_2CH_2CH_3$ 2-Methylpentan-2-ol (**M**)

6. **a.** $CH_2{=}CHCH_3 + H_2 \rightarrow CH_3CH_2CH_3$ Propane

   b. $CH_2{=}CHCH_3 + Cl_2 \rightarrow \underset{Cl}{CH_2}\underset{Cl}{CH}CH_3$ 1,2-Dichloropropane

   c. $CH_2{=}CH_2 + Br_2 \rightarrow \underset{Br}{CH_2}\underset{Br}{CH_2}$ 1,2-Dibromoethane

   d. $CH_2{=}CH_2 + HBr \rightarrow CH_3CH_2Br$ Bromoethane

   e. $CH_2{=}CHCH_3 + Br_2 \rightarrow \underset{Br}{CH_2}\underset{Br}{CH}CH_3$ 1,2-Dibromopropane

   f. $CH_2{=}CH_2 + H_2O(g) \rightarrow CH_3CH_2OH$ Ethanol (**M**)

7. Ethene decolorises bromine water and acidified potassium permanganate, but ethane does not (ethene burns with a sooty flame, ethane with a cleaner flame).
8. **a.** Some double or triple carbon–carbon bonds in the molecule. (*A*)

   b. i. Take a known volume of the fat and dissolve it in a non-polar solvent. Add the bromine until the orange colour no longer disappears when the two solutions are mixed. Record the volume of the bromine added. The fat solution which requires the greater volume of bromine before the orange colour persists has the more double bonds to add the bromine molecules, so has the greater degree of saturation. (*E*)

   ii. If a polar solvent is used, this will not mix with the fat.

9. 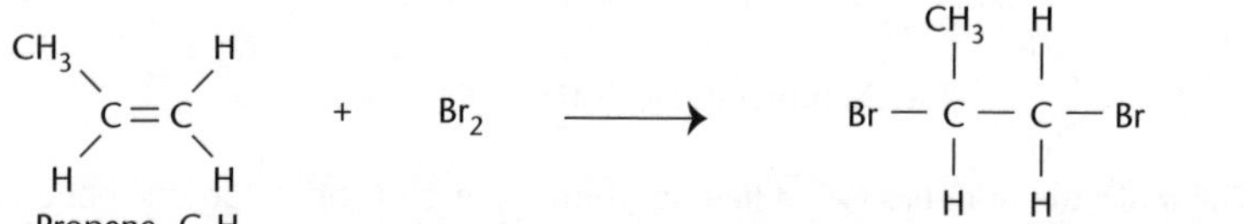 $(CH_3)(H)C{=}C(H)(H)$ + Br_2 ⟶ $Br{-}\overset{CH_3}{\underset{H}{C}}{-}\overset{H}{\underset{H}{C}}{-}Br$

   Propene, C_3H_6 1,2-Dibromopropane, $C_3H_6Br_2$

   (**M** – correct placement of symbols; **E** – name and formula of product correct)

Unit 12.4 Activity 4D: Polymers (page 222)

1. **a.** The process by which a large molecule is formed by joining together many small molecules or 'monomers'. (***M***)

b. **i.** $CH_2{=}CCl_2$ **ii.** $CH_2{=}CHCH_2CH_3$

2. **a.**

```
CH3     H
   \   /
    C=C
   /   \
  H     H
```

b.

```
F     F
 \   /
  C=C
 /   \
F     F
```

(***M***)

3. **a.**

```
     H   H   H   H
     |   |   |   |
···—C—C—C—C—···
     |   |   |   |
     H   Cl  H   Cl
```

b.

```
     H   H   H   H
     |   |   |   |
···—C—C—C—C—···
     |   |   |   |
    CH3 CH3 CH3 CH3
```

c.

```
     H   H   H   H
     |   |   |   |
···—C—C—C—C—···
     |   |   |   |
     H   CN  H   CN
```

4. **a.** **vi.** The product of a polymerisation process.

b. **iv.** A hydrocarbon that contains one double bond.

c. **ii.** A simple molecule that can be polymerised.

d. **v.** The process of making a polymer from a monomer.

e. **i.** Two atoms joined together in a molecule using four electrons.

f. **iii.** A material that can be easily shaped.

(***A*** – three terms correctly matched; ***M*** – six terms correctly matched)

5. **a.** **i.** Plastic film, tubing.

ii. Plastic film, rope, chairs, packaging.

b. **i.** Polystyrene.

ii. PVC.

iii. Teflon.

(***A*** – one answer correct in each of **a**. and **b**.; ***M*** – five answers correct)

Unit 12.4 Activity 4E: Alkynes (page 225)

1. **a.** C_4H_6 **b.** $C_{10}H_{18}$ **c.** $C_{12}H_{22}$ (***A***)

2. **a.** Ethyne. **b.** Propyne. **c.** 4-Bromo-4-methylpent-2-yne (***A***)

3. **a.** The orange-brown bromine water will decolorise in the presence of the alkyne.

b.

```
Br     Br            Br  Br
  \   /              |   |
   C=C            H—C—C—H
  /   \              |   |
 H     H             Br  Br
```

1,2-Dibromoethene 1,1,2,2-Tetrabromoethane (***M***)

Unit 12.4 Activity 5A: Naming alcohols (page 229)

1. **a.** Ethanol. **b.** Butan-2-ol. **c.** Methylpropan-2-ol.

d. Butan-1-ol. **e.** Propan-2-ol. **f.** Propan-1-ol.

g. 2,2-Dimethybutan-1-ol. (***A***)

2. **a.**  CH_3 (branch)
CH_3CHCH_2OH

b. 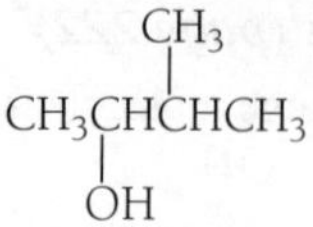 CH_3 (branch)
$CH_3CHCHCH_3$
OH

c. CH_3OH

d. 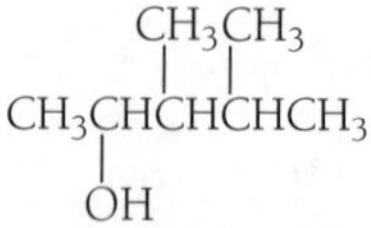 $CH_3\ CH_3$ (branches)
$CH_3CHCHCHCH_3$ (***A***)
OH

3. **a.** Both have the –OH group. **b.** Both names end in *ol*. (***A***)

Unit 12.4 Activity 5B: Isomers and physical properties of alcohols (page 231)

1. $CH_3—CH_2—CH_2—CH_2—OH$

Butan-1-ol

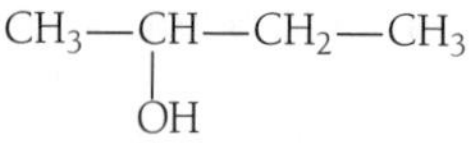

$CH_3—CH—CH_2—CH_3$
OH

Butan-2-ol

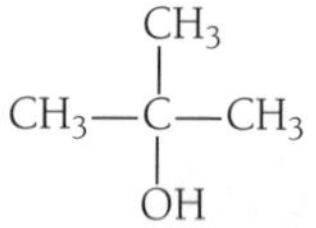

CH_3
$CH_3—C—CH_3$
OH

Methylpropan-2-ol

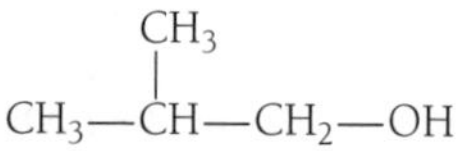

CH_3
$CH_3—CH—CH_2—OH$

Methylpropan-1-ol (***M***)

2. **a.**

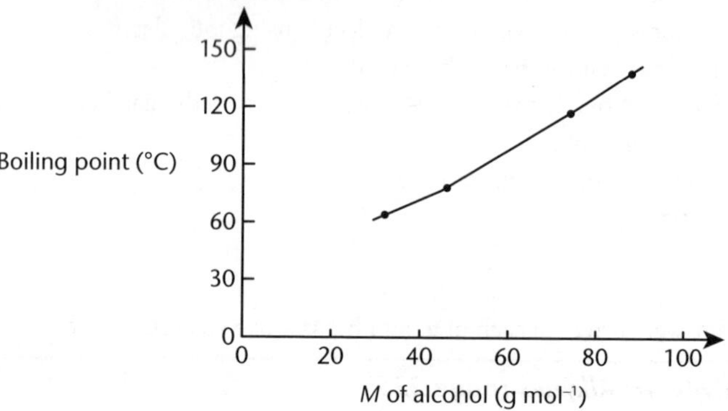

b. 95–99°C

c. 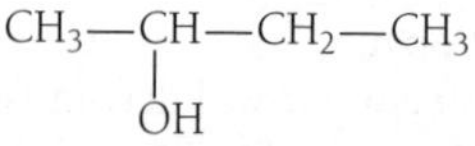 $CH_3—CH—CH_2—CH_3$ (***A***)
OH

Butan-2-ol

d. As *M* increases, the boiling point increases due to increasing number of intermolecular forces in larger molecules with a greater number of electrons. (***M***)

3. **a.** Secondary. **b.** Primary. (***A***)

4. **a.** Propan-l-ol, primary. **b.** Pentan-2-ol, secondary.
c. 2-Methylpropan-2-ol, tertiary. **d.** 2-Methylpropan-l-ol, primary.
e. 3-Methylbutan-2-ol, secondary. (***A***)

5. **a.** Primary alcohols: **i**, **v**; secondary alcohols: **ii**, **iii**, **vi**; tertiary alcohols: **iv**, **vii**, **viii**. (***A***)
   b. **iii**, **iv**, **v**, **vi** and **ii**, **vii**. (***A***)
6. $CH_3CH_2CH_2CH_2OH$ butan-1-ol, $CH_3CH(CH_3)CH_2OH$ 2-methylpropan-l-ol. (***M***)

Unit 12.4 Activity 5C: Reactions of alcohols (page 234)

1. **a.** $CH_3OH(\ell) + \frac{3}{2}O_2(g) \rightarrow CO_2(g) + 2H_2O(\ell)$
   b. $CH_3CH_2OH(\ell) + 3O_2(g) \rightarrow 2CO_2(g) + 3H_2O(\ell)$
2. **a.** Colour change from purple to colourless. (***A***)
   b. Colour change from orange to green. (***A***)
3. **a.** $CH_3CH_2CH_2COOH$
   b. $CH_3CH{=}CH_2$
   c. $CH_3CH{=}CHCH_2CH_3$
   d. HCOOH (***M***)

Unit 12.4 Activity 5D: Alcohols – multiple choice (page 234)

1. C **2.** D **3.** C **4.** B **5.** A

Unit 12.4 Activity 6A: Carboxylic acids – structure and naming (page 237)

1. **a.** $H{-}CH_2{-}CH_2{-}C(=O){-}OH$ (structural formula with all H atoms shown) (***A***)
   b. The structural formula of ethanoic acid is: $H{-}CH_2{-}C(=O){-}O{-}H$ (***A***, ***M***)
   Propanoic acid differs from ethanoic acid by one $–CH_2–$ group, so the carboxylic acids are a homologous series.
2. **a.** $CH_3CH_2CH_2C(=O)OH$
   b. $CH_3CH_2CH_2CH(CH_3)C(=O)OH$
   c. $CH_3CH_2CH(CH_2CH_3)C(=O)OH$ (***A***)
3. **a.** Ethanoic acid.
   b. 2,2-Dimethylpropanoic acid.
   c. 4,4-Dimethylpentanoic acid.
   d. 2-Methylbutanoic acid. (***A***)

Unit 12.4 Activity 6B: Properties of carboxylic acids (page 239)

1. Sodium propanoate.
2. **a.** $HCOOH(\ell) + H_2O(\ell) \rightleftharpoons HCOO^-(aq) + H_3O^+(aq)$
   b. $Zn(s) + 2CH_3COOH(aq) \rightarrow Zn^{2+}(aq) + 2CH_3COO^-(aq) + H_2(g)$ (***M***)

3. **a.** and **b.** Acid donates proton to base.
d. Acid with metal produces hydrogen gas. (**M**)

4. $CH_3COOH(aq) + NaOH(aq) \rightarrow \underset{\text{sodium ethanoate}}{CH_3COONa(aq)} + H_2O(\ell)$

Unit 12.4 Activity 6C: Esters (page 242)

1. **a.** $CH_3\underset{\underset{O}{\|}}{C}OCH_2CH_2CH_3$ **b.** $CH_3CH_2\underset{\underset{O}{\|}}{C}OCH_2CH_3$

c. $H\underset{\underset{O}{\|}}{C}OCH_2CH_2CH_2CH_3$ **d.** $CH_3\underset{\underset{O}{\|}}{C}OCH_2CH_3$

2. **a.** $HCOOH + CH_3OH \rightleftharpoons H{-}\underset{\underset{O}{\|}}{C}{-}O{-}CH_3 + H_2O$

methyl methanoate + water (**A**)

b. $CH_3OH + CH_3CH_2\underset{\underset{O}{\|}}{C}OH \xrightleftharpoons{H_2SO_4} CH_3{-}O{-}\underset{\underset{O}{\|}}{C}{-}CH_2{-}CH_3 + H_2O$

methyl propanoate + water (**A**)

3. **a.** $CH_3CH_2CH_2\underset{\underset{O}{\|}}{C}{-}O{-}CH_3$

b. $CH_3CH_2\underset{\underset{O}{\|}}{C}{-}O{-}CH_2CH_2CH_3$

c. $H\underset{\underset{O}{\|}}{C}OCH_2CH_2CH_2CH_2CH_3$ **d.** $CH_3\underset{\underset{O}{\|}}{C}OCH_2CH_2CH_2CH_3$

e. $H{-}\underset{\underset{O}{\|}}{C}{-}O{-}CH_2CH_3$ **f.** $CH_3CH_2CH_2\underset{\underset{O}{\|}}{C}OCH_2CH_3$ (**A**)

4. **a.** Methyl ethanoate; methanol and ethanoic acid.
b. Methyl methanoate; methanol and methanoic acid.
c. Propyl ethanoate; propanol and ethanoic acid.
d. Ethyl propanoate; ethanol and propanoic acid.
e. Ethyl pentanoate; ethanol and pentanoic acid.
f. Ethyl ethanoate; ethanol and ethanoic acid. (**A**)

5. **a.** Sweet smelling, volatile.
b. Methyl ethanoate and ethyl methanoate

$$H{-}\overset{\overset{H}{|}}{\underset{\underset{H}{|}}{C}}{-}\underset{\underset{O}{\|}}{C}{-}O{-}\overset{\overset{H}{|}}{\underset{\underset{H}{|}}{C}}{-}H \qquad\qquad H{-}\underset{\underset{O}{\|}}{C}{-}O{-}\overset{\overset{H}{|}}{\underset{\underset{H}{|}}{C}}{-}\overset{\overset{H}{|}}{\underset{\underset{H}{|}}{C}}{-}H$$

(**M**)

c. Pentanoic acid and ethanol. (**A**)
d. Reflux the pentanoic acid and ethanol with concentrated sulfuric acid then add sodium carbonate to remove any remaining acid. (**M**)

6. A sweet 'fruity' smell would be produced and a water-insoluble liquid. (***A***)

7. **a.** $CH_3COONa + CH_3CH_2CH_2CH_2OH$ **b.** $CH_3CH_2COOH + CH_3OH$
c. $CH_3CH_2CH_2OH + HCOOH$ **d.** $CH_3CH_2OH + CH_3CH_2COO^-$

Unit 12.4 Activity 6D: Fatty acids and soaps (page 246)

1. Fats are solid at room temperature and contain saturated fatty acids from animals.Oils are liquid at room temperature and contain unsaturated fatty acids from plants. (***A***)

2. $CH_3(CH_2)_7CH{=}CH(CH_2)_7COOH + NaOH \rightarrow CH_3(CH_2)_7CH{=}CH(CH_2)_7COONa + H_2O$ (***M***)

3. **a.** Ester. (***A***)

b. 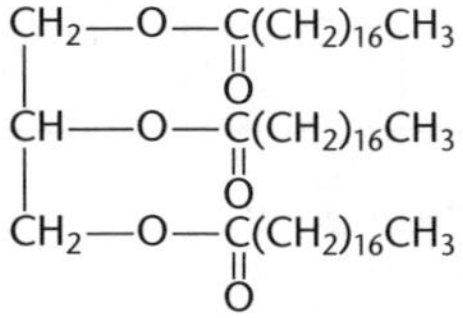
```
CH2—O—C(CH2)16CH3   (M)
|     ||
|     O
CH—O—C(CH2)16CH3
|     ||
|     O
CH2—O—C(CH2)16CH3
      ||
      O
```

c.
```
           O
           ||
CH2—O—C—C(CH2)16CH3                     CH2—OH
|          O                             |
|          ||                            |
CH—O—C—C(CH2)16CH3  + 3NaOH → CH—OH  + 3CH3(CH2)16CONa
|          O                             |                       ||
|          ||                            |                       O
CH2—O—C—C(CH2)16CH3                     CH2—OH
                                        glycerol        sodium stearate (M)
```

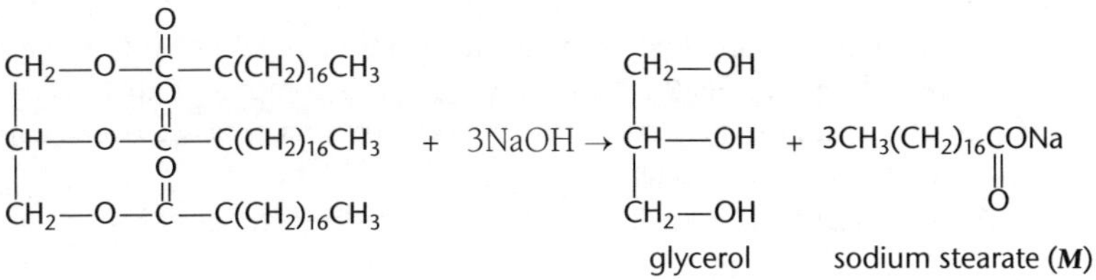

d. Saponification/hydrolysis. (***A***)

4. **a.** Ester. (***A***)

b. **i.** $CH_3(CH_2)_{14}COONa$ (***M***)

ii. 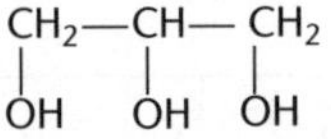
```
CH2—CH—CH2      (M)
|    |    |
OH   OH   OH
```

c. The 'acid' part of the molecule will have some degree of unsaturation. (***M***)

Unit 12.4 Activity 7A: Methanol and ethanol – structure of molecules and commercial production (page 248)

1. **a.** $C_nH_{2n+1}OH$. (***A***)

b. CH_4O.

c. 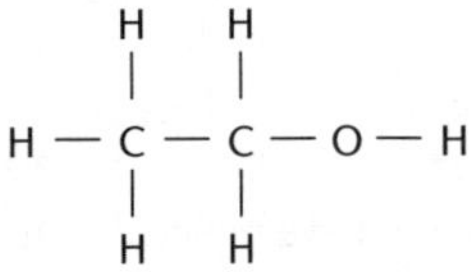
```
    H   H
    |   |
H — C — C — O — H
    |   |
    H   H
```

(***M*** – **a.** and **b.** correct; ***E*** – all three correct)

2.

	a.	b.	c.
CH_4	H | H — C — H | H	Methane	Carbon – 4 Hydrogen –1
CH_4O	H | H — C — O — H | H	Methanol	Carbon – 4 Oxygen – 2 Hydrogen – 1

(**A** – both names correct; **M** – **A** plus both structural formulae correct; **E** – **M** plus all valencies correct)

3. **a.** Stage 1 – natural gas is mixed with steam to produce synthesis gas (a mixture of carbon monoxide and hydrogen) at a temperature of 800–850°C in the presence of a nickel catalyst. Stage 2 – the synthesis gas is compressed to 100 atmospheres (10 MPa) pressure at a temperature of 200–300°C and passed over a copper catalyst to produce methanol. (***E***)

   (***A*** – two stages recognised; ***M*** – products in both stages given)

   b. Starch, in the form of a grain, is converted to maltose by the enzyme amylase. The maltose is then converted to glucose by the enzyme maltase. The glucose is converted to ethanol by the enzyme zymase.

   (***A*** – details on one stage; ***M*** – details of two stages; ***E*** – all stages given, equations not needed.

Unit 12.4 Activity 7B: Properties of alcohols (page 251)

1. **a.** Liquid. (***A***)

   b. The melting point and boiling point of the alcohol are both greater than room temperature. (***A***)

2. **a.** Both alcohols are liquids.

   b. Butanol has the stronger forces because it has a higher boiling point. The boiling point of a liquid is an indication of the energy needed to break the forces between the molecules of the substance. The higher the boiling point, the greater the forces needed.

   (***A*** – **a.** correct; ***M*** – butanol recognised as having greater forces between its molecules, and boiling point linked to energy to break forces between molecules.)

3. Both **a.** (water) and **b.** (ethanol). (***A***) Water and ethanol have molecules that are polar due to the presence of an oxygen atom in the molecule. (***M***)

4. **a.** Ethanol + oxygen → carbon monoxide + *water*.

   b. $C_2H_5OH(\ell) + 3O_2(g) \rightarrow 2CO_2(g) + 3H_2O(\ell)$.

   c. $C_2H_5OH(\ell) + O_2(g) \rightarrow 2C(s) + 3H_2O(\ell)$.

   d. $CH_3OH(\ell) + O_2(g) \rightarrow CO(g) + 2H_2O(\ell)$.

   (***A*** – **a.** correct; ***M*** – **a.** and **b.** correct; ***E*** – all equations correct)

5. **a.** Ethanol. (***A***)

   b. Complete combustion is producing a blue flame that is difficult to see in strong light. (***M***) The large surface area of the pudding allows complete combustion.

6. **a.** *Any two of*: Manufacture of synthetic petrol; fuel; added to methylated spirits; manufacture of other chemicals.

b. *Any two of*: Beverage; fuel; manufacture of other chemicals; solvent.

(**A** – any two uses correct for **a.** or **b.**; **M** – two uses correct for each of **a.** and **b.**)

7. *Either of*: pyridine – to give a foul taste and smell *or* purple dye – to warn people that the liquid is not water.

(**A** – one substance named; **M** – reason for adding the substance)

8. Ethanol is a solvent. Ethanol has a low boiling point which allows it to evaporate easily.

(**A** – one reason; **M** – two reasons)

Unit 12.4 Activity 7C: Ethanoic acid – manufacture and properties (page 252)

1. 3–6%. (**A** – any percentage less than 6)

2. White vinegar is obtained by the conversion of ethanol to ethanoic acid in wine. Malt vinegar is obtained by the conversion of ethanol to ethanoic acid in beer. Malt vinegar is brown due to the presence of the malt. (**A** – different sources named for the conversion of ethanol to ethanoic acid)

3. In both cases, the liquors are poured through casks drilled with air holes and filled with beech shavings on which bacteria adhere. The bacteria, *Mycocarderma aceti*, enable oxygen to convert the alcohol present into ethanoic acid. (**A** – use of oxygen/air;
M – recognition of the need for bacteria; **E** – bacteria named)

4. **a.** $C_2H_4O_2$. (**A**)

b. CH_3COOH or

```
      H
      |
H — C — C — O — H
      |   ||
      H   O
```

(**M** – shortened structural formula correct; **E** – complete structural formula)

c. Propanoic acid. (**M**)

Unit 12.4 Activity 8A: Names and structures (page 255)

1. **a.**

```
        O
       //
 I - C
       \
        H
```

b.

```
  H  O  H  H  H  H
  |  ||  |  |  |  |
 -C-C-C-C-C-C-H
  |     |  |  |  |
  H     H  H  H  H
```

c. $C_6H_{12}O$

```
   H    H    H    H         O
   |    |    |    |        //
 - C  - C  - C  - C  - C
   |    |    |    |        \
   H    H    H    CH3       H
```

d. $C_6H_{12}O$

```
   H    H    H    O    H
   |    |    |    ||   |
 - C  - C  - C  - C  - C  - H
   |    |    |         |
   H    H    CH3       H
```

(**A**)

2. Pentan-2-one. (**A**)

3. **a.** and **b.** Butanal $CH_3CH_2CH_2CHO$, Methylpropanal $CH_3CH(CH_3)CHO$, Butanone $CH_3CH_2COCH_3$. (**M**)

Unit 12.4 Activity 8B: Bonding, properties and uses (page 256)

1. Both molecules have intermolecular attractive forces due to instantaneous dipole-dipole attractions, but aldehydes also have weak intermolecular forces of attraction due to the polarity of the carbonyl group. Aldehydes therefore have higher boiling points. (**M**)
2. **a.** As the mass increases, the size of the non-polar alkyl group increases which does not bond to polar water molecules and the aqueous solubility decreases. (**M**)

   b. Yes – ketones also contain both a polar carbonyl group and non-polar alkyl groups. As the length of the alkyl chains increases, the overall polarity of the molecule decreases. The molecule becomes much less soluble in water, but its solubility in non-polar solvents would increase. (**M**)
3. Ethanol contains intermolecular hydrogen bonds which are stronger attractions than the intermolecular attractive forces between polar ethanal molecules. Stronger intermolecular attractive forces result in higher boiling points. (**M**)

Unit 12.4 Activity 8C: Preparing aldehydes and ketones (page 258)

1. **a.** Propanoic acid. **b.** Fractional distillation.
   c. Propanone is not easily oxidised. (**A**)
2. Because there is no hydrogen bonding between molecules of ethanal, ethanol has a lower boiling point than the other reagents, which can have hydrogen bonding between molecules. Ethanal therefore distils first. (**M**)
3. **a.** $CH_3CH_2CH_2OH$, propan-1-ol. **b.** $CH_3CH_2CHOHCH_3$, butan-2-ol.
   c. $CH_3CH(CH_3)CH_2OH$, 2-methylpropan-1-ol. (**A**)

Unit 12.4 Activity 8D: Distinguishing aldehydes and ketones (page 260)

1. **a.** Oxidation. **b.** Blue – copper(II) ions (complexed with tartrate ions).
   c. Copper(I) oxide – orange-red precipitate. (**A**)
2. **a.** Tollens' reagent.
   b. i. No reaction. **ii.** No reaction.
   iii. Grey solid precipitated, possibly as a silver mirror. (**A**)
3. **a. i.** $H_2O + CH_3CH_2CHO \rightarrow CH_3CH_2CO_2H + 2H^+ + 2e^-$
   $Ag^+ + e^- \rightarrow Ag$
   $H_2O + CH_3CH_2CHO + 2Ag^+ \rightarrow 2Ag + CH_3CH_2CO_2H + 2H^+$
   ii. $H_2O + CH_3CHO \rightarrow CH_3CO_2H + 2H^+ + 2e^-$
   $2e^- + H_2O + 2Cu^{2+} \rightarrow Cu_2O + 2H^+$
   $2Cu^{2+} + CH_3CHO + 2H_2O \rightarrow Cu_2O + CH_3CO_2H + 4H^+$
   iii. $CH_3CH(CH_3)CHO + H_2O \rightarrow CH_3CH(CH_3)CO_2H + 2H^+ + 2e^-$
   $6e^- + 14H^+ + Cr_2O_7^{2-} \rightarrow 2Cr^{3+} + 7H_2O$
   $8H^+ + 3CH_3CH(CH_3)CHO + Cr_2O_7^{2-} \rightarrow 3CH_3CH(CH_3)CO_2H + 4H_2O + 2Cr^{3+}$ (**M**)

   b. i. Colourless solution forms a silver mirror on the surface of the container, or a grey solid.
   ii. Blue solution of Cu^{2+} forms an orange-red precipitate of Cu_2O.
   iii. Orange solution of $Cr_2O_7^{2-}$ goes green due to formation of Cr^{3+}. (**M**)

Unit 12.4 Activity 9A: Organic reactions summary (page 262)

1. **A** Ester.
   B Carboxylic acid.
   C Haloalkane (chloroalkane).
   D Alkene.
   E Alcohol. (***A***)

2.

Molecule	Functional group	Name
i.	Alkene	But-1-ene
ii.	Carboxylic acid	Propanoic acid
iii.	Alkyne	Pent-2-yne
iv.	Alcohol (tertiary)	Methyl propan-2-ol
v.	Ester	Ethyl butanoate

(***M***)

3. **a.** Primary. (***A***)
   b. Oxidation. (***A***)
   c. Bubbles of gas produced. (***A***)
   d. **B** will react rapidly to decolorise bromine. **E** will react slowly in the presence of light to decolorise bromine. (***M***)
   $CH_3CH{=}CHCH_3 + Br_2 \rightarrow CH_3CHBrCHBrCH_3$ addition
   $CH_3CH_2CH_2CH_3 + Br_2 \rightarrow CH_3CH_2CH_2CH_2Br$ substitution
   e. Compounds with the same molecular formula but a different arrangement in space. Examples are **B** and **F**, which have the molecular formula C_4H_8 but are arranged differently. Also **C** and **E**. (***E***)
   f. 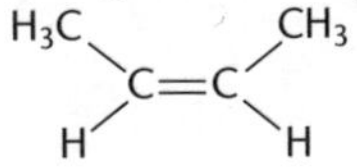
   H_3C and H on one C, $=$ C with CH_3 and H (same side): *cis*-But-2-ene 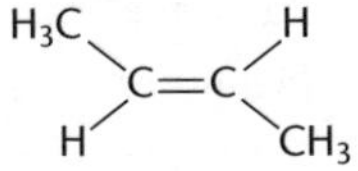
   H_3C and H on one C, $=$ C with H and CH_3 (opposite side): *trans*-But-2-ene (***M***)
   g. **A** and **E** will be water soluble but **C** will not, so add water to identify **C** (which will form two layers when water is added). **E** is a carboxylic acid, so will fizz when added to sodium carbonate. The remaining liquid is **A**. (***E***)

4. **a.** **i.** 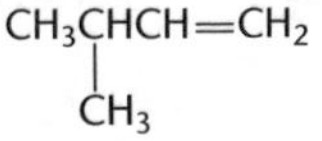 $CH_3CH(CH_3)CH{=}CH_2$ (CH₃ branch on the second carbon)
   ii. 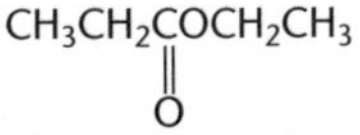 $CH_3CH_2C(=O)CH_2CH_3$
   iii. 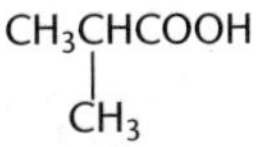 $CH_3CH(CH_3)COOH$
   b. Acidified (potassium) dichromate solution – will turn from orange to green.

5. **a.** **A** Ester.
   B Alkene.
   C Carboxylic acid. (***A***)
   b. **B** Alkene. (***A***)
   c. Fat/oil. (***A***)

6. **a.** **i.** $CH_3CH_2CH_2CH_2Cl$

ii. $CH_3CH{=}CH_2$

iii. $CH_3CH_2\underset{\mid}{\underset{OH}{CH}}{-}\underset{\mid}{\underset{OH}{CH_2}}$

iv. $CH_3CH_2\underset{\|}{\underset{O}{C}}OH$

v. $CH_3CH_2CH_2\underset{\|}{\underset{O}{C}}OCH_2CH_3$ (***M***)

b. **i.** Substitution.

ii. Elimination.

iii. Oxidation.

iv. Oxidation.

v. Esterification. (***A***)

7.

X	$CH_3CH_2CH_2CH_2OH$	Butan-1-ol
Y	$CH_3CH_2CH_2COOH$	Butanoic acid
Z	$CH_3CH_2CH_2\underset{\|}{\underset{O}{C}}OCH_2CH_2CH_2CH_3$	Butyl butanoate
W	$CH_3CH_2CH{=}CH_2$	But-1-ene
V	$CH_3CH_2CH_2CH_2Cl$	1-Chlorobutane
V′	$CH_3CH_2\underset{\mid}{\underset{Cl}{C}}HCH_3$	2-Chlorobutane

V and **V′** may be reversed.

8.

A	$CH_3CH_2CH{=}CH_2$	But-1-ene
B	$CH_3CH_2\underset{\mid}{\underset{OH}{C}}HCH_3$	Butan-2-ol
C	$CH_3CH_2CH_2\underset{\|}{\underset{O}{C}}OH$	Butanoic acid
D	$CH_3CH_2CH_2\underset{\|}{\underset{O}{C}}OCH_2CH_3$	Ethyl butanoate

E $Cr_2O_7^{2-}/H^+$

F NaOH(aq) (***E***)

9. **a.** $CH_3CH_2CH_2CH_3$

b. $CH_3\overset{\overset{CH_3}{\mid}}{C}{=}CH_2$

c. $H\underset{\|}{\underset{O}{C}}OCH_2CH_2CH_2CH_3$

d. $CH_3\underset{\mid}{\underset{CH_3}{C}}HCH_2\underset{\|}{\underset{O}{C}}OH$

e. $CH_3\underset{\mid}{\underset{OH}{C}}H{-}\underset{\mid}{\underset{OH}{C}}HCH_3$ (***M***)

10. **a.** **i.** Add bromine water to each sample.

ii. Cyclohexene will decolorise bromine, but propyl ethanoate will not react. Propene is an alkene, so will add bromine across the double bond.

b. **i.** Use blue litmus or sodium carbonate or magnesium metal.

ii. Ethanoic acid will turn blue litmus red or will form bubbles with sodium carbonate or magnesium metal. Only one answer is needed. 1-chlorobutane will not react with these reagents. Ethanoic acid reacts because of its acid properties.

c. **i.** Use acidified dichromate solution, $Cr_2O_7^{2-}/H^+$.

ii. Butan-1-ol is an alcohol, so will be oxidised by the acidified dichromate to form butanoic acid. The colour of the dicromate solution changes from orange to green. Hexane does not react with dichromate.

d. **i.** Use bromine water.

ii. Cyclohexene reacts rapidly to decolorise the bromine. Cyclohexane reacts slowly in the presence of light. Cyclohexene is an alkene, so the bromine adds across the double bond. The alkene is more reactive than the alkane. Cyclohexane substitutes a bromine atom for a hydrogen atom. (***E***)

11. **A.** $CH_3CH_2CH{=}CH_2$ But-1-ene

B. $CH_3CH_2CH_2CH_2OH$ Butan-1-ol

C. $CH_3CH_2CHCH_3$ Butan-2-ol
|
OH

D. $CH_3CH_2CH_2COH$ Butanoic acid.
‖
O

CH_3
|
E. $CH_3CH_2CH_2COCHCH_2CH_3$
‖
O

12. **F.** $CH_3CH_2CH_2OCCH_2CH_3$ Propyl propanoate
‖
O

G. $CH_3CH_2CH_2OH$ Propan-1-ol

H. CH_3CH_2COH Propanoic acid
‖
O

I. $CH_3CH{=}CH_2$ Prop-1-ene

J. $CH_3CH{-}CH_3$ Propan-2-ol (***E***)
|
OH

13. **a.** Ethanol is a primary alcohol, so will be oxidised to a carboxylic acid by acidified dichromate, $Cr_2O_7^{2-}/H^+$, or acidified permanganate, MnO_4^-/H^+. The dichromate changes from orange to green. The permanganate changes from purple to colourless.

Only one reagent is needed.

Chloroethane does not react.

b. Butanoic acid will turn blue litmus red, because of its acid properties. Hexane will not react to litmus. (Butanoic acid will produce a gas with sodium carbonate or magnesium ribbon.)

14. **a.** **Z** is H_2O/H^+ (***A***)

b. **A** $CH_3\underset{\underset{O}{\|}}{C}OH$ carboxylic acid ethanoic acid

B $CH_3CH_2O\underset{\underset{O}{\|}}{C}CH_3$ ester ethyl ethanoate

C CH_3CH_3 alkane ethane

D CH_3CH_2Br haloalkane bromoethane (***M***)

c. $CH_3CH{=}CH_2 + HBr \rightarrow CH_3\underset{\underset{Br}{|}}{C}HCH_3$ and $CH_3CH_2CH_2Br$

major minor

15. **a.** Methane + steam → carbon monoxide + *hydrogen*.

b. $CO(g) + 2H_2(g) \rightarrow CH_3OH(\ell)$.

c. Starch + *water* $\xrightarrow{\text{amylase}}$ maltose.

d. $C_{12}H_{22}O_{11}(aq) + H_2O(\ell) \xrightarrow{\text{maltose}} 2C_6H_{12}O_6(aq)$.

(***A*** – **a.** correct; ***M*** – **a.** and **b.** correct; ***E*** – all equations correct)

16. Alkane molecules are not attracted to water molecules whereas ethanol molecules are attracted. The alkane molecules are not polar whereas the ethanol molecule, due to the presence of an oxygen atom, is polar. For two liquids to mix, the molecules of each liquid must attract each other strongly.

(***A*** – alkane molecules not attracted to water molecules; ***M*** – ethanol and water molecules attract each other because they are both polar)

17. Carbon dioxide and water. (***A***)

18. **a.** Different. (***A***)

b. Ethanoic acid is a weak acid, ie it has a lower acidity than sulfuric acid of the same concentration. (***M***)

19. The high concentration of ethanol destroys the bacteria that operate with oxygen to convert the ethanol to ethanoic acid. (***A***)

20. **A** $CH_3\underset{\underset{CH_3}{|}}{C}H{-}CH_2CH_2Br$ 1-Bromo-3-methylbutane

B $CH_3\underset{\underset{CH_3}{|}}{C}H\underset{\underset{Br}{|}}{C}HCH_3$ 2-Bromo-3-methlybutane

C $CH_3\underset{\underset{CH_3}{|}}{C}HCH_2CH_2O\underset{\underset{O}{\|}}{C}CH_3$ 3-Methylbutyl ethanoate

D $CH_3\underset{\underset{CH_3}{|}}{C}HCH_2\underset{\underset{O}{\|}}{C}ONa$ Sodium 3-methylbutanoate

E $Cr_2O_7^{2-}/H^+$ (***E***)

21. Bromine water will decolorise 1-pentene. Propanoic acid is soluble in water, so will mix with bromine water, but the colour (brown) will remain. Ethyl ethanoate is not soluble in water and will not react with bromine – two layers will form when mixed with bromine water. One layer will be coloured. (**E**)

22. **a.** **A** $CH_3CH_2CH_2CH_3$ **B** Br_2 (in the presence of light)

b. **C** $CH_3CH{=}CH_2$ **D** HCl

c. **E** $CH_3C(=O)OH$ **F** $CH_3CH_2CH_2OH$ (**M**)

23. **a.** Acidified potassium dichromate or acidified potassium permanganate solution. (**A**)

b. Diagram **iii**. shows the process of refluxing. Heating the flask increases the rate of reaction. In this set-up, liquid in the flask evaporates on heating, but the condenser attached to the flask results in the vapour being condensed and returned to the flask. This means the organic molecules do not escape from the flask. (**M**)

c. **i.** CH_3CH_2CHO (structure: CH_3CH_2C with $=O$ and $-H$) propanal. (**M**)

ii. Change to a setup like diagram i., with the thermometer replaced by a dropping funnel. The dichromate and alcohol mixture would be slowly added to hot acid. The aldehyde product would be distilled off as it forms, since it has a lower boiling point than the alcohol. This prevents any further oxidation of the aldehyde to the corresponding carboxylic acid. (**E**)

Unit 12.5 Activity 1A: The carbon cycle (page 270)

1. Word equation: glucose + oxygen → carbon dioxide + water + energy
Chemical equation: $C_6H_{12}O_6 + 6O_2 \rightarrow 6CO_2 + 6H_2O$ + energy

2. Word equation: carbon dioxide + water + energy → glucose + oxygen
Chemical equation: $6CO_2 + 6H_2O$ + energy (light) → $C_6H_{12}O_6 + 6O_2$

3. Chlorophyll:

- is a green photosynthetic pigment found in most plants, algae, and cyanobacteria.
- is vital for photosynthesis, which will not occur in the absence of chlorophyll.
- is found in chloroplasts, where it is associated with pigment protein complexes called photosystems, which are embedded in the thylakoid membranes of chloroplasts.
- absorbs light and transfers the light energy to a specific chlorophyll pair in the reaction centre of the photosystems.

Chlorophyll has a similar structure to haemoglobin – a porphyrin ring coordinated to a central atom, which is magnesium in chlorophyll (iron in haemoglobin). Chlorophyll a and b have different R groups.

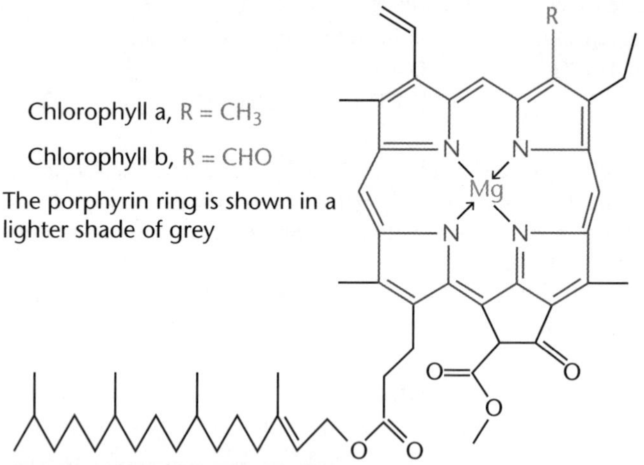

Chlorophyll a, R = CH_3

Chlorophyll b, R = CHO

The porphyrin ring is shown in a lighter shade of grey

Chlorophyll a (R = CH_3) and chlorophyll b = CHO

4. The ocean is an important carbon 'sink' because it takes up more carbon from the atmosphere than it gives up.
   - CO_2 from the atmosphere dissolves in the surface waters of the ocean.
   - Some CO_2 stays dissolved, but much of it gets converted into organic matter in photosynthesis by tiny marine plants (phytoplankton) in the sunlit surface waters.
   - Some carbon gets converted to calcium carbonate, a building material of shells and skeletons. Other chemical processes produce calcium carbonate in the water. As CO_2 is removed from the water for biological and chemical processes, more CO_2 enters the water from the atmosphere.
   - Marine organisms move carbon from the atmosphere, to surface waters, then to the deeper ocean and eventually into rocks. This process is often called a biological pump.
   - When organisms die, their dead cells, shells and other parts sink into deep water. As they decay CO_2 is released into this deep water.
   - Some material sinks to the bottom, where it forms layers of carbon-rich sediments, which eventually may turn into rocks. This part of the carbon cycle can lock up carbon for millions of years.

Unit 12.5 Activity 1B: Coal (page 271)

1. Coal is a combustible black or brownish-black sedimentary rock, composed of carbon hydrocarbons, nitrogen, sulfur, oxygen and water. It usually occurs in rock strata in layers or veins called coal beds or coal seams.
2. Coal is mainly burned as a fossil fuel for the generation of electricity and heat. It is also used industrially for refining metals. Destructive distillation of coal gives gaseous fuels, coke, coal tar and ammonia.

3. MX + R → M + RX, ie the reducing agent R combines with the non-metal X in the metallic compound MX (metal oxide, metal sulfide or metal chloride) via a substitution reaction to yield the metal M and RX.

4. Blast furnaces are used for metals of moderate melting point and low vapour pressure. The furnace is a circular steel shaft about 30 m tall and 9 m in diameter. Ore, limestone and coal are added to the top of the blast furnace. Hot air travels up from the bottom.

   Reduction of iron oxide in a blast furnace:

   - at 400–700°C, CO is produced by incomplete combustion of coal ($CO_2 + C \rightarrow 2CO$). CO reduces Fe_2O_3 to FeO ($Fe_2O_3 + CO \rightarrow FeO + CO_2$)
   - at 700–1000°C, FeO is reduced to Fe ($FeO + C \rightarrow Fe + CO$) and Fe melts at 1150°C
   - at above 1500°C, slag floats on liquid Fe, which is extracted and solidifies into sand or clay moulds.
   - escaped CO is trapped and reused for preheating the blast furnace or gas engines for power generation.

Unit 12.5 Activity 1C: Oil (petroleum) (page 272)

1. Propane is C_3H_8; butane is C_4H_{10}. Both are alkanes and are gases at room temperature.
2. **a.** Fuel oils are a variety of liquid mixtures obtained from the distillation of petroleum; they are used for burning as fuel.

   b. Paraffin wax is most commonly used is in solid form as a lubricant; in candle making; as a food preservative on cheese, fruits, vegetables and sweets; in fertilisers.

   c. Bitumen is used in construction, particularly in paving and roofing, and in roads

   d. Kerosene is used as a fuel oil for lighting, heating and cooking.
3. Jet fuel is a mixture of many different C_5–C_8 hydrocarbons derived from oil, of boiling range 20–150°C. Kerosene is a mixture of longer chain C_9–C_{16} hydrocarbons of boiling range 175–275°C.

Unit 12.5 Activity 1D: Processes in oil refining (page 275)

1. Waxes that are crystalline substances are separated from non-crystalline materials by cooling a solution or a melt and then filtering, ie crystallisation.
2. In linearity characteristic separation, molecules with linear structures are separated from branched ones. In aromaticity separation, aromatic molecules are separated from linear ones.
3. Methane, ethane, propane, butane.

Unit 12.5 Activity 1E: Conversion methods (page 278)

1. Catalytic reforming converts low-octane compounds to high-octane compounds (reformates) by rearranging some hydrocarbons and breaking others into smaller molecules. Catalysts are Pt or Pt–Re on alumina and the process that occurs requires heat and high pressure.
2. The octane rating (or octane number) of a fuel is a measure of its performance. It refers to the tendency of some hydrocarbons to ignite spontaneously in an engine.

It is a measure of a fuel's tendency to burn in a controlled manner, rather than exploding in an uncontrolled manner. The higher the rating, the more the fuel can be compressed before it ignites, which could damage the engine.

The octane rating is measured in a test engine and is defined by comparison with the mixture of 2,2,4-trimethylpentane (isooctane) and heptane that would have the same anti-knocking capacity as the fuel under test. The octane number is assigned by comparing the sample of fuel to mixtures of heptane = 0 and isooctane =100.

3. **a.** Isobutene $H_3C(CH_3)C{=}CH_2$

b. Isobutane $H_3C–CH(CH_3)–CH_3$

c. 2,2,4-Trimethylpentane $H_3C–C(CH_3)2–CH_2–CH(CH_3)–CH_3$

4. **a.** C_7H_{14} (+ Pt–Re catalyst) $\rightarrow C_6H_5–CH_3 + 3H_2$

b. $C_8H_{18} \rightarrow H_3C–CH(CH_3)–CH_2–CH_2–CH(CH_3)–CH_3$

Unit 12.5 Activity 1F: Petroleum in PNG (page 280)

1. It has a low sulfur content.

2. Kutubu crude has a low percentage of mixed naptha (11.4%) and diesel (38.0%). Hence, Napa Napa Refinery needs to import crude from Australia or Asia to supply the PNG market.

3. Yes. Kutubu crude yields a high percentage of fuel gas (0.4–0.6%), LPG (3.9%), unleaded petrol (10.7%) and low-sulfur waxy residue (15.7%). Hence, refineries from Australia and Asia should consider importing crude oil from PNG.

Unit 12.5 Activity 1G: Liquefied natural gas (page 281)

1. Liquefied natural gas is natural gas (predominantly methane, CH_4) that has been converted to liquid form for ease of storage or transport. Liquefied natural gas takes up about 1/600th the volume of natural gas in the gaseous state. It is odourless, colourless, non-toxic and non-corrosive.

2.

```
     H
     |
 H—C—H
     |
     H
```

Methane

3. Natural gas is used as a fuel in its gaseous state. LNG is natural gas that has been cooled below its boiling point to –160°C to a liquid for storage and transport.

4. As a liquid, LNG is not explosive. LNG vapour will only explode in an enclosed space as a 5–15% mixture in air.

Unit 12.5 Activity 1H: LNG refining (page 282)

1. LNG is transported as a liquid in special double-hulled ships designed to handle low temperatures.

2. LNG is stored in insulated storage tanks that keep the temperature low and reduce evaporation. Full containment systems use two tanks – an inner tank that contains the LNG, and an outer tank to contain any leaks from the inner tank.

3. As a liquid, LNG is not explosive. LNG vapour will only explode in an enclosed space as a 5–15% mixture in air.

4. LNG is very convenient to transport. It takes up about 1/600th the volume of natural gas in the gaseous state.
5. Refuse haulers, local delivery (grocery trucks), transit buses. Typically vehicles that are classified as Class 8 (above about 15 000 kg, gross vehicle weight) use LNG.

Unit 12.5 Activity 2A: Addition polymerisation (page 284)

1. 2,2-Dichloroethene, $CH_2{=}CCl_2$
2. Carbon–carbon double bond (C=C).
3. **a.** $-[CH_2-C(CHCN)]_n-$
   b. $-[CH_2-C(CHCH_3)]_n-$
   c. $-[CF_2-CF_2]_n-$
   d. $-[CH_2-CHCl]_n-$

Unit 12.5 Activity 2B: Condensation polymerisation and polyesters (page 285)

1. Two organic molecules join together to form a larger one.
2. Removal of a simple molecule in the process of joining monomers.
3. Hydrogen chloride, sodium hydroxide.
4. $[-HO(CH_2)_3-COOH]_n$

Unit 12.5 Activity 2C: Polyamides (page 286)

a. Pentanedioic acid $HOOC(CH_2)_3-COOH$; 1,5-diaminopentane: $H_2N(CH_2)_5NH_2$
b. $H_2N(CH_2)_5-NH-OC(CH_2)_3-COOH$
c. $-[H_2N(CH_2)_5-NH-OC(CH_2)_3-CO]_n-$

Unit 12.5 Activity 2D: Proteins (page 287)

1. Proteins are formed when amino acids join together. Two amino acids join together to form a dipeptide. Many amino acids join together to form a polypeptide.
2. $H_2N(CH_2)_2CONHCH(CH_3)CH_2COOH$, $H_2NCH(CH_3)CH_2CONH\ CH_2C\ H_2COOH$
3. The sequence of amino acid in protein does not repeat itself.
4. The acidic end of the molecule easily loses the proton to the amine end.

Unit 12.5 Activity 2E: Plastics (page 290)

1. Any of a wide range of synthetic or semisynthetic organic solids that can be moulded. All plastics are soft and mouldable during their production.
2. Acquiring the raw material or monomer. Synthesising the basic polymer. Compounding the polymer into a material that can be used for fabrication. Shaping the plastic into its final form.
3. Extrusion moulding: softened plastic is forced through a shaped die; driving force is supplied by a screw, which provides constant pressure. Compression moulding: plastic forced into shapes by pressure; plastic is put into one half of a two-piece mould and the two halves are brought together and the plastic is melted under high pressure.

Unit 12.5 Activity 2F: Rubber (page 292)

1. South America.
2. Monomer: isoprene $CH_2=C(CH_3)–CH=CH_2$
   Polymer: polyisoprene $–[CH_2C(CH_3)=CH(CH_2)]_n–$
3. Indonesia, Thailand and Malaysia.
4. China.
5. Indonesia (from 1630 to 1810 tonnes).
6. Neglect of the rubber industry by government and private industry.
7. Oil and petroleum.

Unit 12.5 Activity 3A: Gold in PNG (page 294)

1. Some operating gold mines are Ok Tedi Mining Ltd (Western Province), Barrick (PNG) Ltd (Enga Province), Lihir Gold Ltd (New Ireland Province), Tolukuma (Central Province), Kainantu (Eastern Highlands Province). Other deposits are in Hidden Valley and Simberi Island.
2. Positive impacts: increased economic activity around the mine area, enhanced cash flow and money circulation and increased level of affluence. Negative impacts: threats to social fabric of communities around the area leading to collapse of traditional systems and values.

Unit 12.5 Activity 3B: Properties and chemistry of gold (page 295)

1. Historical accounts as early as 400 BCE show that gold was used extensively by the Egyptians, Babylonians and other ancient civilisations. (Students are encouraged to research history of use of gold.)
2. Malleability and durability.
3. Gold is less reactive than copper and thus deposits of gold complexes tend to be found underneath deposits of copper.
4. Traditional PNG used an economic system based on the bartering of perishable items such as fish and sago used in the Hiri-Motu trade between the Motuans and the Gulf people. Similar trading arrangements have been operational in other parts of PNG. Kina and toea shell money were more durable and were used for special situations such as bride price.
5. Dentistry, treatment for rheumatoid arthritis, potential cancer treatment, surgical implants, surgical instruments.
6. **a.** $AuCl_4^-(aq)$ – tetrachloroaurate(III), oxidation number of gold is +3, ie $(x + 4^* - 1 = -1)$.
   b. $[Au(CN)_2]^-(aq)$ – cyanoaurate(I) oxidation number of gold is +1, ie $(x + 2^* - 1 = -1)$.

Unit 12.5 Activity 3C: Extraction and processing of gold (page 296)

1. Pyrometallurgy processes use high temperature to carry out reduction and smelting reactions, while hydrometallurgy processes use aqueous solutions and inorganic solvents to achieve extractive reactions (leaching chemistry).
2. Both methods use cyanide solution to form the gold complex. The difference is the method of reduction. In cyanidation, the cyanide–gold complex is reduced by zinc to

gold metal. In the carbon-in-pulp method, the cyanide–gold complex gets reduced by carbon to gold metal.

3. Frothing agents, collecting agents and air bubbles are used to attract gold minerals to the surface. The gold is carried in the bubbles to the surface and skimmed off to be refined.
4. Mercury liquid and its vapour are poisonous to humans, so miners who use mercury in gold extraction are at risk.

Unit 12.5 Activity 3D: Copper in PNG (page 297)

1. The copper mines are Ok Tedi Mining Ltd (Western Province) and BCL Panguna (Bougainville). Other copper deposits are in Hidden Valley.
2. Positive impacts: increased economic activity around the mine area, enhanced cash flow and money circulation and increased level of affluence. The PNG government derived 20% of its income from the mine alone. Port Moresby continued to grow with contributions from BCL. Negative impacts: threats to social fabric of communities around the area leading to collapse of traditional systems and values. The biggest impact was the 10-year civil war, which claimed an estimated 15 000 lives and crippled the PNG economy.
3. Copper iron sulfide, chalcopyrite (or copper pyrite).
4. Copper(II) oxide.
5. **a.** Cu^{2+} is central to O^{2-} and OH^- arranged in an octahedral shape with $CO_3{}^{2-}$ forming chain links between the octahedrons.
   b. Cu^{2+} is central to O^{2-} and OH^- arranged in a square planar shape with $CO_3{}^{2-}$ forming chain links between the square planar shapes

Unit 12.5 Activity 3E: Applications of copper (page 298)

1. Copper is an excellent electrical conductor (better than any metal, except silver) and thermal conductor, is tough, non-magnetic, corrosion resistant, an attractive reddish colour, antibacterial, easy to alloy, recyclable, ductile and a good catalyst.
2. Copper is more reactive than gold and thus deposits of copper oxides tend to be found above deposits of gold.

Unit 12.5 Activity 3F: Extraction and processing of copper (page 299)

1. To convert sulfides to oxides, thus separating the iron and sulfur. The reaction is:
   $4CuFeS_2(s) + 3O_2(g) \rightarrow 2CuS_2(s) + 2FeO(s) + 2FeS(s) + 2SO_2(g)$
2. Converts the matte to copper when air is blown through the molten matte via a series of chemical reactions such as:
   $2CuS_2(s) + 3O_2(g) \rightarrow 2Cu_2O(s) + 2SO_2(g)$
   $2Cu_2O(s) + CuS_2(s) \rightarrow 6Cu(s) + SO_2(g)$
3. Blister copper is the product of the bessemerisation process. During the process SO_2 bubbles escape from the material leaving blisters on the surface of the product. Blister copper contains 97–98% copper.
4. The secondary and final stages of the copper extraction industry have not been developed in PNG due to lack of government policy (and enforcement) on downstream processing.

Unit 12.5 Activity 3G: Nickel and cobalt in PNG (page 299)

1. Madang Province.
2. Metal Corporation of China (MCC) Ramu NiCo Limited has the largest interest and contribution of 85%, followed by Ramu Nickel Limited (8.5%). The other partners are Mineral Resource Limited (4%) and Mineral Resource Madang Limited (2.5%).

Unit 12.5 Activity 3H: Properties and chemistry of nickel (page 301)

1. Nickel is used primarily as alloys in applications such as stainless steel and many other corrosion-resistant alloys, coinage, armour plating, protective coatings, catalysts for hydrogenating vegetable oils (to form margarine), ceramics, magnets and batteries.

   Electronic configuration of Ni (28e): $1s^22s^22p^63s^23p^64s^23d^8$ or [Ar] $4s^23d^8$.
2. **a.** Nickel hydroxide.

   b. Hexaaqua nickel(II) ion.
3. **a.** Linear structure: HO–Ni–OH

   b. Octahedral structure: Ni at centre surrounded by 6 H_2O molecules (4 planar, 2 axial).
4. Shape-memory alloys are metals that 'remember' their original shape. Nickel–titanium alloys have been found to be the most useful of all shape-memory alloys.

Unit 12.5 Activity 3I: Properties and chemistry of cobalt (page 302)

1. Cobalt is similar to nickel in its properties. It is a brittle, hard, ferromagnetic, silver-grey transition metal. Cobalt salts colour glass a beautiful deep-blue colour.

   The isotope cobalt-60 (^{60}Co) is an artificially produced isotope used as a source of gamma (γ) rays (its high-energy radiation is useful for sterilisation in medicine and of foods).
2. **a.** Cobalt hydroxide.

   b. Cobalt arsenide – smaltite.
3. **a.** Cobalt sulfide – cobaltite. Each Co atom is central and connected to 6 S atoms in an octahedral shape. Each S atom is coordinated to one other Co atom. Thus three Co atoms share 4 S atoms.

   b. Structure is based on corrin ring similar to the porphyrin ring found in heme and chlorophyll. The central metal ion is cobalt.

Unit 12.5 Activity 3J: Extraction of nickel and cobalt (page 303)

1. Nickel is recovered through extractive metallurgy. Nickel is extracted from its ores by conventional roasting and reduction processes that yield a metal of greater than 75% purity. In hydrometallurgical processes, nickel sulfide ores undergo flotation (differential flotation if Ni/Fe ratio is too low) and then smelted. The purity of nickel is enhanced. Solvent extraction is used to separate the cobalt and nickel, with the final nickel concentration greater than 99%.

   Cobalt is extracted through smelting, using heat from a blast furnace and a metal reducing agent, such as carbon, to change the ore's oxidised state. The carbon removes the ore's oxygen and leaves the metal cobalt. The cobalt is taken out to attain 99.9% pure cobalt.

Cobalt is extracted from the nickel or copper by solvent extraction, which is separating compounds based on their solubility in two immiscible liquids, such as an organic solvent and water. Hence, it extracts the cobalt substance from one liquid into another liquid phase.

2. The Mond process increases the nickel concentrate to greater than 99.99% purity. The highly pure nickel produced by this process is known as carbonyl nickel. First, nickel is reacted with carbon monoxide at 40–80°C to form nickel tetracarbonyl in the presence of a sulfur catalyst. The reaction is exothermic. Since the reaction is exothermic and reversible, a low temperature and high pressure is maintained to ensure higher yield of the nickel tetracarbonyl which boils at 42°C.

   Next, the nickel tetracarbonyl is decomposed in a smaller chamber via the reverse endothermic reaction at 230°C to create fine nickel powder. The resultant carbon monoxide is recirculated and reused.

Unit 12.5 Activity 3K: Chemistry and applications of silver (page 304)

1. $AgNO_3 \rightarrow Ag + NO_3$
2. $AgNO_3(aq) + HCl(aq) \rightarrow AgCl(s) + HNO_3(aq)$
3. Take a portion of solution to be tested and add several millilitres of HCl solution. Formation of white precipitate (AgCl(s)) confirms presence of Ag^+ ions.

   Reaction: $Ag^+ + Cl^- \rightarrow AgCl(s)$ (white precipitate)
4. Silver bromide, a silver halide, is a light-sensitive compound used in film emulsion. Other silver halides used in film include silver chloride and silver iodide.
5. The antibiotic property and non-toxicity of silver has made it useful in medicine for thousands of years. It is added to bandages and wound dressings, catheters and other medical instruments and is used in X-rays.

Unit 12.5 Activity 3L: Extraction and processing of silver (page 305)

1. Chalcopyrite $CuFeS_2$; sphalerite (Zn,Fe)S.
2. Argentite Ag_2S; pyragyrite Ag_3SbS_3.
3. Removal of unwanted constituents by transfer into a slag phase because the slag phase is more oxidising than the metal phase. Slag refining is employed for less reactive metals such as silver, gold and copper.

Unit 12.5 Activity 3M: Ocean mining in PNG (page 306)

1. Site is not too deep and is well shielded from rough seas by both New Ireland and New Britain.
2. Bismarck Sea has known deposit and is well shielded from high seas.
3. To ensure easy withdrawal of mining activities should anything go wrong.

Unit 12.5 Activity 3N: Hydrothermal vents (page 307)

1. The deposits were laid down over thousands of years around underwater hot springs, or hydrothermal vents. They occur at depths of 1–2 km and can range in mass from several thousand to 100 million tonnes.
2. Cracks in the sea floor appear and cold seawater enters. The water is heated to around 400°C by magma or 'hot rocks'. The seawater becomes acidic, low in oxygen and highly concentrated in metals dissolved from the surrounding rocks – most of

which are in the form of sulfide. The superheated water (or vent fluid) becomes buoyant and rushes up through the seabed – chimney-like towers are created. Vent fluid continues to gush out of their hollow centres, depositing more sulfide materials.

3. The Mid-Ocean Ridge and Mid-Atlantic Ridge are volcanically active chains of mountain ranges. The Mid-Ocean Ridge is an 84 000 km chain of volcanic mountains that extends throughout the world's oceans. The Mid-Atlantic Ridge is located along the floor of the Atlantic Ocean.
4. Chemosynthetic bacteria, animals such as fish, tubeworms, mussels, clams, snails, shrimps and crabs, the blind Atlantic vent shrimp *Rimacaris*, worms and polychaetes.

Unit 12.5 Activity 4A: Beer (page 310)

1. Alcohol is the intoxicant present in fermented and distilled alcoholic beverages such as beer, wine and spirits. The most commonly used alcohol is ethanol, C_2H_5OH, with the ethane backbone. It is a clear flammable liquid that boils at 78.4°C, which is used as an industrial solvent, car fuel and raw material in the chemical industry.
2. Ethanol C_2H_5OH, methanol CH_3OH.
3. Beer is a fermented beverage that is flavoured with hops and malted with sugar.
4. The four main families of beer styles are determined by the variety of yeast used in their brewing: ale (top-fermenting yeasts), lager (bottom-fermenting yeasts), beers of spontaneous fermentation (wild yeasts), beers of mixed origin.

Unit 12.5 Activity 4B: Brewing (page 313)

1. B
2. **a.** Spent grains are filtered out and the wort is ready for boiling. Decisions will be made about the flavour, colour and aroma of the beer.
   b. Barley is prepared for brewing. Each step of the malting process unlocks the starches hidden in the barley.
   c. The finely ground malt is converted into a sweetened liquid. Mashing converts the starches that were released during malting to sugars that can be fermented.
   d. The brewer selects a type of yeast and adds it to the fermentation tank. The yeast consumes the sugar in the wort and turns it into alcohol and carbon dioxide.
   e. The beer is filtered and carbonated – carbon dioxide is injected to give beer sparkle.
3. **a.** Barley grains.
   b. Carbon dioxide and wort.
   c. Wort and carbon dioxide.
   d. Hops.
4. Ethanol, C_2H_5OH, $H_3C–CH_2–OH$

Unit 12.5 Activity 4C: Fermentation and distillation (page 314)

1. Sugar, ethanol and spirits (rum).
2. Molasses.
3. **a.** Molasses is diluted with water to reduce the sugar content. The yeast that is added converts the available sucrose to ethanol, carbon dioxide and energy. This process takes about 30 hours. The fermentation reaction is:
   molasses + yeast → ethanol + carbon dioxide + energy

b. After fermentation the liquid ('dead wash') is distilled and the vapours condensed to produce ethanol. Distillation separates and concentrates the alcohol component of the liquid mixture.

4. Live wash is the mixture formed from diluting molasses in water and adding yeast at the beginning of fermentation process. Dead wash is the liquid left at the end of the fermentation process.

Unit 12.5 Activity 4D: Ethanol (page 315)

1. Ethanol burns with a smokeless blue flame that is not always visible in normal light. Ethanol is slightly more refractive than water, having a refractive index of 1.36242 (at $\lambda = 589.3$ nm and 18.35°C). The triple point for ethanol is 150 K at a pressure of 4.3×10^{-4} Pa.
2. Ethanol is a versatile solvent, miscible with water and many organic solvents such as acetic acid, acetone and benzene. Ethanol is also miscible with light hydrocarbons such as pentane and hexane, and chlorides, eg trichloroethane and tetrachloroethylene. Ethanol–water mixtures have less volume (1.92) than the sum of their individual components at the given fractions. The reaction is exothermic, up to 777 J mol^{-1} being released at 298 K.
3. 1.92×50 mL = 96 mL

Unit 12.5 Activity 4E: Effects of ethanol (page 316)

1. Ethanol is mostly used as a social drug for relaxation at low dosage.
2. It is fatal at high dosages. Intoxication is responsible for many social problems in society.
3. The ethanol content is not monitored, which could lead to death.
4. Transportation fuel, waste water treatment, fuel cells, chemical feedstock.

Unit 12.5 Activity 5A: Oil from plants (page 318)

1. There are more than 40 plant species with edible kernels in PNG. 11 species such as galip, karuka, okari and aila are used in village agriculture.
2. Locally: Cut nut (*Barringtonia procera*), galip (*Canarium indicum*), karuka (*Pandanus julianettii*), okari (*Terminalia kaernbacchii*), aila (*Inocarpus fagifer*), talis (*Terminalia catappa*). International: In coconut oil, palm oil and peanut oil.
3. Oil palm has been recently introduced to PNG and the PNG Government has embarked on an exercise to secure customary land for cultivation through its Special Agricultural Business Lease (SABL) program. Other plant species can also be cultivated.

Unit 12.5 Activity 5B: Coconut oil (page 319)

1. In margarine, soap, cosmetics, non-dairy cream, snack foods, industrial lubricants, cold rolling of steel strip.
2. Lauric acid 45%, myristic acid 17%, palmitic acid 8%, caprylic acid 8%.
3. As a practical activity, observe the factory processes.

Unit 12.5 Activity 5C: Extracting and processing palm oil (page 320)

1. Threshing, sterilisation or boiling, digestion of fruits, pressing, clarifying and drying.
2. Cell debris, fragments from seeds, fibrous material, non-oily solids, fragments of sand, stone and other impurities.
3. Cooking eg frying food.
4. As a practical activity, observe the factory processes.

Unit 12.5 Activity 5D: Saponification (page 320)

1. A salt of fatty acids produced from hydrolysis of esters with an alkali during a saponification reaction.
2. Fats and oils (triesters of glycerol) and bases (NaOH, KOH).
3.

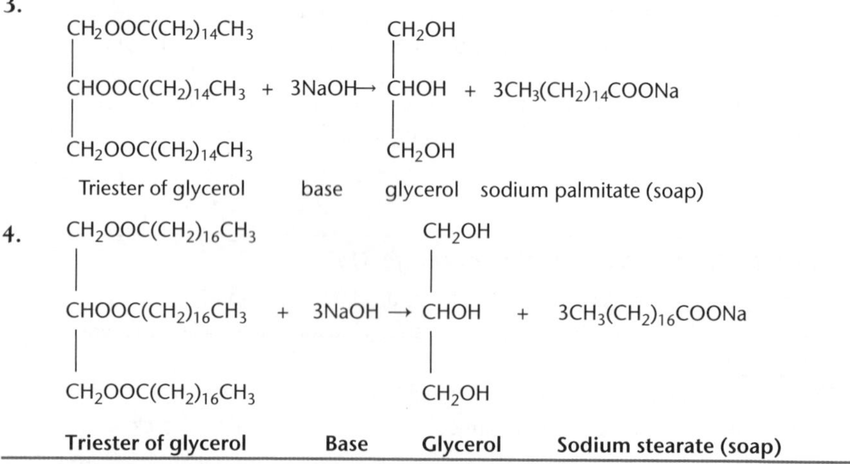

Unit 12.5 Activity 5E: Soap and detergent (page 322)

1. Colgate Palmolive Ltd and K.K. Kingston Ltd.
2. Receiving raw materials, such as coconut oil, palm oil or animal fat from cows and pigs; saponification; salting; removal of excess salt and sodium hydroxide; addition of colours and perfumes; production of bars or powder; packaging and distribution.
3. A soap molecule has two parts – a polar end and a non-polar hydrocarbon chain. The organic non-polar end is attracted to dirt while the polar end is attracted to water, thereby removing the dirt from dirty clothes or body.
4. Soap is less effective in hard water, which contains magnesium and calcium ions. Detergents are designed to soften hard water. Detergents do not contain sodium and potassium salts of fatty acids. They contain long chains sodium dodecylbenzene sulfonate and sodium dodecylbenzene sulfate. The cleansing function of synthetic detergent molecule is similar to that of soap. The long molecules in detergents are by-products of the petroleum industry.

Unit 12.5 Activity 5F: Biodiesel (page 324)

1. The fuel obtained by reacting vegetable or animal fats and oils with short-chain alcohols such as methanol or ethanol in base-catalysed esterification reactions.
2. Triglycerides + ethanol → ethyl esters + glycerol + soap + water + excess ethanol

3. They are catalysts used to increase the rate of the reaction but are not used up in the reaction process.
4. If water was present during the base-catalysed transesterification it would hydrolyse the triglycerides, giving salts of the fatty acids (soaps) instead of producing biodiesel.
5. Cetane number is a measure of combustion quality. The higher the number, the smoother the engine will run. Biodiesel has a higher cetane rating than diesel.
6. **a.** True
   b. False
   c. True
   d. True
   e. True

Unit 12.5 Activity 6A: Lime production (page 326)

1. Individual student response.
2. Limewater turns a milky colour when carbon dioxide reacts with the calcium hydroxide to form a white precipitate of calcium carbonate.
3. In New Hanover, lime powder is traditionally used for body decoration during traditional singsing.
4. Betel nut fruit contains chemical compounds called alkaloids. The main stimulant present in the fruit is an alkaloid called arecoline. When mixed with lime (calcium hydroxide), the arecoline is converted to arecaidine. This has various effects on the central nervous system. Both arecoline and arecaidine interfere with the action of some of the nerves in the brain, which results in the stimulant effects. The action of these alkaloids might also explain why betel nut affects the appetite. Tannins in the fruit may help treat the symptoms of dysentery.
5. Advantage: betel nut chewing has been proposed to protect the teeth and gums. Disadvantage: betel nut chewing has been associated with mouth cancer and spread of tuberculosis in PNG.

Unit 12.5 Activity 6B: Salt production (page 326)

1. Individual student response.
2. No hard work is required. It is cheap. It is more convenient to buy readily available product and saves time.

Unit 12.5 Activity 6C: Dyes and paint production (page 327)

1. Individual student response.
2. There has not been any serious research and development done on this.

Unit 12.5 Activity 6D: Traditional medicine (page 328)

1. Individual student response.
2. Alkaloids are produced as secondary metabolites by a large variety of organisms, including bacteria, fungi, plants and animals. Alkaloids are a group of natural products that contain carbon, hydrogen and nitrogen. They may also contain oxygen, sulfur and more rarely other elements such as chlorine, bromine, and phosphorus. The nitrogen is usually basic but may also be neutral and even weakly acidic. Alkaloids can be poisonous or therapeutic in small doses.

They often have pharmacological effects (eg cocaine) and are used as medications (eg morphine), as recreational drugs (eg caffeine, nicotine, psilocin), or in traditional rituals.

3. Individual student response.
4. A real possibility exists for such arrangements.

Unit 12.5 Activity 6E: Researching traditional medicine (page 328)

1. Individual student response.
2. Individual student response.

Unit 12.5 Activity 7A: Mining and the environment (page 330)

1.
   - Destroys wetlands, which are natural habitats for birds, fish and other organisms.
   - Destroys forests when trees are removed for access to ore.
   - If dams are not built properly massive pollution can occur downstream overflow.
   - Underground gold or coal mining can require the removal of almost an entire layer of material deep under the surface.
   - Dispersal of heavy metals, such as lead, into the atmosphere. Cyanide spills, mercury contamination have been experienced. This can have serious health effects, including mental retardation in children.
2. Underground mining can require the removal of almost an entire layer of material deep under the surface, weakening the structures on the surface. Open-cut mining leaves large gaping holes in the ground, making land unusable.

Unit 12.5 Activity 7B: Environmental effects of mining (page 333)

1. Individual student response.
2. Individual student response.
3. Individual student response.
4. Individual student response.
5. Individual student response.
6. Hydrothermal vents are volcanic hot springs that occur on the sea floor at depths of 1–6 kilometres. Around these vents, cold oxygen-rich seawater is forced down through the porous sea bed by very high pressures. The heat and high pressure causes the seawater to become acidic, low in oxygen and highly concentrated in metal sulfides dissolved from the surrounding rocks. The metal-rich, superheated water (or vent fluid) becomes buoyant and rushes up through the seabed, forming the hydrothermal vents and mixing with cold water from the ocean bottom. This causes the minerals in the vent fluid to crystallise, forming massive sulfide deposits. Some of these form below the seabed in the cracks and crevices of the vents, while above the seabed, uniquely sculptured, chimney-like towers are formed.
7. Seafloor volcanic systems and deep-sea ecosystems are still under-researched and not well understood. Only 5% of Earth's oceanic ridges have been surveyed in any detail.

Unit 12.5 Activity 7C: Petroleum and gas mining (page 334)

1. Carbon dioxide, sulfur dioxide, increase in the volume of hazardous materials and chemicals contaminating land and oceans. Over time the ageing oil and gas installations become pollutants.

2. Solar energy, mostly in the form of short-wavelength visible radiation, penetrates the atmosphere and is absorbed by Earth's surface. The heated surface then radiates some of that energy into the atmosphere in the form of longer-wavelength infrared radiation. Although some of this radiation escapes into space, much of it is absorbed by greenhouse gases in the lower atmosphere, which in turn re-radiate a portion back to Earth. This is similar to how the glass in a greenhouse acts.

   The greenhouse effect is essential to life on Earth; increased levels of greenhouse gases in the atmosphere due to the burning of fossil fuels is increasing the effect and contributing to global warming.

3. Greenhouse gases absorb and emit radiation. Examples include water vapour, carbon dioxide, methane, nitrous oxide, ozone and chlorofluorocarbons (CFCs).

4. Humans have been releasing extra quantities of greenhouse gases into the atmosphere, which trap more heat, enhancing the natural greenhouse effect. The enhanced greenhouse effect is the direct result of human activities. Processes such as the burning of fossil fuels, industrial operations and forest-clearing release carbon dioxide, methane and nitrous oxide into the atmosphere. CFCs are also potent greenhouse gases and they also destroy the ozone layer.

   Human emissions of carbon dioxide, more than any other greenhouse gas, have contributed most to the enhancement of Earth's natural greenhouse effect.

Unit 12.5 Activity 7D: Combustion of fossil fuels (page 337)

1. $C_3H_8(g) + 5O_2(g) \rightarrow 3CO_2(g) + 4H_2O(l)$ + heat

2. Water vapour – evaporation from bodies of water; carbon dioxide – burning fossil fuels; methane – rice growing, cattle, natural gas, mining; nitrous oxide – agriculture and industry; ozone – produced in atmosphere by reaction of pollution with sunlight; chlorofluorocarbons (CFCs) – refrigerants, aerosol propellants. CFCs are being phased out.

3. Humans have been releasing extra quantities of greenhouse gases into the atmosphere, which trap more heat, enhancing the natural greenhouse effect. The intensification of the greenhouse effect due is considered to be the main contributing factor to global warming.

4. Acid rain refers to the deposition of wet (rain, snow, sleet, fog, cloud water and dew) and dry (acidifying particles and gases) acidic components.

5. Sulfurous acid is formed by the reactions:

   $SO_2(g) + H_2O(l) \rightarrow H_2SO_3(aq)$

   $SO_3(g) + H_2O(l) \rightarrow H_2SO_4(aq)$

   Carbonic acid is formed by the reaction:

   $CO_2(g) + H_2O(l) \rightarrow H_2CO_3(aq)$

6. Natural source – lightning. Manmade source – combustion of fossil fuels.

Unit 12.5 Activity 7E: Persistent organic pollutants (page 339)

1. A pollutant derived from use of agro-industrial chemicals. POPs produce very adverse effects on both the ecosystem and human health due to their very high toxicity and their prolonged persistence.

2. Dichlorodiphenyltrichloroethane.

Cl Cl Cl Cl Cl

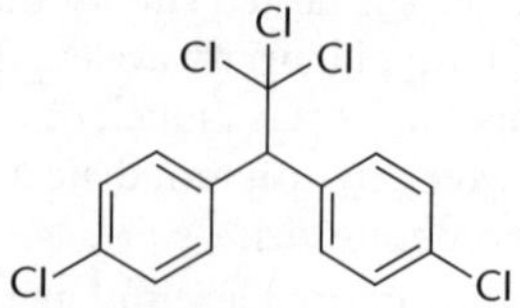

DDT (dichlorodiphenyl trichloroethane)

3. Chlorofluorocarbon. These range from the simple trichlorofluromethane (CCl_3F).

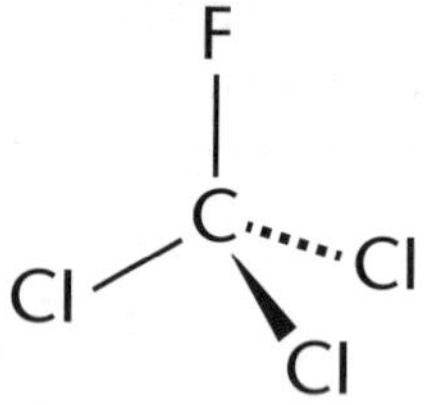

F C Cl Cl Cl

to the more complicated 1,3-dichloro-1,2,2,3,3-pentafluoropropane ($CClF_2CF_2CHClF$).

4. Individual student response.
5. Electrical appliances consume energy, which uses fuel, liberates heat and carbon dioxide. Turning off unused appliances such as lights helps lower energy consumption, reduces carbon dioxide output and reduces the contribution to the enhanced greenhouse effect.
6. The rainforest will absorb the carbon dioxide generated by industrialised countries and thus contribute to slowing down global warming.

Glossary/Index

Note: Numbers in brackets refer to pages on which term is defined or discussed.

acetaldehyde (253): common, non-systematic name for ethanal, CH_3CHO.

acetic acid (233): common name for ethanoic acid.

acetone (254): common, non-systematic name for propanone, CH_3COCH_3.

acid (75): a substance in solution that will turn blue litmus paper red and has a pH of less than 7.

acid (Brønsted) (75): a species that is able to donate a proton (to a base).

acid dissociation constant, K_a (100): the equilibrium constant for the dissociation of a weak acid in which the concentration of water is considered as constant; also known as dissociation constant and acidity constant.

acid rain (336): rain that is more acidic than normal due to the presence of sulfur dioxide, nitrogen dioxide and other acidic gases formed in industrial areas.

acid–base titration (1, 85): an experimental procedure where a neutralisation reaction is used to determine the concentration of an acid or a base.

acidic (76): neutralises bases or alkalis.

acidic oxide (160): an oxide that reacts with a basic solution but not with an acidic solution.

acidic salt (86): a salt that contains an ion that reacts with water to form $H_3O^+(aq)$.

acidified (127): has had acid added, or occurring in acidic solution.

acidity constant (100): see *acid dissociation constant.*

activity series (164): an order for the arrangement of metals, with the most active metal at the top of the series; for Level 1 NCEA Chemistry, the metals listed are, in order of activity: sodium > lithium > calcium > magnesium > aluminium > zinc > iron > lead > copper > silver > gold.

addition polymerisation (221): alkene monomers joining together to form a polymer.

addition polymer (283): a polymer formed by addition reactions.

addition reaction (217): reaction in which a small molecule such as hydrogen chloride is added across a double (or triple bond) in an organic compound.

aerosol (338): a dispersion of very small drops of liquid in a gas, eg air; forming part of what is commonly referred to as 'smog'.

alcohol (188, 227): a compound of carbon formed by replacement of a hydrogen atom by a hydroxyl group in the molecule of an alkane; general formula $C_nH_{2n+1}OH$.

aldehyde (253): a compound with a terminal –CHO group; systematic IUPAC name, alkanal.

aliquots (88): fixed volumes of liquid delivered by a pipette.

alkali (78): a soluble base that will turn red litmus paper blue and has a pH greater than 7.

alkaline (78): a solution with pH greater than 7; may also be referred to as a basic solution.

alkanal (253): systematic IUPAC name for aldehyde, R–CHO.

alkane (188): a compound of hydrogen and carbon, in a series, joined by single bonds only, with the general formula, C_nH_{2n+2}.

alkanol (227): systematic IUPAC name for alcohol, R–OH.

alkanones (254): systematic IUPAC name for ketones, R–CR′O.

alkene (188, 211): a compound of hydrogen and carbon, containing at least one carbon–carbon double bond, in a series with the general formula, C_nH_{2n}.

alkyl group (189, 197): a group of carbon and hydrogen atoms, eg methyl CH_3 and ethyl C_2H_5 that can be represented by R, R′ or R″.

alkyne (188, 223): a compound of hydrogen and carbon, which contains a carbon–carbon triple bond, in a series with the general formula, C_nH_{2n-2}.

allotrope (171): the existence of an element in two or more different forms in the same physical state.

alloy (295): two or more metals dissolved in each other (*Note:* carbon, a non-metal, can form an alloy with iron (called steel)).

alpha (α) particle (48): nucleus of helium atom.

alumina (165): aluminium oxide, obtained from the mineral bauxite and used to produce aluminium metal.

amalgam (296): an alloy formed between a metal and mercury.

amide (285): a compound containing the amide functional group, $–CONH_2$, –CONH– or –CON(R)–; systematic IUPAC name, alkanamide.

amine (286): a compound containing the amine functional group, eg RNH_2, RNHR′, group, RN(R″)R′; systematic IUPAC name, aminoalkane.

amino acid (286): monomer unit for proteins, containing both a carboxylic acid group, –COOH, and an amine group, $–NH_2$.

amphiprotic (81): substances whose particles can act as both proton donors and proton acceptors.

amphoteric (7, 82): metal oxide or hydroxide that can be acidic (react with bases) or basic (react with acids).

analysis (29): separation of compounds/mixtures into their constituents and/or properties of constituents.

anhydrous (34, 42): without water – used to describe crystals that have lost their water of crystallisation.

anion (56): a negatively charged ion.

anode (136): electrode at which oxidation occurs; the positive terminal in an electrolysis process to which anions are attracted.

antifreeze (229): substance used in car radiators in cold climates to prevent water freezing; 1,2-ethanediol is an important component.

antioxidant (288): another name for a reducing agent, or reductant; in terms of food preservation, the antioxidant removes oxygen, and thus prevents bacteria surviving to cause food to decay; also added to plastics to protect against ozone or oxygen.

aqueous (37): dissolved in water.

aqueous solution (37, 182): a solution formed by dissolving a substance in water.

arene (188): aromatic hydrocarbon.

atomic number (Z) (48): the number of protons in the nucleus of an atom.

Avogadro's number (18): 6.02×10^{23} particles.

balanced (23): number of each type of atom on both sides of a chemical equation is the same.

base (75): the oxide, hydroxide, carbonate or hydrogen carbonate of a metal; a substance that will neutralise an acid.

base (Brønsted) (75): species that is able to accept a proton (from an acid).

base dissociation constant, K_b (106): equilibrium constant for the reaction of a weak base and water.

basic (78): can neutralise an acid.

basic oxide (159): any oxide which reacts with an acidic solution but not a basic solution.

basic salt (79): an ionic compound, eg Na_2CO_3, that dissolves to form a solution with an increased concentration of hydroxide ions.

battery (150): commercially produced electrochemical cell or collection of electrochemical cells.

bauxite (165): a mineral containing oxides of aluminium and iron; used as a source of aluminium metal.

Benedict's solution (259): reagent used to test for the presence of aldehydes; when added to an aldehyde, the blue solution (due to the presence of complex ions of copper(II)) forms an orange-red precipitate ($Cu_2O(s)$).

benzene (188): simplest arene C_6H_6.

biodegradable (289): breaks down in the environment.

biodiesel (323): a biofuel made from vegetable or animal fats and oils and alcohols.

boiling point, bp (71): the temperature at which a liquid changes to a gas or vapour.

bond (59): a force between two atoms which creates a molecule, a part of a molecule, or an ionic substance; or a force between two molecules responsible for the liquid or solid state of a substance.

bond angle (66): the angle between three specified atoms in a molecule, eg H–C–Cl in $CHCl_3$.

bonding pairs (66): electron pairs that are shared between two atoms and are attracted to both atomic nuclei.

bonding electrons (60): shared electrons in a bond.

branch chain (196): a smaller chain of carbon atoms joined to the parent chain.

branched-chain structure (196): structure of organic compounds in which the carbon atoms are *not* all joined to each other in a single 'line'.

brewing (312): a process in beer production.

brittle (178, 301): a property of a solid that causes it to break easily into irregular-shaped fragments.

bromination (225): reaction involving replacement of an atom in a molecule with bromine or adding bromine (Br_2) across a double or triple bond.

buckminsterfullerene (171, 174): the first fullerene compound to be discovered, commonly called a buckyball, formula C_{60}.

burette (85): a long vertical tube with a tap at the bottom used in titrations to measure the volume of titre required.

by-product (299, 303, 314, 322, 324): a secondary product of a chemical reaction other than the main product.

calibrated (86): set to a specific value or measurement.

cap rock (176): a hard rock overlying a deposit of petroleum, natural gas, coal, etc.

carbon cycle (175, 269): the natural process by which carbon moves through the living world, mostly by the release (respiration) and absorption (photosynthesis) of carbon dioxide.

carbonyl (253): the C=O group in an organic molecule.

carboxylic acid (188, 235): a compound containing a carboxyl (–COOH) functional group; general formula, RCOOH.

catalyst (160): a substance that changes the rate of a chemical reaction without itself being used up.

catalytic reforming (277): process in petroleum production that rearranges some hydrocarbons and breaks others to form low-octane petrol.

cathode (136): electrode at which reduction occurs; the negative terminal in an electrolysis process to which cations are attracted.

cation (55): a positive ion that is attracted to the cathode in an electrolysis process.

cell diagram (137): symbolic representation of an electrochemical cell or half-cell.

cell potential (E^0_{cell}) (138): the potential, measured in volts, of an electrochemical cell to push and pull electrons through a circuit; symbol, E^0_{cell}.

CFC (338): chlorofluorocarbon compounds used as propellants but now banned in most countries as they catalyse the decomposition of ozone in the protective ozone layer.

charcoal (177): an impure form of carbon obtained by heating wood in the absence of air.

chemical bonds (59): strong attractive forces between atoms, molecules or ions.

chemical change (161): change that results in a new substance and that is difficult or impossible to reverse.

chemical equation (23): a brief and accurate way of summarising what happens in a chemical reaction.

chlorination (218): reaction involving replacement of an atom in a molecule with chlorine or addition of chlorine (Cl_2) across a double or triple bond.

chrome plating (167): when a thin layer of chromium is electroplated onto a metal object.

***cis* isomer (215):** geometric isomer in which identical groups are located on the same side of a double bond.

coal (271): rock composed of carbon, hydrocarbons, nitrogen, sulfur and water; formed from the remains of ancient plants.

coke (164, 176): produced from coal that has been heated in the absence of air; oils and gases are expelled from the coal to leave a porous solid that is mainly carbon; used in the reduction of metal ores.

collagen (286): complex, fibrous insoluble protein.

colorimeter (6): a device used in colorimetry to measure the visible light absorbed by a solution.

combustion (204): a chemical reaction involving oxygen that produces heat and light and sometimes sound.

complete combustion (177, 204): a form of combustion in which the maximum amount of oxygen is used; commonly applied to the combustion of hydrocarbons in which the products are carbon dioxide and water.

complex ion (155, 300): central metal ion bonded to a specific number of ligands.

concentration, *c*(X) (3, 37): quantity of a substance, X, dissolved in a quantity of solvent (units mol L^{-1}, g L^{-1} or g/100 g solvent); concentration of an individual species, X, in solution or a gas in a volume is shown as [X], where the units of concentration are mol L^{-1}.

concordant (3, 89): consistent; within small (acceptable) differences of other values.

condensation polymer (283): a polymer produced by condensation reactions.

condensation reaction (240): reaction that produces water as a by-product.

condensed structural formula (191): a simpler form of the structural formula, used especially for larger molecules.

conical flask (86): a cone-shaped flask used in the laboratory for heating and in titration reactions.

conjugate acid (82): a species having one more proton (H^+) than its conjugate base.

conjugate acid–base pair (82): two species – molecules or ions – that differ in their structure by a single proton (H^+).

conjugate base (82): a species having one less proton (H^+) than its conjugate acid.

copper plating (167): when a thin layer of copper is electroplated onto a metal object.

corrode (148): natural oxidation of a metal; when iron is being oxidised (corroded), the process is called rusting.

corrosion (148): a chemical process by which a metal is 'eaten' away or the surface of the metal is changed in appearance.

covalent bond (59, 60): a bond formed between two atoms by the sharing of electrons from the outer energy level of each atom; the bond is due to the nucleus of each atom attracting the outer electrons of the other atom.

cryolite (166): a substance (Na_3AlF_6) added to alumina to reduce the melting point during the electrolytic process for the production of aluminium.

cyclic (compound) (188): organic compounds whose molecules are made up of three or more carbon atoms joined in a ring structure; cyclic structures may be saturated or unsaturated molecules.

cycloalkane (201): an alkane that contains a ring structure; general formula C_nH_{2n}.

DDT (338): dichlorodiphenyl trichloroethane, a pesticide now banned in many countries.

decanting (117): separating a liquid from a solid in a vessel by pouring the liquid carefully from the vessel to leave the solid at the bottom of the vessel.

dehydrate (117): the removal of water that is chemically bonded in a compound.

dehydration reaction (234): an elimination reaction in which a molecule of water is 'removed' from a molecule.

deliquescent (45, 118): a substance that forms a solution as it absorbs water from the air.

detergent (322): alkylaryl sulfonates used as an alternative to soap in areas with hard water.

deuterium (49): an isotope of hydrogen with a mass number of 2.

diamond (171, 172): an allotrope of carbon which is a three-dimensional network.

diatomic (57): a polyatomic ion made up of two atoms.

dicarboxylic acid (285): organic compound with two carboxyl (–COOH) groups.

dilute (7): contains small amounts ($< 10^{-3}$ mol) of a species per litre of solution.

dipeptide (286): a dimer consisting of two amino acid monomers.

dipole (63): a species with two oppositely charged regions.

discharge (150): the release of electrical energy from a battery on connecting the circuit.

disproportionation (156): simultaneous oxidation and reduction of the same element during a chemical reaction.

dissociation (76): process whereby a species is broken down into smaller species; degree of dissociation is the extent to which a substance is dissociated in aqueous solution.

double bond (61): a covalent bond formed by two atoms, each sharing two of their electrons with another atom – four electrons are used to form the bond.

dry ice (183): solid carbon dioxide.

drying agent (302): a substance that will absorb water – commonly used to dry gases.

ductile (295): a property of metallic substances – can be drawn out into a thin wire.

***E*° (139):** see *standard reduction potential*.

effervesce (182): release gas during a reaction involving a liquid; the gas causes the liquid to froth.

effloresce (118): lose water from a hydrated substance so that a crystalline substance crumbles to a powder.

electrical conductivity (297): the ability of a material to conduct electricity without undergoing chemical change.

electrochemical cell (135): a cell in which oxidation and reduction reactions occur, often in separate compartments; an example is a galvanic cell in which the reaction is spontaneous.

electrodes (135): solid bodies that carry current to or from an electrolyte in an electrolytic cell.

electrolysis (161): using an electric current to decompose an electrolyte to produce chemicals at electrodes.

electrolyte (161): a liquid which conducts electricity because it contains ions which are free to move.

electrolytic cell (150, 161): a cell that uses a supply of electrical energy to bring about a non-spontaneous chemical reaction.

electromotive force, EMF or E°_{cell} (138): the potential difference across a voltage source when no current is flowing.

electron (47): a subatomic particle of negligible mass and carrying a unit negative charge.

electron configuration (51): the positioning of electrons according to energy levels of the electrons around a nucleus of an atom in the ground state.

electron dot diagram (52): also known as a *Lewis diagram*; a diagram showing bonding between two atoms; the dots represent electrons from the outer energy level of each atom; different-coloured dots, or dots and crosses, may be used to differentiate between the electrons in the outer energy level of each atom.

electronegative (63): having a negative charge.

electronegativity (63): a measure of the pull of an atom on the electrons it shares in a covalent bond.

electroplating (167): the use of electrolysis to coat one metal onto another.

electrostatic attraction (59): attraction between positively charged and negatively charged particles.

elimination reaction (233): two atoms or small groups are removed from a molecule forming a double (or triple) covalent bond in that molecule.

empirical formula (29): a formula that gives the simplest whole number ratio of the atoms or ions in a compound.

endpoint (87): the point in a titration at which an indicator changes colour.

endothermic (176): refers to a chemical process in which heat energy is absorbed.

energy level (51): a term used to define the status/position of an electron in an atom, outside the nucleus of the atom; the term 'energy shell' is sometimes used.

enhanced greenhouse effect (185): intensification of natural greenhouse effect caused by human emissions of greenhouse gases, especially CO_2.

enzyme (247): shape-specific, biological catalyst made of protein.

equilibrium (8): the state of reversible reactions where the rate of the forward reaction equals the rate of the backward reaction.

equilibrium constant, K_c (100): equilibrium constant based on concentrations of all species.

equilibrium equation (100): the stoichiometric equation showing the species involved in an equilibrium reaction.

equivalence point (86): the point in a titration or a reaction at which the correct amounts of reactants have been added to complete the reaction according to the stoichiometric equation.

essential amino acid (286): an amino acid that an organism cannot synthesise and which must be present in the diet.

esterification (240): condensation reaction involving an alcohol group (–OH) and acid group (–COOH) reacting to form an ester group.

esters (188, 239): organic compounds containing a –COO– functional group.

ethanol (309, 314): 'alcohol', ethyl alcohol found in fermented and distilled alcoholic beverages.

ethene (211): C_2H_4, first member of the alkenes; non-systematic name, ethylene.

ethylene glycol (273, 314): non-systematic name for ethane-1,2-diol (ethanediol).

exothermic (176): refers to a chemical process in which heat energy is released.

external circuit (135): the conducting wire and electrical devices which connect the terminals of a battery or some other power source.

extractive metallurgy (294): the process of extracting and purifying metals from their ores.

face (172): a side of a crystal.

fats (244): biological triesters consisting of a glycerol unit (propane-1,2,3-triol) and three large fatty acid units.

fatty acids (244): carboxylic acids with long hydrocarbon chains containing even numbers of carbon atoms.

Fehling's solution (259): reagent used to test for the presence of aldehydes; when added to an aldehydes, the blue solution (due to the presence of complex copper(II) ions) forms an orange-red precipitate ($Cu_2O(s)$).

fermentation (247, 309): the natural process whereby carbohydrates, eg starch or glucose, can be converted, by enzymes, to ethanol and carbon dioxide.

ferromagnetic (300, 301): can be magnetised; ferromagnetic materials, such as iron, contain unpaired electrons that will align with each other in response to a magnetic field.

filtration (165): a separation technique using filter paper to separate a solid from a solution.

flotation (296): a process of separating minerals according to their tendency to float or sink in a liquid.

formal equations (113): show all species associated with a reaction – include spectator ions.

formaldehyde (253): non-systematic name for methanal, HCHO.

formalin (255): solution of about 40% methanal and 60% water used as a preservative and disinfectant.

formula (191): the simplest unit of an element or compound shown in symbol form with the type and number of atoms of each element indicated.

forward reaction (104): the initial or left-to-right reaction of an equilibrium.

fossil fuels (187, 270): coal, oil and gas.

fractional distillation (194, 273): the process of separation of two volatile liquids by careful heating of the mixture of liquids so that the more volatile liquid is separated as a vapour, which then condenses into a receiver. The less volatile liquid remains in the distillation flask.

fuel (176): a substance that can be used domestically or commercially to react with oxygen in the air and the reaction releases heat energy.

functional group (188): the characteristic arrangement of atoms and bonds which governs the specific chemistry of that organic compound; the part of the molecule where reaction is likely to occur.

galvanic cell (150): electrochemical cell that converts chemical energy into electrical energy.

galvanise (149): to coat a surface with zinc to prevent or reduce rusting or corrosion.

general formula (188): a formula used to represent the atom composition of a molecule of any member of a homologous series.

geometric isomers (215): different isomers (*cis* and *trans*) due to the presence of non-rotating bonds, eg double C=C bonds, or C–C bonds in cyclic compounds.

GER (124): an acronym (a word formed from the initial letters of a phrase) to help remember that 'gain of electron is reduction'.

global warming (185): the raising of the temperature of the Earth's atmosphere attributed to an increase in the concentration of gases, eg carbon dioxide and methane, in the atmosphere.

glucose (247): a simple sugar of molecular formula, $C_6H_{12}O_6$.

glycerol (229): 1,2,3-propanetriol, also known as glycerine, a component of triglycerides.

gram (18): a unit of mass.

graphite (137, 171): an allotrope of carbon which has a two-dimensional structure.

greenhouse effect (185): the term used to explain global warming; global warming is related to gases, eg carbon dioxide and methane, in the atmosphere retaining heat that would otherwise be lost to outer space.

greenhouse gas (335): a gas that contributes to global warming.

group (54): vertical column of the periodic table; atoms of the elements in a group have the same number of valence electrons but different numbers of energy levels.

half-cell (135): a single electrode in a solution containing ions.

half-equation (126): an equation written to illustrate an oxidising or a reducing process involving electron transfer; adding a half equation (eg illustrating oxidation) to the half equation that is complementary (eg illustrating reduction) produces an overall equation.

halide (133): compound containing a group 17 element (chlorine, bromine, etc).

haloalkane (188, 208): organic molecule formed when a halogen atom or atoms are substituted for a hydrogen atom or atoms.

halogen (133): a group 17 element.

halogenation (217): an addition reaction where halogen atoms add across a double or triple bond.

homologous series (188): a series of compounds containing the same elements with a common difference in their formulae; the alkanes are an example of an homologous series with a general formula of C_nH_{2n+2}, ie each member of the series differs from the next member by the group CH_2.

hydrate (34): a compound formed by the addition of water to another compound; the common example is the addition of water to an anhydrous salt to produce a hydrated salt that contains water of crystallisation.

hydrated (34): containing chemically bonded water molecules.

hydration (72, 148, 218): an addition reaction where a water molecule is added across a double or triple bond.

hydrocarbon (186): a compound that contains carbon and hydrogen only.

hydrogen bond (230): intermolecular attraction between an H atom attached to O, N or F atoms in one molecule and the non-bonding electrons on O, N or F of another molecule.

hydrogen carbonate (9): the radical with the composition HCO_3, forming the hydrogen carbonate ion, HCO_3^-.

hydrogen ion (75): a proton; the aqueous hydrogen ion, $H^+(aq)$ is usually referred to as the hydronium ion, $H_3O^+(aq)$.

hydrogenation (217): addition reaction in which a hydrogen molecule (two hydrogen atoms) adds across a double (or triple) carbon-to-carbon bond.

hydrolysis (242): the splitting of a molecule when it undergoes a chemical reaction with water.

hydronium ion (75): the name of the hydrated hydrogen ion, H_3O^+; formed from $H^+ + H_2O$.

hydrothermal vents (306): volcanic hot springs on the sea floor.

hygroscopic (42): absorbs water from the air.

hypochlorite (303): ClO^- ion; oxidant found in bleaches.

incomplete combustion (176, 205): a form of combustion in which less than the maximum amount of oxygen is used; commonly applied to the combustion of hydrocarbons in which the products are carbon monoxide and/or carbon plus water.

indicator (acid–base) (86): weak acid that shows a colour change when dissociating to form the conjugate base; eg methyl red indicator (pK_a = 5.1) or phenolphthalein indicator (pK_a = 9.3).

inert (59): unreactive to all chemical reagents and changes in physical conditions; the term applies to gases such as helium, neon and argon.

inert gas (59): helium, neon, argon, krypton or xenon, forming group 18 of the periodic table.

insoluble (73): a substance that does not dissolve in a solvent such as water.

intramolecular bonding (98): covalent bonding between the atoms in a molecule.

intermolecular forces of attraction (221): forces of attraction between molecules.

ion (55): an atom, or group of atoms, that has a positive or negative charge; the ion is formed by the loss or gain of electrons from the atom or group of atoms.

ion–electron equation (126): equation describing an oxidation or reduction reaction showing ions involved and the electrons lost or gained by the reactant(s).

ionic bond (59): a bond formed between two atoms by electron(s) being donated by one atom and accepted by the other atom.

ionic compound (19): compound formed from elements whose atoms combine by loss and gain of electrons, producing charged particles called ions. Ionic compounds have the characteristics of high melting point, electrically conducting when molten, and crystalline in form.

ionic equation (7): equation that includes only the ions involved in a reaction; spectator ions are omitted.

ionic product for water, K_w (104): K_w = $[H_3O^+(aq)][OH^-(aq)]$ (= 10^{-14} at 25°C).

isomers (188): species – usually organic molecules – which have the same formula but the atoms are arranged in a different structure.

isotope (49): an atom of an element that has a different number of neutrons in its nucleus than other atoms of the same element.

IUPAC (190): International Union of Physical and Applied Chemists.

K (100): general symbol for any equilibrium constant.

K_a (100): acid dissociation constant or acidity constant; equilibrium constant for a reaction in which a substance donates a proton to a water molecule.

K_b (106): base dissociation constant; equilibrium constant for a reaction in which a substance accepts a proton from a water molecule.

K_c (103): equilibrium constant based on all relevant concentrations of reactants and products.

ketone (253): a compound containing a carbonyl (C=O) functional group; also known as alkanone.

kilogram (18): 1000 g; SI unit of mass.

K_w (104): *ionic product for water*; equilibrium constant for proton transfer reaction between two water molecules.

lattice (34): stable structure in which anions and cations attract each other electrostatically so that each is surrounded by a group of oppositely charged ions arranged in a regular order.

lattice energy (73): energy needed to break ionic bonds in a lattice to produce hydrated ions or molten substances.

lautering (312): a process in beer production.

lead–acid battery (150): electrochemical cell containing lead plates in sulfuric acid; also known as a lead-accumulator battery.

LEO GER (124): Loss of Electrons Oxidation; Gain of Electrons Reduction.

Lewis diagram (52): an atom diagram showing the electron arrangement in the outer energy level; two or more atom diagrams showing the bonding between the atoms as electron sharing, or as electron donation and acceptance.

Lewis structure (52, 60): diagram of molecules and ions in which all valence shell electrons are shown as lines for bonding electrons and dots for unbonded electrons (eg lone pairs); also called *Lewis diagram*.

LHE (137): abbreviation for left-hand electrode.

ligand (295): a small molecule or ion with one or more lone pairs of electrons which can form a coordinate bond with a central positive ion.

lime (325): calcium oxide (CaO).

limewater (111, 180): a solution of calcium hydroxide used for detecting carbon dioxide gas.

linolenic acid (318): a polyunsaturated acid found in coconut oil.

liquefied natural gas (280): natural gas that has been cooled below its boiling point.

litmus (80): a substance used in solution, or absorbed on paper, that has a red colour in acid and a blue colour in alkali.

lone pair (of electrons) (66, 71): see *non-bonding pair*.

macromolecule (220, 283): a single molecule consisting of hundreds of thousands even millions of atoms covalently bonded to neighbouring atoms.

malleable (300, 304): a property of metallic substances – they bend easily.

malting (310): a process in beer production that releases the starch in barley.

Markovnikov's rule (219): during an addition reaction, the hydrogen atom from H–Cl or H–OH will bond to the carbon atom (in the double bond) having the greatest number of hydrogen atoms.

mass (17): the amount of matter in a substance.

mass concentration (39): concentration measured by using the mass of solute in a given volume of solution.

mass number (*A*) (48): the number of protons plus the number of neutrons in the nucleus of an atom.

melting (71): a state change from solid to liquid when enough of the weakest attractive forces between the species making up a substance are broken to destroy the solid lattice structure.

melting point (71): the temperature at which a solid changes to a liquid.

metal (56): an element whose atoms have 1, 2 or 3 electrons in their outer energy level; most metals are silver-grey and shiny in appearance.

methyl orange (87): indicator ($pK_a = 3.7$) used in some acid–base titrations.

mineral (272, 293): naturally occurring substance that is mined, or occurs in the sea, and is commercially valuable.

minor products (218): the lesser product(s) formed when competing organic reactions occur.

miscible (314): the property of two liquids where they can be shaken together to form a homogeneous mixture.

mnemonic (124): aid to memory.

molar mass (g mol^{-1}) (19): the mass in grams of one mole (6.02×10^{23}) of particles of a substance.

mole/mol (3, 18): the amount of substance that contains the same number of particles as there are carbon atoms present in 12 g of carbon-12, and is equal to 6.02×10^{23} particles.

molecular formula (32, 191): the formula of a substance that indicates the number and type of atoms present in one molecule or unit cell of the substance.

molecule (59): the smallest particle of an element or compound that can exist alone; the atoms in the particle are covalently bonded together.

molten (161): the liquid state.

monatomic (56): derived from one atom.

monomer (220, 283): a simple molecule that can be bonded to other molecules of the same, or different, types to form a large molecule – the product is called a polymer.

mother liquor (117): solution remaining after crystals have formed.

natural gas (280): gas formed from decaying remains of ancient animals and plants, CH_4.

negative ion (136): an atom or a radical that has a negative charge because it has more electrons than protons.

neutral (81): neither acidic nor basic; pH of 7 at 25°C.

neutron (47): a sub-atomic particle with a mass of 1.00 amu and no charge; all atomic nuclei, except hydrogen-1, contain one or more neutrons.

non-bonding pair (of electrons) (68): a pair of electrons not involved in a covalent bond; also called lone pair (of electrons).

non-metal (57): an element whose atoms have 4, 5, 6, 7 or 8 electrons in their outer energy level; there are no common properties of non-metals in the way that there are for metals.

non-polar covalent bond (70): covalent bond in which electrons are shared equally between the two bonding atoms.

non-biodegradeable (220): molecules which are unable to be broken down by decomposers; includes most synthetic polymers.

nucleus (47): the central part of any atom which contains protons and neutrons; comprises a very small part of the total size of the atom.

nylon (285): artificial fibre with many uses; a polyamide (containing many amide links, –CONH–) formed by a condensation reaction.

octane rating (275): a measure of a fuel's performance.

octet rule (59): simplistic rule that the atoms in a molecule often form bonds so that they have eight electrons in their outer shell.

OIL-RIG (124): Oxidation **I**s **L**oss of electrons, **R**eduction **I**s **G**ain of electrons.

oil (271): fossil fuel formed from remains of plants and animals.

oils (244): liquid, unsaturated fatty acids bonded to glycerol; found in plants.

oleic acid (244, 318): 9-octadecenoic acid, $CH_3(CH_2)_7CH{=}CH(CH_2)_7COOH$; an unsaturated fatty acid.

open-cut mining (329): a method of extracting minerals located close to the surface of the Earth.

ore (164, 293): naturally occurring mineral from which economically valuable constituents, eg metals, can be obtained.

organic acid (77, 235): a compound that acts as an acid because the hydrogen at the –COOH group can be transferred to a base; another term for carboxylic acid.

organic chemistry (187): the study of the chemistry of the living world; the element carbon is present in all organic compounds.

organic compounds (187): all compounds containing carbon except carbon monoxide, carbon dioxide and the carbonates.

overall equation (128): an equation to illustrate an oxidising process and a complementary reducing process involving electron transfer; formed by adding two half equations, one illustrating oxidation and the other illustrating reduction.

oxidant (oxidising agent) (124): a chemical substance that: can add oxygen to another substance, *or* can remove hydrogen from another substance, *or* possesses particles (atoms or ions) that can remove electron(s) from the particles of another substance.

oxidation (119, 136): a chemical reaction in which: oxygen is added to a substance, *or* hydrogen is removed from a substance, *or* particles (atoms or ions) remove electron(s) from the particles of another substance.

oxidation number (120): a number indicating the extent to which an element is oxidised or reduced in a compound; also known as oxidation state.

oxidation state (121): the 'degree' to which an element has been oxidised or reduced.

oxidation–reduction (reactions) (136): chemical reactions involving the transfer of electrons; also called redox reactions.

oxidation–reduction titration (1): see *titration*.

oxidising agent (124): a reactant which oxidises another reactant.

ozone (288, 335, 338): a gaseous allotrope of oxygen whose molecule contains three atoms of oxygen, O_3.

parent chain (196): the longest chain of carbon atoms in a branched-chain molecule.

particle (17): the term used to describe an atom, or molecule, or ion; a small unit in a chemical substance.

peptide (285, 287): molecule formed by linking a small number of amino acids, eg a dipeptide has two amino acids linked.

peptide link (285): a specific arrangement of atoms, –CO–NH– or –NH–CO–, found in proteins and polyamides indicating where two monomer units have joined together in a condensation reaction.

percentage composition (29): the mass of each element in a substance expressed as a percentage of the total mass.

period (54): horizontal row of the periodic table in which atoms of the elements have the same number of energy levels, but different numbers of valence electrons.

periodic table (54): arrangement of all the 110 known elements according to the number of valence electrons and the number of electron shells; increase in atomic number occurs across the table, columns have elements with similar properties down the table.

persistent organic pollutant (POP) (337): a toxic chemical that was used in industry or agriculture and which moves around the world in the wind or water; resists environmental degradation so persists in the environment.

pH (80): a numerical scale used to indicate acidity or alkalinity in an aqueous solution. The usual range of the scale is 0–14. pH of 0–6 indicates acidity with 0 very high acidity, and 6 extremely low acidity. pH 7 indicates neutrality. pH 8–14 indicates increasing alkalinity.

pH scale (80): a scale formed from the pH of different solutions.

phase (23): different forms of the same substance including states and allotropes.

photosynthesis (176): a process occurring in the leaves of trees whereby carbon dioxide and water, in the presence of sunlight and chlorophyll, produce glucose and oxygen.

physical dissolving (182): a process of dissolving in which the solute does not react chemically with the solvent.

physical property (196, 202, 215): a property that, when displayed, does not change the substance chemically, eg colour, electrical conductivity, ductility.

pigment (270, 326): a substance added to impart colour to another substance.

pipette (86): a glass tube calibrated to dispense an exact volume of solution.

pK_a (102): $-\log K_a$.

plastic (221, 287): the description of a material that is easily moulded, ie the material can be stretched and shaped; the term is also used to describe materials that have these properties, especially if the material is synthetic (man-made).

polar (70): referring to a covalent bond in which the electrons are not uniformly shared between the two atoms, ie the atom at one 'end' of the bond has a small positive charge and the atom at the other end of the bond has a small negative charge; the term also applies to a molecule in which the polar bonds do not cancel, ie the molecule has a positive end and a negative end.

polar bond/polar covalent bond (63): a covalent bond with an associated separation of charge (δ+ and δ–); the bonded pair of electrons is shared unequally between the bonding atoms due to differences in electronegativity.

polarity (70): the property of a covalent bond, or a molecule, that has a positive end and a negative end.

polyalcohol (229): an alcohol that has more than one hydroxyl functional group.

polyamide (285): a polymer containing many amide (–CONH– or –CONR–) links.

polyatomic ion (57): a radical that has a negative charge, by virtue of the number of electrons in the particle exceeding the number of protons by one or more; or a radical that has a positive charge, by virtue of the number of electrons in the particle being less than the number of protons by one or less.

polyester (284): a polymer containing many ester (–COO–) links.

polyethene (221): a polymer made from ethene monomers; structure is that of covalently bonded carbon and hydrogen atoms in linear chains.

polymer (220, 283): a very large molecule made from joining many smaller molecules together.

polymeric ester (242): a synthetic polymer made of monomers linked by ester linkages, such as terylene.

polymerisation (220): the reaction that joins together many smaller molecules (monomers) to make very large molecules.

polypeptide (286): a polymer consisting of several amino acids linked.

polyunsaturated (244, 318): containing several C=C bonds.

polyunsaturated fatty acid (244): a fatty acid that contains more than one double bond.

porous (273): the property of a solid material that allows a liquid to penetrate through it, eg filter paper is porous.

positive ion (117): atom or radical that has a positive charge because it has fewer electrons than protons.

precipitate (259): the insoluble product formed when two solutions are mixed.

precipitation (155): formation of a solid (precipitate) from the combining of ions in solution.

primary (1°) alcohol (230): alcohol in which the –OH group is attached to a carbon atom bonded to only one other carbon atom.

primary (1°) haloalkane (209): haloalkane in which the halogen atom is attached to a carbon atom bonded to only one carbon atom.

primary haloalkane/halide (209): a haloalkane containing a CX group, where X is the halogen atom, with two hydrogen atoms attached to the carbon, eg $R–CH_2Cl$.

primary standard (42): a standard solution where the concentration is determined by weighing a substance and dissolving it in a known volume of solvent.

primary structure (proteins) (286): the order of amino acids in a protein polymer.

product (23): substance produced by a chemical reaction.

protein (286): a fundamental material of plant and animal structure that is composed of amino acids; most commonly observed as muscle in animals.

proton (47): a sub-atomic particle with a mass of 1.00 amu, and a positive charge; all atomic nuclei contain one or more protons.

proton transfer (reactions) (75): reactions in which a proton (H^+) is transferred from an acid to a base.

PVC (222, 283): polyvinylchloride polymer $[CH_2CHCl]_n$; IUPAC name, polychloroethene; monomer for PVC is vinyl chloride (systematic name chloroethene, CH_2CHCl).

qualitative (7): the study of what (*not* 'how much') is produced.

quantised (51): able to have only certain values.

quantitative (17): the study of how much is produced.

quantitative analysis (17): techniques used to determine the amounts of substances present in a mixture.

R (189): simple representation for an alkyl group (R′ and R″ being different alkyl groups).

rate (of a chemical reaction) (217): the speed of a chemical; factors such as temperature, pressure, concentration and presence of a catalyst may alter the speed of a chemical reaction.

reactant (23): a substance or species taking part in a chemical reaction.

reagent (5, 7): a substance or combination of substances that is used for a specific purpose in a chemical reaction.

recharge (150): passing electrical charge through a battery to reverse the spontaneous reactions.

redox (119): describes chemical reactions where one substance is reduced while another is oxidised.

redox couple (135): two species that are the reduction or oxidation product of each other; the oxidised and reduced species in an electron transfer half-reaction.

reducing agent (177): a reactant that reduces another reactant.

reductant (reducing agent) (124): a chemical substance that: can remove oxygen from another substance, *or* add hydrogen to another substance, *or* possesses particles (atoms or ions) that can add electron(s) to the particles of another substance.

reduction (119, 136): a chemical reaction in which: oxygen is removed from a substance, *or* hydrogen is added to a substance, *or* particles (atoms or ions) add electron(s) to the particles of another substance.

reflux (240): return escaped, vaporised, reactants or solvent to the reaction vessel.

repeating units (220, 283): the repeating sequence of atoms and bonds found in a polymer.

respiration (176): a chemical reaction occurring in plants and animals involving oxygen and a fuel to produce energy and, most commonly, carbon dioxide and water as products of respiration.

reversible reaction (79): a reaction that can proceed in the direction of forming products or in the direction of forming reactants.

RHE (137): right-hand electrode.

rough titration (89): initial titration performed quickly to give an estimate of the volume required from the burette.

rubber (290): a natural polymer produced by the rubber tree.

rust (148): the product formed (hydrated iron(III) oxide) when iron is oxidised (corroded).

rusting (148): an electrochemical process involving the corrosion of iron by water and oxygen.

sacrificial corrosion (149): the corrosion of a more reactive metal connected to a less reactive second metal to prevent that second metal from corroding.

salt (109): an ionic compound; formed when the hydrogen in an acid is replaced by a metal.

salt bridge (136): device that allows a flow of ions so that the charge in each half-cell of an electrochemical cell remains balanced.

saponification (245, 320): the process of forming soap from the reaction between triglycerides and sodium hydroxide (or potassium hydroxide).

saturated (188): a term used to describe an organic compound with only single covalent bonds between the atoms in the molecule of the compound.

saturated fatty acid (244): fatty acid which contains single bonds only.

saturated solution (180): a solution in which no more solute can be dissolved in a given volume at a given temperature.

scientific notation (17): a way of writing numbers using powers of 10 so that there is just one digit before the decimal place.

seafloor massive sulfides (307): deposits of metal sulfides that settle around undersea hydrothermal vents.

secondary (2°) alcohol (230): alcohol in which the –OH group is attached to a carbon atom bonded to two other carbon atoms.

secondary (2°) haloalkane (209): haloalkane in which the halogen is attached to a carbon atom bonded to two other carbon atoms.

secondary standard (42): a standard solution where the concentration is determined by titrating with a primary standard.

secondary structure (proteins) (286): the structure due to the formation of hydrogen bonds between sections of the protein molecule.

seed crystals (165): small crystals added to a solution to encourage large crystals of the same material.

semimetals (116): elements with properties similar to both metals and non-metals, also called metalloids.

SHE (139): see *standard hydrogen electrode*.

side chains (201, 286): smaller chains of carbon atoms joined to the parent chain.

significant figures (20): the digits obtained in a measurement (which are then used to determine the number of figures used in the answer to a calculation).

silica (165, 298): a common name for silicon dioxide, SiO_2; a three-dimensional network solid.

smelting (165, 298): extracting a metal from its ore.

soap (245, 320): sodium, or potassium, salt of a fatty acid, eg sodium stearate, $C_{17}H_{35}COONa$.

solid (20): a state of matter where the substance has its own shape; the particles in a solid are closely attached and arranged in a regular pattern.

solubility (182): a measure of the maximum amount of a substance which dissolves in a given amount of solvent to form a saturated solution, at that temperature.

soluble (7): describing the property of substance that will dissolve in a solvent to form a solution.

solute (37): a substance that dissolves in a solvent to make a solution.

solution (37): a homogeneous mixture formed when a solute dissolves in a solvent.

solvent (37): a liquid that dissolves a solute to make a solution.

spectator ions (7, 124): ions present in a solution that are not involved in the reaction occurring in the solution.

spectrophotometer (6): a device used in colorimetry to measure the light absorbed by a solution in various parts of the spectrum.

spontaneous (reaction) (148, 150): a reaction that occurs without any apparent external influence.

stable (42): unreactive in the conditions stated, eg in the presence of air, or when a substance is heated.

stalactite (184): a growth of calcium carbonate in a limestone cave that hangs from the ceiling of the cave; the '*c*' in stalactite reminds you of *c*eiling.

stalagmite (184): a growth of calcium carbonate in a limestone cave that rises from the ground of the cave; the '*g*' in stalagmite reminds you of *g*round.

standard cell diagram (137, 144): symbolic representation of an electrochemical cell, in which certain conventions apply as to the order in which the components of the cell are written, eg Zn(s) | Zn2+(aq) || Cu2+(aq) | Cu(s).

standard conditions (139): 100.0 kPa (1 atmosphere), 25°C and solutions have concentration 1 mol L^{-1}.

standard hydrogen electrode, SHE (139): reference electrode against which electrode potentials are measured.

standard reduction potential, *E*° (139): reduction electrode potential measured in volts under standard conditions, which indicates the ability of a species to gain electrons; may sometimes be called standard electrode potential.

standard solution (4, 42): a solution for which the concentration is accurately known.

standardise (42): to accurately determine the concentration of a solution.

standardising (45): finding the concentration of a secondary standard.

state (23): one of the three forms of matter, which have very different arrangements and speeds of constituent particles, ie gas, liquid and solid.

stearic acid (244): octadecanoic acid, a saturated fatty acid with the formula $CH_3(CH_2)_{16}COOH$.

steel (178): an alloy of iron and carbon; specialist steels also contain other metals.

straight chain (190): a term used to describe an organic molecule where the carbon atoms are bonded to each other in one chain; the carbon atoms at the end of the molecule are primary and all the other carbon atoms in the chain are secondary.

straight-run fractions (274): chemicals collected from first fractional distillation of petroleum.

strong acid (77, 98): a species that almost completely dissociates in water to release hydrogen ions and anions.

strong base (78, 105): a substance that releases (or forms) the maximum amount possible of hydroxide ions in aqueous solution.

structural formula (191): a formula which shows the number of atoms in a molecule and how the atoms are bonded to each other.

structural isomers (196): compounds with the same molecular formula but different structural formulae, and so which have different names and physical properties but similar chemical properties.

subatomic particle (47): a particle smaller than any other atom, eg protons, neutrons and electrons.

sublimation (183): the process whereby a solid changes directly to a gas when heated, or a gas condenses directly to a solid when cooled.

sublime (183): the transformation of a substance from the solid phase to the gaseous phase without going through the liquid phase, or vice versa.

substitution (207): reactions of organic compounds that involve replacement of one atom or group with another.

superconductor (174): a substance that has no resistance to a current of electricity; capable of carrying huge current flows.

synthesis gas (247): a mixture of carbon monoxide and hydrogen produced from a reaction between methane and steam.

synthetic (220): human-made.

Système International (SI) d'Unités (18): a set of units used for convenience for all measurements in science.

tarnish (304): the result of the slow corrosion of a metal in air whereby the appearance of the metal becomes dull.

Teflon (222, 283): polytetrafluoroethene polymer, $(C_2F_4)_n$; also called PTFE.

tertiary (3°) alcohol (230): alcohol in which the –OH group is attached to a carbon atom bonded to three other carbon atoms.

tertiary (3°) haloalkane (209): haloalkane in which the halogen atom is attached to a carbon atom bonded to three other carbon atoms.

tertiary structure (proteins) (287): the complex three-dimensional structure of protein molecules due to interactions between different parts of the molecule.

thermal conductivity (295, 297, 300): the ability of a material to conduct heat without undergoing chemical change.

thermal cracking (276): the use of high temperatures to convert alkanes to alkenes.

thermoplastic (289): a plastic that can be moulded again and again.

thermosetting plastic (289): a plastic that can only melt and take shape once.

titration (45): the process of neutralising an acid with an alkali, using a pipette, a burette, a conical flask and a coloured indicator.

titres (2, 89): in a titration, the volumes of solution needed to be dispensed from the burette in order to reach the equivalence point.

Tollens' reagent (258): reagent used to test for the presence of aldehydes, as shown by the formation of a silver mirror on the inside of the test tube used; contains the diamminesilver(I) complex ion, $[Ag(NH_3)_2]^+$.

tonne (18): an industrial unit of mass, equal to 1000 kg.

***trans* isomer (215):** geometric isomer in which identical groups are located on opposite sides of a double bond.

transesterification (322): a reaction in which an alcohol and an ester react to form another alcohol and another ester.

transition metals (58, 153): metals whose atoms have incomplete d-orbitals; groups 3–11 in the periodic table.

triester (244, 317): the ester formed from glycerol and fatty acids containing three –COO– groups.

triglyceride (242, 317): naturally occurring ester found in fats and oils, formed when one molecule of glycerol joins to three fatty acid molecules.

triple bond (61): a covalent bond formed by two atoms each sharing three of their electrons with another atom – six electrons are used to form the bond.

triple point (315): temperature and pressure at which a substance exists as solid, liquid and gas in equilibrium.

tritium (49): an isotope of hydrogen with a mass number of 3.

underground mining (329): a method of extracting minerals that are too deep for open-cut mining.

universal indicator (80, 112): an acid-base indicator that can be used to determine the pH of a solution because it produces a range of colours corresponding to different pH values.

unsaturated (188): compound that contains a double (C=C) or triple (C≡C) carbon–carbon bond.

unsaturated fatty acid (244): fatty acid which contains one or more double bonds.

unsaturation (217): the term used to describe the situation where the molecule of an organic compound possesses a double or triple bond.

unsymmetrical alkene (218): alkene having different groups attached to at least one of the carbon atoms of the double bond.

upward displacement (180): the process for collecting a gas that is denser than air by delivering the gas to the bottom of gas jar so that the air is displaced upwards.

valence (143, 299): a term used to describe electrons in the outer energy level of an atom that are used for bonding.

valence electrons (52): those electrons available for bonding; usually the outer shell electrons but can also refer to *d*-orbital electrons.

valence level (52): the energy level that valence electrons are found in, also called the outer level.

valency (58): of an element; 1. the number of single covalent bonds formed by one atom of the element; 2. the charge on the ion formed from one atom of the element.

vaporise/vaporisation (240): conversion of liquid to gas.

vapour (183): another term for gas; usually referring to the product formed by a liquid evaporating or boiling.

viscous (229): resistant to flow.

volatile (202): describing a liquid with a low boiling point (<100°C); the liquid vaporises easily.

volumetric analysis (85): quantitative analysis, based on volumes and concentrations, used to determine the concentration of solutions.

volumetric flask (43): long-necked flask used to measure an accurate volume (50.00 mL, 100.0 mL, 250.0 mL etc).

wash-bottle (86): bottle of deionised water used to wash solid or solution from the sides of a container back into a solution.

water of crystallisation (34, 117): water that is chemically bound in a crystal in the formation of a hydrate, eg blue copper sulfate ($CuSO_4.5H_2O$) contains water of crystallisation.

weak acid (77, 99): an acid that only partially dissociates (1–5%) in water.

weak base (7, 79): a base that only reacts to a small extent with water and so produces only a small increase in the concentration of OH^- ions.

word equation (113): a statement of a reaction in words; identifies the reagents and products of a reaction by name only.

Z (63): see *atomic number*.

zwitterion (286): ion having separate positively and negatively charged groups caused by the transfer of a proton from one part of a molecule to another.